Innovative design

Warren J. Luzadder, p.e.

PURDUE UNIVERSITY

PRENTICE-HALL, INC., ENGLEWOOD CLIFFS, NEW JERSEY

Innovative design

with an introduction to design graphics

Library of Congress Cataloging in Publication Data

LUZADDER, WARREN JACOB.
　Innovative design with an introduction to design graphics.

　Bibliography: p.
　1. Engineering graphics.　2. Engineering design.
3. Computer graphics.　I. Title.
T353.L883 1975　　620'.004'2　　74-17146
ISBN　0-13-465641-5

Innovative design
with an introduction to design graphics
by Warren J. Luzadder, P.E.

other books by the author

Basic Graphics for Design, Analysis, Communications and the Computer, 2nd ed., Prentice-Hall, Inc., 1968.

Fundamentals of Engineering Drawing for Design, Communication, and Numerical Control, 6th ed., Prentice-Hall, Inc., 1971.

Fundamentos De Dibujo Para Ingenieros, 2nd ed., Compania Editoria Continental, S. A. Mexico City, 1967.

Graphics for Engineers, Prentice-Hall of India Private Limited, 1964.

Technical Drafting Essentials, 2nd ed., Prentice-Hall, Inc., 1956.

Problems in Engineering Drawing for Design and Communication, 6th ed., Prentice-Hall, Inc., 1971.

Engineering Graphics Problems for Design, Analysis, and Communication, 2nd ed., Prentice-Hall, Inc., 1968.

Problems in Drafting Fundamentals, Prentice-Hall, Inc., 1956.

Purdue University Engineering Drawing Films, with J. Rising et al.

PRENTICE-HALL INTERNATIONAL, INC., *London*
PRENTICE-HALL OF AUSTRALIA, PTY. LTD., *Sydney*
PRENTICE-HALL OF CANADA, LTD., *Toronto*
PRENTICE-HALL OF INDIA PRIVATE LIMITED, *New Delhi*
PRENTICE-HALL OF JAPAN, INC., *Tokyo*

designed by Betty Binns

To my son, Robert Warren Luzadder,
an outstanding author in his own right

Contents

Preface

This completely new text, with its emphasis centered primarily on innovative design, is intended to fulfill a newly recognized need in engineering and technical education for the presentation of the fundamental principles of design to students as early as their freshman year. With this new approach, a beginning course in design graphics can become an engineering orientation course that presents the student with both the basic fundamentals of design graphics and satisfying experiences in creative thinking directed towards solving recognized engineering problems. Design graphics has entered a new era that has been brought about by space age research, the ever increasing use of the computer for design, and the continued improvement and development of numerically controlled machine tools. The new techniques of computer-aided design make possible computer-plotter prepared drawings and graphs as well as quick and ready solutions to complex design problems. These factors have brought about considerable change in both the responsibility and the work assignments of engineers, technologists, and technicians.

Since design graphics and innovative design are interrelated in the total design process, this interrelationship has been adopted as the basis for the preparation of this text. However, the text material has been so arranged that an instructor may either: (1) use the text for one or more traditional design-graphics courses that may be followed by a course in innovative design; (2) use the text for a traditional course accompanied by a design project that parallels the teaching of design graphics; or (3) use the text material for a design course with the student acquiring some basic knowledge of design graphics as needed along the way to completion of the project. The second procedure is now being used at the author's institution with satisfactory results.

Since design graphics is an integral part of the total design process, which in itself is central to engineering, the inclusion of both innovative design and the fundamentals of design graphics in one or more beginning courses is no longer debated. However, the methods of achieving educational objectives are only now beginning to be evaluated. Continued development and experimentation are needed. The presentation of innovative design accompanied by some design graphics in an early course gives the beginning student needed insight into engineering and an opportunity to evaluate the role of the engineer and technologist in society. The student will discover that for engineering problem solving, an engineer or technologist is constantly being called upon to use logical, analytic, and creative thought processes to arrive at conclusions and make necessary decisions. Design experience in the beginning courses makes the fact clear to the student that graphics is important for recording and conveying ideas. He is led to realize that graphics is essential to his education along with mathematics and physics, for he will be using graphical methods for analysis and communication throughout his professional career.

Since students taking basic courses in graphics lack knowledge in depth of the fundamental engineering sciences that are needed for sophisticated engineering problem solving, the instructor teaching introductory design must direct his students' efforts toward the more unsophisticated types of creative work. There is a wide choice of problems at this level of design that will meet engineering objectives. However, the fact must always be borne in mind that the student himself must create the ideas that he must later try to communicate. At the start, it will be discovered that even the design of a bracket for a flag pole offers an opportunity for one to exercise creative ingenuity. To complete this simple task satisfactorily requires the exercise of space visualization, an understanding of how the bracket can function best, some knowledge of the methods available for manufacturing it, and the preparation of the necessary preliminary sketches and the final shop drawings. A relatively simple experience at this level can kindle a first spark of creative and inventive ability. At the same time the motivation level rises because the student feels that he has become involved in the realism of engineering. For more complicated devices, finished detail drawings may be required for only a few selected parts; this offers real learning experiences in the preparation of production drawings. All of the parts need not be drawn.

Upon completion of a report (written and/or oral) and a final set of drawings, the student has enjoyed the experience of disciplined creative thought and has learned much about problem definition. Hopefully, he or she will have come to recognize the importance of graphical communication and will have improved his spatial visualization ability in the process.

An instructor may render final judgment of a design by evaluating all of the reports and drawings submitted. This can be done handily and fairly by using a grading guide that will evaluate the (1) quality of technical work, (2) function and appearance, (3) economic analysis, and (4) manufacturing methods and requirements.

In addition to meeting present day course requirements, a worthwhile text should anticipate trends and include new material that may profitably be presented to students over the several years that must pass before another edition can be justified. Anticipating trends is sometimes difficult but the author has made an attempt to do so by including considerable material covering computer-aided design (Part 4–Chapter 13) and by giving the measurement values for many of the problems in millimeters, centimeters, kilometers, kilograms, and so forth. With American industry already pointing the way, it would appear that the metric system will be widely used within this country by the end of this decade. Facing up to this fact, the major engineering societies have urged the immediate initiation of metric education programs at all levels from kindergarten through college. In engineering education this means that industrial drafting room practice standards must be altered and the ANSI (American National Standards Institute) standards rewritten with metric values being shown. It is for this reason that the presentation of the metric system has been largely limited to an introduction in the first edition of this text. Hopefully, in a second edition many more steps can be taken towards total conversion. In the meantime, it is suggested that an instructor encourage his students to use metric units in preparing drawings, even though the measurements for a problem selected for use are given in the English system. A number of tables will be found in the Appendix for converting to metric.

For convenience this text has been arranged in five main parts: Part 1, Graphics and the Creations of Man; Part 2, The Design Process; Part 3, Design Graphics; Part 4, Computer-aided Design; and Part 5, Design for Production.

Part 2 (Chapters 2–6) provides useful information on innovative design as it has now been included in design-graphics courses in our leading colleges of engineering. By following the suggestions given in these chapters, engineering and technology students can have a meaningful experience in solving an engineering problem. Part 3 (Chapters 7–12) provides essential information on basic graphics, spatial geometry, geometry of design, pictorial presentation, graphs, and nomography. Part 4 (Chapter 13) covers the field of computer-aided design, automated drafting, and numerically controlled machine tools. At all institutions, particularly where computer-aided design systems are available, Chapter 13 should be assigned for reading and discussion. In addition, since practical knowledge of these systems can only be gained by either working experience or by on-the-scene observations, instructors of graphics and design should schedule plotter time and CRT (cathode-ray tube) console time for demonstrations. If this is not possible, several excellent motion pictures are available (see Bibliography) that may be shown in lieu of actual demonstrations. Finally, Part 5 (Chapter 14) presents the graphic standards relating to screw threads, fasteners, springs, bearings, welds, and the practices of dimensioning.

To bring this text abreast of new technological developments, a number of leading industrial organizations have generously assisted the author by supplying appropriate illustrations that were needed in developing specific subjects. Every commercial illustration supplied by American industry has been identified using a courtesy line. The author deeply appreciates the kindness and generosity of these many companies and the busy people in their employment who found the time to select these drawings and photographs that appear in almost every chapter.

The author is grateful to Professors K. E. Botkin, R. L. Airgood, R. P. Thompson, and other members of the graphics staff at Purdue University for their many valuable suggestions in regard to the content and organization of the new section covering innovative design. Their classroom experiences are reflected in the final presentation.

Special appreciation must be expressed for the contributions of Professors W. L. Baldwin, R. H. Hammond, and Byard Houck. Professor Baldwin of Purdue University developed the material on linkages. Professors Hammond and Houck contributed new material to the chapter on computer-aided design. Professor Houck is an authority on the TRIDM program that has been prepared at North Carolina State University at Raleigh, since he played a large role in its development. Not to be forgotten is the fact that there are many persons, some known and others unknown, who have made valuable contributions to this classroom text. The author's indebtedness to these persons is hereby reaffirmed.

To the instructor: Since this text has been prepared also for use in traditional graphics courses (where students may have had very little, if any, training in the preparation of engineering drawings), the knowledge needed to develop the skills essential for writing (drawing) the graphic language have been presented in Parts A,

B, C, and D of the Appendix. These four separate sections covering the drafting skills and techniques are; A – Use of Instruments, B – Freehand Technical Lettering, C – Geometry of Design, and D – Geometry of Developments. Engineers and technologists must have a working knowledge of graphics while the draftsmen and technicians assisting engineers in the development of a design must have a complete command of the language, since they must interpret the design sketches and the layout drawings and from these prepare the final production drawings that must be accurate, clear, and concise. Users of this new and different text will find that sufficient material has been included for the usual courses presented to all of these members of the design team (design engineer, engineering technologist, and technician).

Purdue University   W.J.L.

Part 1

Graphics and the creations of man

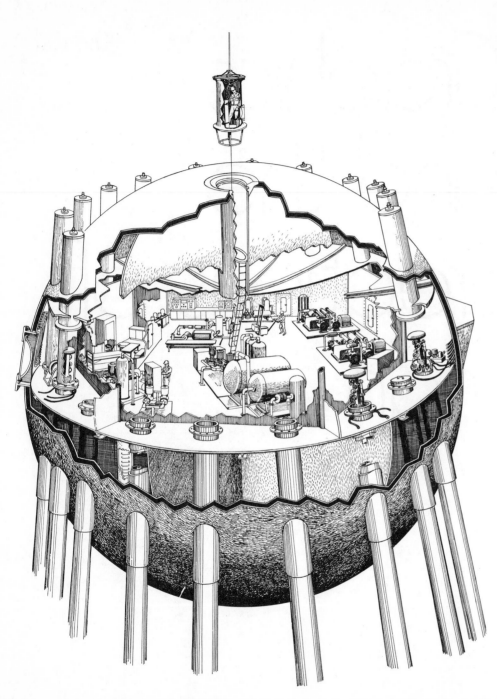

The Underwater Production System illustrated represents a new approach to solving the critical problems of offshore oil recovery. This system is being developed by Ocean Systems, Inc., for rather deep water (over 300 feet). The pictorial representation shows a personnel-transfer capsule being lowered into the 80-foot sphere. In the development of the spherical capsule and its related components, many of the problems involving intersections between geometric shapes may be solved by the graphical methods of projection. (*Courtesy of Design News and Ocean Systems, Inc., an affiliate of Union Carbide Corporation.*)

1.1 □ Our technological age

We of today have inherited a mechanized world, a world in which working machines have reduced man's labor and increased his production in terms of man-hours. This increased production, resulting in lowered costs, has in turn provided better material living and more leisure time for all. Thus, we come ever closer to the great American dream that has existed since that day when the first of our forebears set foot on the shores of this continent and moved into the wilderness to make that dream come true. In this decade, we look to a thirty-hour and three-day work week. All of this did not just happen nor did it result from brawn and sweat alone. This great age, an age in which we travel in outer space (Figs. 1.1 and 1.2), was created by the imaginative minds of men seeking answers to questions and seeking ways to satisfy the needs of man and to protect him from his inhospitable environment, where he has been literally forced to fight for his very existence. Innovative ideas resulting in new machines to lessen man's labors have increased the equality between men, the goal of all. Some critics say, however, that we have over-mechanized and that machines have created new problems that are destroying man and his environment. Also, if it is true as some charge, that we have created more problems than we have solved, then an ever-increasing amount of research and creative designing lies ahead for the engineer and technologist. It is to the technically trained with their creative minds that mankind must look for needed solutions.

1.2 □ Primitive design and metalworking at the dawn of history

The excavation of long-lost cities and tombs has provided evidence

Innovative design: from the cave man to the space man

that various metalworking techniques were in use as early as 4000 B.C. Such items as metal tools, carts with wheels, and plows were being used by early civilizations in the Near East about 2800 B.C. Methods bearing some resemblance to our present-day techniques for working softer metals had been crudely developed and were being used by these ancient peoples to produce items for their community life. As these craftsmen learned metalworking and woodworking methods and became able to form and handle stone, the design process was established. These early people showed an awareness of beauty in producing useful items and ornaments using copper, gold, and silver.

1.3 □ The six fundamental innovations of the ancient world

The sail and five simple machines were utilized by the Greeks of the ancient world. All of these are still used by man in the twentieth century.

The use of the sail is thought to have been the conception of some unknown imaginative genius who lived somewhere at the eastern end of the Mediterranean Sea about 3000 B.C. (Fig. 1.3). There can be little doubt that he was seeking a way to propel himself without having to resort to the backbreaking methods of using either a pole or an oar. In so doing, he harnessed a new source of power for war vessels and ships of commerce that was to be used into the Space Age when men would ride and walk on the moon (Fig. 1.2).

Today we have machines, some numerically controlled, that do many things, and yet even now many of the mechanical devices upon which we depend are largely combinations of these same five simple mechanisms that were created before the birth of Christ, and were widely used at the time of Archimedes (Fig. 1.7). The five fundamental machines are: (1) the lever, (2) the inclined plane and wedge, (3) the wheel and axle, (4) the pulley, and (5) the screw. All of these mechanisms seem simple to us but in their time they were outstanding achievements of man's innovative mind. They changed the course of history and started man along the path toward the technological age in which we now live.

The oldest machine in the world is without any doubt the lever (Fig. 1.4). Its principle was quite possibly discovered during the Paleolithic Age when one of our prehistoric ancestors moved a large stone away from the entrance of his cave by first placing one end of a long timber lever underneath it and then, after having braced his lever against an object that would serve as a fulcrum, moved the unwanted stone by pushing or pulling downward on the free end. The crowbar, boat oar, wheelbarrow, and steam shovel are just a few of the present-day devices that depend upon the principle of the lever.

Fig. 1.1 □ A photograph of Mariner.
The Mariner shown above weighs 910 pounds and with solar panels deployed has a wing span of 19 feet. T.V. cameras and electronic instruments are contained in the pod beneath the octagonal framework. The Mariners have supplied much valuable information about the red planet Mars. (*Courtesy National Aeronautics and Space Administration.*)

Fig. 1.2 □ The Lunar Rover.
The photograph above shows the Lunar Rover taken to the Moon by Apollo 15 in July 1971. Mount Hadley is at the upper right and the most distant lunar feature that can be seen is 25 km away. The Lunar Rover is one of our most noted innovative designs. (*Courtesy National Aeronautics and Space Administration.*)

The second mechanism is the inclined plane and closely related wedge (Fig. 1.5). The basic principles of the inclined plane and wedge are also thought to have been known back in the Paleolithic Age. Present-day applications of the inclined plane and wedge appear in the design of ramps, hand chisels, air hammers, gang plows, and so forth. The medieval battleaxe was a combination of the lever and the wedge.

The wheel and axle (Fig. 1.7) appears to have been conceived about 3000 B.C. Aside from being used on chariots, one of its earliest uses was in connection with the windlass to move something of great weight. It is probable that, with a rope tied to the weight and fastened to the axle, the spokes of the wheel were used to turn the axle. Five thousand years of civilization have been built around the wheel. The wheel and axle has been the most important innovative design of all time. The reader should try to picture, if he can, our world without the wheel, for the wheel along with our spoken and written languages made civilization, as we know it, possible.

The pulley, like the lever, was designed to assist man in lifting a weight too heavy for him to lift alone (Fig. 1.6). The use of a windlass and pulley for drawing water from a well is the oldest known use of the simple fixed pulley. The fixed pulley of this system gave the puller no aid. Today, we use a pulley system for lifting that we call the block and tackle. Also, we use drive pulleys of various types on many of our modern machines. Pulleys have been used since about the eighth century B.C. Archimedes developed the block and tackle in the third century B.C.

The geometry of the spiral helix was worked out by an early Greek mathematician. This made possible the development of the screw as we know it. Archimedes developed some of the most famous screws of antiquity (Fig. 1.7). One of these was designed to raise water; another, it is said, was used to drag a rather large three-masted boat, fully loaded, onto land. Screws are used for raising weights, for clamping, or for fasteners.

1.4 □ The Renaissance—age of Leonardo da Vinci

The innovative creations of the Middle Ages sprang largely from the mind of one man, Leonardo da Vinci (Fig. 1.7). Aside from his work as a great painter, he prepared sketches and drawings of mechanical conceptions that have never been equaled in quality. Many of the concepts shown by his drawings were as much as several hundred years ahead of his time. His work extended through the fields of military equipment, optics, physics, engineering, and even anatomy. He was so far ahead of his age, however, that only a few of his conceptions could be produced, and these few were largely machines of war.

Fig. 1.3 □ The Sail.
The sail harnessed a new source of power to propel men across the seas. It hastened the spread of civilization all along the shores of the Mediterranean.

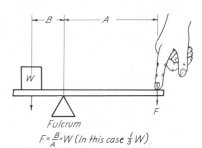

$$F = \frac{B}{A} \times W \text{ (In this case } \tfrac{1}{3}W)$$

Fig. 1.4 □ The Lever.
The lever represents one of the earliest innovations of man. Its use extends back into the past well beyond recorded history.

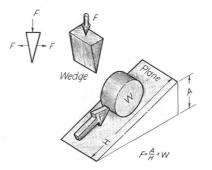

Wedge

Plane

$$F = \frac{A}{H} \times W$$

Fig. 1.5 □ The inclined plane and wedge.
The use of the inclined plane and wedge is thought to have been known as far back in the past as the Paleolithic Age.

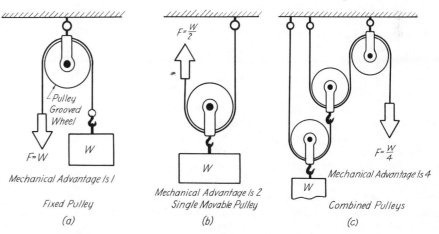

$F = W$

Pulley Grooved Wheel

W

Mechanical Advantage Is 1

Fixed Pulley

(a)

$F = \frac{W}{2}$

W

Mechanical Advantage Is 2
Single Movable Pulley

(b)

$F = \frac{W}{4}$

W

Mechanical Advantage Is 4

Combined Pulleys

(c)

Fig. 1.6 □ Pulleys.
By means of pulley systems man became able to lift objects that were far too heavy for him to lift alone.

1.5 □ The industrial revolution and the age of invention (innovative design)

The industrial revolution was fomented by steam. The interest in steam as a source of power may have been sparked by a report submitted to King Charles II of England in 1685 by Sir Samuel Morland. In this report, he wrote of water evaporated by fire: ''the vapours require a greater space, about 2000 times that occupied by water. And rather than submit to imprisonment it will burst a piece of ordnance. But being controlled . . . it bears its burden peaceably, like good horses, and thus may be of great use to mankind.'' This master mechanic to the king was a man of vision, . . . though the author can find no record that he ever conceived a useful machine that could utilize the power he had observed. In any case, however, this knowledge interested practical men of that time and led to the beginning of an age of invention that has extended across nearly three hundred years to the present. Although James Watt, an instrument maker, developed the condensing steam engine in 1765, Captain Thomas Savery, a military engineer, had already developed a usable steam pump to drain mines in 1698. From this point on, the steam engine was to play a momentous role for more than two centuries in supplying the motive power for industry and transportation. The discovery of steam and the early inventions that harnessed its power fired the imaginations of men with innovative minds and Sir Samuel Morland's thought, '' . . . may be of great use to mankind'' was to be fully realized.

The age of invention did not begin in all countries at the same time. However, once it was well under way, new machines, new products, and new means of transportation were rapidly conceived and developed. In England, Savery's steam pump (1698) was followed by the condensing steam engine developed by James Watt in 1765. This lead to the application of steam to run spinning and weaving machines. From this point on, one new innovative idea bred another in rapid order right down through to the present day, and no let-up appears to be in sight. In Europe during the 19th century, inventors came forth with a steam-powered printing press (1811–Germany), photography (1826–France), the water turbine (1827–France), the internal combustion engine (1860–France), the four-cycle gas engine (1876–Germany), and finally the first automobile (1885–Germany). Michael Faraday (England) discovered the principles of the electric motor in 1822 and developed the first electric generator in 1831. His attempts to utilize this new source of energy . . . opened up a whole new field of invention and brought many new machines into the home. Today there are homes that have fifty or more distinct electric devices to make the life of the occupants easier and more enjoyable.

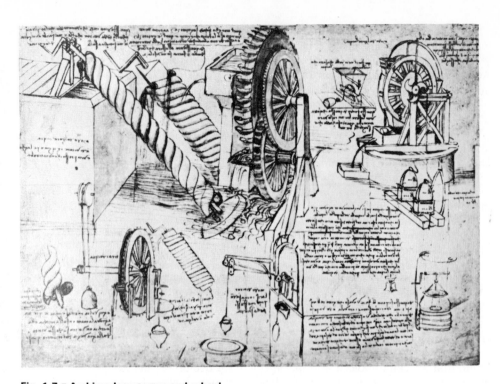

Fig. 1.7 □ Archimedean screw and wheel.
Sketch showing Archimedean screw and wheel by Leonardo da Vinci (1452–1519), engineer, scientist, and painter. (*From Collections of Fine Arts Department, International Business Machines Corporation.*)

In 1825, just six years before Faraday developed his generator, another Englishman, George Stephenson, produced the first successful steam locomotive. This first locomotive had a number of drawbacks, but by 1834 inventors had made sufficient improvements that a train, on its initial run, sped across the English countryside at speeds of from six to ten miles an hour.

In America, innovative people were at work as well and new inventions followed one after another. A few of the American inventions are: the Fulton steamboat (1790), the Whitney cotton gin (1793), the McCormick reaper (1834), the Morse telegraph (1837), the Otis passenger elevator (1857), the Sholes typewriter (1867), the Bell telephone (1876–Fig. 1.8), and the dictaphone (1885). In this burst of inventive activity during the 19th century, the homemaker was not forgotten, for in these years innovative designers brought forth such items for the household as: the washing machine (1851), the vacuum cleaner (1869), the electric iron and electric fan (1882), and the electric stove (1896).

1.6 □ Inventions of the twentieth century

The two world wars of the twentieth century accelerated the rate at which new inventions were pouring forth. This is true even though there was some slowing down during the years of the great depression. This century has brought many great changes in the lives of nearly all peoples. The automobile and the airplane can be cited as the major causes of the greatest change in the dimensions of human existence in all history (Fig. 1.9). People in industrialized countries have become more mobile (Fig. 1.13); we have traveled to new places and have become deeply involved with new people and new problems. As a consequence, we have complicated as well as diversified our lives. This may not be all good or all bad. However, the time has come for inventors and engineers to give more attention to the psychological and sociological impact that the development of a proposed product may have upon the well-being of mankind.

The principal developments of this century include the airplane, radar equipment (1936), the jet-propulsion engine, the laser beam and its applications (1954), the nuclear reactor (1942), the transistor, and, finally, space vehicles for travel to the planets and beyond.

The first airplane flight was taken at Kitty Hawk on December 17, 1903. The distance from take-off to landing was 852 feet. However, two years later the Wright brothers again took to the air and this time flew 24 miles in 38 minutes. Now jet planes have been designed to fly at speeds approaching 1500 mph and most of us will think little about it when we finally travel in such planes. We will be propelled by some type of jet-propulsion engine

Fig. 1.8 □ The telephone.
"My word! It talks," exclaimed Emperor Dom Pedro of Brazil on June 25, 1876 when he listened to the receiver of the 1876 telephone shown at the Philadelphia Centennial Exposition. (*Courtesy American Telephone and Telegraph Company.*)

Fig. 1.9 □ First-class lounge of a Boeing 747 jet airplane.
The full-size mockup left shows the interior of the first-class section of the Boeing 747, a plane that can carry from 350 to 490 passengers. The first of these planes was delivered for use late in 1969. (*Courtesy The Boeing Company.*)

Fig. 1.10 □ The Picturephone, Mod II, is shown in use. (*Courtesy American Telephone and Telegraph Company.*)

that represents the latest design in a progression that started with a turbo-jet engine patented in 1930 by an officer of the British Royal Air Force.

Some of the other outstanding inventions of the century are: the radio telephone (1901), talking motion pictures (1913), television-image pickup tube (1928), the ENIAC electronic computer (1946), NC (numerically controlled) machine tooling systems (1952), the solar battery (1954), and, the commercial gas-turbine automobile (1963). Several innovative creations of a more recent date are shown in Figs. 1.10–1.14.

1.7 □ Mass production (assembly line) and the potential of automation

While Americans with innovative minds were inventing new machines, they were also devising a system that was to become known as the American mass production system. In the late 1700's the Eli Whitney Arms Manufactory was the first factory to adopt a limited mass production method with the interchangeability of component parts. Following this initial effort, the mass production system continued to develop and receive acceptance until in 1913 the Ford Motor Company emerged with the so-called "moving assembly line", a system as we know it today, where the work is brought to the men. Mass production lowers production costs, which in turn lowers the price and leads to an increased demand. The assembly line method set up at the Ford plant represents the synthesis of three basic elements, namely: (1) the complete interchangeability of parts, (2) precision tooling, and (3) a division of labor into separate tasks that can be performed routinely.

In the 1970's the American people are finding that they are facing a new industrial revolution that in all probability will demand new skills and new thinking on man-machine relationships. This new revolution that is being brought about by the computer and the development of numerically controlled (NC) machine tools may change our national economy, work habits, and social structure in ways that even now cannot be fully foreseen. This new industrial revolution, which started in the mid-forties, is now well under way; many new machines that "think" are already being widely used. This movement towards an automated society, in which plants practically operate themselves, will mean dislocation and retraining for some; the nation's political leaders as well will be forced to face the "people problems" that automation will tend to create. It is to be hoped, however, that as the social problems brought about by this new industrial revolution are recognized and solved, there will be more leisure and new freedoms for all. To realize this goal, the technically-trained with their innovative minds

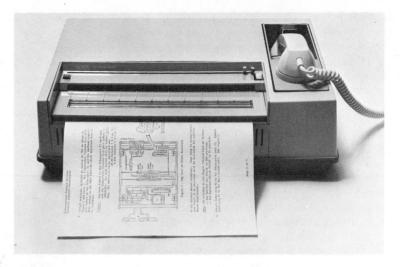

Fig. 1.11 □ The Xerox 400 Telecopier transceiver.
This transceiver marks the beginning of a new era in facsimile transmission. It sends and receives, over regular telephones, a letter-size document in approximately four minutes.(*Courtesy Business Products Group—Xerox Corporation.*)

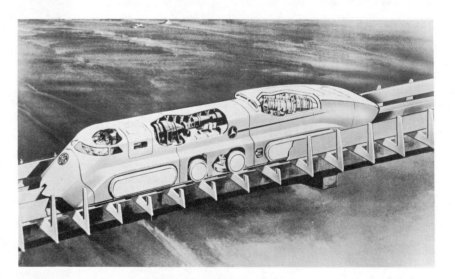

Fig. 1.12 □ Tracked Air Cushion Research vehicle.
The 300-mph TACRV shown could be used to meet the greatly increased intercity transportation needs. It is designed to be supported one-half inch above a channel-shaped guideway. Hundreds of drawings will be required to produce this vehicle. (*Courtesy General Electric Company.*)

must face up to their responsibilities in the design of productive systems. No longer can the production of a needed product be the only goal.

1.8 □ Inventor or scientist?

At the present time, most of our so-called inventors work as part of a design team in the design rooms and laboratories that are maintained by both government and industry. In a sense these are captive inventors who receive substantial salaries for their innovative ideas and the subsequent development work needed for the creation of a new product. Most of these men prefer to be known either as engineers, designers, or scientists rather than as inventors. This may be due to the fact that in the past an inventor was considered to be an eccentric individual who worked alone or with only a few helpers to develop an item that most people were sure could not be constructed.

Salaried inventors working for the nation's leading companies are permitted a vast amount of freedom to develop new ideas. Those working for our large companies are usually allotted the time and given sufficient funds to develop an idea until it can be pronounced either a complete failure or a success. Forward-looking companies realize that an idea of questionable worth can lead to other ideas that may have financial potential.

In spite of the fact that nearly all designers are now employed by corporations or the government, a few independent inventors continue to design and develop items with fame and fortune in mind. Some of these men and women have been successful and have amassed considerable fortunes.

Fig. 1.13 □ The Minigap System.
Creative thinking produces new ways of life for man. The Minigap System shown links cars together into a caravan that is led by a specially built leader vehicle. Computers inside the cars take over control of brakes, accelerator, and steering to free the motorist from the task of driving as long as he remains hooked-up. Caravans can mix in the traffic flow with other vehicles. (*Courtesy Ford Motor Company.*)

Fig. 1.14 □ NEMO on the Ocean Floor.
NEMO is a 66-inch-diameter sphere of acrylic plastic. The sphere is constructed from twelve identical curved pentagons. The capsule with $2\frac{1}{2}$-inch walls is bonded with acrylic adhesive. One of the first uses of NEMO will be as a diver control center at points of Seabee underwater construction sites. In operation, the manned observatory is lowered into the sea from a Navy support vessel with the crewmen flooding the ballast tanks for the descent. NEMO operates independently, controlled by an array of push buttons (41) linked to a solid-state circuitry control system. NEMO has been designed for a normal stay underwater of eight hours. Under emergency conditions it may remain submerged for as long as twenty-four hours. (*Official Navy photograph. Courtesy The Military Engineer Magazine.*)

Part 2

The design process

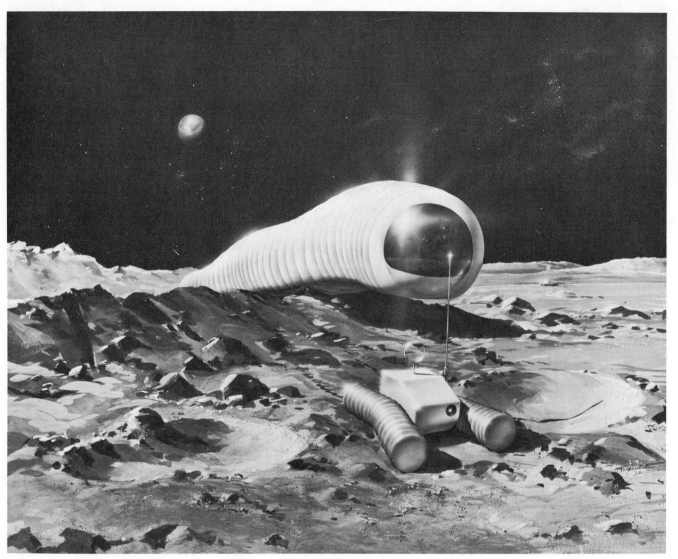

An Artist's Concept of a Lunar Surface Vehicle (Moon Crawler).
The representation shown suggests a type of vehicle that might be used for travel across the rough terrain of the moon. Artist's renderings of this type are needed for an evaluation of a conceptual design when many of the persons involved in the decision making may be government officials and others who do not have a strong scientific and technical background. (*Courtesy Philco-Ford Corp., Aeronutronic Division.*)

Chapter 2

Fundamentals of innovative design: the total design process

2.1 □ Design

In the dictionary, design is defined as follows: (1) to form or conceive in the mind, (2) to contrive a plan, (3) to plan and fashion the form of a system (structure), and (4) to prepare the preliminary sketches and/or plans for a system that is to be produced. Engineering design is a decision-making process used for the development of engineering systems for which there is human need (Fig. 2.1). To design is to conceive, to innovate, to create. One may design an entirely new system, or modify and rearrange existing things in a new way for improved usefulness or performance. Engineering design begins with the recognition of a social or economic need (Fig. 2.2). The need must first be translated into an acceptable idea by conceptualization and decision making. Then, the idea must be tested against the physical laws of nature before one can be certain that it is workable. This requires that the designer have a full knowledge of the fundamental physical laws of the basic sciences, a working knowledge of the engineering sciences, and the ability to communicate ideas both graphically (Fig. 3.7) and orally. The designer should be well grounded in economics, have some knowledge of engineering materials (Chapter 6) and be familiar to a limited extent with manufacturing methods. In addition, some knowledge of both marketing and advertising will prove worthwhile, since usually what is produced must be distributed at a profit. Proficiency in designing can be attained only through total involvement; since it is only through practice that the designer acquires the art of continually providing new and novel ideas. In developing the design, the engineer or engineering technologist must apply his knowledge of the engineering and material sciences, while at the same time taking into account related human factors, reliability, visual appearance, manufacturing methods, and sale price. It may therefore be said that the ability to design is both an art and a science.

A creative person will almost never follow a set pattern of action in developing an idea. To do so would tend to structure his thinking and might limit the creation of possible solutions. The design process calls for unrestrained creative ingenuity and continual decision making by a freewheeling mind. However, the total development of an idea, from recognition of a need to the final product, does appear to proceed loosely in stages that are recognized by authors and educators (Fig. 2.2).

Creative thinking usually begins when a design team, headed by a project leader, has been given an assignment to develop something that will satisfy a particular need. The need may have been suggested by a salesman, a housewife, or even an engineer from another company now using a product or a machine produced by the design team's own company. Most often the directive will come down from top management, as was the case with the development of the electric knife some years ago. Although it is always pleasant for an individual to think about the careers of famous inventors of the past, and dream of the fame and fortune that might await the development and marketing of one of his ideas, the fact is that almost all new and improved products, from food choppers to aircraft engines, represent a team effort.

2.2 □ Design synthesis

The process of combining constituent elements in a new or altered arrangement to achieve a unified entity is known as design synthesis. It is a process that involves reasoning from assumed propositions and known principles to arrive at comparatively new design solutions to recognized problems. The synthesis of systems for simple combinations as well as for complicated assemblies requires creative ability of the very highest order. The synthesis of both parts and systems usually requires successive trials to create new arrangements of old components and new features.

Proof of this point is the Land camera, considered by many people to be an entirely new product, although in reality the camera represents a combination of features and principles common to existing cameras to which Mr. Land added several new ideas of his own, including a new type of film and film pack that for the first time made it possible to develop film and pictures within the camera itself. Mr. Land's design activity no doubt started with a recognition of the need and desire that people had to take pictures that could be seen almost immediately. The early automobile

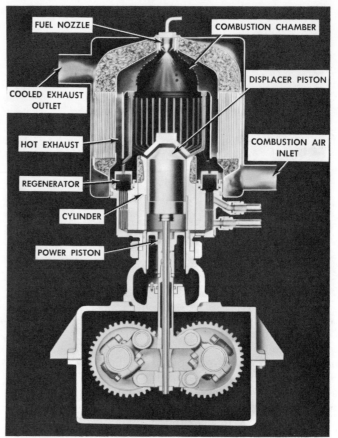

Fig. 2.1 □ The Sterling engine.
The Sterling engine is an interesting engine that can be considered as an alternate engine for an automobile. It is considerably different from the Wankel. The cross-sectional view of a single-cylinder engine shows its complexity. The engine has rather low hydrocarbon and carbon monoxide emissions. It tends to be rather heavy and expensive to produce at this time. (*Courtesy General Motors Corporation.*)

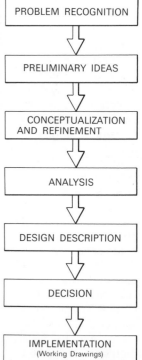

Fig. 2.2 □ Stages of the design process.

is another example that bears out the fact that old established products have features that are used as a starting basis for a new product. For example, at the turn of the century the automobile looked like a horseless carriage; the horse was taken to the barn and a motor was added in its place.

2.3 □ Design of systems and products

In general most design problems may be classified as being either a systems design or a product design, even though it may be quite difficult in many instances to recognize a problem as belonging entirely to one classification or the other. This is due to the fact that there often will be an overlap of identifying characteristics.

A systems design problem involves the interaction of numerous components that together form an operating unit (Fig. 2.3). A complex system such as the climatic-control system for an automobile (heating and cooling), an automatic movie projector, a stadium, an office building, or even a parking lot, represents a composition of several component systems that together form the complete composite system (Fig. 2.4). Some of the component systems of an office building, aside from the structural system itself, are: the electrical system, the plumbing system (including sewers), the heating and cooling systems, the elevator system, and the parking facilities system. All of these component systems, when combined, will meet the needs of a total system; but usually the total system design will involve more than just a technological approach. In the design of composite systems for use by the public, as in the design of a highway, an office building, or a stadium, the designer must adhere to the availability of funds. He must build into his design the safety features that are required by law and, at the same time, give due consideration to human factors, trends, and even present-day social problems. In many cases, a designer will find that there are political and special-interest groups that limit his freedom of decision making. Under such conditions he must be willing to compromise in the best interests of all concerned. This means that a successful designer must have experience in dealing with people. Social studies courses taken along with and in addition to the scientific courses offered in engineering colleges will prove to be helpful.

Product design is concerned with the design of some appliances, systems components, and other similiar unit-type items for which there appears to be a market. Such a product may be an electric lock, a power car-jack, a lawn sprinkler, a food grinder, a toy, a piece of specialized furniture, an electrical component, a valve, or other item that can be readily marketed as a commercial unit. Such products are designed to perform a specific function and to

Fig. 2.3 □ Transmitting drawings and sketches electronically.
Long-distance xerography networks, utilizing microwave, coaxial cable, or special telephone line, provide a practical means for transmitting and receiving electronically any form of communication, typed or drawn, across a plant area or across the continent. The high-speed, high-capicity facsimile transmission and receiving equipment consists of an LDX scanner and LDX printer. The document scanner on the sending end converts images to video transmission signals. At the receiving end the printer electronically restores the transmitted images and produces a black-on-white copy of the original document, sketch, or drawing. The original is simply inserted into the feed slot of the scanner. It is returned to the operator in seconds. The scanner senses colors without exposure adjustment. (*Courtesy Xerox Corporation.*)

satisfy a particular need. Total product design includes not only the design of an item, but the testing, manufacture, and distribution of the product as well. Design does not end with the solution of a problem through creative thinking. The design phase for a product to be sold on the open market ends only when the item has received wide acceptance by the public. Of course, if the item being designed is not for the consumer market, the designer will have less reason to be concerned with details of marketing. This would be true in the case of components for systems and for items to be used in products produced in finished form by other companies.

After initially recognizing a need or desire for a new product, the creative individual enters the next step in the design procedure—the research and exploration phase. During this phase, the possible ways and the feasibility of fulfilling the need are investigated. That is, the question is raised at this point as to whether the contemplated product can be marketed at a competitive price or at a price the public will be willing to pay. Except in the case of the public sector, where a system is being designed under the direction of a government unit (Fig. 2.5), profit is the name of the game. A search of literature and patent records will often show that there have been previous developments with the possibility of infringements on the design and patents of other companies. The discovery that others already may have legal protection for one or more of the possible solutions will tend to control the paths that innovative thinking may take in seeking an acceptable design solution. Although freewheeling innovative thinking is the basic element of successful design, the experienced engineer never fails to consider known restraints when he is brainstorming possible solutions. In addition to avoiding patent infringements, the designer must be fully aware of possible changes in desires and needs and give full consideration to the production processes. Also, not to be overlooked in the decision-making stage, are the visual design (styling aspects) and the price range within which the product must be sold.

The ultimate success of a design is judged on the basis of its acceptability in the marketplace and on how well it satisfies the needs of a particular culture (Fig. 2.6). In these latter decades of the twentieth century, we have added to these judgments a requirement that what has been created must not damage the physical environment around us, an environment which has been rapidly deteriorating because of some past technological advances. New requirements are being prepared and approved in rapid order that set standards to reduce if not eliminate pollution. These restrictions were almost unknown to the creative designers of the first half of this century. Today, it has become almost mandatory for a designer selecting a material or an energy source to ask

Fig. 2.4 □ Mitchell Park Conservatory.
The Flower Domes in Mitchell Park in Milwaukee represent creative design at its best. Each dome encloses 750,000 cubic feet of space. Temperatures and percentages of humidity are kept under rigid control for varied types of plants. (*Courtesy Milwaukee County Park Commission.*)

Fig. 2.5 □ An OGO-D orbiting geophysical observatory satellite in stowed condition.
This satellite may well represent one of man's greatest accomplishments. (*Courtesy TRW Systems group.*)

himself if the material can be recycled and whether any residue discharged into the atmosphere is damaging to life.

2.4 □ Innovative design—individuals and groups

In the past there have been a number of creative individuals, working almost alone, who have created products that have advanced our culture, products for which people still have a great desire. Examples of such products are the printing press, the steam engine, the gasoline engine, the automobile, the telephone (Fig. 2.6), the phonograph, the motion picture camera and projector, the radio, the television, and many more. These individuals were keen observers of their culture and they possessed inquisitive attitudes that led to experimentation. Some called them dreamers. If so, their eyes were wide open, they worked long hours, and they persisted in making a new try after each failure. Their fame and their fortune, although rightly due them, resulted from the fact that each brought forth a new product or system at a time when it would be readily accepted and when it could be more or less mass-produced. These men possessed a good sense of timing and they were not afraid of failure or ridicule. They belonged in the age in which they lived.

At present, it is the practice of large industrial organizations to use group procedures in order to stimulate the imaginations of the individual members of the group and thereby benefit from their combined thinking about a specific problem. Within the group, the innovative idea of one individual stimulates another individual to present an alternative suggestion. Each idea forms the basis for still other ideas until a great number have been listed that hopefully will lead to a workable solution to the problem at hand. This group attack on a problem produces a long list of ideas that would be difficult for a single individual to assemble in so short a time (Fig. 2.7). A single person attempting to solve the problem alone would find it necessary to conduct extensive searches of current literature and patent records and to hold numerous discussions with knowledgeable experts in seeking the guidance needed for making necessary decisions.

The most widely used of the group-related procedures is known as "brainstorming." In applying this procedure to seek the solution to a design problem, a group of optimal size meets in a room where they will be relatively free of interruptions and distractions. The people selected should be knowledgeable and there should be no one assigned to the group who may take a strong negative attitude toward the design problem to be considered. Usually, group size ranges from six to fifteen persons. If the group consists of less than six people, the back-and-forth interchange of ideas is reduced and the length of the list of ideas, which will be the basis for eventual decision making, is shorter. On the other hand, should

Fig. 2.6 □ 1876 Liquid Telephone.
"Mr. Watson, come here; I want you!" These historic words, the first articulate sentence ever spoken over an electric telephone, were uttered by Alexander Graham Bell on the night of March 10, 1876. The receiver was a tuned reed. (*Courtesy American Telephone and Telegraph Company.*)

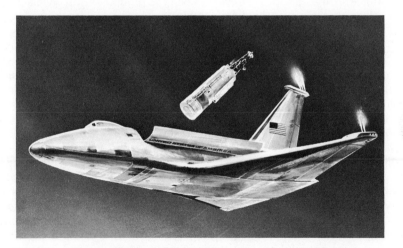

Fig. 2.7 □ The Space Shuttle Orbiter opens cargo doors to place Earth resources and weather satellite into orbit.
This concept was developed by North American Rockwell's Space Division. Manned orbiter will be capable of carrying passengers and a variety of cargo such as spacecraft and satellites to orbit and return to earth. The Orbiter can carry up to 65,000 pounds of payload in its 15 foot by 60 foot cargo bay. (*Courtesy National Aeronautics and Space Administration.*)

the group be larger than fifteen, some individuals who may be capable of making excellent suggestions will have little chance to talk. Also, very large groups tend to be dominated by a few individuals.

During the brainstorming phase of design, no appraisals or judgments should be made nor should criticisms or ridicule of any nature be permitted. The group leader, if he is a capable person, will encourage positive thinking and stimulating comments. He will discourage those who may want to dominate the discussion.

The ideas and suggestions of the members of the group should be listed on a chalk-board and, no matter how long the list, none should be omitted. All ideas should be welcomed and recorded. After the meeting, all the ideas presented may be typed and reproduced for the information of the group and for reference at later discussions.

As the list of suggestions lengthens, several possible solution patterns usually emerge. These in turn lead to still more suggestions for other possible combinations and for improvements to likely solutions to the design problem.

One or two lengthy brainstorming sessions can result in several acceptable design solutions that represent the combined suggestions of the individuals in the group. However, making a group evaluation of the ideas that have evolved from brainstorming is another matter, since no one member feels completely responsible for the results. The major weakness of any group procedure, such as brainstorming, is that individual motivation is dampened to some extent. However, since group procedures are productive and have become widely used in industry, there have been studies made that will hopefully lead to changes that will minimize this recognized weakness and improve group effectiveness. Means must be found to raise the motivation of each participant to the highest possible level. Detailed information on group procedures may be found in the chapter that follows.

2.5 □ The creation of new products

A product or a system may be said to have been produced either through evolutionary change or by what appears to be pure innovation. The word *appears* has been used appropriately, since few, if any, products are ever entirely new in every respect. Most products that appear to have drawn heavily on innovation usually combine both old and new ideas in a new and more workable arrangement. A product of evolutionary change, however, develops rather slowly, over a long period of time, and with slight improvements being made only now and then. Such a product may be reliable and virtually free of design and production errors, but the small amount of design work involved, done at infrequent intervals, will never really challenge a creative person.

In today's competitive world, when products are produced for world-wide consumption, evolutionary change is hardly sufficient to insure either the economic well being or even the survival of those companies that seem to be willing to let well enough alone. Rapid technological changes coupled with new scientific discoveries have increased the emphasis on the importance of new and marketable products that can gain a greater share of the total market than is possible with the product the company may now be promoting. In meeting this need for new and marketable ideas, the designer will find that his innovative ability and his experience and knowledge are being taxed to the limit and, since he may in one sense be stepping into the unknown, he will be taking some risks.

The unusual characteristics that seem to be a part of the general make-up of every outstanding designer are: (1) the ability to recognize a problem, (2) the ability to take a questioning approach toward all possible solutions, (3) the possession of an active curiosity about a problem at hand, (4) the innate willingness to take responsibility for what he has done or may do, (5) the ability to make needed decisions and to defend his decisions in writing and orally, and finally (6) the possession of intellectual integrity.

Mr. William Lear, one of the most prominent designers of the last three decades, has spent his entire working life discovering needs and then finding ways to fulfill them. In the case of the development of the eight-track stereo tape, he was working from economic considerations that required more repertoire on tape without adding more tape. This meant either running the tape slower or adding more tracks. The practical answer from Mr. Lear's viewpoint was to add more tracks. In addition to the Learoscope, an automatic direction-finder for use on airplanes, Mr. Lear is responsible also for the development of the car radio, the automatic pilot, the Lear jet plane, and more than 150 other inventions. A few years ago, he began development work on the problem of steam-powered automobile engines. Mr. Lear's inventing has been done when he was surrounded by many people with considerable know-how. His work involved the gathering of a maze of information and ideas from which he could pick out the salient facts and discard the unimportant ones, while always keeping the goal in mind and solving the problem at the least possible cost.

2.6 □ Background for innovative designing

Designing should be done by people who have a diversified background and who are not entirely unfamiliar with the problem at hand. As an example, even though the design of a product may be thought of as being in the field of mechanical engineering, a designer with a knowledge of electrical applications and controls will find it a distinct advantage since many of our present-day products use electric current as an energy source. In this case, even though an electrical engineer may be a member of the design group, the mechanical engineer and others should have at least some understanding of his suggestions. This added knowledge will enable them to modify their thinking about a product that is largely a mechanical device. Examples of these products include electric locks, electric food choppers, and electric typewriters.

The background required will vary considerably depending upon the field in which the individual works. For example, a person who may be designing small household appliances would probably never need more than the knowledge acquired from his basic engineering courses, while a designer in the aerospace field would need a background based upon advanced study in chemistry, physics, and mathematics. On the other hand, there are respected and competent designers in industry who have had as little as two years of technical education. With this limited training and several years of on-the-job experience, these men and women have become able to design fairly complex machines.

Due to the increasing complexity of engineering, the rapid development of new materials, and the accumulation of new knowledge at an almost unbelievable rate, it has become absolutely necessary for engineering design to become a team effort in some fields. Under such conditions, the design effort becomes the responsibility of highly qualified specialists. A project requiring designers of varied specialized backgrounds might need, for example, people with experience in mechanical, electrical, and structural design and persons with considerable knowledge of materials and chemical processes (Fig. 2.8). If it is decided that styling is important, then one or more stylists must be added to the team. A complete design group for a major project could include pure scientists, metallurgists, craftsmen, and stylists in addition to the designers.

Finally, graphics must not be overlooked when one considers the background needed to become a successful designer. Anyone who hopes to enter the field of design, other than as a specialist or craftsman, must have a thorough training in this area. He must have a working knowledge of all of the forms of graphical expression that are presented in this text and at the same time be capable of expressing himself well both orally and in written form in his preliminary and final reports. The methods used for the preparation of oral and written reports are discussed in Chapters 3–5.

2.7 □ History and background of human engineering

Human engineering is a relatively new area in the field of design; it has gained the recognition that it rightly deserves only within the last few decades. In order to simplify matters we might define

Fig. 2.8 □ A Jet engine.
The ultimate goal of every design must be the production of a product or system useful to man such as the jet engine shown being assembled. (*Courtesy Pratt and Whitney Aircraft Division of United Aircraft Corporation.*)

human engineering as adapting design to the needs of man; that is, we engineer our designs to suit human behavior, human motor activities, and human physical and mental characteristics. The applications of human engineering apply not only to man-machine systems and consumer products but to work methods and to work environments as well. In the early years of its development, human engineering was concerned mainly with the working environment and with the comfort and general welfare of all human beings. As time passed, designers came to realize that more than just safety and comfort should be considered, and that almost all the designs with which they were concerned were in some way related to man's general physical characteristics, his behavior, and his attitudes. At this point, designers began to include these added human factors in their approach to the solution of a design problem so as to secure the most satisfying and efficient man-machine relationship possible.

It is in the area of human engineering that an engineering designer must adhere to the input of specialists who may be involved in a wide range of disciplines. These disciplines may be industrial engineering, industrial psychology, medicine, physiology, climatology, and statistics. Stimulated to a great extent by the space program, scientists from all of these disciplines have, together and separately, become deeply involved in basic research and laboratory experimentation, which has led to a continuous input of new man-machine information into design. Although in the past human engineering has been associated mainly with industrial engineering, industrial psychology, and industrial design, designers in all fields of engineering must now not only be knowledgeable about the principles of human engineering but they must be capable of utilizing these principles and related information whenever they are developing a product that involves human relationship.

Typical body dimensions, representing an average-size person, are used when a product is being designed for general use. The measurements of typical adult males and females, as determined from studies made by Henry Dryfuss, may be used for most designs requiring close adaptation to human physical characteristics. Since many designs involve both foot and hand movements these average body dimensions, as tabulated, include arms, hands, legs, and feet along with other parts of the human body. Data relating to body proportions and dimensions is known as anthropometric data. Information relating to body proportions may be obtained from *The Measure of Man*, Whitney Library of Design.

The first known serious studies of the human body were made by Leonardo da Vinci. To record his studies for his own use and for the use of others, he made some of the finest and most accurate detail sketches and drawings of the human body known to man. These drawings show even the intricate details of muscle forma-

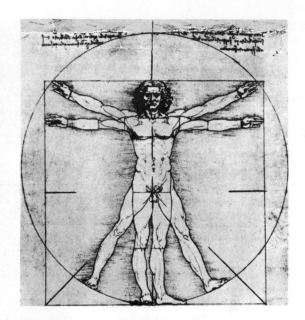

Fig. 2.9 □ Human proportions and body dimensions as illustrated by Leonardo da Vinci (1452–1519).

tion. His work, done in defiance of the laws of his time, is still used today in several textbooks. Da Vinci's studies of the human body mark the beginning of the science of biomechanics. See Fig. 2.9.

From the time of Leonardo da Vinci until early in the twentieth century, very little work was done toward the development of this science. This lack of interest in people and their relationship with equipment and tasks to be performed was due largely to the fact that from the beginning of the Industrial Revolution (which many people say began with invention of the steam engine patented by James Watt in 1769) until the early 1900's, the interest of designers was centered mainly on the creation of new products and the raising of production efficiency to the high levels needed to compete in world markets. Until about 1911, revolutionary change was the order of the day and there was very little time available for consideration of the human anatomical, physiological, behavior, and attitude factors that have now become the basis of our present man-machine-task systems. Even though our computer-programmed numerically controlled machine tools permit us to do almost any task with only minimal human intervention, man is still needed in most of our man-machine-task systems and he must be taken largely as he has been created. In our designs we must not overlook even the possibility of boredom, for should man become sufficiently bored he might just "pull the plug" and everything would stop.

2.8 □ Human engineering in design
There are a number of factors in human engineering, other than human anthropometric measurements, that must be dealt with in design. These factors include motor activities (Fig. 2.10) and body orientation; the five human senses (sight, hearing, touch, smell, and sometimes even taste); atmospheric environment, temperature, humidity, and light; and, finally, accelerative forces if they are exceptional and are likely to cause undue physical discomfort.

During the first half of this century most of the research done in human engineering, aside from the anthropometric studies already mentioned, was directed toward: work areas and the position of controls; physical effort and fatigue; and the speed and accuracy to be expected in the performance of particular tasks. Finally, when it became evident that this was not enough, industrial engineers and industrial psychologists turned their attention to the more complex activities of the average human being. These new studies dealt primarily with receiving information through sight and sound, the making of decisions in response to stimuli, and finally, the performance in direct response to these decisions.

The study of body motion deals primarily with the effective range of operation of parts of the body, usually the arms and legs, and

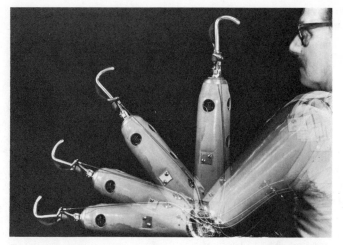

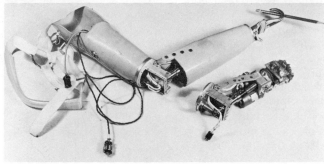

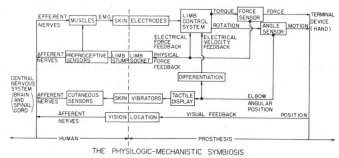

THE PHYSILOGIC-MECHANISTIC SYMBIOSIS

Fig. 2.10 □ The "Boston Arm."
The range of movement of the "Boston Arm" is demonstrated in the upper photograph. The arm, developed as a joint project of Liberty Mutual Insurance Companies, Harvard Medical School, Massachusetts General Hospital, and the Massachusetts Institute of Technology, acts as does a normal arm through thought-impulses transmitted from the brain to existing arm muscles. The design of a product for the handicapped can provide much satisfaction to the designer. (*Courtesy Liberty Mutual.*)

the amount of body force a human being may reasonably be expected to exert in the performance of an assigned task. For example, many time and motion studies have been made of the range of operation of persons performing given tasks while seated at assigned work areas. At the same time, in many of these studies attention has been given to the location and the amount of force required to operate levers and controls in relation to the size and strength of the operator (Fig. 2.11).

Vision is an important factor in all designs where visual gauges or colored lights on control panels are a part of a control system that involves manual operation or, in the case of numerically controlled machines, monitoring for manual intervention at specified times. Control panels for such equipment must be designed to be within the visual range of the average person, and distinctive colors must be selected and used for the colored lights. The colors selected must be easily recognizable and must be capable of quickly attracting the operator's attention. Not to be overlooked is the fact that vision studies have produced much new information for use in highway design. From these same studies have come new ideas for our highway warning and information signs along with suggestions for their placement.

When a designer must consider the working environment as a part of the total design, he should realize that environment includes (1) temperature, (2) humidity, (3) lighting, (4) color schemes, and (5) sound. These are a few of the factors that also deserve full consideration in the design of a large industrial plant, a particular work area in a plant, the cockpit of an airplane (Fig. 2.12), or the cabin of a space vehicle.

The overall environment and the design of the working areas and the living quarters of the undersea laboratory shown in Fig. 2.13 were based on human requirements. The design of any undersea craft or laboratory involves problems that are similar in many ways to those encountered in the design of space vehicles, in that an artificial living environment must be created and maintained for extended periods of time. This requires a self-contained atmosphere. The members of the crew also must have ample space in which to work and live under climatic conditions that duplicate those on land. Crew members must be able to perform their tasks under normal lighting conditions and they must be able to see to the outside. Aside from the design of features and components that are related directly to the performance of the research assignments, the overall development of the undersea craft can be said to be based on the physical needs and the psychological attitudes and reactions of human beings.

In the first years of our space program, the scientists of NASA found it necessary to conduct many physiological tests to learn

Fig. 2.11 □ The Pedipulator.
This 18-foot tall balancing machine was constructed at the General Electric Research and Development Center. Its balance can be controlled easily by the natural balance action of the operator. (*Courtesy General Electric Research and Development Center.*)

Fig. 2.12 □ The cockpit of a Boeing 747—a design group's nightmare.
(*Courtesy The Boeing Company.*)

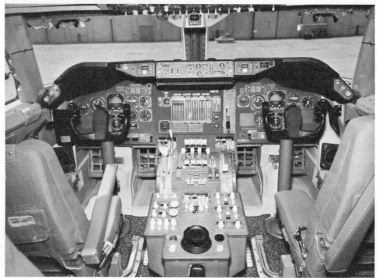

how humans might live and work in a small space capsule and how an astronaut would react to the weightlessness that he would encounter in outer space. It seemed almost certain that we could create a vehicle with a self-maintaining atmosphere and that a more or less comfortable, radiation-resistant space suit could be devised, but we were not absolutely sure that a man could travel in space and finally walk on the surface of the moon without suffering some permanent physical damage.

From the early stages of design of the initial vehicle to final decision making, almost every phase of design was in one way or another subordinated to recognized human factors. Environmental conditions, problems of space limitations, and even the placement of necessary instrumentation were adjusted to the physical and mental abilities of the astronauts as determined by tests. The well-being of the astronauts was always in mind and every effort was made to adapt the design to planned activities and their necessary movements, such as making instrument readings, reaching critical gauges, manually operating certain controls, and making adjustments with ease and efficiency under conditions where speedy action might be necessary. Even though the cabin of the space vehicle was necessarily limited in size, space had to be provided for the several astronauts to float about. No astronaut could be confined to a single position for any length of time and still maintain proper blood circulation. Every action that would be performed by an astronaut in space travel was studied under the environmental conditions that would be experienced. This NASA project represented human engineering at its best.

Human engineering is applied to a wide range of consumer items. Automobiles (Fig. 2.14), refrigerators, furniture, office equipment, lawn mowers, hand tools and other like items have long been designed with human factors in mind.

The automotive industry, which has been designing and re-designing cars for more than seventy-five years with human factors in mind, is now producing cars in accordance with stricter government safety regulations after being accused by a vocal few of having produced American cars for comfort, power, style, and speed with less than full attention being given to the safety of the occupants (Fig. 2.14). Even though the charge in this case is probably unfounded designers must realize that this situation can happen whenever engineers ignore, even to limited extent, some of their professional responsibilities and give in to the whims of stylists and to the suggestions of administrators and sales managers before having made a thorough study of each and every proposal that has been presented.

Under pressure from consumer groups and organizations interested in safety and in protecting our natural environment, there

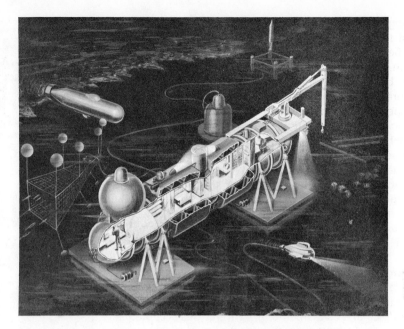

Fig. 2.13 □ An artist's conception of the Atlantis undersea habitat.
The project, which was to be designed and developed as a joint project of the University of Miami and the Space Division of Chrysler Corporation, has been abandoned at this time. The project goal was to explore the continental shelf along our coasts. (*Courtesy Space Division Chrysler Corporation.*)

PINTO OPTIONAL SEAT BELT-STARTER INTERLOCK

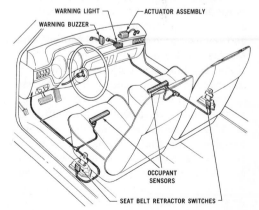

Fig. 2.14 □ An advanced concept of a seat belt-starter interlock system that hopefully will force seat belt usage.
The car cannot be started unless the driver's lap belt and that of his front seat passenger have been fastened. Sensor switches feed information to the circuits that control the interlock system. (*Courtesy Ford Motor Company.*)

has been a rising tide of governmental laws and regulations (Fig. 2.15). Many of these new laws and directives in a sense provide controls in the field of human engineering. In the case of environmental pollution, automobile companies, acting to meet regulations set forth by the Environmental Protection Agency, have developed several new devices to reduce undesirable pollutants at a scheduled percentage rate to produce an almost pollution-free car by 1976. In the interest of safety, some type of passive protection system, such as the air-bag device shown in Fig. 2.16, must be added by a set date.

These government laws and regulations have come into existence because of a growing interest on the part of the general public in human engineering. Designers in the years ahead must be fully cognizant of all such regulations and must be willing to abide by them or seek to have them changed should they appear to be unreasonable or impractical.

2.9 □ Visual design

Visual design includes the use of line, form, proportion, texture, and color to produce the eye-pleasing appearance needed to bring about the acceptance of a consumer item. Without this acceptance there would be no profit, and even though the item might otherwise have been carefully engineered, it would soon disappear from the marketplace. The sketches shown in Fig. 2.17, for the dash panel of an automobile, tastefully combine these visual elements into an attractive design. An illusion of depth has been obtained by means of pencil shading.

Since artistic styling is now recognized as being one of the most important factors in sales, many engineers have come to accept the role of stylists in the development of a consumer product, particularly when they are employed by a company that is small and cannot afford to employ one or more trained stylists. Large companies, such as the Ford Motor Company (Fig. 2.18), General Motors, and the Boeing Company (Fig. 2.19) have styling divisions. Medium-size companies often turn to nationally known organizations to get needed help; the styling is then done under contract agreement. Many books have been written about aesthetics that have proved to be helpful to engineers (see Bibliography). At present, design engineers, who have been trained largely to solve technical problems, are reading more about art in design and they are considering the eye-appeal and the overall appearance of products as part of their engineering interests (Fig. 2.20).

2.10 □ Constructive criticism

There are people who seem to find it easier to criticize than to mix praise with alternative suggestions. In any group meeting a critic should show respect for good ideas and be able to offer

Fig. 2.15 □ Air pollution control system.
A new three unit electrostatic precipitor system (light-colored units at the left) that supplements the air cleaning capacity of the three units on the roof. The complete system removes up to 99 percent of particulate matter given off by four large modern open-hearth furnaces. This total system represents the thinking of creative engineers working together to solve environmental problems. (*Courtesy Republic Steel Corporation.*)

Fig. 2.16 □ The major elements of an air-bag/seat-belt restraint system now under development by the Ford Motor Company.
The air-bag assembly and the seat-belt starter-interlock components are identified. It should be noted that there are separate signal lights for the seat-belt and air-bag systems. (*Courtesy Ford Motor Company.*)

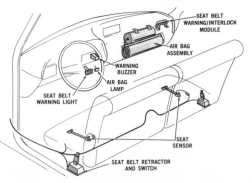

AIR-BAG/SEAT-BELT RESTRAINT SYSTEM WITH STARTER INTERLOCK

constructive suggestions. If there is to be feedback, which will hopefully lead to the introduction of more ideas, the discussion must be free of any harsh criticism. Harsh criticism may cause a sensitive person to assume a defensive position or to withdraw almost entirely from participation in group action. It is the responsibility of the project leader to prevent this from happening.

2.11 □ Recognition of a need

A design project usually begins with the recognition of a need and with the willingness of a company to enter the market with a new product. At other times an idea may be initiated and developed by an individual who either seeks economic benefits for himself or who seeks to solve some social or environmental problem. In either case, the identification of the need in itself represents a high order of creative thinking and the search for a solution to the need requires considerable self-confidence and inner courage. As can be easily observed from reviewing the achievements of our distinguished inventors of the past, those who are closely attuned to life around them become aware of needs or less-than-ideal situations that are worthy of their attention.

It is important that a proposed design activity have clear and definite objectives that will justify the money and effort to be expended in product design and development. The statement defining the objectives should identify the need and state the function the product is to perform in satisfying this need. The identification of the need may be based on the designer's personal observations, suggestions from salesmen in the field, opinion surveys, or on new scientific concepts. The identification of the design problems involved in creating the needed product comes later in the design process and will not be considered at this time.

2.12 □ A formal proposal

The statement covering the recognition of need can be used as a basis for a formal proposal that may be either a few short paragraphs or several typewritten pages. A complete proposal may include supporting data in addition to the description of the plan of action that is to be taken to solve the problem as identified. The report should have the same general form as other technical reports and might include a listing of requirements and possible limitations as then recognized. In preparing the report one should keep in mind that the proposal, when approved, gives the broad general parameters of an agreement under which the project will be developed to its conclusion. It is recommended that students in design groups follow this same procedure and submit a written proposal to their instructor who, in the classroom, takes the place of the administrator.

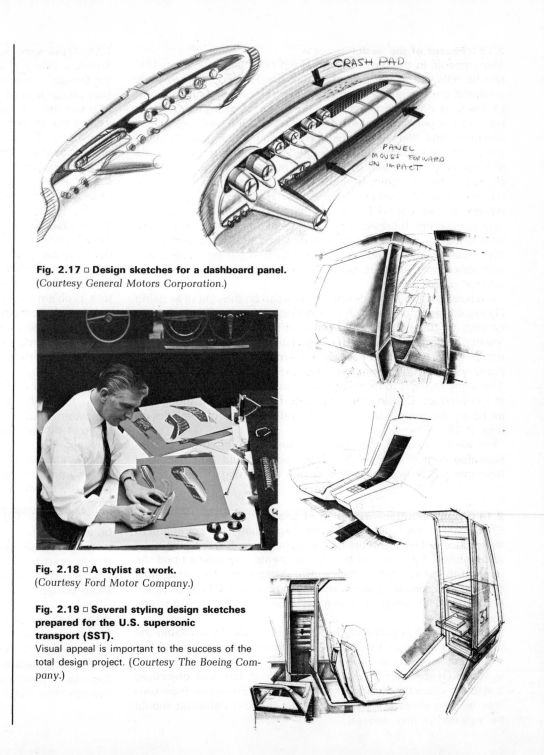

Fig. 2.17 □ Design sketches for a dashboard panel. (*Courtesy General Motors Corporation.*)

Fig. 2.18 □ A stylist at work. (*Courtesy Ford Motor Company.*)

Fig. 2.19 □ Several styling design sketches prepared for the U.S. supersonic transport (SST). Visual appeal is important to the success of the total design project. (*Courtesy The Boeing Company.*)

2.13 □ Phases of the design process

Many people in the past have prepared outlines of the steps that can be followed in the process of design (Fig. 2.21). They have prepared these outlines in order to give some semblance of order to the total design process from the point of recognition of the need to the point of marketing the product. One must recognize, however, that there are actually many combinations of steps in the overall procedure, with no single listing either the best or the one and only combination. The design procedure required in many cases can be very complex and successful designers have found different ways to achieve their goals. However, the phases of design, as recognized by this author, have been listed here in sequential order to provide some degree of direction to the student who is making his first attempt to design a product under a contrived classroom situation. More experienced persons may find it desirable to alter this outline to make it more suitable to their own method of designing.

The basic phases in the design process can be thought of as being (1) identification of need, (2) task definition (goal), (3) task specifications, (4) ideation, (5) conceptualization, (6) analysis, (7) experimental testing (Fig. 2.22), (8) design (solution) description, and (9) design for production. Phase 9 (implementation) usually is not a primary concern of a designer. Also, the designer may or may not give some thought to manufacture, distribution, and consumption of the product. Consideration of these factors may be thought of as being the tenth, eleventh, and twelfth stages of total design (Fig. 2.25).

The identification of need has been previously discussed in considerable detail in Sec. 2.11. The reader should turn back and read this section a second time before continuing.

2.14 □ Task definition—definition of goals

Briefly stated, task definition is the expression of a commitment to produce either a product or a system that will satisfy the need as identified. By means of broad statements, the product and the goals of the project are identified. The statements, as written, must be clear and concise to avoid at least some of the difficulties (often encountered in design) that can be traced directly back to poorly defined goals.

Even though it is probable that the person who has initiated the project has already gathered some pertinent information and has some preconceived ideas, it is desirable that this material not be included. It is better to present the goals in terms of objectives and then allow the designers to pursue the project in their own way, as free of restrictions as possible. The task definition should be included in the proposal.

2.15 □ Task specifications

This is a listing of parameters and data that will serve to control the design. This stage will ordinarily be preceded by some preliminary research to collect information related to the goal as defined. In preparing the task specifications, the designer or design group lists all the pertinent data that can be gathered from research reports, trade journals, patent records, catalogues, and other sources that possess information relating to the project at hand. Included in this listing should be the parameters that will tend to control the design. Other factors that deserve consideration, such as materials to be used, maintenance, and cost may also be noted.

2.16 □ Ideation

The ideation phase of design has been discussed for group procedure situations in Sec. 2.4. It is recommended that the reader review this discussion to refresh his memory regarding brainstorming procedures. It is well to remember that often a lasting solution to a problem has resulted from a creative idea selected from a number of alternative ones. The mathematical likelihood of finding an optimum solution is greatly enhanced as the list of possible alternatives grows longer. Truly great creations are possible when one, acting either alone or in a group, lets his imagination soar with little restraint. If an engineer on a design assignment can set aside his engineering know-how and blind himself, at least temporarily, to traditional approaches he is in the right mood to accept almost any challenge. This is the mood and the mental approach to great discoveries. Some call it ideation; others have called it "imagineering." If we can learn to open up our engineering minds to new approaches to our technical problems as well as to our existing problems of air-pollution, industrial waste, transportation, and even unemployment, there is no limit to what we can accomplish.

This open-minded "imagineering" approach to problems must be tempered with a sense of professional responsibility. It is no longer acceptable to solve an immediate problem using a solution that in years to come can endanger the environment. "Imagineering" is engineering for the total well being of mankind and all other living things.

2.17 □ Conceptualization

Conceptualization follows the preliminary idea (ideation) stage when all the rough sketches and notes have been assembled and reviewed to determine the one or more apparent solutions that seem to be worthy of further consideration. In evaluating alternative solutions, consideration must be given to any restrictions that have been placed on the final design. It is at this stage that the preliminary sketches should be restudied to see that all worthwhile ideas are being included and that none have been inadvertently overlooked. At no time during this phase should the designer become so set in his thinking that he does not feel free to develop still another and almost entirely new and different concept, if necessary. He should realize that it is more sensible to alter or even abandon a concept at this stage than later, when considerable money and time will have been invested in the project. The conceptualization stage of design is that stage where alternative solutions are developed and evaluated in the form of concepts. Considerable research may be necessary and task specifications must be continually reviewed. As activity progresses, many idea sketches are made as alternative approaches are worked out; these approaches are evaluated for the best possible chance of product success. It is not necessary at this stage of the design procedure for any of the alternative solutions to be worked out in any great detail.

2.18 □ Selection of optimum concept

As the design of a product or system progresses, a point is reached in the procedure when it becomes necessary to select the best design concept to be presented to the administrators in the form of a proposal. In making this final selection, a more or less complete design evaluation is made for each of the alternative concepts under consideration. These evaluations may reveal ways that costs can be reduced and value improved; means of simplifying the design to reduce costs may also become apparent.

2.19 □ Design analysis

After a design concept has been chosen as the best possible solution to the problem at hand, it must be subjected to a design analysis; that is, it must be tested against physical laws and evaluated in terms of certain design factors that are almost certain to be present.

Total analysis of a proposed design will include a review of the engineering principles involved and a study of the materials to be used. In addition, there should also be an evaluation of such design considerations as (1) the environmental conditions under which the device will operate, (2) human factors, (3) possible production methods and production problems, (4) assembly methods, (5) maintenance requirements, (6) cost, and (7) styling and market appeal. If the design is based on newly discovered scientific principles, some research may be in order before a final decision can be reached.

It is at this stage in the design procedure that physics, chemistry, and the engineering sciences are utilized most fully. In making

Fig. 2.20 □ Ford styling.
When people think of clay modeling they usually think of the development of an exterior configuration. However, the artistry of the clay sculptor is equally important in interior design. Working from sketches and renderings, sculptors create seats, door panels, instrument panels, and other interior components that are ultimately painted or trimmed to closely simulate real fabric and metal. (*Courtesy Ford Motor Company.*)

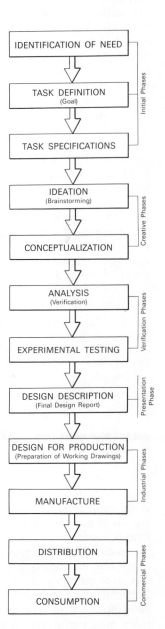

Fig. 2.21 □ The basic phases of the design process.

the usual design analysis, the engineer must depend on the formal training that he has received in school, and although considerable mathematics may be needed in making most of the necessary calculations, he will find it convenient at times to resort to graphical methods. Over the years, graphical methods have proved to be most helpful in evaluating and developing a design. For example, descriptive geometry methods can be employed for making spatial analyses and critical information can be obtained by scaling accurate drawings.

If the design analysis should prove that the design as proposed is inadequate and does not meet requirements, the designer may then either make certain modifications or incorporate into his designs some new concept that might well be a modification of an earlier idea that was abandoned along the way. More will be said in Chapter 4 about testing and analyzing at the more advanced stages of design.

2.20 □ Experimental testing

The experimental testing phase of the design process ranges from the testing of a single piece of software or hardware to verify its workability, durability, and operational characteristics through the construction and testing of a full-size prototype of the complete physical system.

A component of a product should be tested in such a way that the designer can predict its durability and performance under the conditions that will be encountered in its actual use. Needed tests may be performed using standard test apparatus or with special devices that have been produced for a particular test.

There are three types of models that may be constructed for the purpose of testing and evaluating a product. These are: (1) the mockup, (2) the scale model, and (3) the prototype.

2.21 □ Mockups (Fig. 2.22)

A mockup is a full-size ''dummy'' constructed primarily to show the size, shape, component relationships, and styling of the finished design. At this point the designer can see his conception begin to take shape for the first time. Automobile manufacturers customarily produce mockups to evaluate proposed changes in the styling of automobile bodies for new models (Fig. 2.23). Needed modifications in size and body configuration can be determined by studying the mockup and analyzing its overall appearance. Since interior styling is important also, interiors are modeled in clay to reveal the aesthetic appearance the stylist had in mind (Fig. 2.20). In the automobile industry, mockups are made to secure early approval of management for model change. A mockup is more meaningful than a sketch to those whose support and approval is needed. One must realize, however, that numerous

Fig. 2.22 □ A full-size mockup that was constructed in designing the Boeing 747, the world's largest jet airliner.
(*Courtesy The Boeing Company.*)

Fig. 2.23 □ A full-size mockup in clay is prepared to evaluate a proposed body configuration for a forthcoming model.
(*Courtesy Ford Motor Company.*)

sketches and artistic renderings (Fig. 2.19) are made before any work is started on a mockup. These sketches are used as guides. Mockups may be made of clay, wood, plastic, and so forth.

2.22 □ Preliminary and scale models (Fig. 2.24)

Models may be made at almost any stage in the design process to assist the designer in evaluating and analyzing his design. Models are made to strengthen three-dimensional visualization, to check the motion and clearance of parts, and to make necessary tests to clear up questions that have arisen in the designer's own mind or in the mind of a colleague.

The designer may prepare a preliminary model to understand more fully what the shape of a component should be, how well it may be expected to operate, and how it might be fabricated most economically. In some cases, the model might be so simple that it could be made of paper, wood, or clay.

Scale and test models may be constructed either for analysis and evaluation or for the purpose of presenting for approval the design as developed in a more or less refined stage. Scale models may be made of balsa wood, plastic, aluminum, wire, steel, or any other material that can be used to a good advantage. The designer should select a scale that will make the model large enough to permit the movement of parts should a demonstration of movement be desirable.

2.23 □ Prototypes

A prototype is the most expensive form of model that can be constructed for experimental purposes. Yet, since it will yield valuable information that is difficult to obtain in any other way, its cost is usually justified. Since a prototype is a full-size working model of a physical system that has been built in accordance with final specifications, it represents the final step of the experimental stage. In it the designers and stylists see their ideas come to life. From a prototype the designers can gain information needed for mass-production procedures that are to come later. Much can be learned at this point about workability, durability, production techniques, assembly procedures, and, most important of all, performance under actual operating conditions. Since prototype testing offers the last chance for modification of the design, possible changes to improve the design should not be overlooked, nor should a designer ever be reluctant to make a desirable change.

Since a prototype is a one-of-a-kind working model, it is made by hand using general purpose machine tools. Although it might be best to use the same materials that will be used for the mass-produced product, this is not always done. Materials that are easier to fabricate by hand are often substituted for hard-to-work materials.

Fig. 2.24 □ An administrator looks over a model of a new laboratory that is to be built at Ford's Research and Engineering Center.
The new facilities will include two wind tunnels and development and environmental test areas. (*Courtesy Ford Motor Company.*)

In the development of a design, designers deal first with the mockup; next they work out specific problems relating to single features with preliminary models; then they evaluate the design with scale models; and, finally, they may, if desirable, test the whole concept using a prototype. This order in the use of models maintains a desirable relationship between concept and analysis and represents a logical procedure for the total design process.

2.24 □ Design (solution) description—final report
In the solution description the designer describes his design on paper, to communicate his thinking to others. Although the purpose of the final report may be to sell the idea to upper management, it may also be used to instruct the production division on how to construct the product. It usually will contain specific information relating to the product or system. In some cases a process will be described in considerable detail.

A complete design description, prepared as the main part of a formal report, should include: (1) a detailed description of the device or system, (2) a statement of how the device or system satisfies the need, (3) an explanation of how the device operates, (4) a full set of layout drawings, sketches, and graphs, (5) pictorial renderings, if needed, (6) a list of parts, (7) a breakdown of costs, and (8) special instructions to insure that the intent of the designer is followed in the production stage. After the design description is accepted and approved, there remain only the commercial stages of the total design process, namely, implementation, manufacture, distribution (sales), and consumption.

2.25 □ Implementation—design for production
Implementation is that phase of the total design process when working drawings are prepared for the men in the shops who must fabricate the nonstandard parts and assemble the product. See Chapter 5.

2.26 □ Manufacture (Fig. 2.25)
From the time the task specifications are written and through all the stages up to the manufacture, the designer works closely with a production engineer who is familiar with the available shop facilities, production methods, inspection procedures, quality control, and the assembly line. If this is done, the problems encountered in the manufacturing stage will be few in number.

2.27 □ Distribution (Fig. 2.25)
Since a designer usually has little expertise in this area, task problems relating to distribution are passed along to marketing specialists. These specialists have the knowledge and the supporting staff

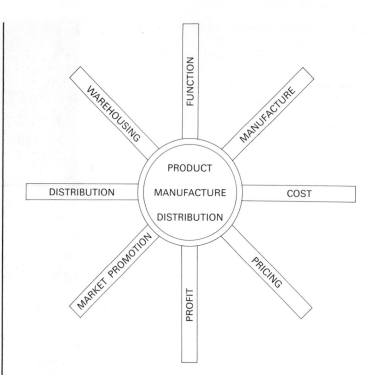

Fig. 2.25 □ Product design-factors relating to manufacturing, marketing, and distribution.

required to decide on the proper release date and to set a competitive price based upon market testing and cost and profit studies. Included among these specialists will be experienced advertising men who prepare the needed advertising and promotional literature. However, these specialists do consult the designer frequently during this stage since he knows more about the product than anyone else. He is the one person who can be depended upon to supply needed technical data and information concerning the capabilities and limitations of the product. Furthermore, the sales promotion people expect ideas from the designer that will lead to a wide and favorable distribution.

2.28 □ Consumption

There should hopefully be a consumer feedback from this last stage of the design process that will prove useful to the designer when it becomes necessary to alter and improve the product at the time of the next model change. Most of this feedback information will come from the sales force, from distributors, and from service departments. Some of this information will be in the form of complaints made by irate users, but much of it will be received as constructive suggestions. These constructive suggestions take the designer back to the need stage for the next model and a new round of the design cycle starts all over again.

The consumption stage represents the goal sought by the designer and this is where the design has its ultimate test. At this stage the design will be pronounced a success or a failure by the users and consumers who are and always will be the final judges.

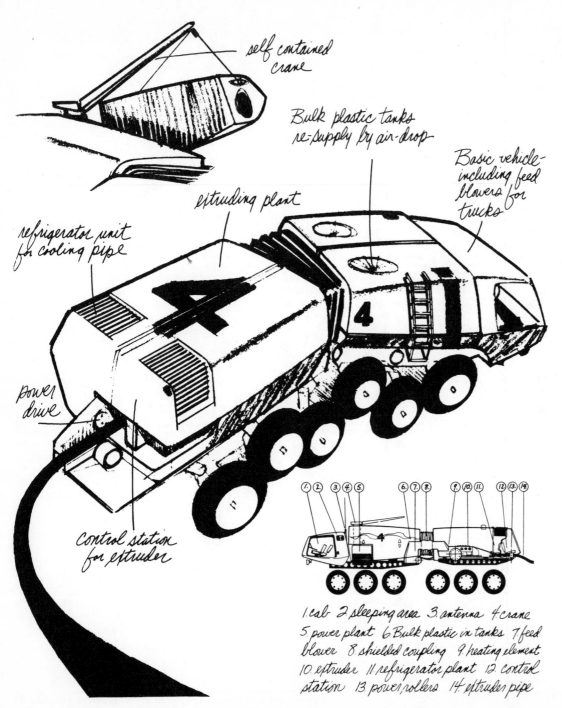

self contained crane

Bulk plastic tanks re-supply by air-drop

extruding plant

refrigerator unit for cooling pipe

Basic vehicle-including feed blowers for trucks

power drive

control station for extruder

1. cab 2. sleeping area 3. antenna 4. crane 5. power plant 6. Bulk plastic in tanks 7. feed blower 8. shielded coupling 9. heating element 10. extruder 11. refrigerator plant 12. control station 13. power rollers 14. extruder pipe

The design of a self-contained pipelayer which could be in use in 1985.
It is intended to facilitate irrigation of large tracts of desert land. The equipment as designed is capable of transporting sufficient bulk plastic to lay approximately two miles of plastic pipe from each pair of storage tanks. Tanks are to be discarded when empty and replaced by air drop. (*Courtesy of Donald Desky Associates, Inc., and Charles Bruning Company.*)

Chapter 3

The creative phases of design: organizing the design effort

3.1 □ Introduction

In Chapter 2 an effort has been made to make the reader aware of the various stages of the total design process. That process will be discussed in more detail in this chapter and in the several chapters that follow. As the reader progresses through the material, he will note that he is once again following the stages of the creative design process in sequence, beginning with a recognition of need. However, from this point on the discussion is directed toward the application of the design process to the production of much-needed items. In this chapter, the reader will find answers to some of the questions he may have about designing, along with some suggestions and recommendations that should assist him when he accepts the challenge to solve a problem and takes his first few faltering steps into the creative design process (Fig. 3.1).

3.2 □ Recognition (identification) of a need

As has been stated previously the creative design process begins with the recognition of a need (Fig. 3.2). This is the common beginning point of all the innovative thinking. Problem recognition may be either of two types: (1) the recognition of a need not heretofore met, such as the need for an electrically operated door lock to provide greater security for homes and apartments or (2) the recognition of a shortcoming or design defect in an existing product or system. The recognition of a problem of the second type usually results from criticism by consumers or from the reports of service departments. Identification of need merely means the recognition of the existence of a need, nothing more. Commitment to a problem (with the setting up of criteria that must be met) comes later, following a fact-finding study and the gathering of pertinent information.

3.3 □ The pre-screening of an idea for a new product

After a designer has become "tuned-in" on an idea he must determine if he is being realistic in thinking that his idea has a real chance for success in the marketplace and also if his company will want to launch a new product. If after due consideration the designer still feels that a real need does exist (Fig. 3.3), that his idea is economically feasible, and that it will win company approval, he is ready to take the next step and prepare the problem statement that identifies the problem and defines the goal. This step is the task definition stage.

3.4 □ Task definition (problem statement)

As has already been stated in Section 2.14, the task definition identifies the need and states in concise yet general terms what the designer or design group proposes to do to satisfy this need. The statement should be rather complete and comprehensive since it will be used later as a part of the proposal to be prepared to win approval for the project (Sec. 3.5). Many projects have been turned down because of poorly stated goals or because the statement itself has been hastily written and is not clear.

3.5 □ The proposal (Fig. 3.4)

A proposal may be a simple one-page letter or a substantial presentation that has been prepared as a technical report; this depends on the complexity of the project. In any case, some form of proposal in writing is usually required to obtain authorization to initiate a project and to substantiate a request for funds and the utilization of company manpower. It is for this reason that the definition of the problem must be presented in simple and concise language that reflects sound thinking. The purpose of the project should be clearly identified. Just a few statements implying that an individual or a group of individuals have a desire to "play around" with a problem is never enough to win the approval of an economy-minded executive who is responsible to the stockholders of the company. The proposal should include much of the preliminary data supporting the initiation of the project, data that has been obtained from literature, field reports, public opinion surveys, and studies of the competitors' products. To assure ease of comprehension, as much of this material as possible should be presented in graphical form and then analyzed, with significant factors being pointed out in written statements.

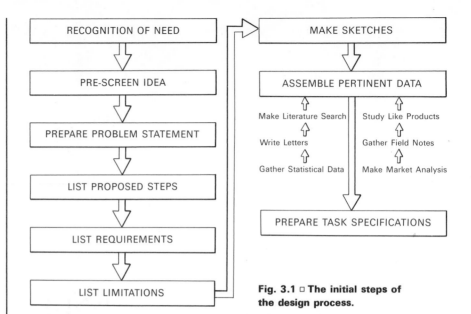

Fig. 3.1 □ The initial steps of the design process.

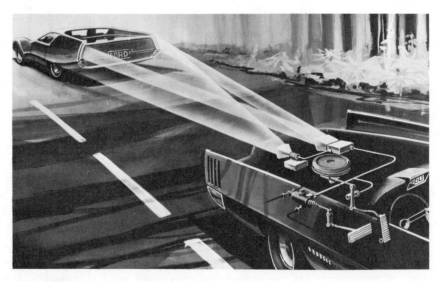

Fig. 3.2 □ An entirely new concept for the electronic control of automobiles is illustrated above by an artist's rendering.

Many additional drawings of a more conventional type will be needed to make the idea a reality. The trailing car is equipped with a transmitter (at left) which projects an invisible beam at the car ahead. Tail lights of the car in front, which does not have to be specially equipped, reflect the beam back to a receiver (right). A computer "reads" the signal and adjusts brakes and accelerator automatically so that a pre-set safe following distance will be maintained. (*Courtesy Ford Motor Company.*)

It is recommended that a formal proposal contain the following elements: (1) problem statement, (2) the plan of approach to a solution, (3) manpower needs, (4) facilities and equipment requirements, (5) budget (a detailed listing of expected expenditures), and (6) a proposed time schedule. If a formal proposal is rather lengthy and somewhat complex, it is usually desirable to add a summary at the end. In the summary, the report should be briefly reviewed and important points emphasized. Experienced designers place tabular data in an appendix as the last item in the report.

3.6 □ Task specifications

Before the task specifications can be prepared, it is essential to collect all pertinent data relating to the goal. This may involve research of all current literature, patent records, and recent reports covering scientific investigations. It may be necessary to study the products of competitors, and the catalogues of suppliers who may have needed components.

In preparing task specifications, one must list all data and parameters that tend to affect or control the design. Included should be a listing of the requirements that are essential for the achievement of the goal set for the design. It may be necessary at this time to write some of the requirements in the form of questions that hopefully can be answered later as the design process continues. The student is urged to use simple statements in listing the requirements, parameters, and unique features that are essential to the design solution. Much care is needed in preparing this list; no feature or requirement should be omitted.

Every design problem will have some limitations that must be listed as negative factors that cannot be overlooked. Such negative limitations may control size, weight, or the emission of specific types of pollutants. Other limitations might relate to environmental factors, to types of materials that must be used, and to types and sources of energy.

3.7 □ Brainstorming-design sketches

Idea sketches (Fig. 3.5) may be prepared on either plain or grid-ruled paper depending on the preference of the student or designer (Fig. 3.6). However, the designer should not become wholly dependent upon grid lines since he may need at times to explain himself by sketching on almost any type of paper, including a paper napkin. Ordinarily all sketches should be made in a permanently-bound notebook that has the pages numbered. These idea (brainstorming) sketches, along with the more fully developed work sheets that follow, form a complete record of a sequence of ideas that a designer may be called upon to present at a later time (at

Fig. 3.3 □ This unique underwater buoyancy transport vehicle moves loads up to 1000 pounds.
Developed by the Naval Undersea Research and Development Center this craft, with a crew of two men, can operate at depths as great as 850 feet. It has its own power, lighting, and built-in propulsion system. Official U.S. Navy photograph by Public Affairs Office. (*Courtesy Military Engineer Magazine.*)

Fig. 3.4 □ The elements of a proposal.

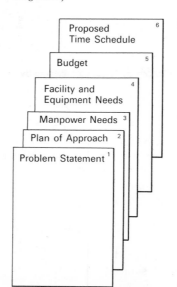

- Proposed Time Schedule ⁶
- Budget ⁵
- Facility and Equipment Needs ⁴
- Manpower Needs ³
- Plan of Approach ²
- Problem Statement ¹

a review session or even at times in a court of law if litigation should arise). The designer should be certain that he has recorded the following information on each and every idea sketch and work sheet: (1) name of the product or system, (2) name of designer or the name of the project leader of a group, (3) date (month, day, and year), and (4) page number or sketch number. Many designers, especially those who have been involved in litigation in the past, will insist upon having a knowledgeable witness add his signature and date to those sketches that represent a more or less final solution to the design problem. He does this because he knows that a design sketch can become an important and valuable record even under ordinary circumstances and at times a witness may be called upon to identify a sketch or a drawing.

Before preparing the sketches required in the design of a product or system (Fig. 3.7), the reader should review the sketching techniques discussed in Chapter 7. Design sketches need not be works of art but they should be prepared well enough to be understood quickly and easily without lengthy explanation. The lettering should be readable and the notes complete enough to reflect the designer's thoughts concerning the design during the period of ideation. If the sketches are incomplete, notes are brief, and relevant features are either carelessly omitted or only partially shown, considerable time may be expended at a later time when it becomes necessary to recall lost or discarded concepts for reconsideration. It is always well to record on a work sheet passing thoughts and general comments pertaining to an idea, since accumulated work sheets are a permanent record of the total thoughts of the designer or the design group.

Idea sketches (Figs. 3.8 and 3.9) result from ideation or idea-generation that some call brainstorming. This approach to the generation of alternative solutions for a given problem has been already presented in considerable detail in Sec. 2.4, and the student is urged to reread this section. It might be noted here, however, that a person who can produce many ideas is likely to find somewhere in the total accumulation a few really significant ones that are worthy of additional consideration as acceptable solutions. To develop the ability to come up with a number of alternate ideas requires constant practice and considerable imagination coupled with an awareness of the environment and alertness to existing conditions. Ideas are born through association, as one idea builds upon another. See also Sec. 5.38, covering the preparation of patent notebooks.

3.8 □ Systems design

Many problems require the design of a system rather than a product. Systems design problems range from the design of simple systems such as the layout of a parking lot or an arrangement

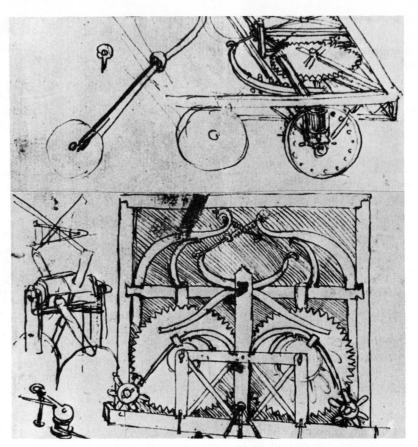

Fig. 3.5 □ An idea sketch prepared by Leonardo da Vinci (1452–1519).
The da Vinci ''automobile'' was to have been powered by two giant springs and steered by the tiller, at the left in the picture, attached to the small wheel. (*From Collection of Fine Arts Department, International Business Machines Corporation.*)

Fig. 3.6 □ A sketch of a simple mechanism.

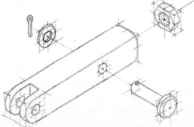

of sidewalks on a campus up to the more complex systems needed to prevent water and air pollution, to transport large masses of people, to move materials, or to maintain life in an artificial environment (Fig. 2.13).

As his initial step the designer should write the problem statement identifying the need. This statement, carefully worded, should make clear why the design effort is justified. The act of writing the problem statement for the records and for the benefit of others who may become involved, will tend to mentally tune-in the inventor to the project. In a classroom situation, a student should write the problem statement directly on his first work sheet. The statement on the sheet should be followed in order by: (1) a statement of justification of the design effort, (2) a list of requirements, and (3) a list of necessary limitations.

Should the designer find that he is listing some of the requirements in the form of questions, as mentioned in Section 3.6, he should recognize that he is writing notes to himself covering requirements that will need to be investigated before there is a complete problem identification.

The limitations vary with the type of system. There may be cost limitations, space limitations, material limitations, safety limitations, and so forth. Usually the primary limitation is the total amount of money available for the construction of the system and, since administrators who allocate funds take a dim view of a design that exceeds the total money available, a designer should make every effort not to exceed this ceiling. For example, if one-hundred-thousand dollars has already been set aside to improve a parking lot, the total cost should not exceed this amount. In the classroom, where the student does not have the background or the time needed to prepare a cost study, it is the usual practice of the instructor to require the student designer or design team to deal with overall costs only in very general terms.

As the design procedure continues, additional work sheets follow the first, one after another. The second work sheet may be used to answer questions raised on the first sheet, the third work sheet to answer questions encountered on the second, and so on until the design has been more or less finalized after considerable research and ideation.

3.9 □ Approach to product design

A product design problem is approached in much the same way as is a systems problem. A problem statement is prepared, the design need justified, and the requirements and limitations listed. However, the information that needs to be collected may be somewhat different, and unit cost usually replaces the overall construction cost of the systems design as a limitation.

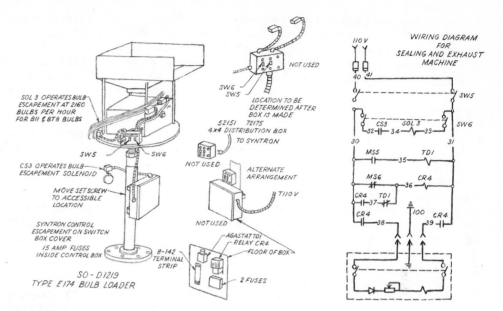

Fig. 3.7 □ A portion of a design sketch.
(*Courtesy of General Electric Company.*)

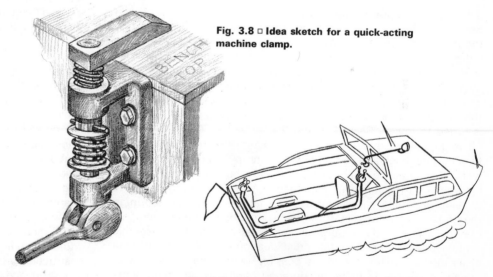

Fig. 3.8 □ Idea sketch for a quick-acting machine clamp.

Fig. 3.9 □ Idea sketch showing remote control system for a motor boat.
(*Courtesy Teleflex, Inc.*)

3.10 □ The individual working independently

A person working alone must perform all the tasks and make all the sketches and notes that would be expected from a group of persons working together as a team. As the only person working on the problem, he must communicate with himself, study alternate solutions through sketches, and record his own notes and explanations. As in the case of the team, he must generate as many ideas as possible for a solution in the hope that from the long list he will find the one best solution he seeks. Without others present to react to his thinking, he must avoid the temptation to accept and develop one of his initial ideas before he has considered numerous alternative solutions. In developing these alternative approaches he must expect to produce many sketches with explanatory notes as he considers each new or modified conception that offers promise. These are quick sketches prepared to record possible ideas that might otherwise be forgotten. Some of these sketches may record ideas that during the ideation stage apparently have little merit as solutions and yet, as is often the case, when one or more of these sketches are altered somewhat they become possible solutions worthy of further consideration. Every idea generated should be recorded in the designers workbook for future reference. See Sec. 3.7.

3.11 □ The design-team approach

Because of the complex technology involved in present-day design, the team approach to the solution has been adopted by American industry. The knowledge and training of many persons in specialized fields (engineers, physicists, chemists, mathematicians, etc.) must be brought together to reach a common goal that requires a pooling of all their talents. Much has already been said about the group approach to design in the previous chapter. The reader is urged to review Secs. 2.4 and 2.6.

It is suggested that the team design effort be organized so as to alternate back-and-forth between group work and individual work. This practice may start at the very beginning, when each member of the team can approach the problem individually to develop preliminary ideas for use at a later time in brainstorming the problem with the rest of the team. In addition, some members may be asked to do some preliminary research and to assemble some needed data prior to the initial ideation session. The inclusion of as much individual work as possible within the framework of the team activity is to be encouraged, since by so doing some individual identity is preserved within the total effort and the main criticism of the group approach, which is that the individual may be completely submerged, is minimized. Work assignments should be made by the project (group) leader who should make every effort to exploit the special training and background of each individual member of the group.

3.12 □ Selecting the project leader

A design team needs a capable leader to organize the group's activities, to assign special tasks to group members, and to lead brainstorming sessions. He should be a person who is capable of encouraging stimulating comments and unique and positive suggestions directed toward a solution to the problem. It is the project leader's responsibility to organize and schedule the design effort to achieve productive results. Finally, since he will be the spokesman for the group in almost all cases, he must be a person who can work well with those people in higher positions of authority in the organization. He must be highly respected by all persons with whom he will work and be able to maintain their confidence. Most often, an administrator will assign the project leader on the basis of his experience, and then the group will be organized to work with him.

3.13 □ Organizing the design effort

Good organization, as well as the proper and orderly scheduling of the design tasks and related activities, is essential. This is probably one of the most important phases of the design effort. Work on the assignment can progress in an orderly manner if the various activities needed to achieve productive results are anticipated. The schedule for the design effort should be drawn up almost immediately after it has been decided that continuation of the design process beyond the recognition-of-need stage is warranted. The preparation of an activities schedule usually follows the preliminary approval of the proposed design project (proposal) by an administrator or engineer who has overall responsibility for the design and development of new products. In the classroom, of course, the instructor or professor in charge will give approval.

A technique of scheduling project work that has been in use now for about twenty-five years is known as PERT (Project Evaluation Review Technique). PERT was developed in 1958 by the Special Projects Office of the Naval Bureau of Ordinance in order to take advantage of the Critical Path Method (CPM) adopted in 1956 by the E. I. duPont de Nemours Company as a management control method for the company's engineering projects. CPM procedures were first applied to the planning and scheduling of several construction projects. With this method, information covering the work sequence and the length of activities was fed to a computer; the computer output was a work schedule. In the application of PERT, major events, milestones in time that must be met to maintain a set schedule, are identified, along with the activities needed to make these events happen. Even though PERT and CPM procedures are most commonly applied for management control of product-oriented design projects (when it is necessary to meet a set schedule called for in a contract), these methods may be used to a limited extent for classroom design projects where a sequence of activities must be performed to arrive at a solution. Necessary limitations placed upon the size of this text make it impossible to present a complete discussion of the PERT and CPM methods for scheduling project activities. Those who may need to know more about the critical-path method (CPM), as it is used in many cases in conjunction with PERT, should consult the texts listed in the bibliography under ''Engineering Design.''

The application of just a few of the basic principles of PERT, as they may be applied to a classroom design project, have been illustrated in Fig. 3.10 where an imaginary events network has been used to determine the critical path for a sequence of activities.

Scheduling is of the utmost importance in planning the activities of design groups. One approach to planning and scheduling the design activities and recording the progress of a group working in a classroom situation is as follows: (1) prepare a design event and activities assignment schedule using a form suitable for listing both the design events and activities and for keeping a record of progress of design work; (2) develop the events network, if the design is sufficiently complex to warrant its preparation; and (3) record the progress of the work as the design develops. A suggested format is shown in Fig. 3.14, at the end of the chapter with the list of suggested design problems.

To prepare a PERT network one must first prepare a list of events and activities and arrange them in a relatively logical sequence. Such a list should be accepted as final only after it has been approved by engineering and administrative personnel. Ordinarily, these people should be in a position to assist in the identification of major events, to make suggestions regarding the activities, and to set the length of time from one major event to the next. See Fig. 3.10. A list should be made and revised until an acceptable outline of events has been obtained.

Final revised list—events

1 Start-Problem Identification
2 Gather Data-Literature Research
3 Gather Data-Statistical
4 Complete PERT Network
5 Prepare Task Specifications
6 Budget Accepted
7 Complete Ideation
8 Select Best Concept
9 Complete Concept Analysis
10 Complete Market Analysis
11 Construct Prototype
12 Prototype Tested
13 Complete Cost Analysis
14 Complete Final Report
15 Give Oral Presentation

An events network is a diagrammatic representation showing the events and activities necessary to achieve a design goal. Each major event, as shown by a circle, is a milestone in the total design program and the line between two events indicates the dependency of one upon the other. The arrowhead indicates the direction of progress of the activity to achieve the next event. Quite obviously, each event must be completed on time along all tracks if the design work is to be completed on schedule. A close study of the diagrammatic network in Fig. 3.10, reveals parallel paths of progress in the design work that should be carried on at the same time. The time allocated for an event may be indicated by numbers placed along the activity line. In Fig. 3.10, a single example of this practice has been shown at the far right along the path between the last two events (14–15). If the activity network shown had been completed, time values would appear along all of the activity lines. It must be noted that time is given in days. In the example, the optimistic time for preparation of the oral report is one day, the likely time is two days, and the maximum time, where unexpected difficulties are encountered, is considered to be three days. It is very unlikely that any student design team would spend more than ten to fifteen hours in preparing for the oral report.

In the illustration the beginning event is event number 1 and the final event of the design stage is number 15. From the events network, it can be easily observed which events take priority and must be completed before work can continue on schedule. For example, event 7 cannot occur until events 1, 2, 3, and 5 have been completed.

The critical path (⟹▷), beginning at the starting event (1) and ending at main event (14), is the longest time path in the network. It sets the activities and identifies the events that are of primary interest and therefore should logically receive the most attention. Other events, along other tracks, are considered to have less priority. To achieve the final goal, all the events must be completed in sequence along the critical path.

3.14 □ Obtaining ideas and information—
literature research and the use of consultants

It is essential to make a careful study of all similar products or systems already on the market and to review patent records of designs for which patents have been granted or are pending (Fig. 3.11). A review of such patent records and manufacturers' brochures may indicate the need for considerable change in the problem identification statement to insure a saleable product. If it is found that a product much like the one the designer has in mind is already available, the designer must either alter his goal drastically or completely abandon his idea and acknowledge that the element of need has been satisfied. To continue the design beyond the investigation stage in such a situation could lead to embarrassment for the designer should a patent infringement suit be lodged against his company. In addition, the designer will be still more unhappy if and when he is called to task for wasting the company's monetary resources. On the positive side, however, a survey of patent records, research reports, manufacturers' brochures, and trade journals can stimulate the mind of the designer and provide him with ideas. This happens because most design projects are related to existing designs and are intended to offer improvements to better satisfy a recognized need. Hopefully, the new product will include the application of known principles in a new way that is patentable.

There are many sources from which the designer can gather information, such as: (1) technical magazines, (2) patent records, (3) published research reports, (4) manufacturers' brochures, and (5) market periodicals. Not to be overlooked are private research reports and the helpful information and guidance that can be obtained from professional consultants.

The libraries of companies and technical schools contain both current and back issues of specialized technical and trade magazines that present recent developments in their fields. Some of the articles in these technical periodicals frequently give complete information on new principles applied to unique designs. Usually, the written material is accompanied by sketches, drawings, and photographs to clarify the explanations. In looking through these journals and magazines, the designer should look for ideas that can be incorporated, with some modification, into his design project. It is also worthwhile for him to scan the advertisements for information on materials and components that will be needed in the final product. Since these advertisements often contain little information, it is usually necessary to write to the manufacturer

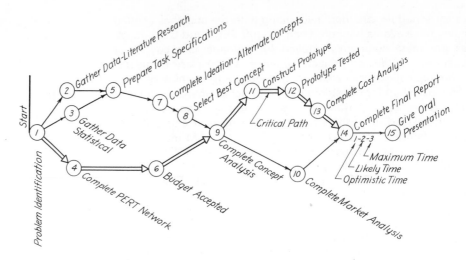

Fig. 3.10 □ A list of numbered events and an imaginary events network with the critical path indicated.
(*The time estimates are in days.*)

Fig. 3.11 □ Recognition of a need can lead to the design of a machine such as the T-600A Rock Cutter shown above.
(*Courtesy Vermeer Manufacturing Company.*)

The creative phases of design: organizing the design effort □ **40**

requesting additional information and technical specifications.

Every manufacturer of such an advertised product has prepared literature that fully describes the component that is being offered for sale on the commercial market. Some brochures are quite large and elaborate and have so much technical information that they can be considered as handbooks. This is particularly true in the case of technical publications from bearing manufacturers. These contain considerable design information. Both students and designers will find manufacturers willing to supply brochures without charge upon written request. Literature obtained in this manner should be retained in the designer's files for future reference. The student who hopes to have a career in design should start early to build his file of such material.

Not to be overlooked by either the student or the designer are the scientific and engineering society magazines (AAAS, ASME, AIChE, ASTM, ASCE, SAE, etc.) which publish the results of recent testing and research. Significant new developments that might be incorporated in a design often appear in these publications.

Many ideas can be gained from a review of existing patents, and since a patent when issued becomes a part of the public record, anyone may purchase a single copy of a specified patent for fifty cents. An order for copies of patents should be addressed to the U.S. Patent Office, Washington, D.C. 20231. The reader will find additional information on patents in Chapter 5.

If a design is at all complex, a designer will usually find it necessary to seek the assistance of specialists on phases of the design outside of his area of training and experience (Fig. 3.12). The person that the designer or the design team consults may be a metallurgist, chemist, or physicist associated with the company's research division. People with these specialized scientific backgrounds can assist with the inclusion of new discoveries into the design, can provide guidance in the application of known principles, and can minimize research. A consultant may also be a technician, another engineer, a sales representative from the designer's own company, or a representative from another company or from a division of government having an interest in the final product. If needed assistance is not available within the company, a professional consultant or a consulting firm may be employed to assist with the project and evaluate possible alternative solutions when the time comes to make decisions.

The members of a student design team can usually obtain needed assistance in the form of technical information from their own classroom instructor and from instructors teaching other courses. Not to be overlooked are fellow students who have backgrounds that enable them to provide some suggestions and needed information that will be helpful in developing a workable solution. At the author's institution each student involved with a project is

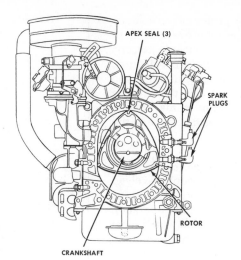

APEX SEAL (3)

SPARK PLUGS

ROTOR

CRANKSHAFT

Fig. 3.12 □ The Wankel engine.
This rotary combustion engine could be the powerplant of the future for passenger car use. Because of its different mechanical configuration, a variety of new and different mechanical problems have been encountered that have been subject to extensive research and development. Emissions are very similar to those of the piston engine. (*Courtesy General Motors Corporation.*)

supplied with a long list of names of major professors in various departments of engineering and technology who are willing to serve as consultants. At other schools, this list may include persons associated with local and regional industries who are willing to supply information. Also included are off-campus consultants, who may be based several hundred miles away, yet visit the campus at scheduled times to consult with student design teams. Help is usually available, and a student should never hesitate to seek the information that he needs.

3.15 □ Formulation and tentative evaluation of ideas (concepts)

It is at this point that the designer first becomes deeply involved in true inventive activity with the end product firmly fixed in his mind. This phase of the design process, with its rapid generation of alternate solutions (concepts) for the stated problem, is the most creative period of total design (Fig. 3.13). Real innovative thinking takes place, ideas are generated subject to few restraints, and concepts are recorded, mainly in the form of freehand sketches. The several alternatives are not developed to any extent at this stage; rather they are left to be worked out in more detail at a later time. However, it is advisable now and then to test the most likely alternatives against task specifications and to make tentative evaluations of their chances for success. In doing this, one must realize that any evaluations made at this point are tentative; real decisions are made in the next stages of design. Preliminary decision making, refinement, and design analysis are discussed in Chapter 4.

All preliminary sketches and notes that result from tentative evaluations of concepts must be carefully maintained in the project notebook so that they will be readily available for reference during the decision-making and preliminary design stages. It is not unlikely that one or more of the ideas that have been lightly regarded at this stage may at a later time be revived and incorporated into the final design or into a related design.

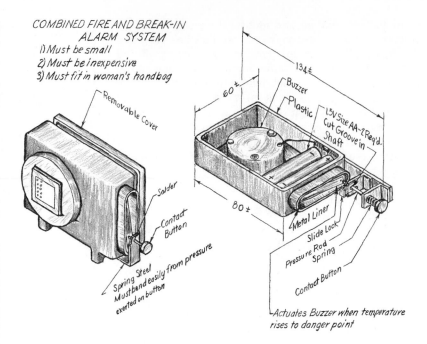

Fig. 3.13 □ Idea sketches for a small portable safety alarm that will give warning for both fire and an attempted break-in.
(Dimension values are in millimeters.)

Design projects

Design projects

The following problems are offered as suggestions to stimulate creativity and give some additional experience in both pictorial and multiview sketching. Students who have an inclination to design useful mechanisms should be encouraged to select a problem for themselves, for the creative mind works best when directed to a task in which it already has some interest. However, the young beginner should confine his activities to ideas for simple mechanisms that do not require extensive training in machine design and the engineering sciences.

It is suggested that the group leader for the design project prepare a design event and activities schedule using a form that is similar to the one shown in Fig. 3.14. A carefully prepared record should be kept of the progress of the design work.

DESIGN SCHEDULE & PROGRESS REPORT

Project:_____ Work Periods:_____

Design Team:_____

Leader:_____ **Due Dates**

Member:_____ Investigation Report:_____

_____ Proposal:_____

_____ Preliminary Report:_____

_____ Final Report:_____

ASSIGNMENT		EST. HRS	START DATE	PERCENT COMPLETE				ACTUAL HOURS
	Members			25%	50%	75%	100%	
1								
2								
3								
4								
5								

Fig. 3.14 □ Design schedule and progress record.

Suggested projects

1 Prepare design sketches (both pictorial and multiview) for an open-end wrench to fit the head of a bolt (regular series) having a body diameter of 1 in. Give dimensions on the multiview sketch and specify the material.

2 Prepare design sketches (both pictorial and multiview) for a wrench having a head with four or more fixed openings to fit the heads of bolts having nominal diameters of $\frac{5}{8}$, $\frac{3}{4}$, $\frac{7}{8}$, and 1 in. Dimension the multiview sketch and specify the material.

3 Prepare a series of design sketches for a bumper hitch (curved bumper) for attaching a light two-wheel trailer to a passenger automobile. Weight of trailer is 295 lb.

4 Prepare sketches for a hanger bracket to support a $\frac{1}{2}$-in. control rod. The bracket must be attached to a vertical surface to which the control rod is parallel. The distance between the vertical surface and the center line of the control rod is 4 in.

5 Prepare sketches for a mechanism to be attached to a two-wheel hand truck to make it easy to move the truck up and down stairs with a heavy load.

6 Prepare sketches for a quick-acting clamp that can be used to hold steel plates in position for making a lap weld.

7 Prepare a pictorial sketch of a bracket that will support an instrument panel at an angle of 45° with a vertical bulkhead to which the bracket will be attached. The bracket should be designed to permit the panel to be raised or lowered a height distance of 4 in. as desired.

8 Prepare sketches for an adjustable pipe support for a $1\frac{1}{2}$-in. pipe that is to carry a chemical mixture in a factory manufacturing paint. The pipe is overhead and is to be supported at 10-ft intervals where the adjustable supports can be attached to the lower chords of the roof trusses. The lower chord of a roof truss is formed by two angles $2\frac{1}{2} \times 2\frac{1}{2} \times \frac{5}{16}$ that are separated by $\frac{3}{8}$-in.-thick washers.

9 Prepare sketches and working drawings for an easy-to-operate fast-release glider hitch. It is suggested that the student talk with several members of a local glider club to determine the requirements for the hitch and what improvements can be made for the hitch that is being used. Follow the stages of design listed in Sec. 2.13.

10 Prepare design sketches for a camera mounting that may be quickly attached and removed from any selected surface on an automobile, boat, or other type of moving vehicle. The device is to sell for not more than $9.95. Standard parts are to be used if possible. Follow through the several stages of design listed in Sec. 2.13 as required by the instructor. Make either a complete set of working drawings or drawings of selected parts.

11 Design a tire pump that will be more efficient and easier to operate than the ones now on the market.

12 Design an electric door lock that can be opened by pushing buttons in a special pattern.

13 Design a thermostatically controlled, electrically heated sidewalk upon which snow and ice will not accumulate.

14 Design an automatic automobile theft alarm.

15 Design a weather-controlled window-closing system.

16 Design a child's toy.

17 Design an automatic pet food dispenser.

18 *Class Projects.* Each student in the class is to prepare a description (on a single sheet of paper) of an innovational idea of a needed design that he deems to be suitable for a group project. From among the ideas collected several of the most worthy will be assigned to the class for development, one idea to each group. Each group of students is to be considered as a project group and shall be headed by one student member who may be thought of as being the project engineer. The instructor will assume the role of coach for all of the groups. About midway in the total time period assigned for the development of the design, each group is to submit a written preliminary design report that shall be accompanied by sketches. A final design report (see Sec. 5.2) will be due when the project has been completed. Each report will be judged on the following: (1) evidence of good group organization, (2) quality of technical work, (3) function and appearance, (4) economic analysis, (5) manufacturing methods and requirements, and (6) over-all effectiveness of communication (written and/or oral). NOTE: The instructor will decide whether or not finished shop drawings are to be made for all of the parts. Information needed for the preparation of finished drawings may be found in Part V (Chapter 14) of this text.

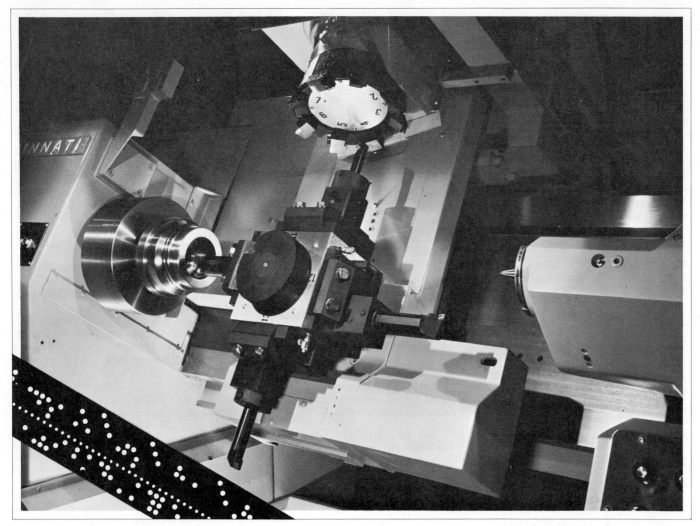

**With the coming of numerical control
(NC) American industry stands at
the threshold of a new era.**
In some industries, numerical control now
starts at the product design level where engineers and technologists utilize the computer for creating mathematical models that
may ultimately become line drawings prepared on a plotter (see Chapter 13). Under
a total system concept, now a practical reality, numerical control will be utilized from
the very early stages of the design through
drafting, manufacturing, and quality control. We are entering the electronic
age—an age that may be identified in history as the ''second industrial revolution.''
Graphic methods have now assumed new
importance and the graphic language has
become a means for communication with a
computer.* (*Courtesy Cincinnati Milling
Machine Company.*)

4.1 □ Preliminary design

After a number of design concepts have been generated in the brainstorming phase, feasibility studies must be made to determine which of the alternative designs (usually two or more) seem to be potential solutions. This preliminary decision-making step requires the designer to select (largely on the basis of feasibility studies and his own experience) the designs that seem to be superior to the others and that come closest to meeting the requirements of the task specifications. Until the preliminary design work has been undertaken, the designer usually will have only a hazy picture of these potential solutions and will not be in any position to accept or reject any of the designs, either in part or as a whole. This being the case, the design task must be pursued through the design-refinement stage where needed development work (Fig. 4.1) can be carried on that will lead to the selection of the most acceptable solution. This preliminary design work, performed largely for decision making, can be one of the more expensive phases of the design process, depending upon the extent of the drafting work and analysis required to make the needed evaluations.

In the design refinement stage those solutions that seem to have the most merit are developed to a point that will permit the intelligent selection of the one best solution. In design refinement, it

*A complete chapter covering numerically controlled machine tools may be found in the author's text, *Fundamentals of Engineering Drawing for Design, Communication, and Numerical Control*, 6th ed. (Englewood Cliffs, N.J.: Prentice-Hall, 1971).

Preliminary design and design analysis

is usually necessary to prepare design layout drawings using a scale that will permit a check of critical dimensions and clearances (Fig. 4.2). Layout drawings clarify ideas and component relationships that up to this point have appeared only in sketches that can often be very misleading. It is during this period also that the designer begins his analysis and makes some early decisions.

The design layouts (see Fig. 4.2) made at this stage must be accurate and must show all necessary detail. Sufficient information must be included for the physical development of one or more of the design concepts at a later time. It will be from one of these layouts that detailers, model makers, mockup builders, and those who may build the prototype will take the information that they will need for their work.

The beginner, of course, will wonder what is meant by *best* and what factors or design criteria make one design more acceptable than another. A few of the design criteria are: reliability of performance, durability, low cost, simplicity of maintenance, and visual attractiveness. With his criteria for the project established, the designer can resort to the use of some type of design decision chart that will more or less force him to make a detailed evaluation of each of the alternative solutions in relation to the design criteria. In applying this technique, each design factor should be weighed against its importance to the total design goal as defined in the task specifications. In tabulating the values representing the designer's evaluation of each of the criteria for the several alternatives, the alternatives might be placed at the top of the chart as headings for vertical columns, with the design criteria being listed at the left.

In selecting and defining design criteria, care must be taken to see that restrictions have not been set that are so severe that new and unique ideas will be almost completely eliminated.

Even with the use of some form of decision chart, the selection of the best design still will depend largely upon the designer's engineering experience and upon factual information that has been accumulated.

Selecting the one or more design solutions to be considered in the next stage of design could be the responsibility of a single designer, but usually the decision is made by the design team leader and his associates and, on occasion, some administrators. In any case, the designer or project leader should have all the research material, technical information, idea sketches, layout drawings, graphs, and notes that have been accumulated up to this stage readily available for reference.

4.2 □ Design analysis

Design analysis begins to some extent at the very beginning of

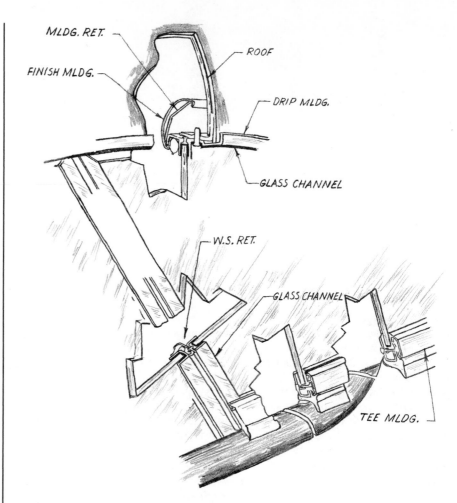

Fig. 4.1 □ **A preliminary design sketch showing a detail of construction as proposed.** (*Courtesy General Motors Corporation.*)

a design project, when the information gathered is analyzed. For example, during initial research it is often necessary to analyze the design of existing products that may resemble the end product of the proposed design. Then, as design work progresses, considerable feedback information can be obtained by analyzing the design at several points in the design procedure. As has already been pointed out, design analyses are made for each proposal of merit in the process of selecting the one best design proposal. The design analysis can bring about the overall improvement of the design since it provides information that can lead to a simpler design, greater value at less cost, a better choice of materials, and at times even new or improved methods of production.

In general, *analysis* means separating the whole into its constituent elements (opposite of synthesis). The analysis process is the method of studying the nature of something to determine its essential features and their relationship. In making a design analysis at this stage, the design analyst reviews and evaluates the design from the standpoint of function, strength, physical quantities, market requirements, production costs, consumer cost, human factors, and visual appeal (Fig. 4.3). In addition, he should test the design against physical laws utilizing mathematics, physics, chemistry, and the engineering sciences that have been the core studies of his formal education. Engineering analysis is used more at this stage than in any other phase of the design process because from the start of work through conceptualization most designers are usually content to give only passing attention to physical and scientific limitations because they know that in the analysis phase the design will be thoroughly checked against scientific and physical principles. Analysis in itself offers little if any opportunity for creative work, although it must be pointed out that analysis often requires that a concept be altered by innovation and then be reanalyzed. This being true, analysis can be said to form an iteration loop with conceptualization, in that any design proposal being analyzed may be looped back into the conceptualization phase for a revision of concept and then may be returned along the loop for reanalysis.

Should the design analysis reveal that the design proposal is wholly inadequate from the standpoint of the design specifications and that more than minor changes are needed, the designer must either reconsider a concept that may have been passed over or he must repeat the entire design process from the very beginning. This would be an unusual situation that could have been avoided if the designer or design team had been alert to potential trouble during the earlier phases of the design process.

If the analyst is someone other than the designer, he must first attempt to recreate the approach to the design that has been taken

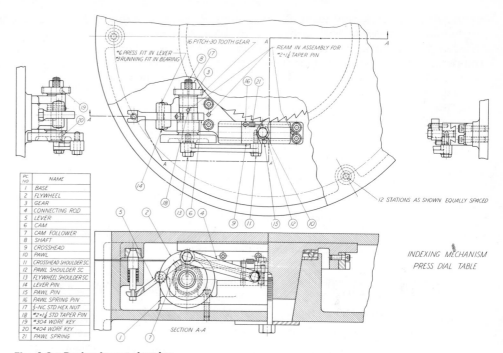

PC NO.	NAME
1	BASE
2	FLYWHEEL
3	GEAR
4	CONNECTING ROD
5	LEVER
6	CAM
7	CAM FOLLOWER
8	SHAFT
9	CROSSHEAD
10	PAWL
11	CROSSHEAD SHOULDER SC.
12	PAWL SHOULDER SC.
13	FLYWHEEL SHOULDER SC.
14	LEVER PIN
15	PAWL PIN
16	PAWL SPRING PIN
17	$\frac{3}{8}$-NC STD HEX NUT
18	#2×1$\frac{1}{4}$ STD TAPER PIN
19	#304 WDRF KEY
20	#404 WDRF KEY
21	PAWL SPRING

INDEXING MECHANISM PRESS DIAL TABLE

SECTION A-A

Fig. 4.2 □ Design layout drawing.

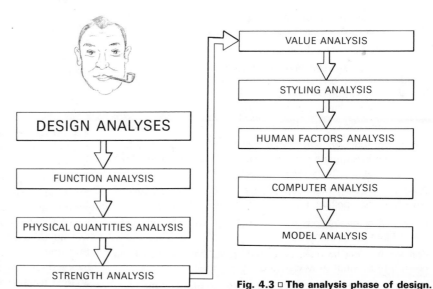

Fig. 4.3 □ The analysis phase of design.

by the designer or design team. Then he must determine if the product, as proposed, will function with high efficiency and whether or not it has the desirable features required by today's standards. Finally, he must determine if it can be produced and sold at an acceptable price and if it will have the visual appeal (styling) needed to be marketable.

A general list of factors that must be taken into account in analyzing almost every design follows. The list has been presented in the form of questions that must be answered by the analyst.

1 Is there a real need for the product?

2 Does the design satisfy this need? How well does it do so?

3 Has the function or purpose of the product been clearly determined?

4 Will the product, as designed, perform its function at a minimum cost?

5 What level of performance should be expected? Does the design insure this level?

6 Will the product, as designed, operate with highest efficiency?

7 Has the product been "value engineered"?

8 Have proper materials been specified? Has the best choice been made?

9 Have the critical properties of these materials been considered? Will these critical properties affect performance?

10 Will the product have durability?

11 Will the product be reliable under the environmental conditions in which it will operate? Will the operation be adversely affected by extremes in temperature, air pollution, chemical exposure, high altitude, and so forth?

12 Have human engineering factors been given full consideration?

13 Will the product be safe to operate? Have needed safety features been incorporated into the design? Does the product, as designed, satisfy governmental safety regulations?

14 Does the product meet federal, state, and local regulations covering noise pollution and environmental pollution?

15 Have critical dimensions been recognized and needed clearances been properly noted?

16 Will the product be mass-produced? If so, has this fact been taken into account in developing the design?

17 Have the methods most likely to be used to produce the components and assemble the product been considered in the design? Might the design be modified to permit the use of new and unique production methods and new materials?

18 Does the design provide for ease of maintenance?

19 Has the product been designed to be attractive to the customer? Have aesthetic values been incorporated into the design? Are there decorative features that will increase the prestige of the product?

20 Will the major features satisfy the consumer? Do these features insure maximum saleability?

21 Are there unique features that enhance the product and make it more marketable? If not, can one or more such features be added? Should they be added?

22 Does the design meet present-day expectations in regard to miniaturization, disposability, portability, and ease of operation?

4.3 □ Function analysis

Even though it is highly unlikely that experienced designers will design a product that will hardly function at all, it is still necessary to evaluate the functional capabilities of a design to determine how well the product will perform the tasks for which it was designed. The analyst must determine whether or not the functional capabilities of the design will meet the requirements of intended use under all conditions that may arise during operation. The function analysis can be considered to be the most important analysis that the analyst makes, since a product that will not perform its intended tasks satisfactorily and efficiently will prove to be unmarketable and that design should be abandoned. Always, the best design is the one that results in a product that best performs the intended tasks. Therefore, in making a function analysis of a design proposal, the analyst must determine whether the product, when constructed in accordance with the proposed design, will work as well as, or better than another design.

4.4 □ Physical quantities

In almost every design there are physical quantities that are critical to the design, and to ignore these important physical factors is to destroy the practicality of the product. For example, if a product must be portable to achieve maximum usage, it must be lightweight and of rather small size. If it is designed with too much weight and bulk, the product can have only limited usage and will be difficult to market. The important physical characteristics that need to be determined and evaluated in the case of almost every design are overall size, shape, weight, and materials. The analyst must evaluate these properties in light of known requirements while keeping in mind that in nearly every case good design is characterized by the achievement of maximum strength and durability with a minimum of weight and size, all at low cost.

4.5 □ Strength analysis

Strength evaluation should follow the analysis of function and physical quantities since strength is closely associated both with function and with size and weight. If the design is not adequate to take the stresses imposed by the loading, the design will result in a product that will not function as required to satisfy the need that exists. For example, the design of a truss that will not support the maximum load expected is an unacceptable design for that truss. Strength is also interrelated with the intelligent selection of optimum materials that will satisfy design requirements at minimum cost. Optimization as it relates to material selection is discussed in Sec. 4.6.

A strength analysis usually should be made by an engineer or technologist who has acquired considerable knowledge of materials and their characteristics and who also is capable of determining the nature and magnitude of the stresses that may be imposed.

Graphical methods are quite commonly used to determine stress in the components of a structural system. Explanations of the principles underlying these methods as they are applied to stress analysis may be found in Chapter 9. Graphical methods have proved to be adequate to analyze most structural systems for their behavior under stress induced by loading and by external forces.

Information on materials and their characteristics may be found in Chapter 6.

4.6 □ Value analysis (optimization)

Optimization is important to good design. In essence, optimization is the attempt to achieve, to as great a degree as possible, desired product characteristics. For example, the designer tries to get the best quality while utilizing the least material; the most power output for the least weight (engines-motors); and the best appearance (see styling) at the lowest cost. Analysis of a project design must include an overall evaluation to determine whether the design is an optimum design giving maximum performance at minimum cost.

Value analysis programs provide a means for the determination of the cost of a product in relation to its function with the purpose of identifying and removing unnecessary costs that inadvertently may be incorporated into the design. The use of value analysis procedures represents the acceptance of the responsibility to obtain a dollar of value for each dollar spent for production. In other words, value analysis seeks the achievement of greater product value for less money.

The systematic application of value analysis techniques is now called for in most Defense Department contracts so as to assure maximum value with the reliable performance of essential functions at the lowest possible cost to the taxpayer. The following definition has been set forth by the Department of Defense: "Value analysis is the technical analysis of an item by an engineer or a team of specialists to determine whether the function of the item can be achieved less expensively without compromising quality."

Basically there are two types of values that may be considered in making a value analysis of a design. These are: (1) use (function) value, and (2) prestige value. The prestige value analysis is relevant mainly to consumer products that face strong competition. Maximum use value is obtained by providing for the acceptable performance of a function at the lowest possible cost.

It is because of the experience that industrial companies have gained in making required value analyses under government contracts that value analysis techniques and methods have been extended and are now being applied to the designs of many of the more complex consumer products and systems.

4.7 □ Styling analysis

During the last few decades it has become evident that the marketing of products has become more and more dependent upon outward appearance. The year-by-year change in body styles by automobile companies is one of the best examples of the use of styling to promote sales (Fig. 4.4). In the past, styling has been done by stylists, industrial designers, artists, and to a limited extend by engineers. Now engineers are becoming more deeply involved in the task of making the product that they have designed more pleasing in its outward appearance and therefore more marketable. The best possible styling of a product usually results when the function is reflected in beauty of form, with little if any artificial ornamentation being added.

Even though the reaction of the public to the engineer's or stylist's effort may not be known until feedback has been received through the sales department, the initial effort should be analyzed to determine whether the design for appearance is either outmoded or too far advanced for ready acceptance by present-day society. A properly styled design should be for the present, and not directed to the past or to the distant future. Finally, since modern design includes product appearance as well as product function, the relationship of inner parts, and production methods, today's engineer should have the background needed to participate in all phases of total design.

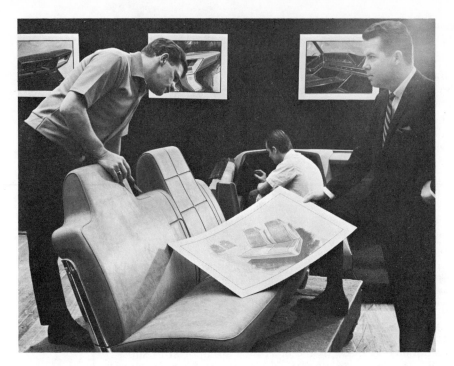

Fig. 4.4 □ Styling the interior of an automobile combines both science and art. All automotive finishes must undergo long and rigorous tests. (*Courtesy Ford Motor Company.*)

4.8 □ Human engineering analysis

Almost every product and system that is designed will in one way or another serve man and it is to be hoped that he ultimately will be better off due to its existence and use. Therefore, the analyst must evaluate the use of the product from the standpoint of human requirements, both mental and physical. He must also analyze the design and investigate the operation of the product to determine whether or not it can be operated safely and without causing undue fatigue. Some other questions that may require an answer, particularly in the case of man-machine operations, are: (1) can the recording dials be easily and accurately read, (2) are the controls in the best position for ease of operation, (3) has the work area been correctly located, (4) is the design realistic in regard to human motor activities and human ability to respond to stimuli calling for the performance of needed tasks with speed and accuracy, (5) has the design included all possible features that will make the operation of the machine or system as nearly fool-proof as possible, and finally, (6) have all needed safety features been incorporated into the design?

Human engineering has been discussed previously in Chapter 2. The reader is urged to return to Sec. 2.8 to review the factors important in analyzing a product or system from the standpoint of human engineering.

4.9 □ Computer analysis (Fig. 4.5)

Each year the computer is utilized more and more to solve engineering problems. This practice leaves the engineer with added time that can be used for the creative work that is unique to the domain of the mind of man. Only man can think and arrange, then rearrange and test features in creative combinations to produce products to satisfy the needs of man.

There have been phenomenal advances in the development of computers since the early days when their use was limited primarily to obtaining quick answers to programmed equations. At that time, problems were solved mainly in terms of numbers, with some output in the form of graphical curves represented by a printout of calculated points. Even with such limitations, computers played a big part in creative design in the early 1960's, while work continued to harness their potential capabilities for the use of scientists and engineers. At the present time, the computer and plotter are everyday useful design tools. Engineers may now sketch a proposed design with a light-pen on the face of a sensitized cathode-ray tube (Fig. 13.5), test it, alter it, and have the computer system accept it as input data to be stored in memory units for recall at a later time. If he so desires, the engineer can obtain photographically a permanent copy of his design as displayed on

Fig. 4.5 □ A CRT graphic display of an engineering problem.

the cathode-ray tube. This procedure as applicable to creative design is explained in Chapter 13.

At the CRT console a designer can design an electronic circuit and rearrange and analyze components. Should he wish to see what would happen if a component is changed in the circuit displayed, he can obtain an immediate answer by entering the new value. In a very short interval of time, the waveform that would result from the change will be displayed. A graphic display of an engineering problem is shown in Fig. 13.8.

Straight lines, circle arcs, and even exotic curves may be drawn using an X-Y plotter such as the one that was shown in Fig. 4.6. Statements in a computer language are fed to the computer by means of punched cards. The computer in turn processes the information supplied and produces either a punched or magnetic tape that is used to direct the plotter. Plotters may also be run on-line with the computer. Using a general plot program, a designer can show scientific and test data graphically for analysis. Present-day plotters are capable of producing fully annotated linear and logarithmic charts in quick time. For analysis purposes, a mathematical structure may first be plotted as an unloaded configuration and then theoretical loads may be applied through the computer to different parts of the structure (Fig. 13.10). New plotting tapes may then be generated and replots made of the new load structure.

In highway work, perspective drawings may be prepared on a plotter from the same data as that used in the earthwork program. These driver-eye views, which have been taken at intervals along the highway as designed, enable the design engineer to spot points of potential hazard that might cause driver error.

Other pictorial graphics programs have been developed to provide designers with three-dimensional pictures of their design. For example, a series of three-dimensional perspective scenes may be produced to simulate a pilot's view as he approaches an airport runway for a landing, or a fighter pilot's view of the deck of a carrier during a landing operation as the carrier moves through heavy seas. Such scenes enable designers to analyze their designs from the standpoint of man–machine relationships. Other human factors have been programmed into computer graphic systems to permit an analysis of man–machine work relationship patterns. At least one such program permits an analysis of an average man moving into various positions. Statements fed into the computer direct the simulation and the man can be seen in positions that he will assume as he performs specified tasks. The computer program permits the human figure to be placed in an infinite number of positions for analysis and study.

Fig. 4.6 □ Analysis of highway interchanges has been one of the principle applications of X-Y plotters.
(Extracted from Computer Yearbook and Directory. Courtesy American Data Processing, Inc.)

4.10 □ Scale models

As stated in Sec. 2.22, scale models may be prepared either for analysis of operation and movements or for presentation of the design for acceptance and approval. In both cases, a scale should be selected that will produce a model sufficiently large that all movements can be clearly seen and analyzed. If these requirements are satisfied, the model will be large enough to be demonstrated. Many models are made using balsa wood.

The use of mockups to analyze proposals for suggested styling has been discussed in Sec. 2.21. Prototypes are covered in Secs. 2.23 and 5.8.

4.11 □ Analysis report

After completing the design analysis the designer should prepare a design analysis report. First, the product as designed should be fully described. This should be followed by a clear description of the purpose and function of the item including the functions of its major parts. Special features should be mentioned and the factors that were considered in an attempt to bring about design optimization should be listed and described. In the main body of the report, the results of the analyses should be given in considerable detail to show to what degree the task specifications and design factors were fulfilled. The report should close with suggested improvements to the design. Preparing an analysis report at this stage of the design process is not a waste of time, since much of the report can be abridged and included in the final design report that follows.

4.12 □ Progress reports

Progress reports will be required by management executives who find it necessary to keep in close touch with the progress of a design project. Such reports offer periodic reviews of the status of a project and show whether or not there has been any wide variation or deviation from the time and budget schedule. In some cases, a letter directed to the project manager or even a memorandum circulated to interested persons may suffice as a progress report. However, more complex and costly projects may require a formal technical report that reviews in considerable detail the progress being achieved.

4.13 □ The final design report

With the completion of preliminary design work and the preparation of an analysis report, the designer must now select the best design and make his recommendations in a written design report. The preparation of a final report is discussed in detail in Chapter 5.

The scene shows the Boeing 747 final assembly line.
At the Boeing Company's plant in Everett, Washington, three 747's in final stages of completion can be seen in the line. From the factory, completed airliners are towed to the preflight line where they are prepared for production test flights and delivery to the airlines. (*Courtesy The Boeing Company.*)

Finalization of the design: approval and product development

A □ Design (final) report

5.1 □ Design reports

A designer must be fully capable of communicating with people using words (written and spoken), symbols, and drawings (Fig. 5.1). This means that he must use his verbal skills to prepare the written reports and oral presentations needed to clarify ideas that may already have been recorded by preliminary graphs, rough sketches, and drawings. Although the design (final) report discussed here is prepared mainly to sell a design concept and to present the designer's intent to management and engineering executives, it must be written with production and marketing personnel in mind as well, since it may be used by people in these areas for planning and for the preparation of advertising.

5.2 □ The design (final) report (Fig. 5.4)

The final report is prepared when all of the development work on the design project has been completed and the time has come for making a decision as to whether or not to continue the project into the design-for-production stage. Since it will be filed along with the design proposal as a permanent record of the project, the report should be well prepared and should have numerous drawings and contain a considerable amount of detail (Figs. 5.2 and 5.3). As he organizes the report the writer should not forget that the design and the data may have a strong influence on other projects, and that the most important elements of the report are the conclusions and recommendations. These, of course, are now much more reliable than those presented in the proposal and in the intermediate progress reports, since they are based on analyses, testing, and the collection of data and information over the total design development and refinement period. The main body of the report should contain a full and complete explanation of the design solution and a description of how the device or system works. Also included in this section of the report should be a listing of the advantages and disadvantages of the design; a cost breakdown; and a projected forecast of the sales potential against competitive products now on the market.

A final report should be composed so as to sell the concept of the design to persons in engineering, management, or possibly an outside customer or distributor. Since these people are apt to have widely varying backgrounds, the report should be worded in such a way that each reader may be able to gain some understanding of the operation and the potential worth of the product.

A well-prepared final report is likely to contain most but not necessarily all of the following elements (Fig. 5.4):

1 *Cover.* An inexpensive type of report folder with the company name, the company address, and the type of the company's business on the front face.

2 *Title Page.* This first page of the report should show the name of the product or system, the name of the company for which the design was developed, the name of the designer and/or designers, the date, and so forth.

3 *A Letter of Transmittal.* This letter is usually addressed either to the executive of the designer's own company or to an administrator of a client company who has been in one way or another responsible for the development of the product or system. This letter should be very brief.

4 *Table of Contents, List of Illustrations, List of Tables.*

5 *Preface.* On this page there should be a few short statements that point out the need for the product and list the goals set forth to guide the designer in developing the design.

6 *Abstract* (abridged description). This page should contain a few brief statements that summarize the description of the design solution appearing in the main body of the report.

7 *Main Body.* This part of the report covers in considerable detail the approach to the problem, and the methods and procedures used in developing the design. Graphs and drawings should be used to present data and to describe the design (Fig. 5.5). This portion of the report should be well illustrated

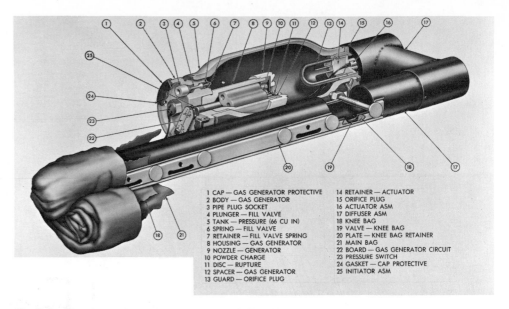

1 CAP — GAS GENERATOR PROTECTIVE	14 RETAINER — ACTUATOR
2 BODY — GAS GENERATOR	15 ORIFICE PLUG
3 PIPE PLUG SOCKET	16 ACTUATOR ASM
4 PLUNGER — FILL VALVE	17 DIFFUSER ASM
5 TANK — PRESSURE (66 CU IN)	18 KNEE BAG
6 SPRING — FILL VALVE	19 VALVE — KNEE BAG
7 RETAINER — FILL VALVE SPRING	20 PLATE — KNEE BAG RETAINER
8 HOUSING — GAS GENERATOR	21 MAIN BAG
9 NOZZLE — GENERATOR	22 BOARD — GAS GENERATOR CIRCUIT
10 POWDER CHARGE	23 PRESSURE SWITCH
11 DISC — RUPTURE	24 GASKET — CAP PROTECTIVE
12 SPACER — GAS GENERATOR	25 INITIATOR ASM
13 GUARD — ORIFICE PLUG	

Fig. 5.1 □ Shown are the components of the Allied Chemical Corporation air-bag modules to be used in the air-bag/seat-belt restraint system now under development by the Ford Motor Company.
The system has a main bag (cushion) and a small knee bag. (*Courtesy Ford Motor Company.*)

EXPERIMENTAL CATALYTIC CONVERTER

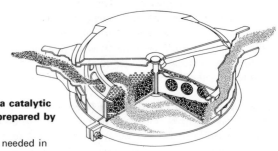

Fig. 5.2 □ This pictorial representation of a catalytic converter appeared in a technical report prepared by General Motors.
Pictorial drawings such as this one are often needed in final design reports. Some engineers think that the utilization of a complex combination of components along with a catalytic converter in the exhaust line will prove to be the most effective method of treating hydrocarbons and carbon monoxide. (*Courtesy General Motors Corporation.*)

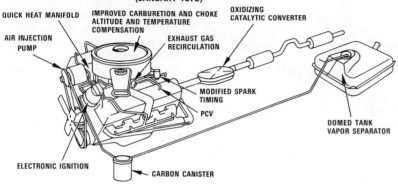

**ADVANCED EMISSION CONTROL SYSTEM
(JANUARY 1972)**

QUICK HEAT MANIFOLD

IMPROVED CARBURETION AND CHOKE
ALTITUDE AND TEMPERATURE
COMPENSATION

OXIDIZING
CATALYTIC CONVERTER

AIR INJECTION
PUMP

EXHAUST GAS
RECIRCULATION

MODIFIED SPARK
TIMING

PCV

DOMED TANK
VAPOR SEPARATOR

ELECTRONIC IGNITION

CARBON CANISTER

Fig. 5.3 □ **Items other than the catalytic converter that are a part of the total package of emission control are shown in the pictorial representation of the advanced emission control system under development at G.M.**
This presentation was extracted from a technical report prepared by G.M. engineers.
(*Courtesy General Motors Corporation.*)

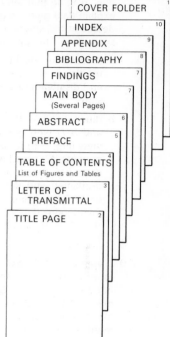

COVER FOLDER 1
INDEX 10
APPENDIX 9
BIBLIOGRAPHY 8
FINDINGS 7
MAIN BODY 7
(Several Pages)
ABSTRACT 6
PREFACE 5
TABLE OF CONTENTS 4
List of Figures and Tables
LETTER OF 3
TRANSMITTAL
TITLE PAGE 2

Fig. 5.4 □ **Contents of a design (final) report.**

Fig. 5.5 □ **The drawing below showing the internal configuration of a proposed manned undersea station appeared as one of three drawings in the Atlantis Program Project report (see Fig. 2.13).**
This was to have been a joint research and development program in cooperation with the University of Miami for ocean resources development. It is an excellent example of a type of drawing that is commonly prepared for final design reports. (*Courtesy Space Division Chrysler Corporation.*)

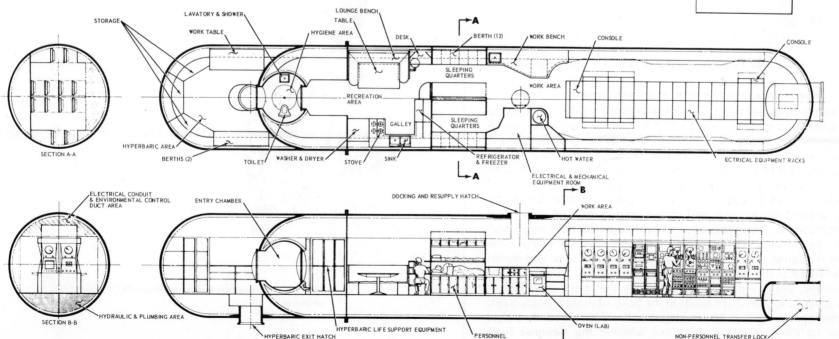

STORAGE

LAVATORY & SHOWER

LOUNGE BENCH
TABLE

WORK TABLE

HYGIENE AREA

DESK

BERTH (12)

WORK BENCH

CONSOLE

CONSOLE

SLEEPING
QUARTERS

WORK AREA

RECREATION
AREA

SECTION A-A

HYPERBARIC AREA

BERTHS (2)

TOILET

WASHER & DRYER

STOVE

SINK

GALLEY

SLEEPING
QUARTERS

REFRIGERATOR
& FREEZER

HOT WATER

ELECTRICAL & MECHANICAL
EQUIPMENT ROOM

ECTRICAL EQUIPMENT RACKS

ELECTRICAL CONDUIT
& ENVIRONMENTAL CONTROL
DUCT AREA

ENTRY CHAMBER

DOCKING AND RESUPPLY HATCH

WORK AREA

SECTION B-B

HYDRAULIC & PLUMBING AREA

HYPERBARIC EXIT HATCH

HYPERBARIC LIFE SUPPORT EQUIPMENT

PERSONNEL
LOCKERS

OVEN (LAB)

NON-PERSONNEL TRANSFER LOCK

(Fig. 5.9) so that long written explanations are not needed to explain data given in graphs and diagrams or to explain minor details of the solution shown by drawings and pictorial representations. This part of the report is likely to contain:

a Discussion—text material covering: (1) a description of how the device or system works, (2) findings, (3) the advantages and disadvantages of the design described, and (4) conclusions.

b Illustrations—(1) pictorial sketches and drawings (Figs. 5.6 and 5.7) with possibly one drawing showing an exploded view (Fig. 5.8) to reveal the assembled relationship of components, (2) a well-executed design layout drawing that fully describes the design solution in sufficient detail that the draftsmen and detailers can use it to prepare the needed detail (production) drawings at a later time with only brief instructions (Fig. 4.2), and (3) charts, graphs, and diagrams (Fig. 5.9) to supplement the written discussion.

8 *Bibliography.*

9 *Appendix.* This section can be used for data and supporting information that supplement the material given in the main body of the report. The student may include in this section many of the preliminary sketches and drawings that he has made so that he can have a complete record of the development of his design.

10 *Index.*

B □ Presentation for acceptance

5.3 □ Securing the approval of a proposed design

When the designer has finished the final report, his work is not yet completed since the design that he has selected must still be approved by a group of company associates and administrators before it will be assigned for further development in the design-for-production phase (see Sec. 5.7). To gain the approval that he must have, he will first present each member of this group of ''experts'' with a copy of the final report some days before he is to present his findings and recommendations orally. This oral presentation is important, since even though the final report presents the design in considerable detail, there are still many questions that arise. The answers to these questions may well determine whether the design is accepted or rejected. In anticipating questions, the designer should be aware that some members of this group, composed of experts in the fields of design, production, marketing, advertising, and management, will be extremely critical if for no other reason than to see whether the designer feels strongly that he has a good solution and is willing to defend it.

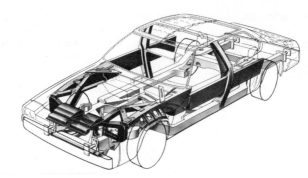

Fig. 5.6 □ This pictorial representation that appeared in a technical report shows the experimental safety vehicle (ESV) that has been built and is being tested by GM.
This car has been built solely as an engineering research project to provide needed information. (*Courtesy General Motors Corporation.*)

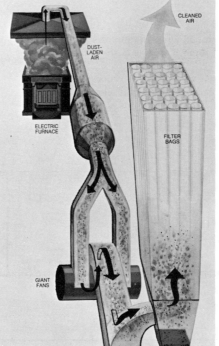

Fig. 5.7 □ Shown is a schematic pictorial representation to illustrate the use of a bag-house to control emissions from electric steelmaking furnaces.
(*Courtesy Bethlehem Steel Corporation.*)

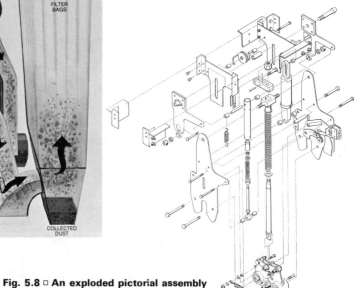

Fig. 5.8 □ An exploded pictorial assembly drawing.
(*Courtesy Lockheed Aircraft Corporation.*)

Finalization of the design: approval and product development □ **58**

The designer or design project leader should prepare a set of prompter cards (Fig. 5.10) for use during the delivery of the presentation and should have at hand a selection of illustrations (drawings) for an aide to project using a slide projector, an opaque projector, or an overhead visual cast. Students giving presentations often use all three in addition to a movable blackboard and a flip-chart mounted on an easel.

Since the principal purpose of giving the presentation is to sell the design, a designer should show enthusiasm for his design solution and be willing to defend it. In defending his solution, however, he should not attempt to cover up the disadvantages of the design while presenting only its advantages. During the delivery, he should accept criticism with grace and make every effort to temper his responses with quiet respect. Finally, he should remain in firm control of the presentation, even during the heated discussions that may arise between members of the group during the question and answer period.

5.4 □ Planning and organizing the presentation

In organizing the presentation, the designer must not overlook the composition of the group. Should it be composed almost entirely of the designer's professional associates, the presentation may be more technically oriented than would be the case with a mixed group of associates, administrators, and laymen for whom a public project may have been undertaken and whose opinion will carry the most weight in making a final decision. In planning the presentation, the designer must be aware of the fact that administrators will be primarily interested in the economic feasibility of producing the product or system, while the function and soundness of the design will be the main concern of the designer's associates. If the product manager and advertising manager are present, the former will be interested in the materials used and the ease with which the parts may be fabricated while the latter is likely to center his attention on visual appeal, special features, and whether or not the product will be competitive.

The presentation may be organized on 3-by-5-inch prompter cards using the final written report as a guide. Each statement that will be supplemented by showing a drawing, graph, or photograph should be on a separate card along with a rough sketch, if needed, as shown in Fig. 5.10. The speaker can add one or more related statements to each card and then arrange the complete set of cards in sequence for use as a guide during the delivery of his oral presentation. Before being sequenced for use as prompter cards, however, all cards having drawings requiring the preparation of transparencies and overlays for overhead projection should be placed in one pile while cards that call for drawings to be shown using an opaque projector should be stacked in a second pile.

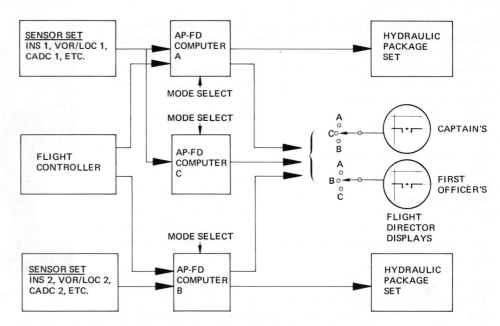

Fig. 5.9 □ A schematic diagram is often needed to clarify a portion of a design report.

The diagram shown was used in a report prepared on the design of the Boeing 747. (*Courtesy The Boeing Company.*)

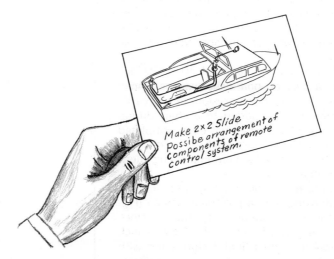

Fig. 5.10 □ A prompter card.

Next, all cards calling for the preparation of slides should be assembled together. Then finally, if a flip-chart is to be used, all cards requiring its use should be assembled in a fourth pile. Then all the work necessary to produce each type of visual aid can be done at one time, with reduced effort and expense. For example, if all illustrations that are to be presented by means of slides are photographed and processed at one time there will be a considerable saving in set-up time.

5.5 □ Visual aids for the presentation

The communication of ideas and information can usually be enhanced through the utilization of one or more types of visual aids. Visual aids both clarify and supplement the ideas being presented and ensure greater understanding by those persons who may not have had the time to thoroughly study the written report but still must be involved in making a decision on the acceptability of the design.

No slide should ever contain a mass of tabular data or long statements to be read and all slides should be prepared so that their projected views on the screen are clearly readable from every point in the room where they will be shown. When feasible, color should be used to attract attention and hold the interest of the viewer. In a flip-chart, the graphical views and the lettering should be large enough so that drawings may be readily understood and so that all statements may be read with ease.

The designer should select the visual media best suited to his presentation. For example, he may choose to use slides and a movie film; slides and transparencies utilizing an overhead projector; or a flip-chart presentation and a mockup or model. Ordinarily, it is advisable to limit the media employed and not attempt to use more than two or three methods.

A flip-chart (see Fig. 5.11) consists of a series of 30-by-40-inch sheets of paper mounted on a backing board that can be placed on an easel for general viewing. A flip-chart presentation is easily prepared and additional items may be added to a chart during the discussion. When the discussion of a chart has been completed it is flipped over the back of the easel to reveal the next chart. A flip-chart presentation is best suited for communicating ideas to a small group of no more than forty people.

An opaque projector is often used to project a page from a book, a regular photograph, or a drawing onto a projection screen (Fig. 5.12). This type of projector is the choice of many students since it enables them to use the $8\frac{1}{2}$-by-11-inch and 11-by-17-inch drawings and sketches directly from their final design report with no need to prepare slides or transparencies. An opaque projector must be used in a completely darkened room.

Photographic slides, usually 2-by-2 inches in size, are commonly

Fig. 5.11 □ The use of a flip-chart.

used for making presentations to large groups. Slides are easy to file and transport but they are time-consuming to prepare and must be shown in a darkened room. However, aside from the preparation of needed artwork, slides may be produced at a relatively low cost, even in color. The use of a combination of colors for slides adds variety to the presentation and serves to increase the level of interest of the viewers. For an effective presentation, the first slide should show either a pictorial assembly or an exploded view of the product and should be accompanied by a discussion of the product's principal function. All too often the person making a presentation starts with details, leaving each member of the audience to decide for himself how the product works to satisfy a need. This will not happen if the slides are arranged in accordance with a thoughtfully prepared script.

Transparencies are used with an overhead projector (visual-cast) that may be placed close to the screen and in front of the group. The person making the presentation can then stand near the projector in the semi-lighted room in a position where he can indicate important items on the transparency by pointing them out with a pencil or some type of pointer. The image of the pointer will appear on the screen. Whenever it is desirable, an idea may be developed in sequential steps through the use of overlays. The basic transparency is prepared first and then up to three overlays can be hinged to the mount so that they can be positioned when needed as the visual display is being developed in support of the discussion.

The operation of moving parts may be demonstrated on the screen by means of a two-dimensional plastic model placed directly on the stage (glass plate) of the overhead projector. Component parts may be moved to illustrate their operational relationship. Fixed members may be held firmly in place with tape.

5.6 □ The presentation

When all visual aids have been prepared and organized in accordance with a presentation outline, the designer should determine a time for the meeting that will be acceptable to those who must attend. He must then select a location and direct his attention to the physical facilities and the arrangement of the visual-aid equipment. Although the task of obtaining the equipment and arranging it in the room may be delegated to an experienced projectionist, the designer should check to see that all of the needed equipment is in the room and that the seating arrangement for the group is satisfactory.

In delivering his presentation, the designer should move along at a moderate pace that is not so slow as to be boring nor so rapid that some members of the group become confused. Proper timing can be determined by several rehearsals. Under no circumstance

Fig. 5.12 □ A screen and projector are used for showing charts and drawings to relatively large groups.
An opaque projector is in use. A slide projector is available. The flip-chart can be seen to the left of the screen.

should the designer appear before the critique group unrehearsed.

If the presentation has been properly organized, it will follow closely the design report that has been in the hands of each member of the group for at least several days. The emphasis in the discussion can then be placed on the significant features of the design and not on minor detail. Weaknesses in the design should be noted as the advantages are being pointed out. At the same time, acknowledgement should be made of the trade-off decisions made to achieve the optimum for one or more aspects of the design. It should be pointed out that hardly ever can all of the factors of a design be totally effective and the best that can be expected is an optimum design that has the best possible trade-off relationships. In closing the discussion, the designer should repeat the conclusions and recommendations that he has made in his technical report.

At the conclusion of the report there should be a question and answer period to resolve any questions from the group members, many of whom have diverse backgrounds and varied interests in the project. In answering these questions, the designer should refer to the written design report and supplementary data, supporting the answer whenever it is possible to do so.

In the classroom, either the presentation may be given by an appointed leader of a design group or the task may be divided among members of the group with each member presenting a portion of the report. Each design group should be allowed approximately ten minutes for their presentation with five additional minutes permitted for questions and answers.

To assist the classroom instructor in arriving at a grade for each of the design reports, class members should be requested to evaluate each presentation using a form that permits an indication of strengths and weaknesses by means of check marks. Each student should prepare a separate report for each presentation, starting with a new and unmarked form each time. This same form can be used by practicing engineers from industry if any have been included along with the students as members of the critique group. The lower portion of the form should be reserved for general comments on the delivery and content of the report.

C □ Steps of implementation

5.7 □ Steps of implementation (Fig. 5.13)

A complete set of working drawings, both detail and assembly, are needed to permit the manufacture of a product. Even though in theory the production design phase directly follows the preliminary design phase and the acceptance of the design recommended

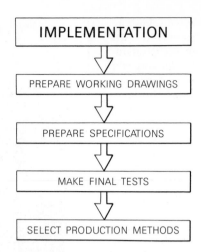

Fig. 5.13 □ The steps of implementation, the final stage of the design process that leads to production.

in the final report, in actual practice there is often no clear dividing line between these two phases. This is because detailed drawings of some components may have been started and to some extent completed well-back in the preliminary design stage, along with one or more design layout assemblies (see Fig. 4.2). Furthermore, it is not unusual for detail design drawings to be made for two or more likely solutions in the design-for-production stage and the final selection of the one best solution delayed until the information derived from these detail design drawings and related assemblies can be used in making the final decision. Such a delay in decision making tends to cause the preliminary design and the production design phases to appear to melt into one another. However, there is still a division line even though the designers themselves may not recognize this fact.

The making of a final decision should never be delayed longer than is necessary, since the design-for-production stage is to be thought of as actually starting with the acceptance of a concept and extending through the complete design of the components (hardware). Because detail and assembly drawings (working drawings) are prepared mainly for the manufacture of the needed component parts and their assembly, the preparation of these drawings is the final step that leads directly to the transformation of the designer's concept into a physical reality. To unnecessarily put aside the task of selecting the best concept will delay completion of the detail design phase and ultimately the production and distribution of the product itself. The conclusion of this stage and the completion of all needed detail and assembly drawings hopefully will lead to successful production (manufacture).

To prepare a set of working drawings (see Part D of this chapter) that will satisfy the requirements of the fabrication shops, one must have a working knowledge of fastening elements, know the rules of dimensioning, and be capable of preparing needed specifications as well. Fastening elements and dimensioning practices are covered in detail in Chapter 14.

In large engineering departments, detail design work is usually performed by draftsmen and detailers, under the direction of a drafting design supervisor who makes assignments and attempts to keep the drafting and detailing work on the schedule set by the project manager (Fig. 5.14). The supervisor responsible for the preparation of the working drawings should be somewhat familiar with most aspects of the design even though he will be expected to work in close contact with the design project leader so as to become fully aware of the complete scope of the design project. The ultimate success of the entire project may well depend to a great extent on how well these two people can work together in making critical decisions. It must be noted here that letting a draftsman alone make decisions, other than very minor ones, can

Fig. 5.14 □ A draftsman discusses part of the design of the TF-41 jet engine with his supervisor.
(*Courtesy Allison Division, General Motors Corporation.*)

be disastrous and can result in an unsuccessful product. However, the production design team as a whole, working closely with the designer, is expected to take a sound functional design and convert it into a producible design for release to the production division in the form of detail and assembly drawings.

Since most products are scheduled for either an in-stock or a delivery date, the detail design supervisor must prepare his work schedule by working back from a date set by the project manager who has the responsibility of coordinating product development work and process development for production. More will be said about process development and process management in Sec. 5.10.

5.8 □ Final testing and design modification (Fig. 5.15)

During the design for production stage, prototype studies continue as the design team makes final material selections, final alterations of geometric configurations, and assembly decisions (prototypes have been discussed in detail in Sec. 2.23). At this time some components may be fabricated and tested to predict service life and level of performance to be expected, since dependability must be achieved in good design. In setting up programs, it is often necessary to reproduce the temperature and climatic conditions under which the product will operate.

For testing the final design, the prototype can yield a great amount of useful information that otherwise might be almost impossible to obtain (Fig. 5.16). With the prototype the designer can test his conception in a final physical form; he and others can observe and analyze workability, durability, reliability and actual performance under operating conditions. Up to this time tests and analyses of the prototype have been made mainly to determine the feasibility of the design. The manufacturing engineer can study the prototype to determine appropriate fabrication techniques and assembly methods. Finally, even though a working scale model may have been constructed at an earlier time to prove the real worth of the design concept and to provide some data for analysis, the prototype as such offers an opportunity to obtain additional data under operating conditions and a one last chance for change in the design concept (Fig. 5.17). Prototypes are expensive to construct and the time and effort expended in their preparation may not be justified for many cases, particularly where the design is not a complex operational design. Scale models may serve as well under ordinary circumstances for evaluating a design and providing data for analysis. Models may be made by students at little expense.

In making a final evaluation of a prototype (Fig. 5.18), the designer should never be reluctant to make any changes that seem to be desirable even though he may consider himself to be face-to-

Fig. 5.15 □ A full-size prototype of the Pioneer Satellite A-B-C-D series.
A prototype provides the best means for an evaluation of a complex design. (*Courtesy TRW Systems Group.*)

Fig. 5.16 □ Two prototypes of the SST supersonic transport were constructed by the Boeing Company at Seattle, Washington.
This craft would have been built almost entirely of titanium. The 300-passenger production version shown above would have been 298 feet long with a wing span of 143 feet 4 inches. (*Courtesy The Boeing Company.*)

face with the finished product. The inner reluctance to make changes at this late stage of the total design process is difficult to overcome. However, it must be remembered that if a needed change is not made to correct a recognized weakness that has been brought to the attention of the designer through operation of the prototype, it may be only a matter of a few months before some word will be coming back concerning this fault either through the sales division or the repair department.

5.9 □ Final technical report and release of the production drawings

When the prototype has been tested and it has become apparent that a potentially successful product has been fully developed, the designer must prepare a technical report and make final preparations for the release of the production drawings to the manufacturing section of the production division. In doing this it may be necessary to submit the final drawings and the report covering prototype testing to those in engineering management who may be expected to give their final approval before the drawings are released for use. These executives may review the final design report (see Sec. 5.2), which was prepared at the end of the preliminary design phase, along with the report on prototype testing to determine if there are any design weaknesses that should be corrected before the product is mass produced.

5.10 □ Process development and management

Process development and process management are engineering responsibilities that are usually under the direction of a manufacturing engineer who has been assigned to the project by the product manager. Upon the release of the design to the production division, the manufacturing engineer, in consultation with the designer or leader of the design group, selects the production methods, machine tools, and other equipment that will be best suited for fabricating the product and controlling its quality. This same manufacturing engineer may have been consulted by the designer during the earlier stages of the design process when questions arose as to what forms some components should have to be produced inexpensively and what materials might be used that could be readily secured and easily processed at a low cost.

Although the manufacturing engineer will be deeply involved, it is usually the product manager's responsibility to coordinate the interaction of material flow and production operations within the plant. Material flow consists of procurement, receiving, storing, and handling of all materials needed to manufacture the finished product.

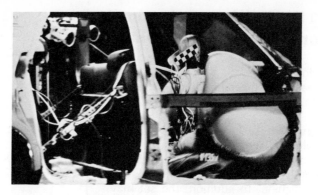

Fig. 5.17 □ The air-cushion (bag) concept may be the answer for high speed impacts.
Presently, the leader among passive restraint systems for use on automobiles and it has undergone the most development. However, numerous problems, characteristic of any new product design, have arisen. (*Courtesy General Motors Corporation.*)

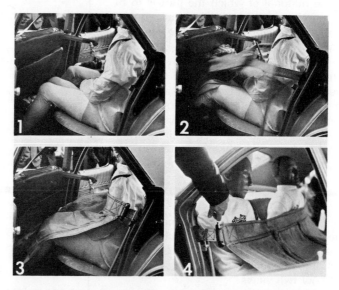

Fig. 5.18 □ A passive restraint system for rear seat passengers now being developed by Firestone is designed to provide accident protection in one-thirtieth of a second.
It is shown above in a stop-action sequence. (1) The rear passenger section with safety blanket stored in a compartment built into the back of the front seat. (2) The blanket about to be deployed. (3) The blanket in place for restraint. (4) This shows how the blanket may be released so that the passengers can leave the car. (*Courtesy The Firestone Rubber Company.*)

D □ Design for production: preparation of working drawings

5.11 □ Classes of production drawings
There are two recognized classes of machine drawings: detail drawings and assembly drawings.

5.12 □ Set of working drawings
A complete set of working drawings for a machine consists of detail sheets giving all necessary shop information for the production of individual pieces and an assembly drawing showing the location of each piece in the finished machine. In addition, the set may include drawings showing a foundation plan, piping diagram, oiling diagram, and so on.

5.13 □ Detail drawing
A detail drawing should give complete information for the manufacture of a part, describing with adequate dimensions the part's size. The title should give the material of which the part is to be made and should state the number of the parts that are required for the production of an assembled unit of which the part is a member. Commercial examples of detail drawings are shown in Figs. 5.19 and 5.20.

Since a machinist will ordinarily make one part at a time, it is advisable to detail each piece, regardless of its size, on a separate individual sheet. In some shops, however, custom dictates that related parts be grouped on the same sheet, particularly when the parts form a unit in themselves. Other concerns sometimes group small parts of the same material together thus: castings on one sheet, forgings on another, special fasteners on still another, and so on.

5.14 □ One-view drawings
Many parts, such as shafts, bolts, studs, and washers, may require only one properly dimensioned view. In the case of each of these parts, a note can imply the complete shape of the piece without sacrificing clearness. Most engineering departments, however, deem it better practice to show two views.

5.15 □ Detail titles
Every detail drawing must give information not conveyed by the notes and dimensions, such as the name of the part, part number, material, number required, and so on. The method of recording and the location of this information on the drawing varies somewhat in different drafting rooms. It may be lettered either in the record strip or directly below the views (Fig. 5.20).

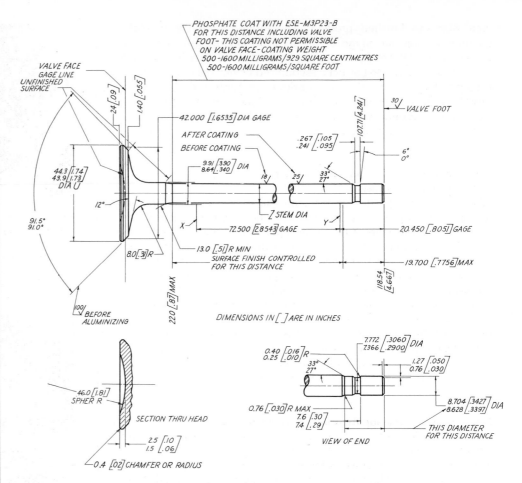

Fig. 5.19 □ A detail drawing showing the use of dual-dimensioning.
The reader should note the procedures used when both inch measurements and millimeters are shown. Many industrial drawings are now being dimensioned in this manner. See Chapter 14. (*Courtesy Ford Motor Company.*)

If all surfaces on a part are machined, a title note, "FINISH ALL OVER," is sometimes added to the detail title.

5.16 □ Title blocks and record strips

The purpose of a title or record strip is to present in an orderly manner the name of the machine, name of the manufacturer, date, scale, drawing number, and other drafting-room information.

Every commercial drafting room has developed its own standard title forms, whose features depend on the processes of manufacture, the peculiarities of the plant organization, and the established customs of particular types of manufacturing. In large organizations, the blank form, along with the borderline, is printed on standard sizes of drawing or tracing paper.

A record strip is a form of title extending almost the entire distance across the bottom of the sheet. In addition to the usual title information, it may contain a section for recording revisions, changes, and so on, with the dates on which they were adopted (Fig. 5.20).

5.17 □ Contents of the title (Figs. 5.20 and 5.21)

The title on a machine drawing generally contains the following information:

1 Name of the part.

2 Name of the machine or structure. (This is given in the main title and is usually followed by one of two words: *details* or *assembly.*)

3 Name and location of the manufacturing firm.

4 Name and address of the purchasing firm, if the structure has been designed for a particular company.

5 Scale.

6 Date. (Often spaces are provided in the preparation of the drawing for the date of completion of each operation. If only one date is given, it is usually the date of completion of the drawing.)

7 Initials or name of the draftsman who made the pencil drawing.

8 Initials of the checker.

9 Initials or signature of the chief draftsman, chief engineer, or another in authority who approved the drawing.

10 Drawing number. This generally serves as a filing number and may furnish information in code form. Letters and numbers may be so combined to indicate departments, plants, model, type, order number, filing number, and so on. The drawing number is sometimes repeated in the upper-left-hand corner (in an upside-down position), so that the drawing may be quickly identified if it should become reversed in the file.

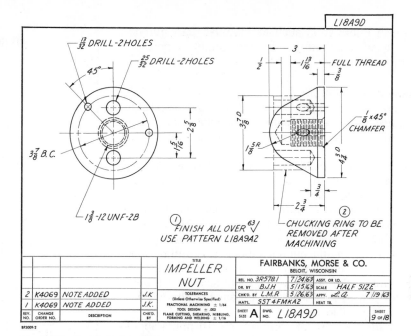

Fig. 5.20 □ A detail drawing.
(*Courtesy Fairbanks, Morse & Co.*)

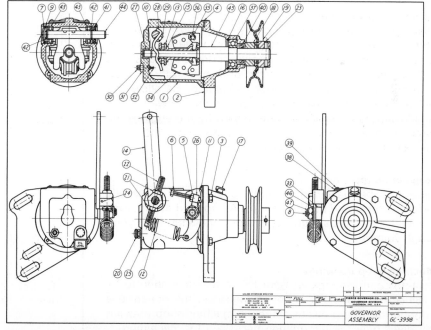

Fig. 5.21 □ An assembly drawing.
(*Courtesy Pierce Governor Co., Inc.*)

Some titles furnish information such as material, part number, pattern number, finish, treatment, estimated weight, superseded drawing number, and so on.

5.18 □ Corrections and alterations

Alterations on working drawings are made either by cancellation or by erasure. Cancellations are indicated by parallel inclined lines drawn through the views, lines, notes, or dimensions to be changed.

Superseding dimensions should be placed above or near the original ones. If alterations are made by erasure, the changed dimensions are often underlined.

All changes on a completed or approved drawing should be recorded in a revision record that may be located either adjacent to the title block (Fig. 5.20) or at one corner of the drawing. This note should contain the identification symbol, date, authorization number, character of the revision, and the initials of the draftsman and checker who made the change. The identification symbol is a numeral or letter placed in a small circle near the alteration on the body of the drawing (Fig. 5.20).

If changes are made by complete erasure, record prints should be made for the file before the original is altered. Many companies make record prints whenever changes are extensive.

5.19 □ Pattern-shop drawings

Sometimes special pattern-shop drawings, giving information needed for making a pattern, are required for large and complicated castings. If the pattern maker receives a drawing that shows finished dimensions, he provides for the draft necessary to draw the pattern and for the extra metal for machining. He allows for shrinkage by making the pattern oversize. When, however, the draft and allowances for finish are determined by the engineering department, no finish marks appear on the drawing. The allowances are included in the dimensions.

5.20 □ Forge-shop drawings

If a forging is to be machined, separate detail drawings usually are made for the forge and machine shops. A forging drawing gives all the nominal dimensions required by the forge shop for a completed rough forging.

5.21 □ Machine-shop drawings

Rough castings and forgings are sent to the machine shop to be finished. Since the machinist is not interested in the dimensions and information for the previous stages, a machine-shop drawing frequently gives only the information necessary for machining.

PIERCE GOVERNOR ASSEMBLY GC-3998 PARTS LIST			
Key No.	Part Name	Part No.	Quantity
1	Governor Body	G-9042-16	1
2	Governor Flange	G-9138-3	1
3	Sems Fastener	X-1784	4
4	Gasket	X-1425	1
5	Hex. Nut	X-977	1
6	Hex. Head Screw	X-890-4	1
7	Welch Plug	X-2019	1
8	Shoulder Stud	G-9799	1
9	Washer	X-2307	2
10	Snap Ring	X-1923	1
11	Stop Bracket	G-9556	1
12	Governor Spring	SN-1304	1
13	Thrust Bearing	X-1336-A	1
14	Throttle Lever Assembly	A-6325	1
15	Thrust Sleeve	G-10813	1
16	Snap Ring—Internal	X-1921	1
17	Oil Cup	X-2053	1
18	Spacer	G-12614-1	1
19	Governor Pulley	G-10908-1	1
20	Oil Lever Check	X-2054	1
21	Hex. Nut	X-1011-1	2
22	Adj. Screw Eye	G-12306	1
23	Roll Pin	X-2620	1
24	Roll Pin	X-2602	1
25	Oil Lever Tag	X-1945	1
26	Spring Adj. Lever	G-5715	1
27	Bushing	X-2721	1
28	Yoke	G-9838	1
29	Sems Fastener	X-1687	2
30	Bumper Screw	G-5113-1	1
31	Hex. Nut	X-246-4	1
32	Bumper Spring	SN-1481	1
33	Spacer	G-11886-1	1
34	Laminated Weight Assembly	A-2446	4
35	Weight Pin	G-14007	4
36	"E" Retaining Ring	X-2996	4
37	Ball Bearing	X-310	1
38	Name Plate	X-581	1
39	Escutcheon Pin	X-455	2
40	Oil Seal	X-652	1
41	Rocker Shaft Oil Seal	A-6118	1
42	External Snap Ring	X-1904	2
43	Ball Bearing	X-328	2
44	Rocker Shaft	G-11698	1
45	Spider and Shaft Assembly	A-6637	1
46	Washer	X-2026-4	1
47	Elastic Stop Nut	X-1845	1

Fig. 5.22 □ Parts list—governor assembly.
(*Courtesy Pierce Governor Co., Inc.*)

5.22 □ Assembly drawings

A drawing that shows the parts of a machine or machine unit assembled in their relative working positions is an assembly drawing (Fig. 5.21). There are several types of such drawings: design assembly drawings, working assembly drawings, unit assembly drawings, installation diagrams, and so on, each of which will be described separately.

5.23 □ Working assembly drawings

A working assembly drawing, showing each piece completely dimensioned, is sometimes made for a simple mechanism or unit of related parts. No additional detail drawings of parts are required.

5.24 □ Subassembly (unit) drawings (Fig. 5.23)

A unit assembly is an assembly drawing of a group of related parts that form a unit in a more complicated machine. Such a drawing would be made for the tail stock of a lathe, the clutch of an automobile, or the carburetor of an airplane. A set of assembly drawings thus takes the place of a complete assembly of a complex machine.

5.25 □ Bill of material or parts list

A bill of material is a list of parts placed on an assembly drawing just above the title block, or, in the case of quantity production, on a separate sheet (Figs. 5.22 and 5.24). The bill contains the part (item or key) number, descriptive name, material, quantity (number) required, and so on, of each piece. Additional information, such as stock size, pattern number (castings), and so forth, is sometimes listed.

Suggested dimensions for ruling a bill of material are shown in Fig. 5.24.

When listing standard parts in a bill of material, the general practice is to omit the name of the materials and to use abbreviated descriptive titles. A pattern number may be composed of the commercial job number followed by the assigned number one, two, three, and so on. It is suggested that parts be listed in the following order: (1) castings, (2) forgings, (3) parts made from bar stock, and (4) standard parts.

Sometimes bills of material are first typed on thin paper and then blueprinted. The form may be ruled or printed (Fig. 5.22).

5.26 □ Title

The title strip on an assembly drawing usually is the same as that used on a detail drawing. It will be noted, when lettering in the block, that the title of the drawing is generally composed of the name of the machine followed by the word *assembly* (Figs. 5.21 and 5.24).

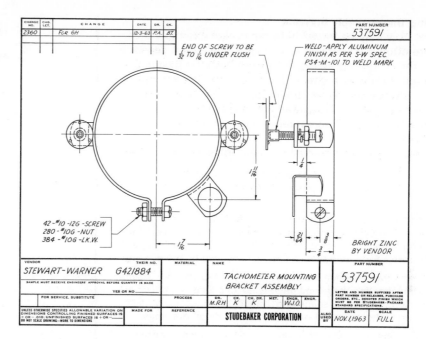

Fig. 5.23 □ A unit assembly drawing.
(*Courtesy Studebaker Corporation.*)

9	$\frac{1}{2} \times 1\frac{1}{16}$ PLAIN WASHER	1	
8	$\frac{3}{8}$-24x$\frac{1}{2}$ SLOTTED DOG PT. SET SC.	1	
7	#10-24x$\frac{3}{4}$ FLAT HD. MACH. SC.	6	
6	BALL	2	C.R.S.
5	HANDLE	1	C.R.S.
4	VISE SCREW	1	C.R.S.
3	JAW PLATE	2	C.R.S.
2	JAW PATT. NO. 19742-2	1	C.I.
1	BASE PATT. NO. 19742-1	1	C.I.
ITEM	NAME	NO PER UNIT	MATERIAL

LIMITS, UNLESS OTHERWISE NOTED: FRACTIONAL $\pm\frac{1}{64}$ DECIMAL $\pm$.010 ANGULAR $\pm\frac{1}{2}°$

TITLE OF UNIT

VISE ASSEMBLY

SCALE FULL SIZE
APPROVED WJL

LAFAYETTE, INDIANA

TRACED BY J.H.D   CHECKED BY J.H.P   DATE 12-10-61
DRAWN BY DOE, JOHN H.   CODE WJL-E-15   DRAWING NO. 19742

Fig. 5.24 □ A bill of material.

5.27 □ Making the assembly drawing

The final assembly may be traced from the design assembly drawing, but more often it is redrawn to a smaller scale on a separate sheet. Since the redrawing, being done from both the design and detail drawings, furnishes a check that frequently reveals errors, the assembly always should be drawn before the details are accepted as finished and the blueprints are made. The assembly of a simple machine or unit is sometimes shown on the same sheet with the details.

Accepted practices to be observed on assemblies are as follows:

1 *Sectioning.* Parts should be sectioned using the ANSI symbols shown in Fig. 8.58. The practices of sectioning apply to assemblies.

2 *Views.* The main view, which is usually in full section, should show to the best advantage nearly all the individual parts and their locations. Additional views are shown only when they add necessary information that should be conveyed by the drawing.

3 *Hidden Lines.* Hidden lines should be omitted from an assembly drawing, for they tend merely to overload it and create confusion. Complete shape description is unnecessary, since parts are either standard or are shown on detail drawings.

4 *Dimensions.* Overall dimensions and center-to-center distances indicating the relationship of parts in the machine as a whole are sometimes given. Detail dimensions are omitted, except on working-assembly drawings.

5 *Identification of Parts.* Parts in a machine or structure are identified on the assembly drawing by numbers that are used on the details and in the bill of material (Fig. 5.21). These should be made at least $\frac{3}{16}$ in. high and enclosed in a $\frac{3}{8}$-in. circle. The centers of the circles are located not less than $\frac{3}{4}$ in. from the nearest line of the drawing.

5.28 □ Installation assembly drawings

An installation drawing gives useful information for putting a machine or structure together. The names of parts, order of assembling parts, location dimensions, and special instructions for operating may also be shown.

5.29 □ Outline assembly drawings

Outline assembly drawings are most frequently made for illustrative purposes in catalogs. Usually they show merely overall and principal dimensions. Their appearance may be improved by the use of line shading.

5.30 □ Exploded pictorial assembly drawings for design reports and instruction manuals

Exploded pictorial assembly drawings are used frequently in the parts lists sections of company catalogs, in instruction manuals

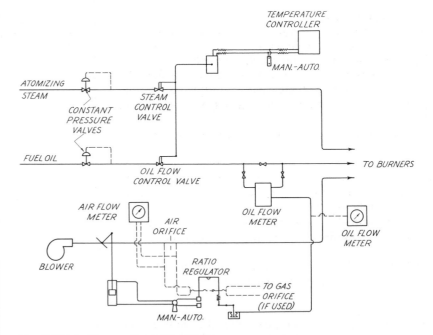

Fig. 5.25 □ A diagram assembly drawing.
(*Courtesy Instruments Magazine.*)

and final design reports. Drawings of this type are easily under-stood by those with very little experience in reading multiview drawings. Figure 5.8 shows a commercial example of an exploded pictorial assembly drawing that could have been prepared for a design report.

5.31 □ Diagram assembly drawings

Diagram drawings may be grouped into two general classes: (1) those composed of single lines and conventional symbols, such as piping diagrams, wiring diagrams, and so on (Fig. 5.25); and (2) those drawn in regular projection, such as an erection drawing, which may be shown in either orthographic or pictorial projection.

Piping diagrams give the size of pipe, location of fittings, and so on. To draw an assembly of a piping system in true orthographic projection would add no information and merely entail needless work.

A large portion of electrical drawing is composed of diagrammatic sketches using conventional electrical symbols (Fig. 5.26). Electrical engineers therefore need to know the American National (ANSI) Standard wiring symbols given in the Appendix.

5.32 □ Chemical engineering drawings

In general, the chemical engineer is concerned with plant layouts and equipment design. He must be well informed on the types of machinery used in grinding, drying, mixing, evaporation, sedimentation, and distillation, and must be able to design or select conveying machinery.

It is obvious that the determining of the sequence of operations, selecting of machinery, arranging of piping, and so on, must be done by a trained chemical engineer who can speak the basic language of the mechanical, electrical, or civil engineer with whom he must cooperate. To be able to do this, he must have a thorough knowledge of the principles of engineering drawing.

Plant layout drawings, the satisfactory development of which requires numerous preliminary sketches (layouts, scale diagrams, flow sheets, and so on), show the location of machines, equipment, and the like. Often, if the machinery and apparatus are used in the manufacturing of chemicals and are of a specialized nature, a chemical engineer is called on to do the designing. It may even be necessary for him to build experimental apparatus.

5.33 □ Electrical engineering drawings

Electrical engineering drawings are of two types: machine drawings and diagrammatic assemblies (Fig. 5.26). Working drawings, which are made for electrical machinery, involve all of the principles and conventions of the working drawings of the mechanical engineer. Diagrammatic drawings have been discussed in Sec. 5.31.

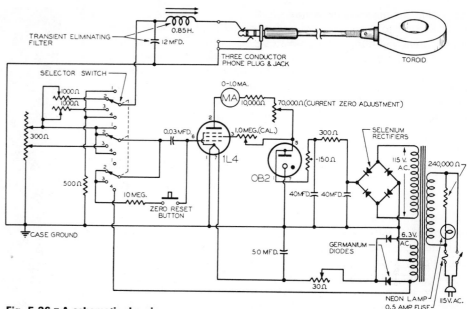

Fig. 5.26 □ A schematic drawing.
(*Courtesy General Motors Corporation.*)

5.34 □ Civil engineering drawings

The civil engineer is concerned with a broad field of construction and with civic planning. The drawings prepared for civil engineers may be in the nature of maps for city, state, and nationwide planning for streets, water systems, sewerage systems, airports, highways, railroads, harbors and waterways or they may be design, fabrication, and erection drawings for concrete and steel structures, as in the case of buildings and bridges.

E □ Process models for complete systems

5.35 □ Piping models (Fig. 5.27)

Plant models with extensive piping installations usually portray complete process systems. The use of piping models varies widely from company to company. At times, some companies build models only after complete sets of design drawings, including piping assembly drawings, have been prepared. These conventional drawings show the piping along with related mechanical equipment, structural steel components, electrical conduit, and so forth. This usage makes the model an instrument that provides the means for review and complete checking of the entire system before final confirmation of the total design. However, due to somewhat different design and construction habits, other companies build models without first preparing assembly piping drawings. In this case, the piping system is designed directly on the model with the aid of flow diagrams. When this practice is followed every pipeline and all piping accessories, such as metering instruments, valves, traps, strainers, and so forth, must be placed in proper location. When almost all design work has been done directly on the model and only a few drawings, flow charts, and sketches have been prepared, the model with all related systems shown ultimately becomes the means for communicating design information to the construction crews in the field (Fig. 5.28).

Since in most cases it is difficult to coordinate an entire processing system that involves complicated piping, all phases of the total system should be shown on the model along with the necessary piping. If all of the related mechanical and electrical equipment, along with minor structural components, platforms, ladders, and supports, are properly positioned with related items such as ducts, chutes, instruments, electrical conduits, and lights, and each pipe is traced out and checked to see that requirements are met, it is most likely that all phases of the total system will fit together at the construction site.

Fig. 5.27 □ A typical equipment layout model showing processing.
(*Courtesy Procter and Gamble.*)

When the model is nearing completion, the design should be examined critically for inaccuracies and faulty representation. In reviewing the model at this stage those responsible for the project should determine whether the proposal fulfills the requirements of the process and whether the most economical design has been adopted.

Piping models are built to scales that vary from $\frac{1}{4}$ in. to the foot to 1 in. to the foot. When conventional design and working drawings have been prepared and accuracy of the model is not important, a small-size model prepared to a scale of either $\frac{1}{4}$ or $\frac{3}{8}$ in. to the foot proves to be satisfactory. However, if the design is to be worked out on the model, the scale should be at least $\frac{3}{4}$ in. to the foot.

When the representation of each piece of equipment has been mounted in its proper location on the basic model, consisting essentially of a baseboard and a replica of the structural design, the piping can be installed through and around the formed background (Fig. 5.27). The two techniques that are commonly employed for the piping are known as (1) the "fine-wire" method and (2) the "true-scale" technique. When the fine-wire method is used, a fine wire with fibre discs or rubber sleeves mounted along its length is used to represent the pipe. The wire, normally $\frac{1}{16}$ in. in diameter, represents the center line of the pipe, while the outside diameter of the discs or sleeves indicates the outside diameter of the pipe.

Because fine-wire models sometimes prove unsatisfactory due to built-in inaccuracies and poor representation, many companies are now using the true-scale technique for building so-called true-scale models (Fig. 5.28). In building these models, plastic rod or plastic tubing, having outside diameters equal to the scaled diameters of the various sizes of standard pipe, are used for the piping. A full line of true-scale piping and piping components of plastic may be purchased ready for use. Plastic, as a model material, is easy to work, light in weight, and may be readily joined using solvent cements. It is for these reasons that it is so widely used for models.

After a model has been reviewed and approved it is color-coded, tagged, and made ready for shipment to the plant construction site. On models that have been prepared primarily for confirmation and approval, pipelines are coded to indicate particular systems to which the lines belong. That is, water lines might be shown in blue, while lines carrying a specified chemical might be presented in green. In the case of models for which there are few, if any, conventional drawings, pipelines are coded to indicate their construction specification.

Three-dimensional models enable designers to discover errors, detect interferences, and determine the most economical design

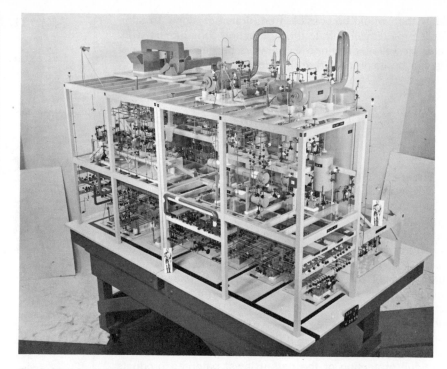

Fig. 5.28 □ A complete true scale production engineering model ready for use at the construction site.

(Courtesy E. I. du Pont de Nemours Company.)

more readily than would be possible from drawings alone. In addition, models prove invaluable to those directing construction work in the field. However, drawings will always be needed since not all details can be incorporated and indicated on a model.

F □ Patents and patent office drawings

5.36 □ A patent

A patent, when granted to an inventor, excludes others from manufacturing, using, or selling the device or system covered by the patent anywhere in the country for a period of seventeen years. In the patent document, in which the invention is fully described, the rights and privileges of the inventor are set forth and defined. Upon the issuance of the patent, the inventor has the right to either manufacture and sell the product himself with a protected market or he can assign his rights to others and charge them for the manufacture and sale or use of his invention. After the seventeen-year period has expired, the inventor no longer has any protection and the invention becomes public property and may be produced, sold, and used by anyone for the good of all.

To ensure full protection of the patent laws, patented products must be marked *Patent* with the patent number following. Even though an invention is not legally protected until a patent actually has been issued, a product often bears a statement that reads: *Patent Pending*.

In accordance with patent law "any person who has invented any new and useful process, machine, manufacture or composition of matter, or any new and useful improvement thereof, may obtain a patent," subject however, to restrictions, conditions, and requirements imposed by patent law. To be patentable an invention must be new and original and be uniquely different; it must perform a useful function; and finally the invention must not have been previously described in any publication anywhere nor have been sold or in general use in the U.S. before the applicant made the invention. It should be noted that an idea in itself is not patentable, since it is a requirement that a specific design and design description of a device accompany an application for a patent.

5.37 □ The patent attorney

When an inventor has decided that he has a patentable device, he should engage a patent attorney who will help prepare the necessary application for a patent. The inventor should depend upon this attorney to advise him as to whether or not the product

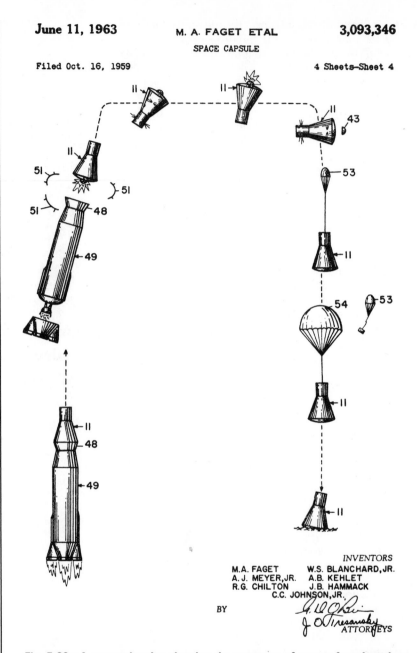

June 11, 1963   M. A. FAGET ETAL   3,093,346
SPACE CAPSULE

Filed Oct. 16, 1959   4 Sheets—Sheet 4

INVENTORS
M.A. FAGET   W.S. BLANCHARD, JR.
A. J. MEYER, JR.   A.B. KEHLET
R.G. CHILTON   J.B. HAMMACK
C.C. JOHNSON, JR.
BY
ATTORNEYS

Fig. 5.29 □ A patent drawing showing the sequence of events from launch to landing.

or process that he has conceived may infringe upon the rights of others. Since most patent attorneys are also graduate engineers, the inventor will find that he has allied himself with a person who can guide his application through the searching, investigating, and processing that takes place before a patent is granted. Generally, an application remains pending for as long as four or more years before a patent is finally granted. In this period the pending patent application may well be amended to include engineering changes that have been made in the device or system.

If after some investigation, the attorney-engineer thinks that the device is novel and therefore patentable, an application for a patent should be prepared and filed. A patent application includes a formal portion consisting of the petition, a power of attorney, and an oath or declaration. This is followed by a description of the invention, called the specifications, and a list of claims relating to it. If the device can be illustrated, one or more drawings should be included (Fig. 5.29).

In selecting a patent attorney, the inventor should remember that this man can be of service to him over many years going well beyond the time when the patent is issued. The attorney or his firm may be retained to prepare all agreements covering the sale and leasing of patent rights, to assist in negotiations relating to these rights, and if the need should arise, to handle charges of alleged infringements.

Many of the large corporations have patent divisions that operate as a section of the home office legal staff. An additional office to deal with patent matters may also be maintained in Washington, D.C. In this case, both offices, staffed with patent attorneys, will have all activities coordinated by a director of patents.

5.38 □ The role of the inventor in obtaining a patent
Even though an inventor must rely almost entirely upon a patent attorney to take the final steps to secure a patent, the inventor plays a vital role up to the time of filing the application. Very often what he has done or not done determines whether his rights to his invention can be safeguarded by his attorney.

Since court decisions in patent suits usually depend upon the inventor's ability to prove that certain design events happened on a specific date, it is always advisable to keep the design records in a hardcover permanently-bound notebook so that there will be no question as to whether or not pages have been added (as could be charged if a loose leaf notebook had been used). A well-kept patent notebook is a complete file of information covering a design. It can serve as the basis of project reports; it can spare the designer and his company the unnecessary expense of repeating portions of the experimental work; and finally, it furnishes indisputable proof of dates of conception and development.

To prepare a legally effective patent notebook the designer should:

1 Use a bound notebook having printed page numbers.

2 Make entries directly using either a black ink pen or an indelible pencil.

3 Keep entries in a chronological order. Do not add retroactive entries.

4 Add references to sources of information.

5 Describe all procedures, equipment, and instruments used for development work.

6 Insert photos of instrument and equipment setups. Add other photos showing models, mockups, and so forth.

7 Sign and date all photos.

8 Have a qualified witness sign and date every completed page. The witness cannot be a co-inventor nor can he have a financial interest in the development of the device. The witness must be a person who is capable of understanding the construction and operation of the device or system. He must also be experienced in reading drawings and understand the specifications.

9 Have the witness write the words "Witnessed and Understood" and then write and date his signature.

10 Have two or more witnesses sign and date every page after reading it.

11 Avoid having blank spaces on any page.

12 Have all new entries witnessed at least once a week.

13 Have the notebook evaluated periodically by a patent attorney. Follow his suggestions.

It takes a considerable time and effort to adhere to these thirteen recommendations for the preparation of an effective patent notebook. However, if an inventor has once become entangled in an infringement lawsuit he realizes that a little extra effort pays off.

5.39 □ Patent drawings
A person who has invented a new machine or device, or an improvement for an existing machine, and who applies for a patent is required by law to submit a drawing showing every important feature of his invention. When the invention is an improvement, the drawing must contain one or more views of the invention alone, and a separate view showing it attached to the portion of the machine for which it is intended.

Patent drawings must be carefully prepared in accordance with the strict rules of the U.S. Patent Office. These rules are published in a pamphlet entitled *Rules of Practice in the United States Patent Office*, which may be obtained, without charge, by writing to the Commissioner of Patents, Washington, D.C.

In the case of a machine or mechanical device, the complete application for a patent will consist of a petition, a ''specification'' (written description), and a drawing. An applicant should employ a patent attorney, preferably one who is connected with or regularly employs competent draftsmen capable of producing well-executed drawings that conform with all of the rules. Ordinary draftsmen lack the skill and experience necessary to produce such drawings. Two sheets of drawings for a patent for a radiation protection system are shown in Figs. 5.30 and 5.31.

Several U.S. Patent Office publications that are available to the general public have been listed in the bibliography.

5.40 □ Rules

The following rules (49–55) are quoted verbatim from the pamphlet, *Rules of Practice in the United States Patent Office:*

49. The applicant for a patent is required by law to furnish a drawing of his invention whenever the nature of the case admits of it.

50. The drawing may be signed by the inventor or one of the persons indicated in rule 25, or the name of the applicant may be signed on the drawing by his attorney in fact. The drawing must show every feature of the invention covered by the claims, and the figures should be consecutively numbered, if possible. When the invention consists of an improvement on an old machine, the drawing must exhibit, in one or more views, the invention itself, disconnected from the old structure, and also in another view, so much only of the old structure as will suffice to show the connection of the invention therewith.

51. Two editions of patent drawings are printed and published—one for office use, certified copies, etc., of the size and character of those attached to patents, the work being about 6 by $9\frac{1}{2}$ inches; and one reduction of a selected portion of each drawing for the official Gazette.

52. This work is done by the photolithographic process, and therefore the character of each original drawing must be brought as nearly as possible to a uniform standard of excellence, suited to the requirements of the process, to give the best results, in the interests of inventors, of the office, and of the public. The following rules will therefore be rigidly enforced, and any departure from them will be certain to cause delay in the examination of an application for letters patent:

(a) Drawings must be made upon pure white paper of a thickness corresponding to two-sheet or three-sheet Bristol board. The sur-

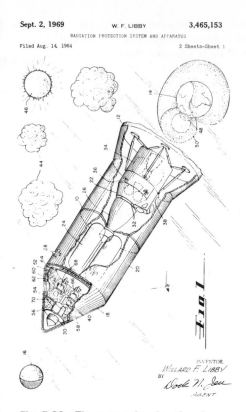

Fig. 5.30 □ The patent drawing showing a radiation protection system.
See also Fig. 5.32.

face of the paper must be calendered and smooth. India ink alone must be used, to secure perfectly black and solid lines.

(*b*) The size of a sheet on which a drawing is made must be exactly 10 by 15 inches. One inch from its edges a single marginal line is to be drawn, leaving the "sight" precisely 8 by 13 inches. Within this margin all work and signatures must be included. One of the shorter sides of the sheet is regarded as its top, and measuring downwardly from the marginal line, a space of not less than $1\frac{1}{4}$ inches is to be left blank for the heading of title, name, number, and date.

(*c*) All drawings must be made with the pen only. Every line and letter (signatures included) must be absolutely black. This direction applies to all lines, however fine, to shading, and to lines representing cut surfaces in sectional views. All lines must be clean, sharp, and solid, and they must not be too fine or crowded. Surface shading, when used, should be open. Sectional shading should be made by oblique parallel lines, which may be about one-twentieth of an inch apart. Solid black should not be used for sectional or surface shading. Free-hand work should be avoided wherever it is possible to do so.

(*d*) Drawings should be made with the fewest lines possible consistent with clearness. By the observance of this rule the effectiveness of the work after reduction will be much increased. Shading (except on sectional views) should be used only on convex and concave surfaces, where it should be used sparingly, and may even there be dispensed with if the drawing be otherwise well executed. The plane upon which a sectional view is taken should be indicated on the general view by a broken or dotted line, which should be designated by numerals corresponding to the number of the sectional view. Heavy lines on the shade sides of objects should be used, except where they tend to thicken the work and obscure letters of reference. The light is always supposed to come from the upper left-hand corner at an angle of 45°.

(*e*) The scale to which a drawing is made ought to be large enough to show the mechanism without crowding, and two or more sheets should be used if one does not give sufficient room to accomplish this end; but the number of sheets must never be more than is absolutely necessary.

(*f*) The different views should be consecutively numbered. Letters and figures of reference must be carefully formed. They should, if possible, measure at least one-eighth of an inch in height, so that they may bear reduction to one twenty-fourth of an inch; and they may be much larger when there is sufficient room. They must be so placed in the close and complex parts of drawings as not to interfere with a thorough comprehension of the same, and therefore should rarely cross or mingle with the lines. When necessarily grouped around a certain part they should be placed at a little distance, where there is available space, and connected by lines with the parts to which they refer. They should not be placed upon shaded surfaces, but when it is difficult to avoid this, a blank space must be left in the shading where the

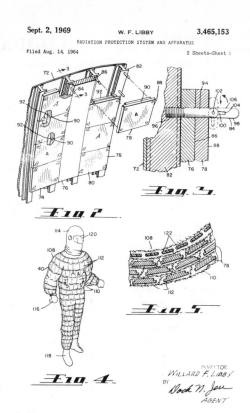

Sept. 2, 1969   W. F. LIBBY   3,465,153

RADIATION PROTECTION SYSTEM AND APPARATUS

Filed Aug. 14, 1964   2 Sheets-Sheet 2

INVENTOR.
WILLARD F. LIBBY
BY
AGENT

Fig. 5.31 □ Drawings showing the details that are related to the claims.

letter occurs, so that it shall appear perfectly distinct and separate from the work. If the same part of an invention appears in more than one view of the drawing, it must always be represented by the same character, and the same character must never be used to designate different parts.

(g) The signature of the applicant should be placed at the lower right-hand corner of each sheet, and the signatures of the witnesses, if any, at the lower left-hand corner, all within the marginal line, but in no instance should they trespass upon the drawings. The title should be written with pencil on the back of the sheet. The permanent names and title constituting the heading will be applied subsequently by the office in uniform style.

(h) All views on the same sheet must stand in the same direction and must, if possible, stand so that they can be read with the sheet held in an upright position. If views longer than the width of the sheet are necessary for the proper illustration of the invention, the sheet may be turned on its side. The space for heading must then be reserved at the right and the signatures placed at the left, occupying the same space and position as in the upright views and being horizontal when the sheet is held in an upright position. One figure must not be placed upon another or within the outline of another.

(i) As a rule, one view only of each invention can be shown in the Gazette illustrations. The selection of that portion of a drawing best calculated to explain the nature of the specific improvement would be facilitated and the final result improved by the judicious execution of a figure with express reference to the Gazette, but which must at the same time serve as one of the figures referred to in the specification. For this purpose the figure may be a plan, elevation, section, or perspective view, according to the judgment of the draftsman. All its parts should be especially open and distinct, with very little or no shading, and it must illustrate the invention claimed only, to the exclusion of all other details. (See specimen drawing.) When well executed it will be used without curtailment or change, but any excessive fineness or crowding or unnecessary elaborateness of detail will necessitate its exclusion from the Gazette.

(j) Drawings transmitted to the office should be sent flat, protected by a sheet of heavy binder's board; or should be rolled for transmission in a suitable mailing tube, but should never be folded.

(k) An agent's or attorney's stamp, or advertisement, or written address will not be permitted upon the face of a drawing, within or without the marginal line.

53. In reissue applications the drawings upon which the original patent was issued may be used upon the filing of suitable permanent photographic copies thereof, if no changes are to be made in the drawings.

54. The foregoing rules relating to drawings will be rigidly enforced. A drawing not executed in conformity thereto may be admitted for purposes of examination if it sufficiently illustrates the

United States Patent Office

3,465,153

Patented Sept. 2, 1969

1

3,465,153
RADIATION PROTECTION SYSTEM AND APPARATUS
Willard F. Libby, Los Angeles, Calif., assignor to McDonnell Douglas Corporation, Santa Monica, Calif., a corporation of Maryland
Filed Aug. 14, 1964, Ser. No. 389,734
Int. Cl. G21f 1/12, 3/02, 7/00
U.S. Cl. 250—108   16 Claims

My present invention relates generally to astronautics, the science of space flight, and more particularly to a system and apparatus and method for the protection of astronauts from the hazards of suddenly encountered radiation fields of extreme intensity in space.

Manned space flights have now been successfully achieved by both the United States and the Soviet Union. Such flights will be followed by manned space probes including lunar and interplanetary missions for the manned exploration of the Moon, Mars and Venus. The manned space program of the United States is directed towards manned exploration first of the moon and then initially only of the two planets Mars and Venus of the solar system since all of its other planets appear to be barren and lifeless. These and other probes will, of course, eventually lead to interstellar journeys over vast distances to other stellar systems for the purpose of conducting explorations aimed at discovering new worlds which are susceptible to colonization by the human race. The propulsion systems require f

2

by using a greater number of stages are offset by the additional complexity involved, and it is very difficult to increase the propellant-weight ratio much beyond a certain value in the present concepts of vehicle systems.

The only feasible alternative remaining is to reduce the inert weight which is not useful for propulsion in the vehicle system so that a greater payload weight can be obtained without the need to increase initial launch weight of the vehicle system. This is an important consideration since any unnecessary inert weight in the various stages of the vehicle system imposes a heavy, additional demand on required engine thrust which is functionally related to launch weight. In a large, three stage booster system to be used on a lunar flight, for example, any change in weight of the final stage will be reflected in a similar change in the total launch weight multiplied, however, by a growth factor which may easily number in the hundreds.

In undertaking manned, space exploration missions, the astronauts may be exposed to radiation fields of high intensity in space. Biological damage is done by the ionization produced by radiation and high energy charged particles which pass through the tissues of the astronauts. A lethal action arises when the radiation dosage is excessive such that changes in living cells result in their death when they attempt division. Of course, extremely high radiation dose rates which may be lethal to an astronaut after a relatively short exposure period are encountered except rarely in ordinary

Fig. 5.32 □ A portion of the first of six sheets of specifications (petition) for the radiation protection system illustrated in Figures 5.30 and 5.31.
Copies of patents may be purchased from the Patent Office (50¢ each).

invention, but in such case the drawing must be corrected or a new one furnished before the application will be allowed. The necessary corrections will be made by the office, upon applicant's request and at his expense.

55. Applicants are advised to employ competent draftsmen to make their drawings.

The office will furnish the drawings at cost, as promptly as its draftsmen can make them, for applicants who cannot otherwise conveniently procure them.

The air-cushion vehicle shown skims along over the water on a cushion of air and docks on land.
It could not have been designed successfully without considerable knowledge of engineering materials. At the present time ACV's are sweeping across the English Channel, three times faster than regular ships, carrying more than 250 passengers and 30 cars at each crossing. (*Courtesy Aluminum Corporation of America.*)

6.1 □ Design and engineering materials

At several stages in the design of a product or system, consideration must be given to the selection of the materials from which the product or the components of the system are to be made. The designer must ask questions about the requirements of the complete design; he must remember that the materials selected for all components must be suitable for all known environmental conditions of operation and for the computed loads and stresses that will be imposed on some of the machine and structural parts. It is often the case that material properties and limitations are controlling factors in a design. Frequently the methods of manufacture, the shape of individual parts, the methods to be used for fastening the parts together, and even the aesthetic appearance of the finished product depend to a great extent on the materials used. Environmental conditions under which a system is to be operated must not be overlooked. For example, one should determine whether the unit will be operated in a country that has a hot and humid climate that will cause parts to deteriorate or corrode. Even worse conditions are encountered by components for aircraft and space vehicles, where units start operation on the ground at 100°F and then move into high altitudes where the temperature may drop to −40°F or lower. Since operational characteristics change radically where temperature extremes are as great as mentioned, the materials for components must be carefully selected and the components may have to be insulated. Often some means for maintaining a nearly constant temperature may be required as a part of the total design; this is true in the case of delicate measuring instruments. In all cases materials must be

Engineering materials: selection and use

selected and treated so that parts will not rust, pit, corrode or wear so rapidly that maintenance becomes costly. A designer should always select those materials that insure the quality of the product and at the same time permit sales within a competitive price range. A poor maintenance record coupled with a high price tag, both of which may be due largely to a poor selection of materials, has led to the disappearance from American and world markets of many otherwise innovative products.

The designer must recognize that the selection of the materials and the methods of fabrication are so closely interrelated that they must be considered simultaneously at several stages in the total design process. In following this practice, the selection of a material for one or more components may lead either to a minor change in the overall design or to a change in the shape of several related components.

In a text that has been written for students, space is not available for a complete presentation of detailed information on all types of materials commonly used. To cover only several characteristics of steel (such as strength, hardness, machinability, wear-resistance, corrosion, heat treatment, carburizing, short time effects of high temperatures, and creep at high temperatures) and then to explain the numbering systems for both carbon steels and alloy steels would take many pages. Only brief descriptions of the most common types of ferrous metals, nonferrous metals, and plastics, along with their principal characteristics, will be given to serve as a guide for their use. More information may be found in handbooks such as *Metals Engineering-Design* (ASME), *Metal Properties* (ASME) and the *SAE Handbook*, as well as in standards published by ANSI, ASTM, ASME, and other technical organizations. A good designer realizes that it is necessary at times to seek the help of a professional metallurgist to insure the best choice of material.

6.2 □ Ferrous metals
Iron is one of the oldest and most used of all metals, having been used by man for weapons and tools for more than 3000 years. The modern conversion of iron into steel has made possible our great technological age. Although iron of very high purity can be made commercially, most of the varieties used by industry are alloys of the metal with varying amounts of other elements such as carbon, manganese, silicon, and sulfur. In some cases, other metals, such as nickel, chromium, vanadium, and tungsten are added to the combination to produce desired metal characteristics.

6.3 □ Cast iron
Cast iron (gray) is widely used by industry as a material for cast shapes. It is characterized by rather high percentages of carbon

(2–4 percent by weight) and silicon. Cast iron is cheap; can be readily cast; can be easily machined; can be welded if carefully preheated; and is resistant to corrosion. It should be noted that while cast iron is strong in compressions, it is weak in tension and lacks ductility. Alloying elements such as nickel, chromium, copper, and molybdenum are sometimes added to cast iron to increase hardness and strength.

Malleable cast iron is characterized by high strength and a somewhat greater ductility and, since it is tough and easily machined, it is used for conditions where ordinary gray cast iron would be too brittle. It is produced from white iron by a special annealing treatment.

6.4 □ Cast steel
Steel castings are widely used in heavy machinery. They are produced from low-carbon steel, medium-carbon steel, and alloy steels. Low-carbon steel castings compete with ordinary castings made from either gray or malleable iron. Carbon steels for castings have good machinability and can be readily welded. These steels are used where high strength and toughness are needed and where resistance to wear and fatigue are important. Most high-carbon steels have somewhat greater strength than low-carbon steels used for castings. However, they have less ductility.

Alloy steels are used for castings requiring high strength, high hardness, and high resistance to fatigue and wear. High temperature properties are better than for the carbon steels. It is the usual practice to resort to heat treatment to secure the full benefit of the added alloys. Alloy steels having high nickel and chromium content are used for high temperature service and for applications where high resistance to corrosion is needed.

6.5 □ Steel
Steel is an alloy of iron and carbon (approximately 0.5 percent carbon by weight); it may contain some alloying elements such as nickel, chromium, molybdenum, silicon, or vanadium. The steels used in machine construction may or may not be heat treated. Those steels that are not heat treated are: (1) low-carbon steels, either hot rolled or cold drawn; (2) free-cutting steels that are intended for easy machining of nuts, fasteners, collars, and dowels; and, (3) low-carbon sheets. The heat-treated steels are: (1) carburized (either low-carbon or low-alloy steels); and (2) quenched and then tempered (either medium- and high-carbon steels or low- and medium-alloy steels).

Since so many different types of carbon and alloy steels are used for the construction of machinery, the SAE (Society of Automotive Engineers) adopted an identifying numbering system that has been widely used for many years. This numbering system provides a means whereby any type of steel can be specified by a simple combination of four digits. The first two digits identify the type and alloy characteristics. The first digit denotes the type of the particular steel. For example, the numeral 1 indicates plain carbon steel; the numeral 2 indicates nickel steel; the numeral 3, a chromium-nickel steel; 4, a molybdenum steel; 5, a chrome steel; and the numeral 6 indicates a chromium-vanadium steel. The second digit denotes the approximate percentage of the principal alloying element, and the last two digits indicate the approximate carbon content in points (one-hundredths of one percent). Since a plain carbon steel is indicated by a 10, the specification 1045 would denote that a plain-carbon steel containing 0.45 percent carbon is to be used for the component. A 2315 combination specifies a nickel steel of approximately 3 percent nickel and 0.15 percent carbon.

Examples:

Plain-carbon steel 1xxx, 10xx

Free-cutting screw stock 11xx

Manganese steels 13xx

Nickel steels 2xxx, 23xx, 25xx

Nickel-chromium steels 3xxx, 31xx, 32xx, 33xx

Molybdenum steels 4xxx, 41xx, 43xx, 46xx, 48xx

Chromium steels 5xxx, 51xx, 52xx

Chromium-vanadium steels 6xxx, 61xx

Silicon-manganese steels 9xxx, 92xx

6.6 □ Plain-carbon steels

A steel is classified as being a plain-carbon steel when the only alloying element is carbon. Within this classification a wide range of strengths and hardness may be secured by varying the percentage of the carbon and by heat treating when the carbon content is at least 0.30 percent. Heat treatment, which consists of heating to above the upper critical temperature and then quenching, gives added hardness and strength, and increases resistance to wear. Hardening grades are available as hot- or cold-rolled bars, rods, and mill shapes. Carbon steels have good forging qualities. Machining qualities are good to excellent. Carbon steel is used for structural shapes, wire, tubing, machine parts, springs, crane hooks, axles, gears, metal cutting tools, and dies. Welding requires special practices for the heat-treating grades. Carbon steels are usually easier to machine than alloy steels.

Carbon steels that are suitable for carbonizing are available in all mill grades. Forging qualities are good but machining qualities vary from very poor to very good. The carbonizing process leaves a soft, strong core for added strength while producing a hard outer surface that is capable of resisting wear and abrasion. The process consists of packing the parts in a carbonizing material and heating in a furnace for a number of hours. For the carbon to be absorbed by the "skin" layer the temperature must be within the austenitic range. Satisfactory results depend greatly upon the heat treating that follows carbonizing.

6.7 □ Alloy (high speed) steels

These steels have been developed to meet the requirements of industry. They are tougher, stronger, and harder than the plain carbon steels. Alloy steels, which are made by alloying chromium, tungsten, molybdenum, nickel, vanadium, and other elements with steel, have made possible the automobile as we know it today. Without these steels man would not now be flying through the air at 600–1500 miles per hour nor would he have walked upon the moon. Of course, iron is the basis of all steel, and carbon is the determinative that is added for hardness and rigidity. To meet particular requirements chromium is added as a refiner of the grain; manganese adds strength and gives depth to hardening; nickel adds strength and toughness; and tungsten and molybdenum add to the hardness. All are widely used singly or in varied combinations.

6.8 □ Aluminum alloys

In recent years the use of lightweight aluminum alloys for components in manufactured products has been increasing at a rapid rate. The greatly increased use of this most versatile metal started when designers in the aircraft industry were forced to drastically reduce the weight of structural parts in order to permit greater payloads; at the same time a metal that offered considerable resistance to damaging corrosion was needed. Aluminum can be rolled into sheets, forged, or extruded and in most of its alloys it is available as plates, bars, tubes, and common structural shapes. The so-called high-strength alloys have about the same strength as mild steel; thus a part made of aluminum will not be as stiff or as strong as it would be if it were made of one of the better grades of steel.

All aluminum alloys have a high resistance to atmospheric corrosion, due to an aluminum oxide layer which forms naturally on surfaces exposed to air. To secure even better protection, an oxide layer may be built up by anodic oxidation. It is interesting to note that the high-strength alloys are not as resistant to corrosion as pure aluminum. Although it takes considerable skill to weld aluminum, it may still be said that the machinability and weldability of aluminum is good. Aluminum alloys are used for making both sand-mold and die castings and for fasteners, pins, rivets, and other such items that can be made as automatic screw-machine products. It is much easier to machine the alloys than the pure soft metal.

Aluminum is used for electrical conductors, chemical equipment, pipe and pipe fittings, bus and truck bodies, automobile trim, cooking utensils, and, as has been mentioned previously, for automatic screw-machine products, die castings, and structural components in aircraft.

6.9 □ Magnesium alloys

Magnesium is another material that is suitable for the design of light-weight structural components. Most of the alloys are available in sheet form and as rods, bars, and extrusions. These alloys are used for both sand-mold and die castings. Because of lightness and strength, magnesium parts are used for space vehicles, missiles, and aircraft as well as for cameras, office machines, appliances, lawn mowers, sporting goods, and luggage. Magnesium is easy to machine and it can be forged using hydraulic presses; although it has relatively good resistance to atmospheric corrosion, it is attacked by sea water and salt-water atmosphere unless surfaces are finished in some manner.

6.10 □ Copper and copper alloys

Copper is a versatile material that may be rolled into sheets or drawn into wire and tubing. It is outstanding for its thermal and electrical conductivity properties. Copper may be used in almost pure form or in combination with zinc and tin and sometimes with lead and other alloys. The most important copper-base alloys, the ones most of us are familiar with, are bronze and brass. In general terms, it can be said that brass is a copper-zinc alloy containing approximately two-thirds copper and one-third zinc, while bronze is a copper-tin alloy made up of nine-tenths copper and one-tenth tin. A so-called red brass consists of 85 percent copper and 15 percent zinc. Other commercial grades of brass have less copper and more zinc; the yellow brasses have as much as 35 percent zinc. With the addition of a small amount of lead ($1\frac{1}{2}$ to 3 percent) a low-leaded brass called clock-brass is obtained. Clock-brasses are used for components in automatic machines and meters and for parts

in clocks. Low-leaded Muntz metal, which is often used for bolts, nuts, sheathing, and wire, is composed of 60 percent copper and 40 percent zinc. Brass and bronze are used commercially in both cast and wrought forms. Bronze is usually considered to be the superior metal for most applications.

Many different copper alloys are used in industry. Some of the alloying elements that are used (other than the zinc, tin, and lead that have already been mentioned) include aluminum, nickel, chromium, manganese, phosphorus, silicon, and beryllium. Aluminum bronze contains as much as 11 percent aluminum; manganese bronze, with about 18 percent manganese, is a white-colored alloy. Common applications of some of these special brass and bronze alloys are: white nickel brass—control brackets and fittings; manganese bronze—gear shifter forks, brackets, and parts for starting motors; nickel phosphorus bronze—cones of synchronizer gears; aluminum bronze—valve seats and guides, gears, and forgings; and phosphorus bronze—bushings. Many other uses are listed in handbooks.

In general the machinability of these alloys varies from fair to good and joining may be done satisfactorily by either soldering, welding, or brazing. All of the alloys have good resistance to sea water and to the industrial atmosphere.

6.11 □ Lead and lead alloys

Soft lead is used for such applications as washers, cable sheathings, coatings, storage batteries, and, as has been mentioned, for alloying. Sheet lead, produced from lead in combination with antimony, is used for roof flashing and for lining tanks and acid containers where the effects of sulfuric, phosphoric, and chromic acids must be resisted. Oddly enough, lead is not resistant to acetic, nitric, and formic acids. Lead has good resistance to both fresh and salt water and to the atmosphere. Lead-based alloys containing a rather high percentage of tin are used for bearing liners subjected to light loads at low speed where heat conditions are not severe.

Antimonial lead (hard lead) is pure lead to which antimony has been added to increase stiffness.

6.12 □ Tin and tin alloys

Tin is used for food can linings and for food handling equipment. The tin-based alloys, commonly known commercially as the babbitt metals, are used mainly in the production of automobile bearings and for parts requiring resistance to corrosion. The alloying elements are usually antimony and copper but lead may be added to the mixture. Pewter, a tin alloy used for tableware and white metal used for jewelry, consists mainly of tin with varying quantities of lead, bismuth, antimony, or copper. Tin is an element in brass casting alloys (Sec. 6.15).

6.13 □ Zinc and zinc alloys

Zinc is widely used for coating (galvanizing) steel, for dry-battery cases, and as an element in the alloying of copper-based metals. Zinc-based alloys are used in the production of die-cast parts; the usual alloying elements are a combination of tin, copper, and aluminum. Zinc and zinc alloys are commercially available in strips, sheets, and extruded shapes and in the form of drawn rods and wire. Zinc alloys may be readily fabricated and easily soldered and welded.

6.14 □ Nickel alloys

There are a number of nickel-base alloys that are used where resistance to corrosion is important. These are known by different trade names. Among these is Monel metal, an alloy composed of approximately two-thirds nickel and one-third copper along with about 3 percent of other elements. Monel metal is a strong, tough, white metal that is malleable, ductile, and corrosion-resistant. It is available in the usual commercial forms and is used for forgings and castings. It is easily workable by all methods of fabrication except extrusion. Monel metal may be welded but Monel rods should be used. Monel 411 and 505 and similar nickel alloys are widely used for food-handling equipment, for tanks and boilers, and for machine parts such as valve seats and bushings. The designer can obtain information about Monel and other trade-name nickel alloys, such as Duranickel, by consulting handbooks or catalogs. Special nickel-based superalloys have been developed in recent years for use in missiles and jet engines, and for similar applications where there are high temperatures and the resistance to corrosion is an important factor in the design.

6.15 □ Alloys for die casting

Alloys of zinc, aluminum, magnesium, and brass are commonly used for die casting. Die castings are formed by forcing a molten alloy between metal dies under high pressure. Die casting, an inexpensive mass-production method for manufacturing small parts, is strictly a machine operation and, since finished products can be made to adhere to rather close tolerances, little if any, machining is required. The process is suitable for parts with thin sections or intricate shapes that would be expensive and difficult to fabricate by any other method. Usually, die castings have very smooth surfaces that are suitable for plating or the application of an organic finish with little or no preparation. The characteristics and properties of the four alloys mentioned are discussed here

briefly. For further information, the designer should consult appropriate handbooks.

Zinc die castings are very widely used because they have high strength and also because the need for only moderate casting temperatures permits a long die life.

An important advantage of aluminum die castings is their light weight. However, since the casting temperature needed is higher than is required for zinc, the die life is much shorter.

Magnesium also offers the advantage of being lightweight, and it has the same disadvantage of aluminum in that the casting temperature is approximately the same. It is usual practice to treat the surface of a magnesium casting to provide a base for the finish as well as to inhibit corrosion.

The principal advantages possessed by die castings produced from brass alloys are great strength and hardness. However, since brass must be cast at a much higher temperature than either zinc, aluminum, or magnesium, die life is considerably shorter. This is an important consideration in making a selection of a material for casting because of the high cost of dies. In the case of either brass or bronze the advantages of strength and hardness must be evaluated against the cost of the needed dies.

6.16 □ Precious and rare metals

In addition to being used for jewelry, the precious metals, namely silver, gold, and platinum, have engineering applications as well.

Gold, a very ductile metal, can be drawn into a very fine wire. When it is alloyed with platinum, palladium, nickel, or zinc, a white metal is obtained. Gold is used for lining chemical equipment, for high-melting solder alloys, and in medicine and dentistry.

Silver is largely used for electroplating the baser metals for protection against corrosion and for improved appearance. It is employed by industry for electrical contacts, for corrosion-resistant equipment, and in silver brazing for joints that need little, if any, finishing. Silver is used also as an alloy in stainless steel.

Platinum is used for linings and surface coverings for chemical and laboratory equipment. In the electrical field it is employed for contact points and electrodes, and in x-ray apparatus. It is also used for dental instruments and dental fillings. Hardness and wearing quality may be increased by alloying with other metals, usually iridium, palladium, or ruthenium.

6.17 □ Ceramics

Some important ceramic products that are widely used by industry include brick, tile, sewer pipe, chemical and electrical porcelain and stoneware, wall tile, china, and porcelain enamel. In general these products are made from some clay-like raw material to which high-temperature treatment is applied during the process of manufacture. The use of ceramic products are many and varied. In industrial plants, ceramic products are employed for lining the shells of portland-cement and steel furnaces, for fire boxes of boilers, for sulfite pulp digesters, and for other similar purposes. A number of ceramic products (refractories) are available that are capable of resisting change in shape, weight, and other physical properties when subjected to very high temperatures. Ceramics have high compression-strength and rigidity.

Porcelain (vitreous) enamels are actually fused silicate coatings applied to a metal such as sheet steel or cast iron.

6.18 □ Rubber—natural and synthetic

There are many everyday products made from natural rubber, an elastic substance that has been used by man for almost a thousand years. In its natural form (latex), it is a white milky substance that is obtained from tropical trees. When coagulated it forms an elastic waterproof material. Although pure natural rubber is still used to a considerable extent for specific applications, the consumption of synthetic rubber now greatly exceeds the amount of natural rubber consumed in the U.S. The growth of production of synthetic rubbers has been phenomenal as various methods have been developed for its production from such materials as natural gas, petroleum, and alcohol. Neoprene, the first successful synthetic, is resistant to gasoline and oils.

Natural rubber, along with butadiene-styrene, is used in soft flexible forms for items such as hose, belts, and gaskets, and for insulation in the electrical industry. Other compositions, both natural and synthetic, are used for tires, insulators, and plugs, and for cases for storage batteries. This versatile material, in both its man-made and natural forms, has many applications. For example, rubber is employed for lining tank cars used for the storage and transportation of acids and other liquids that corrode metal.

Butyl (synthetic) rubber is commonly used for tires, tubes, electrical insulation, and other like items where some degree of flexibility is a required characteristic. For self-sealing tanks, shock mountings, and aircraft hose, butadiene-acrylonitrile is usually selected.

In concluding this brief discussion on rubber, it must be pointed out that since there are so many types and compositions of synthetic rubber now available, most designers will find it desirable and even necessary to consult with a knowledgeable authority before making a final selection.

6.19 □ Plastics

As has been said elsewhere in this text, we may be entering a second industrial revolution, an age of new materials and methods of production directed by computers. It is also quite possible that this new age could be known in the later years as the ''age of plastics'' even though the spectacular creations in the future may depend largely upon new developments of metals and their alloys. Plastics are the miracle materials of the twentieth century, materials that have made it possible for industry to place into the hands of people of low and medium incomes those products that could otherwise even now be available only to the rich and near rich because of high production costs. The designer must never lose sight of the fact that in most cases the item that he is designing must be mass-produced in considerable volume in order to be priced well within the range of competition. The use of plastics, where appropriate, could very possibly be the deciding factor as to whether or not an item will be marketable. Marketability in itself is just as important to the success of a desirable item or unit as are the art and science that must be applied to the design.

Basically, plastics are a group of synthetic resins or natural organic materials which may be shaped when soft and then hardened. Plastics may be either cast, molded, or extruded. This group of versatile materials includes many types of resins, resinoids, polymers, cellulose derivatives, casein materials, and proteins. Designers, who rightly have an eye on marketability, economy, and appearance, use plastic materials in place of wood, glass, and metal for the production of many everyday articles. Additional uses are for decorations, for coatings, and (when drawn into filaments) for weaving strong fabrics that are resistant to atmospheric pollution and to moisture.

There are many types of plastics and their uses are varied. For example, ABS Resins (molded-extruded), which are resistant to phosphoric and hydrochloric acids, alkalis, and salts, are used for appliance housings, housewares, luggage, and piping; alkyds (molded) are used for electronic parts such as resistors, capacitors, and circuit-breaker parts; cellulose acetate (molded-extruded) is suitable for appliance housings, optical parts, and tape; epoxies (cast-molded), which are resistant to alkalis, are used for paints, glues and moldings and for coating electrical parts; melamines (molded) are used for tableware, kitchenware, and lighting fixtures; polyesters (cast) are used for electrical parts and for high-temperature conditions; and polystyrenes (molded-extruded), which are resistant to both alkalis and salts, are used for electrical components, insulators, instrument panels, and instrument housings.

Although the detailed information needed for making a proper selection of a plastic that will satisfy design conditions may be readily obtained from handbooks and from publications prepared by producers of plastic materials, it is usually wise for a designer to consult some knowledgeable person who has had considerable experience with the selection, use, and fabrication of plastics. Often, plastic materials are identified by trademark names such as Bakelite, Vinylite, Lucite, and so forth.

6.20 □ Finishes and coatings

A factor that often influences the selection of a base material is its ability to accept finishing and/or a coating material. Surfaces of base materials are commonly coated or overlaid with a finishing material. This is generally done for one or more of the following reasons:

1 To improve the appearance, as might be the case with a housing that must have some aesthetic appeal.

2 To offer protection and resistance to corrosion.

3 To improve resistance to wear.

Designers should have some knowledge of the various methods that are used to provide coatings. These are: (1) electrodeposited coatings, (2) sprayed metal coatings, (3) hot-dip coatings, (4) vapor-deposited coatings, (5) ceramic and refractory coatings, (6) porcelain-enamel coatings, (7) rust preventative coatings, and (8) hard finish overlays that are applied by a welding process. In addition to these coated finishes, there are several mechanical finishes that are applied to copper, aluminum, and stainless steel. These may be either hammered, buffed, burnished, wire-brushed, or sandblasted finishes.

6.21 □ Selection of materials—summary

A designer should formulate and follow some well-organized method for evaluating design materials. His method for arriving at a final decision must involve such selection factors as: (1) mechanical properties, (2) metallurgical composition, (3) formability, (4) joinability, (5) ease of finishing and coating, and (6) cost. Until a designer has acquired considerable experience, he should prepare some form of table in which he can evaluate these factors quantitatively to get a performance rating for each type of material that he is considering. Some designers use a number system for rating the relative importance of each factor and from the rating derived, they judge the relative merit of each material. This system makes it easy to determine which materials are unsuitable and not worthy of consideration. Materials that remain can then be restud-

ied. This method makes it possible to take advantage of new materials with which the designer may not be familiar. However, the application of this type of method to the selection any material presupposes some knowledge of specific information about the properties of ordinary materials. In the case of a new material, the designer may find it necessary to consult with a producer of this material and, in addition, seek the advice of other designers who may have had some experience with its use. A good designer never relies exclusively upon his own knowledge.

Part 3

Design graphics

central computer

master center transmits programs to classrooms located in Mobile Helio-craft via laser beams in computers with satellite relays

STATION 1	
STATION 2	
STATION 3	
STATION 4	
STATION 5	

PROGRAMED SOFTWARE

CENTRAL COMPUTER

master tracking center is computerized

3D stimuli

laser projector

2D media television

UN 1

hologram

Helio-hover craft

In the future, heli-lifted classrooms may be transported to every part of the earth.
The lessons would be projected in 3-D. Transmittal would be carried by laser beams, relayed by satellite. (*Courtesy Raymond Loewy/William Smith, Inc. and Charles Bruning Company.*)

Chapter 7

A □ Sketching and creative thinking

7.1 □ Introduction

From the earliest times pictorial representations have been the means of conveying the ideas of one person to another person and of one group to another group (Fig. 7.2). There is little doubt that our ancestors traced out in the dust on the cave floor many crude pictures to supplement their guttural utterances. On their cave walls these same primitive men and women drew pictures, which today convey to others the stories of their lives. They used the only permanent means they were aware of at that time.

At present, we have at our command the spoken languages, which no doubt developed from limited semiintelligent throat sounds, and the written languages, graphical and symbolic in form. The descriptive powers of the various forms of presentation may be compared in Fig. 7.1. The use of sign language representation (*a*) is rather easy to learn and may be quickly executed but interpretation is restricted to persons who understand the particular language in which it is presented. The multiview representation, shown in (*b*), may be understood universally by persons who have been trained in its use. However, the given views will prove to be almost meaningless to the many who have not had the advantage of needed training. What then is the one form of representation that can be understood and used by all? It is the pictorial form shown in (*c*).

Fundamentals and techniques of communication graphics

It is necessary that a designer be capable of executing well-proportioned and understandable freehand pictorial sketches, for it is one of the most important modes of expression that he has available for use. Like other means for conveying ideas, when depended on alone, it will usually prove to be inadequate. However, when used in combination with the written or spoken language and related graphical representations, it makes a full understanding by others become sure and not just possible (Fig. 7.2). Each method of expression is at hand to supplement another to convey the intended idea.

The designer should be capable of speaking his own language and possibly one foreign language fluently; he should be able to write to present his ideas clearly and accurately; he should be familiar with the graphical method of presenting shape through the use of multiviews; and, finally, he should be competent to execute well-proportioned and understandable pictorial sketches, which are needed to clarify and insure complete transfer of his ideas to others. (See chapter opening figure.)

The designer is a creative person living in a world where all that he creates must exist in space. He must visualize space conditions, space distances, and movement in space. In addition he must be able to retain as well as alter the image of his idea, which will at the very start exist only in his mind.

As his idea forms the designer nearly always resorts to sketching to organize his thoughts quickly and to more clearly visualize the problems that appear. The first ideas may be sketched in pictorial form as they are visualized. Later, in making a preliminary study, a combination of orthographic design sketches and pictorial sketches may quickly pile up on his desk as problems are recognized and possible solutions are recorded for reference and for conferences with others.

The designer's use of sketches, both pictorial and orthographic (Fig. 7.3), continues throughout the preliminary design stages and into the development and detailing stages. This comes about because the designer is usually called on to serve as both planner and director. Throughout all stages in the development of a product he must solve problems and clarify instructions. Very often a pictorial sketch of some detail of construction will prove to be more intelligible and will convey the idea much better than an orthographic sketch, even when dealing with an experienced draftsman or detailer (Fig. 7.20).

Design sketches may be done in the quiet of the designer's office or amid the confusion of the conference table. To meet the requirement of speed of preparation, one must resist all temptation to use instruments of any type and rely on the pencil alone, for the true measure of the quality of a finished sketch is neatness and

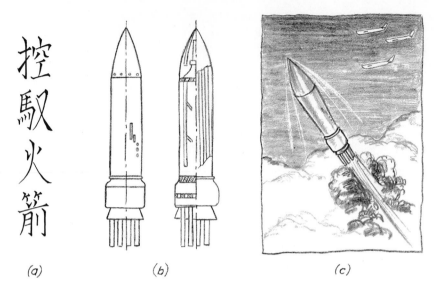

(a)　　　　　　(b)　　　　　　(c)

Fig. 7.1 □ Graphic methods for presenting ideas (symbolic, multiview, and pictorial).

good proportion rather than the straightness of the lines. A pictorial sketch need not be an artistic masterpiece to be useful.

Training for making pictorial sketches must include the presentation of basic fundamentals, as is done with other how-to-do-it subjects. As learning the mechanics of English does not make one a creative writer, so training in sketching will not make one a creative engineer. However, sketching is the means of recording creative thoughts.

Some design sketches drawn by an electrical engineer are shown in Fig. 3.7. These sketches were prepared in making a study of the wiring to the electronic control panel for an automatic machine.

7.2 □ Thinking with a pencil

As an attempt is made to bring actuality to a plan, sketches undergo constant change as different ideas develop. An eraser may be in constant use or new starts may be made repeatedly, even though one should think much and sketch only when it would appear to be worthwhile. Sketching should be done as easily and freely as writing, so that the mind is always centered on the idea and not on the technique of sketching. To reach the point where one can "think with the pencil" is not easy. Continued practice is necessary until one can sketch with as little thought about how it is done as he gives to how he uses a knife and fork at the dinner table.

7.3 □ Value of freehand drawing

Freehand technical drawing is primarily the language of those in charge of the development of technical designs and plans. Chief engineers, chief draftsmen, designers, and squad bosses have found that the best way to present their ideas for either a simple or complex design is through the medium of sketches. Sketches may be schematic, as are those that are original expressions of new ideas (Fig. 7.2), or they may be instructional, their purpose being to convey ideas to draftsmen or shopmen. Some sketches, especially those prepared for the manufacture of parts that are to replace worn or broken parts on existing machines, may resemble complete working drawings (Fig. 7.4).

7.4 □ Projections

Although freehand drawing lacks the refinement given by mechanical instruments, it is based on the same principles of projection and conventional practices that apply to multiview, pictorial, and the other divisions of mechanical drawing. For this reason, one must be thoroughly familiar with projection, in all its many forms, before he is adequately trained to prepare sketches.

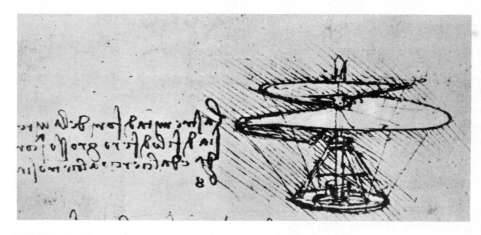

Fig. 7.2 □ An idea sketch of a helicopter prepared by Leonardo da Vinci (1452–1519).
(From Collections of Fine Arts Department, International Business Machines Corporation.)

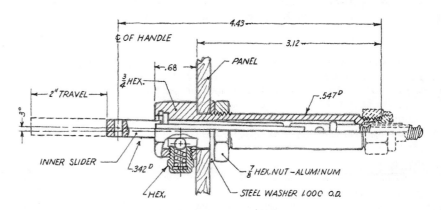

Fig. 7.3 □ A design sketch for a connector of a remote control unit.
(Courtesy Teleflex, Inc.)

B □ Sketching techniques

7.5 □ Sketching materials
For the type of sketching discussed here, the required materials are an F pencil, a soft eraser, and some paper. In the industrial field, men who have been improperly trained in sketching often use straightedges and cheap pocket compasses that they could well dispense with if they would adopt the correct technique. Preparing sketches with instruments consumes much unnecessary time.

For the person who cannot produce a satisfactory sketch without guide lines, cross-section paper is helpful. Ordinarily, the ruling on this paper forms 1-in. squares, which are subdivided into $\frac{1}{8}$-, $\frac{1}{4}$-, or $\frac{1}{10}$-in. squares. Such paper is especially useful when sketching to scale is desirable (Fig. 7.5).

7.6 □ Technique of lines
Freehand lines quite naturally will differ in their appearance from mechanical ones. A well-executed freehand line will never be perfectly straight and absolutely uniform in weight, but an effort should be made to approach *exacting uniformity*. As in the case of mechanical lines, they should be black and clear, not broad and fuzzy (Fig. 7.6).

7.7 □ Sharpening the sketching pencil
A sketching pencil should be sharpened on a file or piece of sandpaper to a conical point. The point then should be rounded slightly, on the back of the sketch pad or on another sheet of paper, to the correct degree of dullness. When rounding the point, rotate the pencil to prevent the formation of sharp edges.

7.8 □ Straight lines
The pencil should rest on the second finger and be held loosely by the thumb and index finger about $1-1\frac{1}{2}$ in. above the point.

Horizontal lines are sketched from left to right with an easy arm motion that is pivoted about the muscle of the forearm. The straight line thus becomes an arc of infinite radius. When sketching a straight line, it is advisable first to mark the end points with light dots or small crosses (Fig. 7.7).

The complete procedure for sketching a straight line is as follows:

  1 Mark the end points.

  2 Make a few trial motions between the marked points to adjust the eye and hand to the contemplated line.

  3 Sketch a *very* light line between the points by moving the pencil in two or three sweeps. When sketching the trial line, the eye should be on the point toward which the movement is directed.

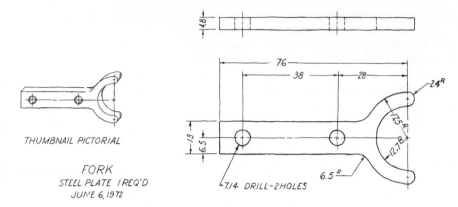

THUMBNAIL PICTORIAL

FORK
STEEL PLATE I REQ'D
JUNE 6, 1972

Fig. 7.4 □ A freehand sketch for the manufacture of a part.
(*Numerical values are in millimeters.*)

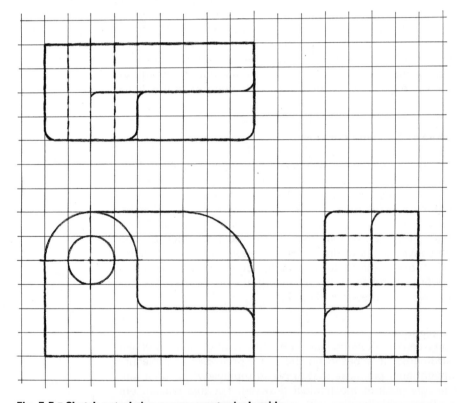

Fig. 7.5 □ Sketch actual size on one-quarter inch grid.

With each stroke, an attempt should be made to correct the most obvious defects of the stroke preceding, so that the finished trial line will be relatively straight.

4 Darken the finished line, keeping the eye on the pencil point on the trial line. The final line, replacing the trial line, should be distinct, black, uniform, and straight.

It is helpful to turn the paper through a *convenient angle* so that the horizontal and vertical lines assume a slight inclination (Fig. 7.8). A horizontal line, when the paper is in this position, is sketched to the right and upward, thus allowing the arm to be held slightly away from the body and making possible a free arm motion.

Short vertical lines may be sketched either downward or upward, without changing the position of the paper. When sketching downward, the arm is held slightly away from the body and the movement is toward the sketcher (Fig. 7.9). To sketch vertical lines upward, the arm is held well away from the body. By turning the paper, a long vertical line may be made to assume the position of a horizontal line and can be sketched with the same general movements used for the latter.

Inclined lines running upward from lower left to upper right may be sketched upward with the same movements used for horizontal lines, but those running downward from upper left to lower right are sketched with the general movements used for either horizontal or vertical lines, depending on their inclination (Fig. 7.10). Inclined lines may be more easily sketched by turning the paper to make them conform to the direction of horizontal lines.

7.9 □ Circles
Small circles may be sketched by marking radial distances on perpendicular center lines. When additional points are needed, the distances can be marked off either by eye or by measuring with a marked strip of paper (Fig. 7.11). Larger circles may be constructed more accurately by sketching two or more diagonals, in addition to the center lines, and by sketching short construction lines perpendicular to each, equidistant from the center. Tangent to these lines, short arcs are drawn perpendicular to the radii. The circle is completed with a light construction line, and all defects are corrected before darkening (Fig. 7.12).

7.10 □ Making a multiview sketch (Fig. 7.4)
When making orthographic working sketches a systematic order should be followed, and all the rules and conventional practices used in making working drawings should be applied. The following procedure is recommended:

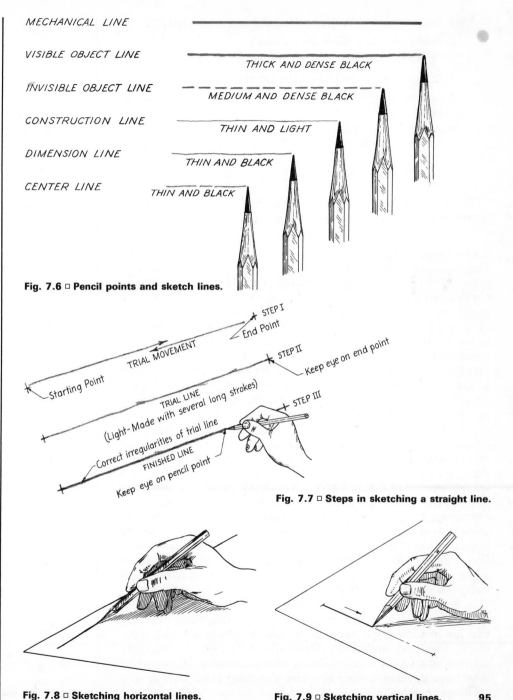

Fig. 7.6 □ Pencil points and sketch lines.

Fig. 7.7 □ Steps in sketching a straight line.

Fig. 7.8 □ Sketching horizontal lines.

Fig. 7.9 □ Sketching vertical lines.

95

1 Examine the object, giving particular attention to detail.

2 Determine which views are necessary.

3 "Block-in" the views, using light construction lines.

4 Complete the detail and darken the object lines.

5 Sketch extension lines and dimension lines, including arrowheads.

6 Complete the sketch by adding dimensions, notes, title, date, sketcher's name or initials, and so on.

7 Check the entire sketch carefully to see that no dimensions have been omitted.

The beginning student should read Part A of Chap. 8 before he attempts to make a multiview sketch.

7.11 □ Proportions

The beginner must recognize the importance of being able to estimate comparative relationships between the width, height, and depth of an object being sketched. The complete problem of proportioning a sketch also involves relating the estimated dimensions for any component parts, such as slots, holes, and projections, to the over-all dimensions of the object. It is not the practice to attempt to estimate actual dimensions, for sketches are not usually made to scale. Rather one must decide, for example, that the width of the object is twice its height, that the width of a given slot is equal to one-half the width of the object, and that its depth is approximately one-fourth the overall height.

To become proficient at sketching one must learn to recognize proportions and be able to compare dimensions "by eye." Until one is able to do so, he can not really "think with his pencil." Some people can develop a keen eye for proportion with only a limited amount of practice and can maintain these estimated proportions when making the views of sketch. Others have alternately discouraging and encouraging experiences. Discouragement comes when one's knowledge of sketching is ahead of his ability and he has not had as much practice as he needs. The many who find it difficult to make the proportions of the completed sketch agree with the estimated proportions of the object may begin by using the graphical method shown in Figs. 7.13(a), (b), and (c). This method is based on the fact that a rectangle (enclosing a view) may be divided to obtain intermediate distances along any side that are in such proportions to the total length as one-half, one-fourth, one-third, and so on. Those who start with this rectangle method as an aid in proportioning should abandon its use when they have developed their eye and sketching skills so that it is no longer needed.

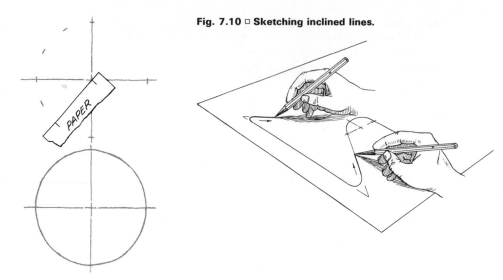

Fig. 7.10 □ **Sketching inclined lines.**

Fig. 7.11 □ **Marking off radial distances.**

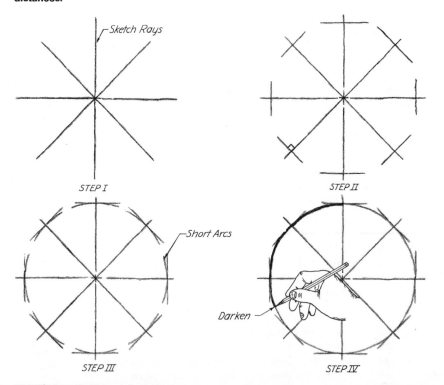

Fig. 7.12. □ **Sketching large circles.**

Sketching must be done rapidly, and the addition of unnecessary lines consumes much valuable time. Furthermore, the addition of construction lines distracts the reader, and it is certain that they do not contribute to the neatness of the sketch.

The midpoint of a rectangle is the point of intersection of the diagonals, as shown in Fig. 7.13(a). A line sketched through this point that is perpendicular to any side will establish the midpoint of that side. Should it be necessary to determine a distance that is equal to one-fourth the length of a side, (say, AC), the quarter-point may be located by repeating this procedure for the small rectangle representing the upper left hand quarter of $ABCD$.

With the midpoint J located by the intersecting diagonals of the small rectangle [representing one-fourth of the larger rectangle $EFGH$, as in (b)], the one-third point along FH may be located by sketching a line from point G through J and extending it to line FH. The point K at the intersection of these lines establishes the needed one-third distance.

To determine one-sixth the length of a side of a rectangle, as in (c), sketch a line from N through point P, as was done in (b), to determine a one-third distance. Point Q at which the line NP crosses the center line of the rectangle establishes the one-sixth distance along the center line.

Figure 7.14 shows how this method for dividing the sides of a rectangle might be used to proportion an orthographic sketch.

The square may be used to proportion a view after one dimension for the view has been assumed. In this method, additional squares are added to the initial one having the assumed length as one side (Fig. 7.15). As an example, suppose that it has been estimated that the front view of an object should be three times as long as it is high. In Fig. 7.15 the height of the view has been represented by the line AB sketched to an assumed length. The first step in making the construction is to sketch the initial square $ABCD$ and extend AC and BD to indefinite length, being certain that the overall length from A and B will be slightly greater than three times the length of AB. Then the center line must be sketched through the intersection of AD and BC. Now BX extended to E locates EF to form the second square, and DY extended to point G locates the line GH. Line AG will be three times the length of AB.

7.12 □ Pictorial sketching

Students may employ pictorial sketches to advantage as an aid in visualizing and organizing problems. Sales engineers may frequently include pictorial sketches with orthographic sketches when preparing field reports on the needs and suggestions of the firm's customers.

With some training anyone can prepare pictorial sketches that

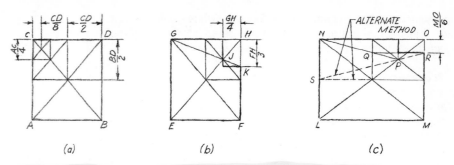

(a) (b) (c)

Fig. 7.13 □ **Methods of proportioning a rectangle representing the outline of a view.**

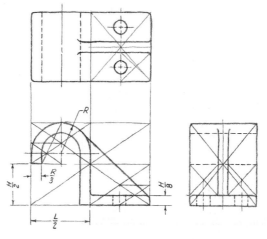

Fig. 7.14 □ **The rectangle method applied in making an orthographic sketch.**

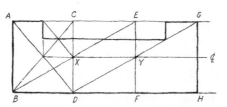

Fig. 7.15 □ **The build-up method.**

will be satisfactory for all practical purposes. Artistic ability is not needed. This fact is important, for many persons lack only the necessary confidence to start making pictorial sketches.

7.13 □ Mechanical methods of sketching

Many engineers have found that they can produce satisfactory pictorial sketches by using one of the so-called mechanical methods. They rely on these methods because of their familiarity with the procedures used in making pictorial drawings with instruments.

The practices presented in Chapter 11 for the mechanical methods, axonometric, oblique, and perspective, are followed generally in pictorial sketching, except that angles are assumed and lengths are estimated. For this reason, one must develop an eye for good proportion before he will be able to create a satisfactory pictorial sketch that will be in no way misleading.

A student having difficulty in interpreting a multiview drawing usually will find that a pictorial sketch, prepared as illustrated in Fig. 7.16, will clarify the form that he is trying to visualize, even before the last lines of the sketch have been drawn.

7.14 □ Isometric sketching

Isometric sketching starts with three isometric lines, called axes, which represent three mutually perpendicular lines. One of these axes is sketched vertically, the other two at 30° with the horizontal. In Fig. 7.16 (step I), the near front corner of the enclosing box lies along the vertical axis, while the two visible receding edges of the base lie along the axes receding to the left and to the right.

If the object is of simple rectangular form, as in Fig. 7.16, it may be sketched by drawing an enclosing isometric box (step I) on the surfaces of which the orthographic views may be sketched (step II). Care must be taken in assuming lengths and distances so that the finished view (step III) will have relatively correct proportions. In constructing the enclosing box (step I), the vertical edges are parallel to the vertical axis, and edges receding to the right and to the left are parallel to the right and left axes, respectively.

Objects of more complicated construction may be "blocked in," as shown in Fig. 7.17. Note that the projecting cylindrical features are enclosed in "isometric" prisms and that the circles are sketched within isometric squares. The procedure in Fig. 7.17 is the same as in Fig. 7.16, except that three enclosing isometric boxes are needed in the formation of the final representation instead of one.

In sketching an ellipse to represent a circle pictorially, an enclosing "isometric square" (rhombus) is drawn having sides equal approximately to the diameter of the true circle (step I, Fig. 7.18).

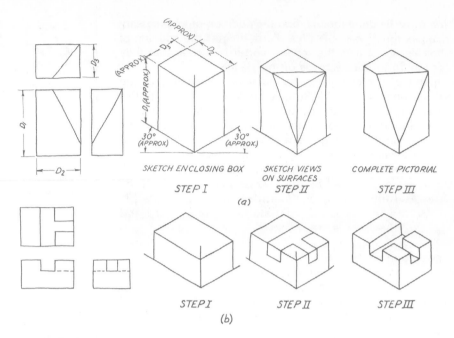

SKETCH ENCLOSING BOX   SKETCH VIEWS ON SURFACES   COMPLETE PICTORIAL

STEP I   STEP II   STEP III

(a)

STEP I   STEP II   STEP III

(b)

Fig. 7.16 □ Steps in isometric sketching.

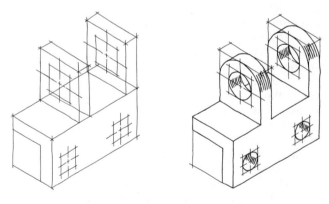

Fig. 7.17 □ Blocking in an isometric sketch.

SKETCH "ISOMETRIC SQUARE" STEP I   SKETCH SHORT ARCS STEP II   COMPLETE ELLIPSE STEP III

Fig. 7.18 □ Isometric circles.

The ellipse is formed by first drawing arcs tangent to the midpoints of the sides of the isometric square in light sketchy pencil lines (step II). In finishing the ellipse (step III) with a dark heavy line, care must be taken to obtain a nearly elliptical shape.

Figure 7.19 shows the three positions for an isometric circle. Note that the major axis is horizontal for an ellipse on a horizontal plane (I).

An idea sketch prepared in isometric is shown in Fig. 7.20.

7.15 □ Proportioning

As stated in Sec. 7.11, one should eventually be able to judge lengths and recognize proportions. Until this ultimate goal has been reached, the graphical method presented in Fig. 7.13 may be used with pictorial sketching (Fig. 7.21). The procedures as used are identical, the only recognizable difference being that the rectangle in the first case now becomes a rhomboid. Figure 7.22 illustrates how the method might be applied in making a sketch of a simple object. The enclosing box was sketched first with light lines, and then the graphical method was applied as shown to locate the points at one-quarter and one-half of the height. To establish the line of the top surface that is at a distance equal to one-third of the length from the end, a construction line was sketched from A to the midpoint B to locate C at the point of intersection of AB with the diagonal. Point C will fall on the required line.

7.16 □ Sketches in oblique

A sketch in oblique shows the front face without distortion, in its true shape. It has this one advantage over a representation prepared in isometric, even though the final result usually will not present so pleasing an appearance. It is not recommended for objects having circular or irregularly curved features on any but the front plane or in a plane parallel to it.

The beginner who is familiar with axonometric sketching will have very little difficulty in preparing a sketch in oblique, for, in general, the methods of preparation presented in the previous sections apply to both. The principal difference between these two forms of sketching is in the position of the axes, oblique sketching being unlike the isometric in that two of the axes are at right angles to each other. The third axis may be at any convenient angle, as indicated in Fig. 7.24.

Figure 7.25 shows the steps in making an oblique sketch using the proportioning methods previously explained for dividing a rectangle and a rhomboid. The receding lines are made parallel when a sketch is made in oblique projection.

The distortion and illusion of extreme elongation in the direction of the receding axis may be minimized by foreshortening to obtain

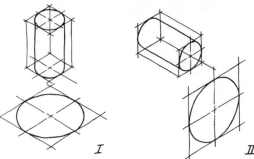
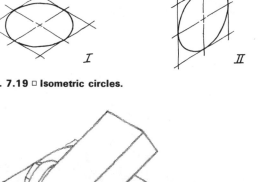

Fig. 7.19 □ Isometric circles.

I II III

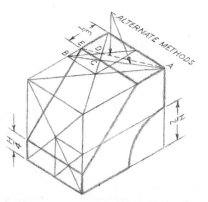

Fig. 7.20 □ An idea sketch in isometric.

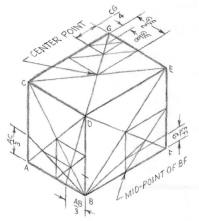

Fig. 7.21 □ Method for proportioning a rhomboid.

Fig. 7.22 □ Proportioning method applied.

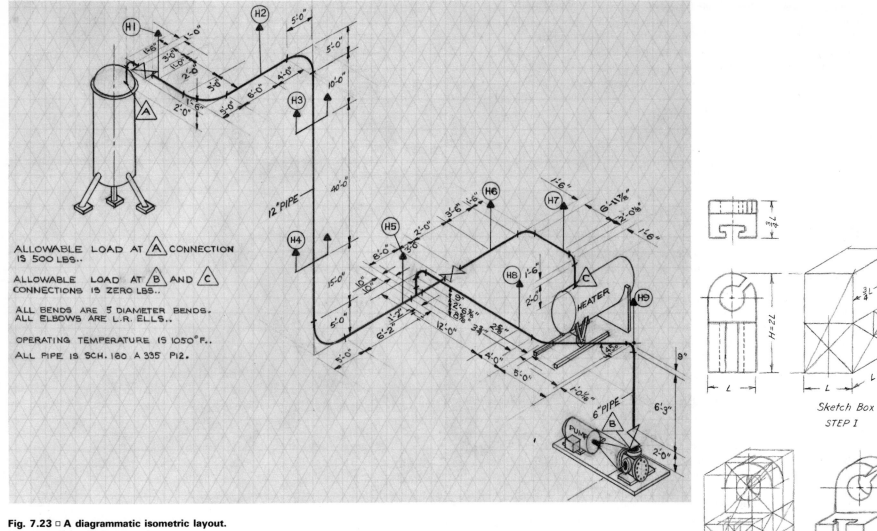

ALLOWABLE LOAD AT △A CONNECTION
IS 500 LBS..

ALLOWABLE LOAD AT △B AND △C
CONNECTIONS IS ZERO LBS..

ALL BENDS ARE 5 DIAMETER BENDS.
ALL ELBOWS ARE L.R. ELLS..

OPERATING TEMPERATURE IS 1050°F..

ALL PIPE IS SCH. 160 A 335 P12.

Fig. 7.23 □ **A diagrammatic isometric layout.**
(Courtesy Grinnell Company.)

Sketch Box
STEP I

Block-in Outline   Complete Pictorial
STEP II   STEP III

Fig. 7.25 □ **Steps in oblique sketching.**

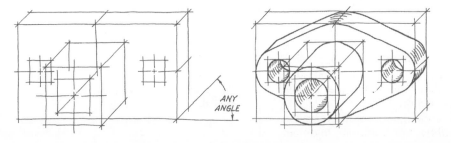

ANY
ANGLE

Fig. 7.24 □ **Blocking in an oblique sketch.**

proportions that are more realistic to the eye and by making the receding lines converge slightly. The resulting sketch will then be in a form of pseudoperspective, which resembles parallel perspective to some extent.

7.17 □ Perspective sketching

A sketch that has been prepared in accordance with the concepts of perspective will present a somewhat more pleasing and realistic effect than one in oblique or isometric. A perspective sketch actually presents an object as it would appear when observed from a particular point. The recognition of this fact, along with an understanding of the concepts that an object will appear smaller at a distance than when it is close and that horizontal lines converge as they recede until they meet at a vanishing point, should enable one to produce sketches having a perspective appearance. In sketching an actual object, a position should be selected that will show it to the best advantage. When the object exists only in one's mind or on paper in orthographic form, then the object must be visualized and the viewing position assumed.

At the start, the principal lines should be sketched in lightly, each line extending for some length toward its vanishing point. After this has been accomplished, the enclosing perspective squares for circles should be blocked in and the outline for minor details added. When the object lines have been darkened, the construction lines extending beyond the figure may be erased.

Figure 7.26 shows a parallel or one-point perspective that bears some resemblance to an oblique sketch. All faces in planes parallel to the front show their true shape. All receding lines should meet at a single vanishing point. Figure 7.27 is an angular or two-point perspective.

As stated previously, a two-point perspective sketch shows an object as it would appear to the human eye at a fixed point in space and not as it actually exists. All parallel receding lines converge (Fig. 7.28). Should these receding lines be horizontal, they will converge at a vanishing point on the eyeline. Those lines extending toward the right converge to a vanishing point to the right (VP_R), and those to the left converge to the left (VP_L). These vanishing points are at the level of the observer's eye (Fig. 7.28). A system of lines that is neither perpendicular nor horizontal will converge to a VP for inclined lines.

In one-point perspective one of the principal faces is parallel to the picture plane. All of the vertical lines will appear as vertical, and the receding horizontal lines will converge to a single vanishing point (Fig. 7.29).

Those interested in a complete discussion of the geometry of perspective drawing should read Secs. 11.25–11.35. The beginner should make two or three mechanically drawn perspectives

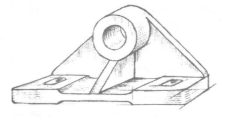

Fig. 7.26 □ A sketch in parallel perspective.

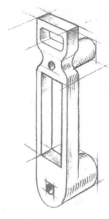

Fig. 7.27 □ A sketch in angular perspective.

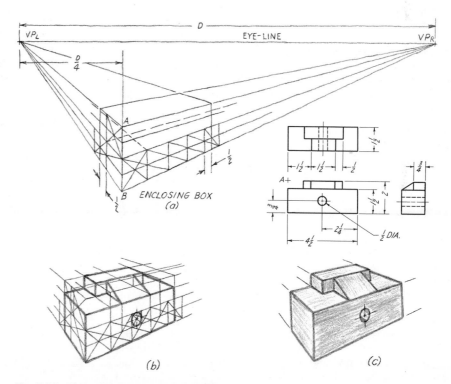

Fig. 7.28 □ Preparing a perspective sketch.

at the start to fix the fundamentals of the methods of perspective projection in his mind, even though there is some difference between sketching what one sees or imagines and true geometrical perspective.

In making a sketch in artist's perspective, several fundamental concepts must be recognized.

First, a circle sketched in perspective will resemble an ellipse (Fig. 7.28). The long diameter of the representation of a circle on a horizontal plane is always in a horizontal direction (Fig. 7.30).

Second, if an object or a component part of an object is above the eyeline, it will be seen from below. Should the object be below the eyeline, it will be seen from above (Fig. 7.28). The farther the object is removed below or above the eyeline, the more one can see of the top or the bottom surface.

Third, the nearest vertical edge of an object will be the longest vertical line of the view, as shown in Fig. 7.28. When two or more objects of the same actual height appear in a perspective sketch, their represented heights will decrease in the view as they near the vanishing point.

7.18 □ Preparing a perspective sketch

The application of proportioning methods to the construction and division of an enclosing box is shown in Fig. 7.28. Read Secs. 7.11 and 7.15. The construction of a required perspective by steps is as follows:

Step I: Sketch the eyeline. This line should be well toward the top of the sheet of sketch paper.

Step II: Locate VP_L and VP_R on the eyeline. These vanishing points should be placed as far apart as possible.

Step III: Assume the position and length for the near front edge AB. The length of this line, along with the spacing of the vanishing points, establishes the size of the finished sketch. The position of AB determines how the visible surfaces are to appear. For instance, if the line AB had been moved downward from the position shown in (a), much more of the top surface would be seen in (b) and (c). Should AB have been moved to the right from its position shown in (a), the left side would have become more prominent. If AB were placed midway between the two vanishing points, then both the front and left-side surfaces would be at 45° with the picture plane for the perspective. As it has been placed in (a), the side face is at 60° to the picture plane, while the front face is at 30°. Before establishing the position for the near front edge, one must decide which surfaces are the most important surfaces and how they may best be displayed in the sketch.

Step IV: Sketch light construction lines from points A and B to each vanishing point.

Step V: Determine the proportions for the enclosing box, in this case $4\frac{1}{2}$, 2, and $1\frac{1}{2}$, and mark off two equal units along AB.

Step VI: Sketch perspective squares, representing the faces of 1-unit cubes, starting at AB and working toward each vanishing point. In cases where an overall length must be completed with a partial unit, a full-perspective square must be sketched at the end.

Step VII: Subdivide any of the end squares if necessary and sketch in the enclosing box (a).

Step VIII: Locate and block in the details, subdividing the perspective squares as required to establish the location of any detail. When circles are to be sketched in perspective by a beginner, it is advisable to sketch the enclosing box first using light lines (b).

Step IX: Darken the object lines of the sketch. Construction lines may be removed and some shading added to the surfaces as shown in (c), if desired.

A sketch in one-point perspective might be made as shown in Fig. 7.29. For this particular sketch the enclosing box was made to the assumed overall proportions for the part. Then the location of the details was established by subdividing the regular rectangle of the front face of the enclosing box and the perspective rectangle of the right side.

7.19 □ Pencil shading

The addition of some shading to the surfaces of a part will force its form to stand out against the white surface of the sketching paper and will increase the effect of depth in a view that might otherwise appear to be somewhat flat.

Seldom are technologists and engineers able to do creditable work in artistic shading, with cast shadows included, as they could many years ago when training in art was part of an engineer's education. It is unfortunate that they lack this training at the present time, for art and design go hand in hand. This is especially true today, for a pleasing and appealing styling sells more products than a good mechanical design.

Within the scope of this chapter, written for beginning students, it will only be possible to present a few simple rules as a guide for those making a first attempt at surface shading. However, continued practice and some thought should lead one to the point where he can do a creditable job of shading and definitely improve a pictorial sketch.

When shading an imaginary part, a designer may consider the

source of the light to be located in a position to the left, above, and in front of the object. Of course, if the part actually exists and is being sketched by viewing it, then the sketcher should attempt to duplicate the degrees of shade and shadows as they are observed.

With the light source considered to be to the left, above, and in front of the object, a rectangular part would be shaded as shown in Fig. 7.30(a). The use of gradation of tone on the surfaces gives additional emphasis to the depth. To secure this added effect by shading, the darkest tone on the surface that is away from the light must be closest to the eye. As the surface recedes the tone must be made lighter in value with a faint trace of reflected light showing along the entire length of the back edge. On the lighted side, the lightest area must be closest to the eye, as indicated by the letter L_1 in (a). To make this lighted face appear to go into the distance, it is made darker as it recedes, but it should never be made as dark as the lightest of the dark tones on the dark surface.

Shading a cylindrical part is not as easy to do as shading a rectangular part but, if it is realized that practically half of the cylinder is in the light and half in the dark and that the lightest light and the darkest dark fall along the elements at the quarter-points of each half, then one should not find the task too difficult (b). The two extremes are separated by lighter values of shade. The first quarter on the lighted side must be made lighter than the last quarter on the dark side. In starting at the left and going counterclockwise there is a dark shade of light blending into the full light at the first quarter-point. From this point and passing the center to the dark line, the tone should become gradually darker. If vertical lines are used for shading, they should be spaced closer and closer together as they approach the dark line. The extreme-right-hand quarter should show the tones of reflected light.

There are two ways that pencil shading may be applied. If the paper has a medium-rough surface, solid tone shading may be used with one shade blending into the other. For the best results, the light tones are put on first over all areas to be shaded. The darker tones are then added by building up lighter tones to the desired intensity for a particular area. For this form of shading, a pencil with flattened point is used.

The other form of shading, and the one that is best suited for quick sketches, is produced with lines of varied spacing and weight. Light lines with wide spacing are used on the light areas and heavy lines that are closely spaced give the tone for the darkest areas. No lines are needed for the lightest of the light areas.

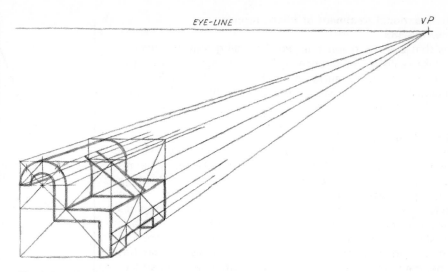

Fig. 7.29 □ **A sketch made in one-point perspective.**

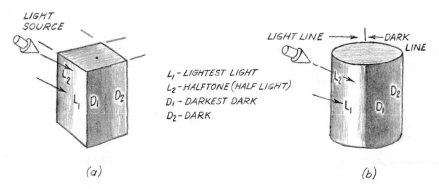

L_1 – LIGHTEST LIGHT
L_2 – HALFTONE (HALF LIGHT)
D_1 – DARKEST DARK
D_2 – DARK

Fig. 7.30 □ **Shading rectangular and cylindrical parts.**

7.20 □ Conventional treatment of fillets, rounds, and screw threads

Sketches that are not given full pencil shading may be given a more or less realistic appearance by representing the fillets and rounds of the unfinished surfaces as shown in Fig. 11.33. The conventional treatment for screw threads is shown in (b) and (c) of the same illustration.

7.21 □ Use of an overlay sheet

In preparing a design sketch, an overlay sheet may be used to advantage in making a sketch that is complicated by many details (Fig. 7.31). In this case, a quick sketch showing the general outline of the principal parts is made first in a rather rough form. Then an overlay sheet is placed over this outline sketch and the lines are retraced. In doing so, slight corrections can be made for any errors existing in the proportions of the parts or in the position of any of the lines of the original rough sketch. When this has been done, the representation of the related minor parts are added. If at any time one becomes discouraged with a sketch (multiview or pictorial) that he is making and feels that he should make a new start, he should use an overlay sheet, for there are usually many features on his existing sketch that may be retraced with a great saving of time.

7.22 □ Illustration sketches showing mechanisms exploded

A sketch of a mechanism showing the parts in exploded positions along the principal axes is shown in Fig. 7.32. Through the use of such sketches those who have not been trained to read multiview drawings may readily understand how a mechanism should be assembled, for both the shapes of the parts and their order of assembly, as denoted by their space relationship, is shown in pictorial form.

Illustration sketches may be made for discussions dealing with ideas for a design, but more frequently they are prepared for explanatory purposes to clarify instructions for preparing illustration drawings in a more finished form as assembly illustrations, advertising illustrations, catalogue illustrations, and illustrations for service and repair charts.

Many persons find that it is desirable, when preparing sketches of exploded mechanisms, to first block in the complete mechanism with all of the parts in position. At this initial stage of construction, the parts are sketched in perspective in rough outline and the principal axes and object lines are extended partway toward their vanishing points.

When the rough layout has been completed to the satisfaction of the person preparing the sketch, an overlay sheet is placed over the original sketch and the parts are traced directly from the sketch

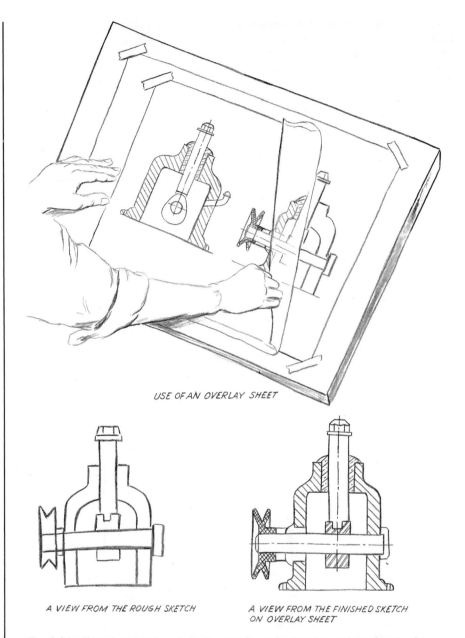

USE OF AN OVERLAY SHEET

A VIEW FROM THE ROUGH SKETCH

A VIEW FROM THE FINISHED SKETCH ON OVERLAY SHEET

Fig. 7.31 □ Use of an overlay sheet for creating a final and complete sketch of a mechanism.

beneath in exploded positions along the principal axes and along the axes of holes. Frequently some beginners make a traced sketch of each individual part and place the sketches in exploded positions before preparing the finished sketch. Others with more experience accomplish the same results by first tracing the major part along with the principal axes and then moving the overlay sheet as required to trace off the remaining parts in their correct positions along the axes.

It should be recognized that in preparing sketches of exploded mechanisms in this manner the parts are not shown perspectively reduced although they are removed from their original place in the pictorial assembly outward and toward the vanishing points. To prepare a sketch of this type with all of the parts shown in true geometrical perspective would result in a general picture that would be misleading and one that would be apt to confuse the nontechnical person because some parts would appear much too large or too small to be mating parts.

7.23 □ Pictorial sketching on ruled paper

Although one should become proficient in sketching on plain white bond paper, a specially ruled paper, shown in Fig. 7.33, can be used by those who need the help of guidelines.

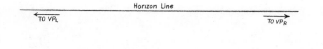

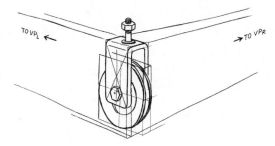

Block-in assembly in outline

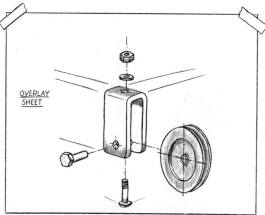

Retrace parts on overlay sheet in exploded positions from assembly sketch

Fig. 7.32 □ A sketch showing the parts of a mechanism in exploded positions.

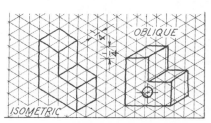

Fig. 7.33 □ Sketches on isometric paper.

Orbiting Laboratory.
Apollo command and service-module craft is shown docked to Earth-orbiting Skylab in concept by North American Rockwell's Space Division. Skylab is the nation's first step in putting a permanent laboratory in space. The Apollo spacecraft, built by Space Division, will ferry crews and supplies to Skylab. Many design sketches, layouts, and shop drawings are needed for the production of the units that make up a total system such as this one.

Chapter 8

A □ Multiview drawing and conventional practices

8.1 □ Introduction

Engineers use the orthographic system of projection for describing the shape of machine parts and structures (Fig. 8.1). Practical application of this method of describing an object results in a drawing consisting of a number of systematically arranged views that reproduce the object's exact shape. These views, positioned in strict accordance with the universally recognized arrangement, must show the three dimensions: width, height, and depth. Although three views (Fig. 8.1) are usually required to describe an ordinary object, only two may be needed for a particularly simple one. A very complicated object may require four or more views. A view projected on an auxiliary plane also may be desirable (Fig. 8.61). Such a view often makes possible the elimination of one of the principal views. Therefore, it is up to the individual to determine the number and type of views needed to produce a satisfactory drawing. He will soon develop a knack for this, if he bears in mind that the number of views required depends entirely on the complexity of the shape to be described.

8.2 □ Definition

Multiview (multiplanar) projection is a method by means of which the exact shape of an object can be represented by two or more separate views produced on projection planes that are usually at right angles to each other.

Basic graphics for design and product development

8.3 □ Methods of obtaining the views

The views of an object may be obtained by either of two methods: (1) the natural method; and (2) the "glass box" method.

Since the resulting views will be the same in either case, the beginner should adopt the method he finds the easiest to understand. Both methods are explained here in detail.

8.4 □ The natural method

In using this method, each of the necessary views is obtained by looking directly at the particular side of the object the view is to represent.

Figure 8.1 shows three of the principal views of an object: the front, top, and side views. They were obtained by looking directly at the front, top, and right side, respectively. In the application of this method, some consider the position of the object as fixed and the position of the observer as shifted for each view; others find it easier to consider the observer's position as fixed and the position of the object as changed for each view. Regardless of which procedure is followed, the top and side views must be arranged in their natural positions relative to the front view.

Figure 8.4 illustrates the natural relationship of views. Note that the top view is *vertically above* the front view, and the side view is *horizontally in line with* the front view. In both of these views *the front of the block is toward the front view.*

8.5 □ The "glass box" method

An imaginary "glass box" is used widely by instructors to explain the arrangement of orthographic views. In using this theoretical approach, it may be considered that planes of projection placed parallel to the six faces of an object form an enclosing "glass box" (Fig. 8.2). The observer views the enclosed object from the outside. The views are obtained by running projectors from points on the object to the planes. This procedure is in accordance with the theory and definition of orthographic projection explained in Sec. 8.2. The top, front, and right side of the box represent the H (horizontal), F (frontal), and P (profile) projection planes.

Since the projections on the sides of the three-dimensional transparent box are to appear on a sheet of drawing paper, it must be assumed that the box is hinged (Fig. 8.3) so that, when it is opened outward into the plane of the paper, the planes assume the positions illustrated in Figs. 8.3 and 8.4. Note that all of the planes, except the back one, are hinged to the frontal plane. In accordance with this universally recognized assumption, the top projection must take a position directly above the front projection, and the right-side projection must lie horizontally to the right of the front projection. To identify the separate projections, engineers call the one on the frontal plane the *front view* or *front elevation,*

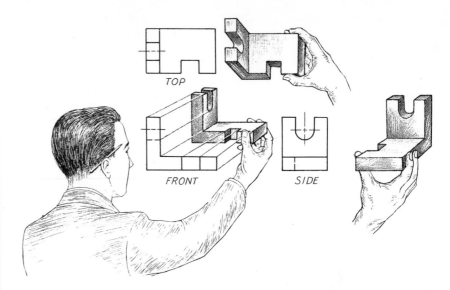

Fig. 8.1 □ Obtaining three views of an object.

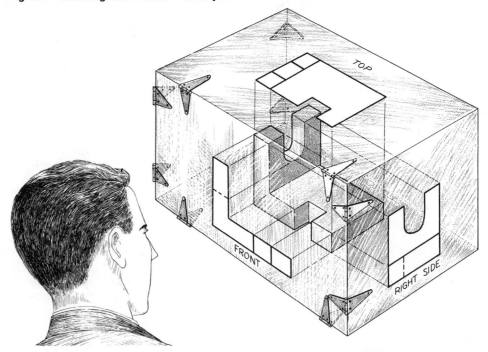

Fig. 8.2 □ The glass box.

the one on the horizontal plane the *top view* or *plan*, and the one on the side or profile plane the *side view, side elevation,* or *end view*. Figure 8.4 shows the six views of the same object as they would appear on a sheet of drawing paper. Ordinarily, only three of these views are necessary (front, top, and right side). A bottom or rear view will be required in comparatively few cases.

8.6 □ The principles of multiview drawing

The following principles should be studied carefully and understood thoroughly before any attempt is made to prepare an orthographic drawing:

1 The front and top views are *always* in line vertically (Fig. 8.5).

2 The front and side views are in line horizontally.

3 The front of the object in the top view faces the front view (Fig. 8.3).

4 The front of the object in the side view faces the front view (Fig. 8.3).

5 The depth of the top view is the same as the depth of the side view (or views) (see Fig. 8.5).

6 The width of the top view is the same as the width of the front view (Fig. 8.5).

7 The height of the side view is the same as the height of the front view (Fig. 8.5).

8 A view taken from above is a top view and *must* be drawn above the front view (Fig. 8.4).

9 A view taken from the right, in relation to the selected front, is a right-side view and *must* be drawn to the right of the front view (Fig. 8.4).

10 A view taken from the left is a left-side view and *must* be drawn to the left of the front view (Fig. 8.3).

11 A view taken from below is a bottom view and *must* appear below the front view (Fig. 8.3).

8.7 □ Projection of lines

A line may project either in true length, foreshortened, or as a point in a view depending on its relationship to the projection plane on which the view is projected (see Fig. 8.6). In the top view, the line projection $a^H b^H$ shows the true length of the edge AB (see pictorial) because AB is parallel to the horizontal plane of projection. Looking directly at the frontal plane, along the line, AB projects as a point ($a^F b^F$). Lines, such as CD, that are inclined to one of the planes of projection, will show a foreshortened projection in the view on the projection plane to which the line is inclined and true length in the view on the plane of projection

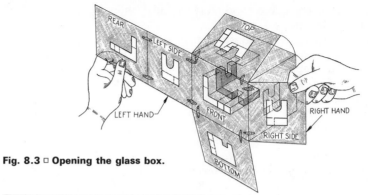

Fig. 8.3 □ Opening the glass box.

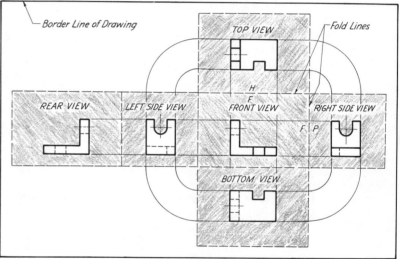

Fig. 8.4 □ Six views of an object on a sheet of drawing paper.

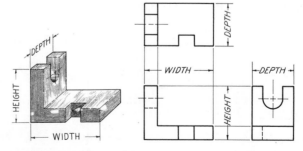

Fig. 8.5 □ View terminology.

to which the line is parallel. The curved line projection $e^F f^F$ shows the true length of the curved edge.

The student should study Fig. 8.7 and attempt to visualize the space position of each of the given lines. It is very necessary both in preparing and reading graphical representations to recognize the position of a point, line, or plane and to know whether the projection of a line is true length or foreshortened, and whether the projection of a plane shows the true size and shape. The indicated reference lines may be thought of as representing the edges of the glass boxes shown. The projections of a line are identified as being on either a frontal, horizontal, or profile plane by the use of the letters F, H, or P with the lowercase letters that identify the endpoints of the line. For example, in Fig. 8.7(a), $a^H b^H$ is the horizontal projection of the line AB, $a^F b^F$ is the frontal projection, and $a^P b^P$ is the profile projection.

It is suggested that the student hold a pencil before him and move it into the following typical line positions to observe the conditions under which the pencil representing a line, appears in true length.

1 *Vertical Line.* The vertical line is perpendicular to the horizontal and will therefore appear as a point in the H (top) view. It will appear in true length in the F (frontal) view and in the P (profile) view.

2 *Horizontal Line* [Fig. 8.7(b)]. The horizontal line will appear in true length when viewed from above because it is parallel to the H-plane of projection and its endpoints are theoretically equidistant from an observer looking downward.

3 *Inclined Line* [Fig. 8.7(c)]. The inclined line is any line not vertical or horizontal that is parallel to either the frontal plane or the profile plane of projection. An inclined line will show true length in the F (frontal) view or P (profile) view.

4 *Oblique Line* [Fig. 8.7(d)]. The oblique line will not appear in true length in any of the principal views because it is inclined to all of the principal planes of projection. It should be apparent in viewing the pencil alternately from the directions used to obtain the principal views, namely, from the front, above, and side, that one end of the pencil is always farther away from the observer than the other. Only when looking directly at the pencil from such a position that the endpoints are equidistant from the observer can the true length be seen. On a drawing, the true length projection of an oblique line will appear in a supplementary A (auxiliary) view projected on a plane that is parallel to the line (Sec. 9.6).

8.8 □ Meaning of lines

On a multiview drawing a visible or invisible line may represent either the intersection of two surfaces, the edge view of a surface,

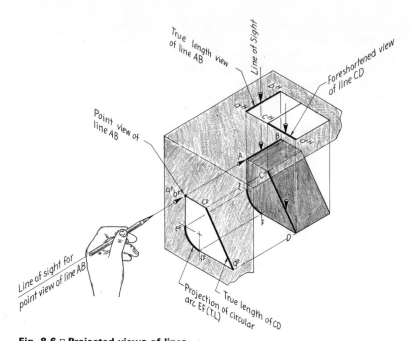

Fig. 8.6 □ Projected views of lines.

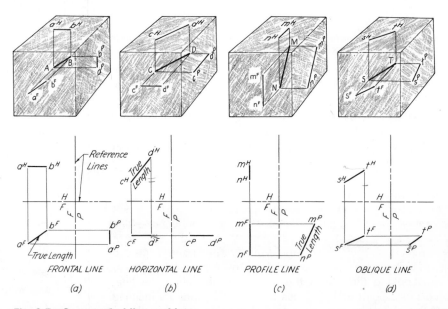

FRONTAL LINE (a)   HORIZONTAL LINE (b)   PROFILE LINE (c)   OBLIQUE LINE (d)

Fig. 8.7 □ Some typical line positions.

or the limiting element of a surface. These three different meanings of a line are illustrated in Fig. 8.8. In the top view, the curved line is an edge view of surface C, while a straight line is the edge view of surface A. The full circle in the front view may be considered as the edge view of the cylindrical surface of the hole. In the side view, the top line, representing the contour element of the cylindrical surface, indicates the limits for the surface and therefore can be thought of as being a surface limit line. The short vertical line in this same view represents the intersection of two surfaces. In reading a drawing, one can be sure of the meaning of a line on a view only after an analysis of the related view or views. All views must be studied carefully.

8.9 □ Projection of Surfaces

The components of most machine parts are bounded by either plane or single-curved surfaces. Plane surfaces bound cubes, prisms, and pyramids, while single-curved surfaces, ruled by a moving straight line, bound cylinders and cones. The projected representations (lines or areas) of both plane and single-curved surfaces are shown in Fig. 8.9. From this illustration the student should note that (1) when a surface is parallel to a plane of projection, it will appear in true size in the view on the plane of projection to which it is parallel; (2) when it is perpendicular to the plane of projection, it will project as a line in the view; and (3) when it is positioned at an angle, it will appear foreshortened. A surface will always project either as a line or an area on a view. The area representing the surface may be either a full-size or foreshortened representation.

In Fig. 8.9 the cylindrical surface A appears as a line in the side (profile) view and as an area in the top and front views. Surface B shows true size in the top view and as a line in both the front and side views. Surface C, a vertical surface, will appear as a line when observed from above.

8.10 □ The selection of views

Careful study should be given to the outline of an object before the views are selected (Fig. 8.19); otherwise there is no assurance that the object will be described completely from the reader's viewpoint. Only those views that are necessary for a clear and complete description should be selected. Since the repetition of information only tends to confuse the reader, superfluous views should be avoided.

Although some objects, such as cylinders, bushings, bolts, and so forth, require only two views (front and side), more complicated pieces may require an auxiliary or sectional view in addition to the ordinary three views.

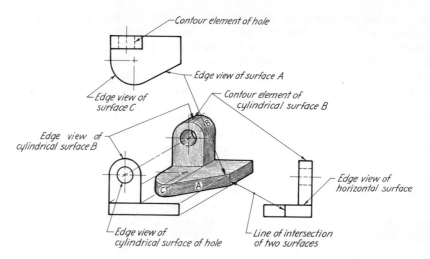

Fig. 8.8 □ The meaning of lines.

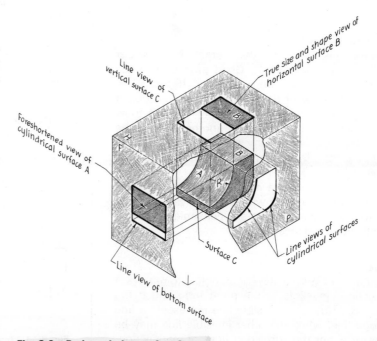

Fig. 8.9 □ Projected views of surfaces.

The space available for arranging the views often governs the choice between the use of a top or side view. The difference between the descriptive values of the two frequently is not great. For example, a draftsman often finds that the views of a long object will have better balance if a top view is used; while in the case of a short object, the use of a side view may make possible a more pleasing arrangement. It should be remembered that the choice of views for many objects is definitely fixed by the contour alone, and no choice is offered as far as spacing is concerned. It is more important to have a set of views that describes an object clearly than one that is artistically balanced.

Often there is a choice between two equally important views, such as between a right-side and left-side view or between a top and bottom view. In such cases, one should adhere to the following rule: *A right-side view should be used in preference to a left-side view and a top view in preference to a bottom view.* When this rule is applied to irregular objects, the front (contour) view should be drawn so that the most irregular outline is toward the top and right side.

Another rule, one that must be considered in selecting the front view, is as follows: *Place the object so as to obtain the smallest number of hidden lines.*

8.11 □ The principal (front) view

The principal view is the one that shows the characteristic contour of the object [see Fig. 8.10(a) and (b)]. Good practice dictates that this be used as the front view on a drawing. It should be clearly understood that the view of the natural front of an object is not always the principal view, because frequently it fails to show the object's characteristic shape. Therefore, another rule to be followed is: *Ordinarily, select the view showing the characteristic contour shape as the front view, regardless of the normal or natural front of the object.*

When an object does have a definite normal position, however, the front view should be in agreement with it. In the case of most machine parts, the front view can assume any convenient position that is consistent with good balance.

8.12 □ Invisible lines

Dotted lines are used on an external view of an object to represent surfaces and intersections invisible at the point from which the view is taken. In Fig. 8.11(a), one invisible line represents a line of intersection or edge line, while the other invisible line may be considered to represent either the surface or lines of intersection. On the side view in (b) there are invisible lines, which represent the contour elements of the cylindrical holes.

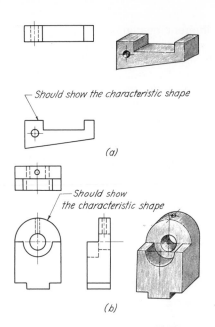

Should show the characteristic shape

(a)

Should show the characteristic shape

(b)

Fig. 8.10 □ The principal view of an object.

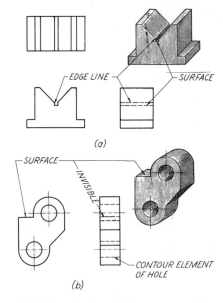

EDGE LINE — SURFACE

(a)

SURFACE — INVISIBLE — CONTOUR ELEMENT OF HOLE

(b)

Fig. 8.11 □ Invisible lines.

8.13 □ Treatment of invisible lines

The short dashes that form an invisible line should be drawn carefully in accordance with the recommendations in Sec. 8.19. An invisible line always starts with a dash in contact with the object line from which it starts, unless it forms a continuation of a visible line. In the latter case, it should start with a space, in order to establish at a glance the exact location of the endpoint of the visible line [see Fig. 8.12(C)]. Note that the effect of definite corners is secured at points A, B, E, and F, where, in each case, the end dash touches the intersecting line. When the point of intersection of an invisible line and another object line does not represent an actual intersection on the object, the intersection should be open as at points C and D. An open intersection tends to make the lines appear to be at different distances from the observer.

Parallel invisible lines should have the breaks staggered.

The correct and incorrect treatment for starting invisible arcs is illustrated at G and G'. Note that an arc should start with a dash at the point of tangency. This treatment enables the reader to determine the exact end points of the curvature.

8.14 □ Precedence of lines

When one discovers in making a multiview drawing that two lines coincide, the question arises as to which line should be shown or, in other words, which line must have precedence if the drawing is to be read intelligently. For example, as revealed in Fig. 8.13, a solid line may have the same position as an invisible line representing the contour element of a hole, or an invisible line may occur at the same place as a center line for a hole. In these cases the decision rests on the relative importance of each of the two lines that can be shown. The precedence of lines is as follows:

Solid lines (visible object lines) take precedence over all other lines.

Dashed lines (invisible object lines) take precedence over center lines, although evidence of center lines may be indicated as shown in both the top and side views of Fig. 8.13.

A *cutting-plane line* takes precedence over a center line where it is necessary to indicate the position of a cutting plane.

8.15 □ Treatment of tangent surfaces

When a curved surface is tangent to a plane surface, as illustrated in several ways on the pictorial drawing in Fig. 8.14, no line should be shown as indicated at A and B in the top view and as noted for the front and side views. At C in the top view the line represents a small vertical surface that must be shown even though the upper and lower lines for this surface may be omitted in the front view,

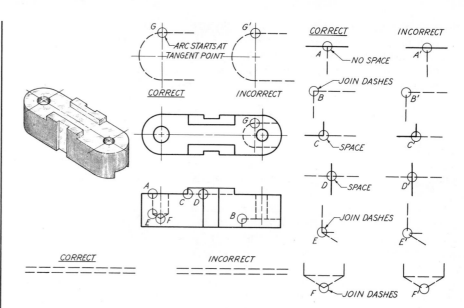

Fig. 8.12 □ Correct and incorrect junctures of invisible outlines.

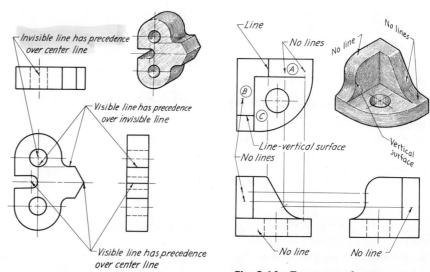

Fig. 8.13 □ Precedence of lines.

Fig. 8.14 □ Treatment of tangent surfaces.

depending on the decision of the draftsman. In the top view a line has been drawn to represent the intersection of the inclined and horizontal surfaces at the rear, even though they meet in a small round instead of a sharp edge. The presence of this line emphasizes the fact that there are two surfaces meeting here that are at a definite angle, one to the other. Several typical examples of tangencies and intersections have been illustrated in Fig. 8.15.

8.16 □ Parallel lines

When parallel surfaces are cut by a plane, the resulting lines of intersection will be parallel, as shown by the pictorial drawing in Fig. 8.16(b), where the near corner of the object has been removed by the oblique plane *ABC*. It can be observed from the multiview drawing in (c) that *when two lines are parallel in space, their projections will be parallel in all of the views*, even though at times both lines may appear as points in one view.

In Fig. 8.16, three views are to be drawn that show the block after the near front corner has been removed [see (b)]. Several of the required lines of intersection can be readily established through the given points *A*, *B*, and *C* that define the oblique plane. For example, $c^F b^F$ can be drawn in the front view and the line through a^F can be drawn parallel to it. In the top view, $a^H b^H$ should be drawn first and the intersection line through c^H should then be drawn parallel to this *H*-view of *AB*. The drawing can now be completed by working back and forth from view to view while applying the rule that a plane intersects parallel planes along lines of intersection that are parallel. The remaining lines are thus drawn parallel to either *AB* or *CB* [see the pictorial in (b)].

8.17 □ Plotting an elliptical boundary

The actual intersection of a circular cylinder or cylindrical hole with a slanting surface (inclined plane) is an ellipse (Fig. 8.17). The elliptical boundary in (a) appears as an ellipse in the top view, as a line in the front view, and as a semicircle in the side view. The ellipse was plotted in the top view by projecting selected points (such as points *A* and *B*) from the circle arc in the side view, as shown. For example, point *A* was projected first to the inclined line in the front view and then to the top view. The mitre line shown was used to project the depth distance for *A* in the top view for illustrative purposes only. Ordinarily, dividers should be used to transfer measurements to secure greater accuracy.

In (b) the intersection of the hole with the sloping surface is represented by an ellipse in the side view. Points selected around the circle in the top view (such as points *C* and *D*) projected to the side view as shown permit the draftsman to form the elliptical outline. It is recommended that a smooth curve be sketched free-

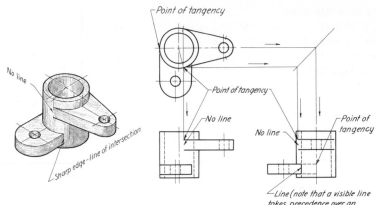

Fig. 8.15 □ Treatment of tangent surfaces.

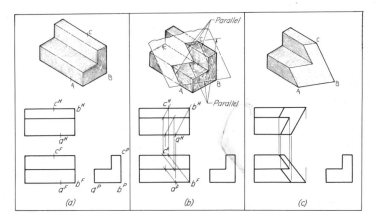

Fig. 8.16 □ Parallel lines.

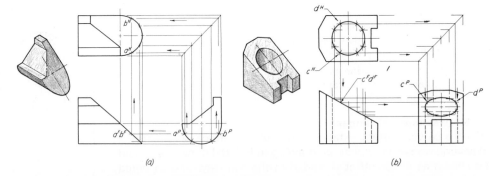

Fig. 8.17 □ Representation of an elliptical boundary.

hand through the projected points before the French curve is applied to draw the finished ellipse, because it is easier to fit a curved ruling edge to a line than to scattered points.

8.18 □ Treatment of intersecting finished and unfinished surfaces

Figure 8.18 illustrates the removal of material when machining surfaces, cutting a slot, and drilling a hole in a small part. The italic *f* on a surface of a pictorial drawing in this text indicates that the surface has been machined. The location of sharp and rounded corners, as illustrated in (*b*) and (*c*), are noted on the multiview drawing. A discussion covering rounded internal and external corners is given in Sec. 8.28.

8.19 □ To make an orthographic drawing

The location of all views should be determined before a drawing is begun. This will ensure balance in the appearance of the finished drawing. The contour view is usually started first. After the initial start, the draftsman should construct his views simultaneously by projecting back and forth from one to the other. It is poor practice to complete one view before starting the others, as much more time will be required to complete the drawing. Figure 8.19 shows the procedure for laying out a three-view drawing. The general outline of the views first should be drawn lightly with a hard pencil and then made heavier with a medium-grade pencil. Although experienced persons sometimes deviate from this procedure by drawing in the lines of known length and location in finished weight while constructing the views, it is not recommended that beginners do so (see Fig. 8.19, step III).

Although a 45° mitre line is sometimes used for transferring depth dimensions from the top view to the side view, or vice versa, as shown in Fig. 8.20(*b*), it is better practice to use dividers, as in (*a*). Continuous lines need not be drawn between the views and the mitre line, as in the illustration, for one may project from short dashes across the mitre line. The location of the mitre line may be obtained by extending the construction lines representing the front edge of the top view and the front edge of the side view to an intersection.

When making an orthographic drawing in pencil, the beginner should endeavor to use the line weights recommended in Sec. A.27 (Appendix). The object lines should be made very dark and bright, to give snap to the drawing as well as to create the contrast necessary to cause the shape of the object to stand out. Special care should be taken to gauge the dashes and spaces in invisible object lines. On ordinary drawings, $\frac{1}{8}$-in. dashes and $\frac{1}{32}$-in. spaces are recommended (Fig. 8.21).

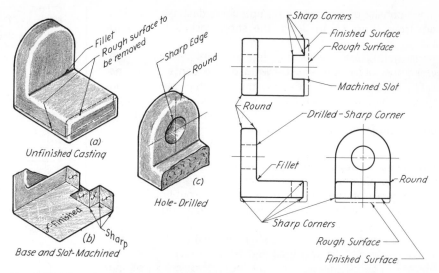

Fig. 8.18 □ Rough and finished surfaces on a casting.

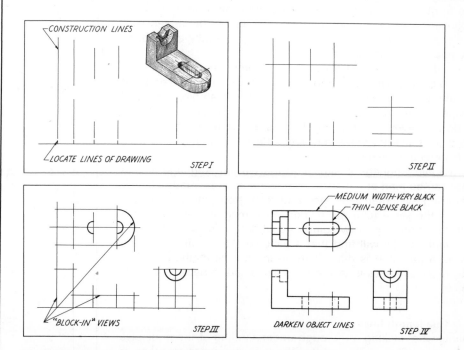

Fig. 8.19 □ Steps in making a three-view drawing of an object.

Center lines consist of alternate long and short dashes. The long dashes are from $\frac{3}{4}$ to $1\frac{1}{2}$ in. long, the short dashes $\frac{1}{8}$ in., and the spaces $\frac{1}{32}$ in. (Fig. 8.21).

When constructing a two-view drawing of a circular object, the pencil work must start with the drawing of the center lines, as shown in Fig. 8.22. This is necessarily the first step, because the construction of the circular (contour) view is based on a horizontal and a vertical center line. The horizontal object lines of the rectangular view are projected from the circles.

8.20 □ Visualizing an object from given views

Most students in elementary graphics courses find it difficult to visualize an object from two or more views. This trouble is largely due to the lack of systematic procedure for analyzing complex shapes.

The simplest method of determining shape is illustrated pictorially in Fig. 8.23. This method of "breaking down" may be applied to any object, since all objects may be thought of as consisting of elemental geometric forms, such as prisms, cylinders, cones, and so on. These imaginary component parts may be additions in the form of projections or subtractions in the form of cavities. By following such a detailed geometric analysis, the student can obtain a clear picture of an entire object by mentally assembling a few easily visualized forms.

It should be realized, when analyzing component parts, that it is impossible ordinarily to determine whether a form is an addition or a subtraction by looking at one view. For example, the small circles in the top view in Fig. 8.23 indicate a cylindrical form, but they do not reveal whether the form is a hole or a projection. If we consult the front view, however, the form is shown to be a hole (subtracted cylinder).

The graphic language is similar to the written language in that neither can be read at a glance. A drawing must be read patiently by referring systematically back and forth from one view to another. At the same time the reader must imagine a three-dimensional object and not a two-dimensional flat projection.

A student usually will find that a pictorial sketch will clarify the shape of a part that is difficult to visualize. The method for preparing quick sketches in isometric is explained in Secs. 7.14 and 7.15.

8.21 □ True-length lines

Students who, lacking a thorough understanding of the principles of projection (Sec. 8.7), find it difficult to determine whether or not a projection of a line in one of the principal views shows the true length of the line, should study carefully the following facts:

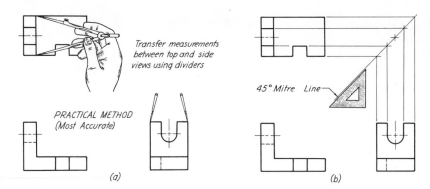

Fig. 8.20 □ Methods for transferring depth dimensions.

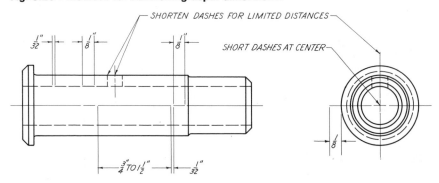

Fig. 8.21 □ Invisible lines and center lines.

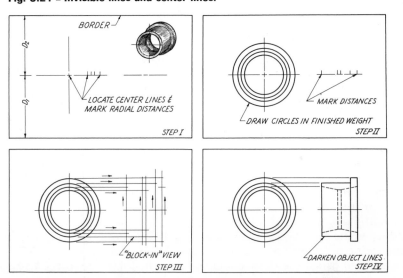

Fig. 8.22 □ Steps in making a two-view drawing of a circular object.

1 If the projection of a line shows the true length of the line, one of the other projections must appear as a horizontal line, a vertical line, or a point, on one of the other views of the drawing.

2 If the top and front views of a line are horizontal, then both views show the true length.

3 If the top view of a line is a point, the front and side views show the true length.

4 If the front view of a line is a point, the top and side views show the true length.

5 If the top and front views of a line are vertical, the side view shows the true length.

6 If the side projection of a line is a point, the top and front views show the true length.

7 If the front view of a line is horizontal and the top view is inclined, the top inclined view shows the true length.

8 If the top view of a line is horizontal and the front view is inclined, the front inclined view shows the true length.

8.22 □ Representation of holes

In preparing drawings of parts of mechanisms, a design draftsman finds it necessary to represent machined holes, which most often are either drilled, drilled and reamed, drilled and countersunk, drilled and counterbored, or drilled and spotfaced. Graphically, a hole is represented to conform with the finished form. The form may be completely specified by a note attached to the view showing the circular contour (Fig. 8.24). The shop note, as prepared by the draftsman, usually specifies the several shop operations in the order that they are to be performed in the shop. For example, in (*d*) the hole, as specified, is drilled before it is counterbored. When depth has not been given in the note for a hole, it is understood to be a through hole; that is, the hole goes entirely through the piece [(*a*), (*c*), (*d*), and (*e*)]. A hole that does not go through is known as a ''blind hole'' (*b*). For such holes, depth is the length of the cylindrical portion. Drilled, bored, reamed, cored, or punched holes are always specified by giving their diameters, never their radii. Drill diameters for number- and letter-size drills are given in Table 30 in the Appendix.

In drawing the hole shown in (*a*), which must be drilled before it is reamed, the limits are ignored and the diameter is scaled to the nearest regular fractional or decimal size. In (*b*) the 30° × 60° triangle is used to draw the approximate representation of the conical hole formed by the drill point. In (*c*) a 45° triangle has been used to draw an approximate representation of the outline of the conical enlargement. The actual angle of 82° is ignored in order to save time in drawing. The spotface in (*e*) is most often

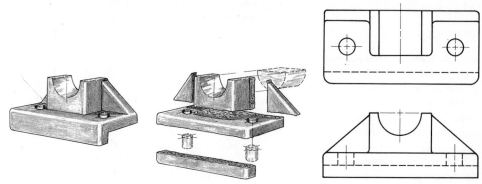

Fig. 8.23 □ ''Breaking down'' method.

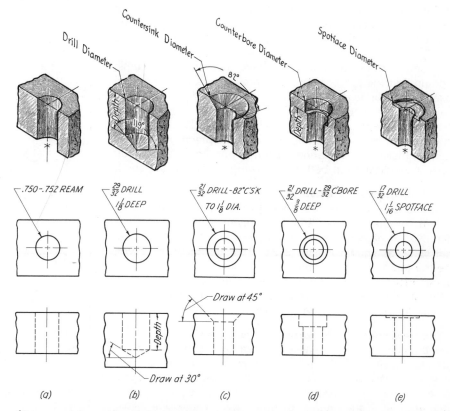

(*a*)   (*b*)   (*c*)   (*d*)   (*e*)

(For representations of threaded holes see chapter covering screw threads and fasteners)

Fig. 8.24 □ Representation of holes.

cut to a depth of $\frac{1}{16}$ in.; however, the depth is usually not specified. Information on the preparation of shop notes for holes may be found in Chapter 14.

8.23 □ Conventional practices

To reduce the high cost of preparing engineering drawings and at the same time to convey specific and concise information without a great expenditure of effort, some generally recognized systems of symbolic representation and conventional practices have been adopted by American industry.

A standard symbol or conventional representation can express information that might not be understood from a true-line representation unless accompanied by a lettered statement. In many cases, even though a true-line representation would convey exact information, very little more would be gained from the standpoint of better interpretation. Some conventional practices have been adopted for added clearness. For instance, they can eliminate awkward conditions that arise from strict adherence to the rules of projection.

These idioms of drawing have slowly developed with the graphic language until at the present time they are universally recognized and observed and appear in the various standards of the American National Standards Institute.

Professional men and skilled workmen have learned to accept and respect the use of the symbols and conventional practices, for they can interpret these representations accurately and realize that their use saves valuable time in both the drawing room and the shop.

8.24 □ Half views and partial views

When the available space is insufficient to allow a satisfactory scale to be used for the representation of a symmetrical piece, it is considered good practice to make one view either a half view or a partial view, as shown in Fig. 8.25. The half view, however, must be the top or side view and not the front view, which shows the characteristic contour. The half view should be the front half of the top or side view. In the case of the partial view shown in (b), a break line is used to limit the view.

8.25 □ The treatment of unimportant intersections

The conventional methods of treating various unimportant intersections are shown in Fig. 8.26. To show the true line of intersection in each case would add little to the value of the drawing. Therefore, in the views designated as preferred, true projection has been ignored in the interest of simplicity. On the side views, in (a) and (b), for example there is so little difference between the descriptive values of the true and approximate representations of the holes that the extra labor necessary to draw the true representation is unwarranted.

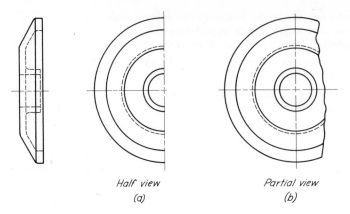

Half view
(a)

Partial view
(b)

Fig. 8.25 □ Half views and partial views.

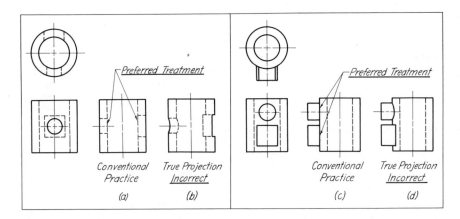

Fig. 8.26 □ Treatment of unimportant intersections.

8.26 □ Aligned views

Pieces that have arms, ribs, lugs, or other features at an angle are shown aligned or "straightened out" in one view, as illustrated in Fig. 8.27. By this method, it is possible to show the true shape as well as the true position of such features. In Fig. 8.28, the front view has been drawn as though the slotted arm had been revolved into alignment with the element projecting outward to the left. This practice is followed to avoid drawing an element—that is at an angle—in a foreshortened position.

8.27 □ Conventional treatment of radially arranged features

Many objects that have radially arranged features may be shown more clearly if true projection is violated, as in Fig. 8.27(b). Violation of true projection in such cases consists of intentionally showing such features swung out of position in one view to present the idea of symmetry and show the true relationship of the features at the same time. For example, while the radially arranged holes in a flange (Fig. 8.29) should always be shown in their true position in the circular view, they should be shown in a revolved position in the other view in order to show their true relationship with the rim.

Radial ribs and radial spokes are similarly treated [Fig. 8.27(a)]. The true projection of such features may create representations that are unsymmetrical and misleading. The preferred conventional method of treatment, by preserving symmetry, produces representations that are more easily understood and that at the same time are much simpler to draw. Figure 8.29 illustrates the preferred treatment for radial ribs and holes.

8.28 □ Representations of fillets and rounds

Interior corners, which are formed on a casting by unfinished surfaces, always are filled in (filleted) at the intersection in order to avoid possible fracture at that point. Sharp corners are also difficult to obtain and are avoided for this reason as well. Exterior corners are rounded for appearance and for the comfort of persons who must handle the part when assembling or repairing the machine on which the part is used. A rounded internal corner is known as a *fillet*; a rounded external corner is known as a *round* (Fig. 8.18).

When two intersecting surfaces are machined, however, their intersection will become a sharp corner. For this reason, all corners formed by unfinished surfaces should be shown "broken" by small rounds, and all corners formed by two finished surfaces, or one finished surface and one unfinished surface, should be shown "sharp" (Fig. 8.18).

Since fillets and rounds eliminate the intersection lines of intersecting surfaces, they create a special problem in orthographic

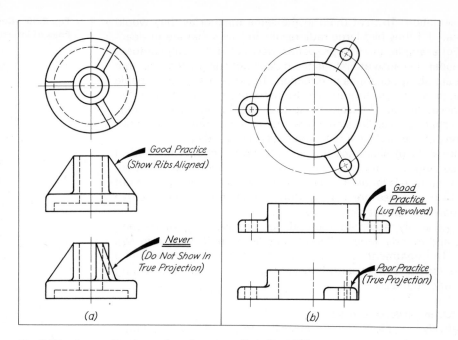

Fig. 8.27 □ Conventional practice of representing ribs and lugs.

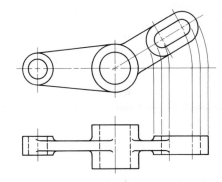

Fig. 8.28 □ Aligned views.

Fig. 8.29 □ Conventional treatment of radially arranged ribs.

representation. To treat them in the same manner as they would be treated if they had large radii results in views that are misleading. For example, the true-projection view in Fig. 8.30(c) confuses the reader, because at first glance it does not convey the idea that there are abrupt changes in direction. To prevent such a probable first impression and to improve the descriptive value of the view, it is necessary to represent these theoretically nonexisting lines. These characteristic lines are projected from the approximate intersections of the surfaces, with the fillets disregarded.

Figure 8.31(b) illustrates the accepted conventional method of representing the "run-out" intersection of a fillet in cases where a plane surface is tangent to a cylindrical surface. Although run-out arcs such as these are usually drawn freehand, a French curve or a bow instrument may be used.

The generally accepted methods of representing intersecting fillets and rounds under different conditions are illustrated in Fig. 8.31. The treatment, in each of the cases shown, is determined by the relationship existing between the sizes of the intersecting fillets and rounds.

8.29 □ Conventional breaks

A relatively long piece of uniform section may be shown to a larger scale, if a portion is broken out so that the ends can be drawn closer together (Fig. 8.32). When such a scheme is employed, a conventional break is used to indicate that the length of the representation is not to scale. The American National Standard conventional breaks, shown in Fig. 8.33, are used on either detail or assembly drawings. The break representations for indicating the broken ends of rods, shafts, tubes, and so forth, are designed to reveal the characteristic shape of the cross section in each case. Although break lines for round sections may be drawn freehand, particularly on small views, it is better to draw them with either an irregular curve or a bow instrument. The breaks for wood sections, however, always should be drawn freehand.

8.30 □ Ditto lines

When it is desirable to minimize labor in order to save time, ditto lines may be used to indicate a series of identical features. For example, the threads on the shaft shown in Fig. 8.34 are just as effectively indicated by ditto lines as by a completed profile representation. When ditto lines are used, a long shaft of this type may be shortened without actually showing a conventional break.

8.31 □ A conventional method
for showing a part in alternate positions

A method frequently used for indicating an alternate position of a part or a limiting position of a moving part is shown in Fig. 8.35.

Fig. 8.30 □ Conventional practice of representing nonexisting lines of intersection.

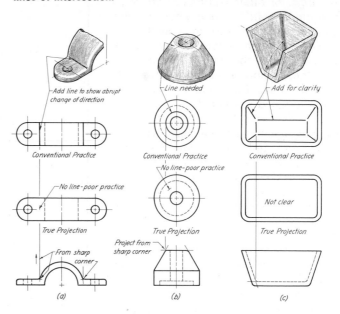

Fig. 8.31 □ The approximate methods of representing run-outs for intersecting fillets and rounds.

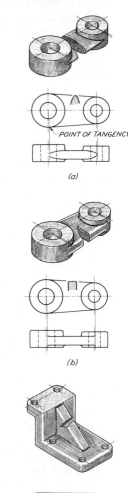

Fig. 8.32 □ A broken-out view.
(*Dimension values shown in [ ] are in millimeters.*)

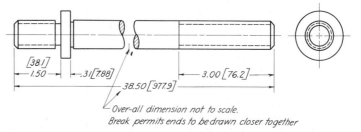

Fig. 8.33 □ Conventional breaks.

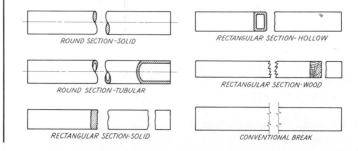

The dashes forming the object lines of the view showing the alternate position should be of medium weight. The phantom line shown in Fig. A.36 (Appendix) is recommended for representing an alternate position.

B □ Sectional views

8.32 □ Sectional views

Although the invisible features of a simple object usually may be described on an exterior view by the use of hidden lines, it is unwise to depend on a perplexing mass of such lines to describe adequately the interior of a complicated object or an assembled mechanism. Whenever a representation becomes so confused that it is difficult to read, it is customary to make one or more of the views ''in section'' (Fig. 8.36). A view ''in section'' is one obtained by imagining the object to have been cut by a cutting plane, the front portion being removed to reveal clearly the interior features. Figure 8.37 illustrates the use of an imaginary cutting plane. The resulting section (front) view, accompanied by a top view, is shown. At this point it should be understood that a portion is shown removed only in a sectional view, not in any of the other views.

When the cutting plane cuts an object lengthwise, the section obtained is commonly called a longitudinal section; when crosswise, it is called a cross section. It is designated as being either a full section, a half section, or a broken section. If the plane cuts entirely across the object, the section represented is known as a *full section*. If it cuts only halfway across a symmetrical object, the section is a *half section*. A *broken section* is a partial one, which is used when less than a half-section is needed. See Fig. 8.38.

On a completed sectional view, fine section lines are drawn across the surface cut by the imaginary plane, to emphasize the contour of the interior (see Sec. 8.39).

8.33 □ A full section

Since a cutting plane that cuts a full section passes entirely through an object, the resulting view will appear as illustrated in Fig. 8.38. Although the plane usually passes along the main axis, it may be offset (Fig. 8.37) to reveal important features.

A full-sectional view, showing an object's characteristic shape, usually replaces an exterior front view; however, one of the other principal views, side or top, may be converted to a sectional view if some interior feature thus can be shown to better advantage or if such a view is needed in addition to a sectioned front view.

The procedure in making a full-sectional view is simple, in that

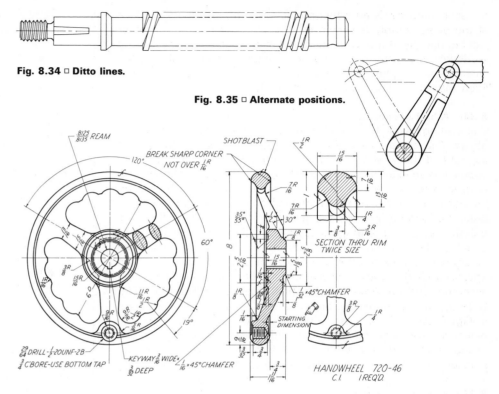

Fig. 8.34 □ Ditto lines.

Fig. 8.35 □ Alternate positions.

Fig. 8.36 □ A working drawing with sectional views.
(*Courtesy Warner and Swasey Company.*)

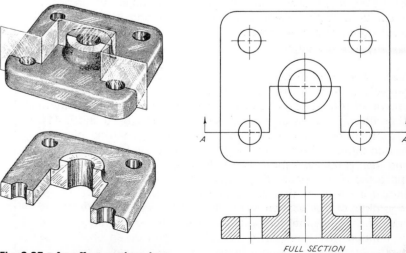

Fig. 8.37 □ An offset cutting plane.

the sectional view is an orthographic one. The imaginary cut face of the object simply is shown as it would appear to an observer looking directly at it from a point an infinite distance away. In any sectional view, it is considered good practice to omit all invisible lines unless such lines are necessary to clarify the representation. Even then they should be used sparingly.

8.34 □ A half section

The cutting plane for a half section removes one-quarter of an object. The plane cuts halfway through to the axis or center line so that half the finished sectional view appears in section and half appears as an external view (Fig. 8.38). This type of sectional view is used when a view is needed showing both the exterior and interior construction of a symmetrical object. Good practice dictates that hidden lines be omitted from both halves of the view unless they are absolutely necessary for dimensioning purposes or for explaining the construction. Although the use of a solid object line to separate the two halves of a half section has been approved by the Society of Automotive Engineers and has been accepted by the American National Standards Institute [Fig. 8.39(a)], many draftsmen prefer to use a center line, as shown in Fig. 8.39(b). They reason that the removal of a quarter of the object is theoretical and imaginary and that an actual edge, which would be implied by a solid line, does not exist. The center line is taken as denoting a theoretical edge.

8.35 □ A broken section

A broken or partial section is used mainly to expose the interior of objects so constructed that less than a half section is required for a satisfactory description (Fig. 8.40). The object theoretically is cut by a cutting plane and the front portion is removed by breaking it away. The ''breaking away'' gives an irregular boundary line to the section.

8.36 □ A revolved section

A revolved section is useful for showing the true shape of the cross section of some elongated object, such as a bar, or some feature of an object, such as an arm, spoke, or rib (Figs. 8.36 and 8.41).
To obtain such a cross section, an imaginary cutting plane is passed through the member perpendicular to the longitudinal axis and then is revolved through 90° to bring the resulting view into the plane of the paper (Fig. 8.42). When revolved, the section should show in its true shape and in its true revolved position, regardless of the location of the lines of the exterior view. If any lines of the view interfere with the revolved section, they should be omitted. It is sometimes advisable to provide an open space for the section by making a break in the object (Fig. 8.41).

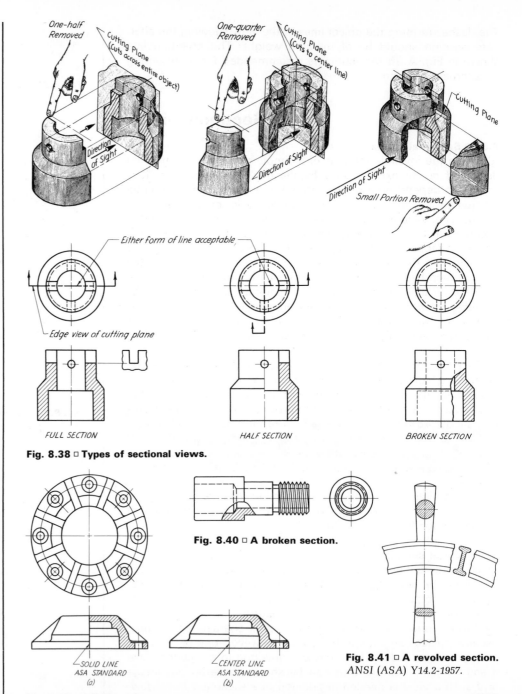

Fig. 8.38 □ Types of sectional views.

Fig. 8.40 □ A broken section.

Fig. 8.41 □ A revolved section.
ANSI (ASA) Y14.2-1957.

Fig. 8.39 □ A half section.

122

8.37 □ Removed (detail) sections

A removed section is similar to a revolved section, except that it does not appear on an external view but instead is drawn "out of place" and appears adjacent to it (Fig. 8.43). There are two good reasons why detail sections frequently are desirable. First, their use may prevent a principal view of an object, the cross section of which is not uniform, from being cluttered with numerous revolved sections (Fig. 8.44). Second, they may be drawn to an enlarged scale in order to emphasize detail and allow for adequate dimensioning.

Whenever a detail section is used, there must be some means of identifying it. Usually this is accomplished by showing the cutting plane on the principal view and then labeling both the plane and the resulting view, as shown in Fig. 8.44.

8.38 □ Phantom sections

A phantom or hidden section is a regular exterior view on which the interior construction is emphasized by crosshatching an imaginary cut surface with dotted section lines (Fig. 8.45). This type of section is used only when a regular section or a broken section would remove some important exterior detail, or, in some instances, to show an accompanying part in its relative position with regard to a particular part.

8.39 □ Section lining

Section lines are light continuous lines drawn across the imaginary cut surface of an object for the purpose of emphasizing the contour of its interior. Usually they are drawn at an angle of 45° except in cases where a number of adjacent parts are shown assembled.

To be pleasing in appearance, these lines must be correctly executed. While on ordinary work they are spaced about $\frac{3}{32}$ in. apart, there is no set rule governing their spacing. They simply should be spaced to suit the drawing and the size of the areas to be crosshatched. For example, on small views having small areas, the section lines may be as close as $\frac{1}{32}$ in., while on large views having large areas they may be as far apart as $\frac{1}{8}$ in. In the case of very thin plates, the cross-section is shown "solid black" (Fig. 8.46).

The usual mistake of the beginning student is to draw the lines too close together. This, plus the unavoidable slight variations, causes the section lining to appear streaked. Although several forms of mechanical section liners are available, most draftsmen do their spacing by eye. The student is advised to do likewise, being careful to see that the initial pitch, as set by the first few lines, is maintained across the area. To accomplish this, he should check back from time to time to make sure there has been no slight general increase or decrease in the spacing. Experienced

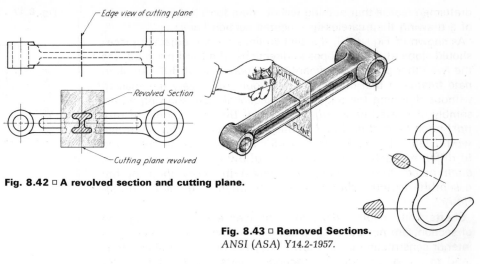

Fig. 8.42 □ A revolved section and cutting plane.

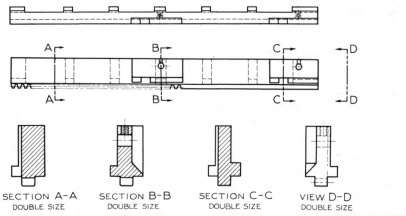

Fig. 8.43 □ Removed Sections.
ANSI (ASA) Y14.2-1957.

SECTION A-A DOUBLE SIZE   SECTION B-B DOUBLE SIZE   SECTION C-C DOUBLE SIZE   VIEW D-D DOUBLE SIZE

Fig. 8.44 □ Removed sections.
ANSI (ASA) Y14.2-1957.

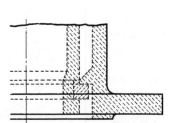

Fig. 8.45 □ A phantom section.

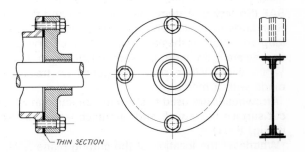

Fig. 8.46 □ Thin sections.

draftsmen realize that nothing will do more to ruin the appearance of a drawing than carelessly executed section lines.

As shown in Fig. 8.47, the section lines on two adjacent pieces should slope at 45° in opposite directions. If a third piece adjoins the two other pieces, it ordinarily is section-lined at 30°. An alternate treatment that might be used would be to vary the spacing without changing the angle. On a sectional view showing an assembly of related parts, *all portions of the cut surface of any part must be section-lined in the same direction, for a change would lead the reader to consider the portions as belonging to different parts. Furthermore, to allow quick identification, each piece (and all identical pieces) in every view of the assembly drawing should be section-lined in the same direction.*

Shafts, bolts, rivets, balls, and so on, whose axes lie in the plane of section, are not treated the same as ordinary parts. Having no interior construction to be shown, they are drawn in full and thus tend to make the adjacent sectioned parts stand out to better advantage (Fig. 8.48).

Whenever section lines drawn at 45° with the horizontal are parallel to part of the outline of the section, it is advisable to draw them at some other angle (say, 30° or 60°).

8.40 □ Outline sectioning
Very large surfaces may be section-lined around the bounding outline only, as illustrated in Fig. 8.49.

8.41 □ The symbolic representation for a cutting plane
The symbolic lines that are used to represent the edge of view of a cutting plane are shown in Fig. 8.50. The line is as heavy as an object line and is composed of either alternate long and short dashes or a series of dashes of equal length. The latter form is used in the automobile industry and has been approved by the SAE (Society of Automotive Engineers) and the American National Standards Institute. On drawings of ordinary size, when alternate long and short dashes are used for the cutting-plane line, the long dashes are $\frac{3}{4}$ in. long, the short dashes $\frac{1}{8}$ in. long, and the spaces $\frac{1}{32}$ in. wide. When drawn in pencil on tracing paper, they are made with a medium pencil.

Arrowheads are used to show the direction in which the imaginary cut surface is viewed, and reference letters are added to identify it (Fig. 8.51).

Whenever the location of the cutting plane is obvious, it is common practice to omit the edge-view representation, particularly in the case of symmetrical objects. But if it is shown, and coincides with a center line, it takes precedence over the center line.

Fig. 8.47 □ Two adjacent pieces.

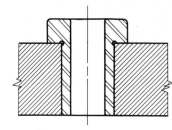

Fig. 8.48 □ Treatment of shafts, fasteners, ball bearings, and other parts.
(*Courtesy New Departure, Division General Motors Corporation.*)

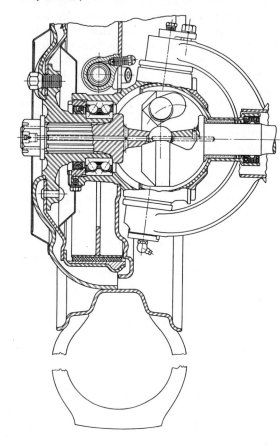

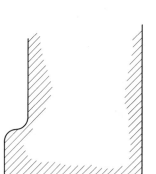

Fig. 8.49 □ Outline sectioning.

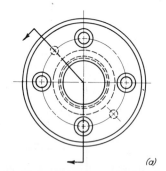

(a)

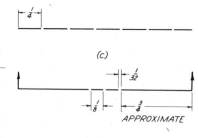

(c)

(b)

APPROXIMATE

Fig. 8.50 □ Cutting plane lines.
(*ANSI Standard.*)

124

8.42 □ Summary of the practices of sectioning

1 A cutting plane may be offset in order to cut the object in such a manner as to reveal an important detail that would not be shown if the cutting plane were continuous (Fig. 8.37).

2 All visible lines beyond the cutting plane for the section are usually shown.

3 Invisible lines beyond the cutting plane for the section are usually not shown, unless they are absolutely necessary to clarify the construction of the piece. In a half section, they are omitted in the unsectioned half, and either a center line or a solid line is used to separate the two halves of the view (Figs. 8.38 and 8.39).

4 On a view showing assembled parts, the section lines on adjacent pieces are drawn in opposite directions at an angle of 45° (Fig. 8.47).

5 On an assembly drawing, the portions of the cut surface of a single piece in the same view or different views always should be section-lined in the same direction, with the same spacing (Fig. 8.48).

6 The symbolic line indicating the location of the cutting plane may be omitted if the location of the plane is obvious (Fig. 8.36).

7 On a sectioned view showing assembled pieces, an exterior view is preferred for shafts, rods, bolts, nuts, rivets, and so forth, whose axes are in the plane of section (Fig. 8.48).

8.43 □ Auxiliary sections

A sectional view, projected on an auxiliary plane, is sometimes necessary to show the shape of a surface cut by a plane or to show the cross-sectional shape of an arm, rib, and so forth, inclined to any two or all three of the principal planes of projection (Fig. 8.52). When a cutting plane cuts an object, as in Fig. 8.52, arrows should show the direction in which the cut surface is viewed. Auxiliary sections are drawn by the usual method for drawing auxiliary views. When the bounding edge of the section is a curve, it is necessary to plot enough points to obtain a smooth one. Section 8.57 explains in detail the method for constructing the required view. A section view of this type usually shows only the inclined cut surface.

8.44 □ Conventional sections (Fig. 8.51)

Sometimes a less confusing sectioned representation is obtained if certain of the strict rules of projection are violated, as explained in Secs. 8.23–8.27. For example, an unbalanced and confused view results when the sectioned view of the handwheel shown in Fig. 8.53 is drawn in true projection. It is better practice to preserve symmetry by showing the spokes as if they were aligned

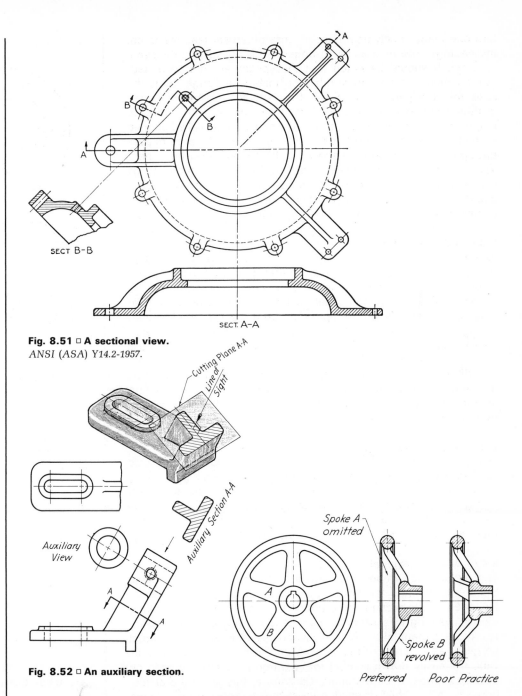

Fig. 8.51 □ A sectional view.
ANSI (ASA) Y14.2-1957.

Fig. 8.52 □ An auxiliary section.

Fig. 8.53 □ Spokes in section.
ANSI (ASA) Y14.2-1957.

into one plane. Such treatment of unsymmetrical features is not misleading, since their actual arrangement is revealed in the circular view. It should be noted that the spokes are not sectioned.

When there are an odd number of holes in a flange, as is the case with the part in Fig. 8.54, they should be shown aligned in the sectioned view to reveal their true location with reference to the rim and the axis of the piece. To secure the so-called aligned section, one usually considers the cutting plane to be bent to pass through the angled hole, as shown in the pictorial drawing. Then, the bent portion of the plane (with the hole) is imagined to be revolved until it is aligned with the other portion of the cutting plane. As straightened out, the imaginary continuous plane produces the preferred section view shown in (a).

8.45 □ Ribs in section

When a machine part has a rib cut by a plane of section (Fig. 8.55), a "true" sectional view taken through the rib would prove to be false and misleading, because the crosshatching on the rib would cause the object to appear "solid." The preferred treatment is to omit arbitrarily the section lines from the rib, as illustrated by Fig. 8.55(a). The resulting sectional view may be considered the view that would be obtained if the plane were offset to pass just in front of the rib (b).

An alternative conventional method, approved but not used as frequently, is illustrated in Fig. 8.56. This practice of omitting alternate section lines sometimes is adopted when it is necessary to emphasize a rib that might otherwise be overlooked.

8.46 □ Half views

When the space available is insufficient to allow a satisfactory scale to be used for the representation of a symmetrical piece, it is considered good practice to make one view a half view, as shown in Fig. 8.57. The half view, however, must be the top or side view and not the front view, which shows the characteristic contour. The half view should be the rear half.

8.47 □ Material symbols

The section-line symbols recommended by the American National Standards Institute for indicating various materials are shown in Fig. 8.58. Code section lining ordinarily is not used on a working (detail) drawing of a separate part. It is considered unnecessary to indicate a material symbolically when its exact specification must be given as a note. For this reason, and in order to save time as well, the easily drawn symbol for cast iron is commonly used on detail drawings for all materials. Contrary to this general practice, however, a few chief draftsmen insist that symbolic section lining be used on all detail drawings prepared under their supervision.

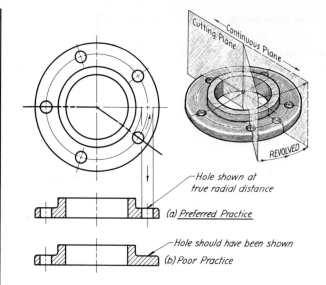

Hole shown at true radial distance

(a) *Preferred Practice*

Hole should have been shown

(b) *Poor Practice*

Fig. 8.54 □ Drilled flanges.

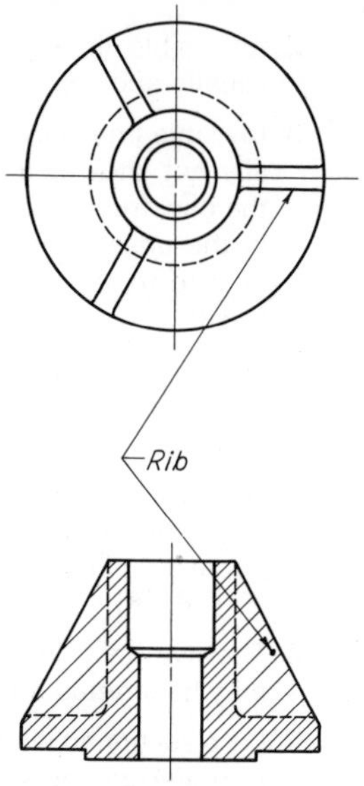

Usually shown through rib

Rib

(a)

Plane Through Center

Plane In Front Of Rib

(b)

Fig. 8.55 □ Conventional treatment of ribs in section.

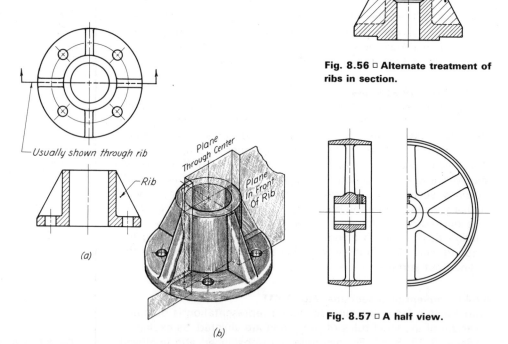

Rib

Fig. 8.56 □ Alternate treatment of ribs in section.

Fig. 8.57 □ A half view.

Code section lining usually is employed on an assembly section showing the various parts of a unit in position, because a distinction between the materials causes the parts to ''stand out'' to better advantage. Furthermore, a knowledge of the type of material of which an individual part is composed often helps the reader to identify it more quickly and understand its function.

C □ Auxiliary views

Primary auxiliary views

8.48 □ Introduction
When it is desirable to show the true size and shape of an irregular surface, which is inclined to two or more of the coordinate planes of projection, a view of the surface must be projected on a plane parallel to it. This imaginary projection plane is called an *auxiliary plane*, and the view obtained is called an *auxiliary view* (Fig. 8.59).

The theory underlying the method of projecting principal views applies also to auxiliary views. In other words, an auxiliary view shows an inclined surface of an object as it would appear to an observer stationed an infinite distance away (Fig. 8.60).

8.49 □ The use of auxiliary views
In commercial drafting, an auxiliary view ordinarily is a partial view showing only an inclined surface. The reason for this is that a projection showing the entire object adds very little to the shape description. The added lines are likely to defeat the intended purpose of an auxiliary view. However, in technical schools, some instructors require that an auxiliary view show the entire object, including all invisible lines (Fig. 8.66). Such a requirement, though impractical commercially, is justified in the classroom, for the construction of a complete auxiliary view furnishes excellent practice in projection.

A partial auxiliary view often is needed to complete the projection of a foreshortened feature in a principal view. This second important function of auxiliary views is illustrated in Fig. 8.72 and explained in Sec. 8.59.

8.50 □ Types of auxiliary views
Although auxiliary views may have an infinite number of positions in relation to the three principal planes of projection, primary auxiliary views may be classified into three general types in accordance with position relative to the principal planes. Figure 8.61

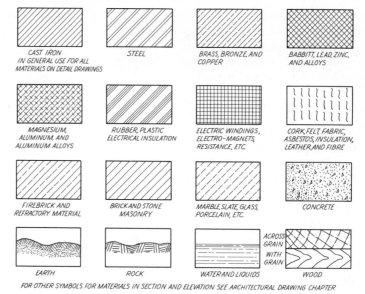

Fig. 8.58 □ Material symbols.
(*ANSI Standard.*)

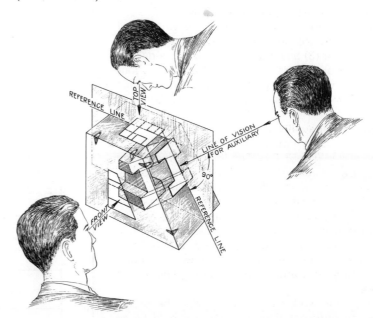

Fig. 8.59 □ Theory of projecting an auxiliary view.

shows the first type, where the auxiliary plane is perpendicular to the frontal plane and inclined to the horizontal plane of projection. Here the auxiliary view and top view have one dimension that is common to both: the depth. Note that the auxiliary plane is hinged to the frontal plane and that the auxiliary view is projected from the front view.

In Fig. 8.62 the auxiliary plane is perpendicular to the horizontal plane and inclined to the frontal and profile planes of projection. The auxiliary view is projected from the top view, and its height is the same as the height of the front view.

The third type of auxiliary view, as shown in Fig. 8.63, is projected from the side view and has a common dimension with both the front and top views. To construct it, distances may be taken from either the front or top view.

All three types of auxiliary views are constructed similarly. Each is projected from the view that shows the slanting surface as a line, and the distances for the view are taken from the other principal view that has a common dimension with the auxiliary. A careful study of the three illustrations will reveal the fact that the inclined auxiliary plane is always hinged to the principal plane to which it is perpendicular.

8.51 □ Symmetrical and unsymmetrical auxiliary views

Since auxiliary views are either symmetrical or unsymmetrical about a center line or reference line, they may be termed (1) symmetrical, (2) unilateral, or (3) bilateral, according to the degree of symmetry. A symmetrical view is drawn symmetrically about a center line, the unilateral view entirely on one side of a reference line, and the bilateral view on both sides of a reference line.

8.52 □ To draw a symmetrical auxiliary view

When an inclined surface is symmetrical, the auxiliary view is "worked" from a center line (Fig. 8.64). The first step in drawing such a view is to draw a center line parallel to the inclined line that represents an edge view of the surface. If the object is assumed to be enclosed in a glass box, this center line may be considered the line of intersection of the auxiliary plane and an imaginary vertical center plane. There are professional draftsmen who, not acquainted with the "glass box," proceed without theoretical explanation. Their method is simply to draw a working center line for the auxiliary view and a corresponding line in one of the principal views.

Although theoretically, this working center line may be drawn at any distance from the principal view, actually it should be so located to give the whole drawing a balanced appearance. If not already shown, it also must be drawn in the principal view showing the true width of the inclined surface.

Fig. 8.60 □ An auxiliary view.

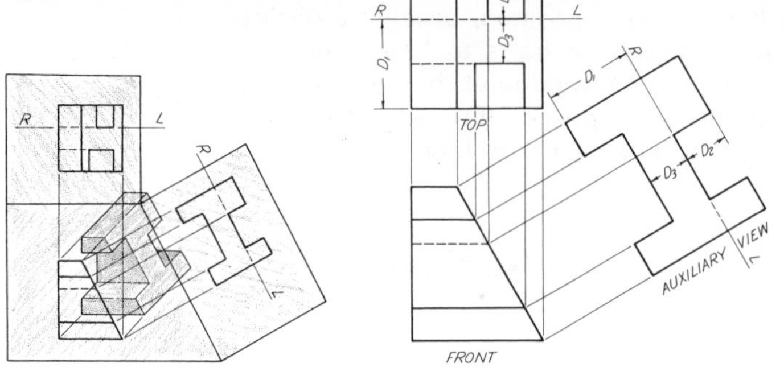

Fig. 8.61 □ Auxiliary view projected from front view.

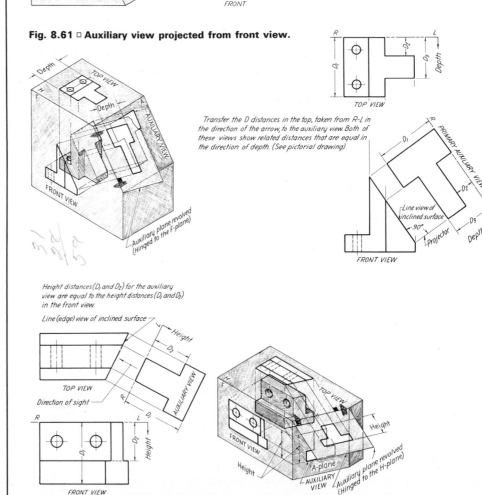

Transfer the D distances in the top, taken from R-L in the direction of the arrow, to the auxiliary view. Both of these views show related distances that are equal in the direction of depth. (See pictorial drawing)

Height distances (D_1 and D_2) for the auxiliary view are equal to the height distances (D_1 and D_2) in the front view.

Fig. 8.62 □ Auxiliary view projected from top view.

128

The next step is to draw projection lines from each point of the sloping face, remembering that the projectors make an angle of 90° with the inclined line representing the surface. With the projectors drawn, the location of each point in the auxiliary can be established by setting the dividers to each point's distance from the center line in the principal view and transferring the distance to the auxiliary view. For example, point X is projected to the auxiliary by drawing a projector from point X in the front view perpendicular to the center line. Since its distance from the center line in the top view is the same as it is from the center line in the auxiliary view, the point's location along the projector may be established by using the distance taken from the top view. In the case of point X, the distance is set off from the center line toward the front view. Point Y is set off from the center line away from the front view. A careful study of Fig. 8.64 reveals the fact that if a point lies between the front view and the center line of the top view, it will lie between the front view and the center line of the auxiliary view, and, conversely, if it lies away from the front view with reference to the center line of the top view, it will lie away from the front view with reference to the center line of the auxiliary view.

8.53 □ Unilateral auxiliary views

When constructing a unilateral auxiliary view, it is necessary to work from a reference line that is drawn in a manner similar to the working center line of a symmetrical view. The reference line for the auxiliary view may be considered to represent the line of intersection of a reference plane, coinciding with an outer face, and the auxiliary plane (Fig. 8.65). The intersection of this plane with the top plane establishes the reference line in the top view. All the points are projected from the edge view of the surface, as in a symmetrical view, and it should be noted in setting them off that they all fall on the same side of the reference line.

Figure 8.66 shows an auxiliary view of an entire object. In constructing such a view, it should be remembered that the projectors from all points of the object are perpendicular to the auxiliary plane, since the observer views the entire figure by looking directly at the inclined surface. The distances perpendicular to the auxiliary reference line were taken from the front view.

8.54 □ Bilateral auxiliary views

The method of drawing a bilateral view is similar to that of drawing a unilateral view, the only difference being that in a bilateral view the inclined face lies partly on both sides of the reference plane, as shown in Fig. 8.67.

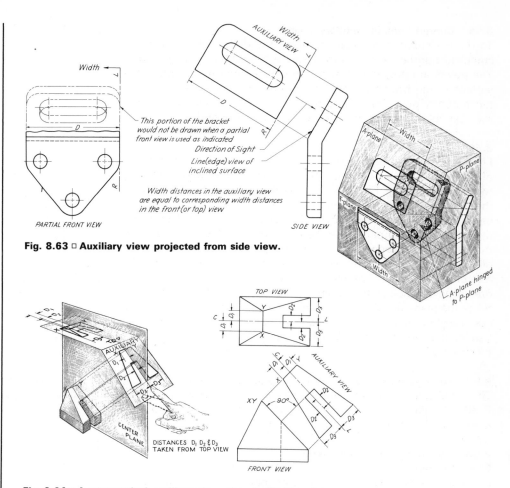

Fig. 8.63 □ Auxiliary view projected from side view.

Fig. 8.64 □ A symmetrical auxiliary view of an inclined surface.

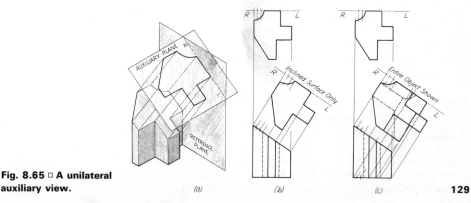

Fig. 8.65 □ A unilateral auxiliary view.

(a)   (b)   (c)

8.55 □ Curved lines in auxiliary views

To draw a curve in an auxiliary view, the draftsman must plot a sufficient number of points to ensure a smooth curve (Fig. 8.68). The points are projected first to the inclined line representing the surface in the front view and then to the auxiliary view. The distance of any point from the center line in the auxiliary view is the same as its distance from the center line in the end view.

8.56 □ Dihedral angles

Frequently, an auxiliary view may be needed to show the true size of a dihedral angle—that is, the true size of the angle between two planes. In Fig. 8.69, it is desirable to show the true size of the angle between the planes forming the V-slot by means of a partial auxiliary view, as shown. The direction of sight (see pictorial) must be taken parallel to the edge lines 1–2 and 3–4 so that these lines will appear as points and the surfaces forming the dihedral angle will project as line views in the auxiliary view. The reference line for the partial auxiliary view would necessarily be drawn perpendicular to the lines 1–2 and 3–4 in the top view. Since the plane on which the auxiliary view is projected is a vertical one, height dimensions were used—that is, distances in the direction of the dimension D in the auxiliary view, were taken from the front view.

✓ 8.57 □ To construct an auxiliary view, practical method

The usual steps in constructing an auxiliary view are shown in Fig. 8.70. The illustration should be studied carefully, as each step is explained in the drawing.

8.58 □ Auxiliary and partial views

Often the use of an auxiliary view allows the elimination of one of the principal views (top or side) or makes possible the use of a partial principal view. The shape description furnished by the partial views shown in Fig. 8.71 is sufficient for a complete understanding of the shape of the part. The use of partial views simplifies the drawing, saves valuable drafting time, and tends to make the drawing easier to read.

 A break line is used at a convenient location to indicate an imaginary break for a partial view.

8.59 □ The use of an auxiliary view to complete a principal view

As previously stated, it is frequently necessary to project a foreshortened feature in one of the principal views from an auxiliary view. In the case of the object shown in Fig. 8.72, the foreshortened projection of the inclined face in the top view can be projected from the auxiliary view. The elliptical curves are plotted by

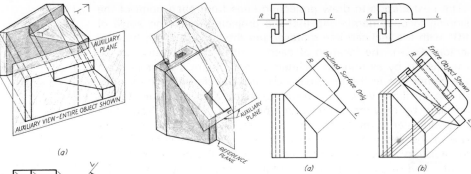

(a)

Fig. 8.67 □ A bilateral auxiliary view.

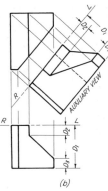

(b)

Fig. 8.66 □ An auxiliary view of an object.

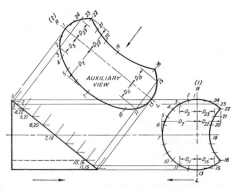

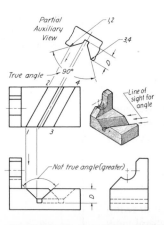

Fig. 8.69 □ To determine the true dihedral angle between inclined surfaces.

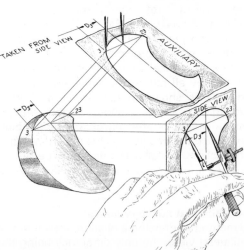

Fig. 8.68 □ Curved line auxiliary view.

projecting points from the auxiliary view to the front view and from there to the top view. The location of these points in the top view with respect to the center line is the same as their location in the auxiliary view with respect to the auxiliary center line. For example, the distance D_1 from the center line in the top view is the same as the distance D_1 from the auxiliary center line in the auxiliary view.

8.60 □ Line of Intersection

It is frequently necessary to represent a line of intersection between two surfaces when making a multiview drawing involving an auxiliary view. Figure 8.73 shows a method for drawing the line of intersection on a principal view. In this case the scheme commonly used for determining the intersection involves the use of elements drawn on the surface of the cylindrical portion of the part, as shown on the pictorial drawing. These elements, such as AB, are common to the cylindrical surface. Point B, where the element pierces the flat surface, is a point that is common to both surfaces and therefore lies on the line of intersection.

On the orthographic views, element AB appears as a point on the auxiliary view and as a line on the front view. The location of the projection of the piercing point on the front view is visible upon inspection. Point B is found in the other principal view by projecting from the front view and setting off the distance D taken from the auxiliary view. The distance D of point B from the center line is a true distance for both views. The center line in the auxiliary view and side view can be considered as the edge view of a reference plane or datum plane from which measurements can be made.

8.61 □ True length of a line

The true length of an oblique line may be determined either by means of an auxiliary view or by revolution of the line. Separate discussions of the procedure to be followed in the application of these methods are given in Chapter 9. To determine the true length of a line by revolution, see Sec. 9.21. To find the true length through the use of an auxiliary view, read Sec. 9.6.

Secondary auxiliary views

8.62 □ Secondary (oblique) auxiliary views

Frequently an object will have an inclined face that is not perpendicular to any one of the principal planes of projection. In such cases it is necessary to draw a primary auxiliary view and a secondary auxiliary or oblique view (Fig. 8.74). The primary auxiliary view is constructed by projecting the figure on a primary auxiliary plane

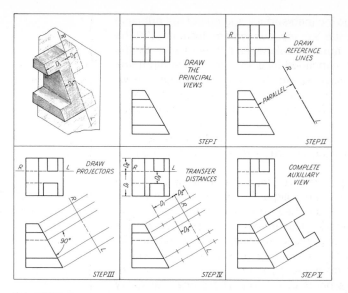

Fig. 8.70 □ Steps in constructing an auxiliary view.

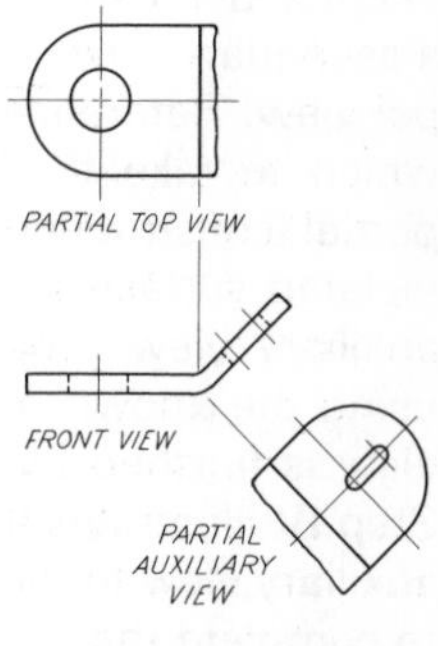

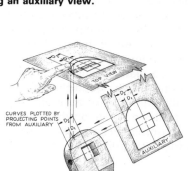

Fig. 8.71 □ Partial views.

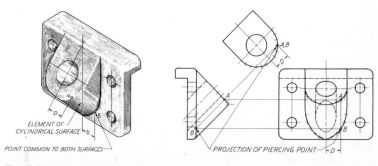

Fig. 8.72 □ Use of auxiliary to complete a principal view.

Fig. 8.73 □ Line of intersection.

131

that is perpendicular to the inclined surface and one of the principal planes. This plane may be at any convenient location. In the illustration, the primary auxiliary plane is perpendicular to the frontal plane. Note that the inclined face appears as a straight line in the primary auxiliary view. Using this view as a regular view, the secondary auxiliary view may be projected on a plane parallel to the inclined face. Figure 8.74(b) shows a practical application of the theoretical principles shown pictorially in (a).

It is suggested that the student read Sec. 9.13 in which the procedure for drawing the normal (true-shape) view of an oblique surface is presented step by step.

Figure 8.75 shows the progressive steps in preparing and using a secondary auxiliary view of an oblique face to complete a principal view. Reference planes have been used as datum planes from which to take the necessary measurements. Step II shows the partial construction of the primary auxiliary view in which the inclined surface appears as a line. Step III shows the secondary auxiliary view projected from the primary view and completed, using the known measurements of the lug. The primary auxiliary view is finished by projecting from the secondary auxiliary view. Step IV illustrates the procedure for projecting from the secondary auxiliary view to the top view through the primary auxiliary in order to complete the foreshortened view of the lug. It should be noted that distance D_1 taken from reference R_2P_2 in the secondary auxiliary is transferred to the top view because both views show the same width distances in true length. A sufficient number of points should be obtained to allow the use of an irregular curve. Step V shows the projection of these points on the curve to the front view. In this case the measurements are taken from the primary auxiliary view because the height distances from reference plane R_1P_1 are the same in both views.

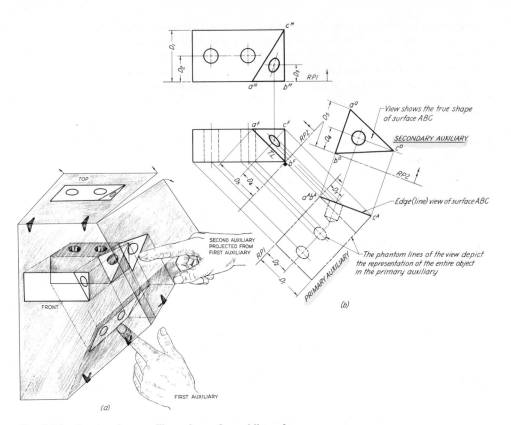

Fig. 8.74 □ A secondary auxiliary view of an oblique face.

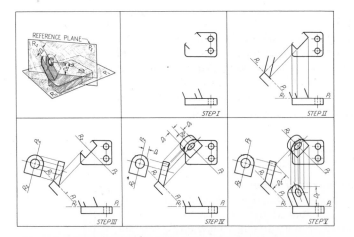

Fig. 8.75 □ Steps in drawing a secondary auxiliary view and using it to complete a principal view.

Problems 1

Multiview representation

The problems that follow are intended primarily to furnish study in multiview projection through the preparation of either sketches or instrumental drawings. Many of the problems in this chapter, however, may be prepared in more complete form. Their views may be dimensioned as are the views of working drawings, if the student will study carefully the chapter covering dimensioning before attempting to record size description (Chapter 14). All dimensions should be placed in accordance with the general rules of dimensioning.

The views shown in a sketch or drawing should be spaced on the paper with aim for balance within the borderlines. Ample room should be allowed between the views for the necessary dimensions. If the views are not to be dimensioned, the distance between them may be made somewhat less than would be necessary otherwise.

Before starting to draw, the student should reread Sec. 8.19 and study Fig. 8.19, which shows the steps in making a multiview drawing. The preparation of a preliminary sketch always proves helpful to the beginner.

All construction work should be done in light lines with a sharp hard pencil.

1 (Fig. 8.76). Draw or sketch the third view for each of the given objects.

2 (Fig. 8.77). Make an orthographic drawing or sketch of the bench stop. The views may be dimensioned. The

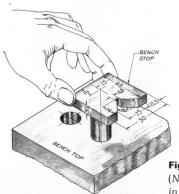

Fig. 8.77 □ Bench stop.
(*Numerical values in millimeters.*)

Fig. 8.76 □ Third-view problems.

shaft portion that fits into the hole in the bench top is 19.05 mm in diameter and 50.8 mm long.

3–19 (Figs. 8.78–8.94). Make multiview drawings of the given objects. The views of a drawing may or may not be dimensioned.

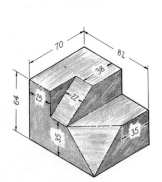

Fig. 8.78 □ Corner block.
(Numerical values in millimeters.)

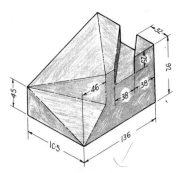

Fig. 8.79 □ Adjustment block.
(Numerical values in millimeters.)

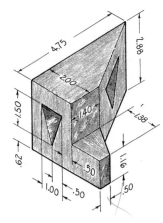

Fig. 8.80 □ Bevel block.

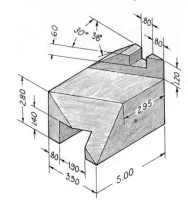

Fig. 8.81 □ End block.

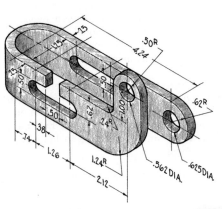

Fig. 8.85 □ Control guide.

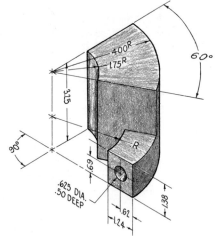

Fig. 8.83 □ Jaw block.

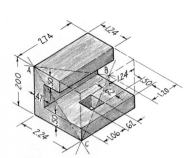

Fig. 8.82 □ Stabilizer block.

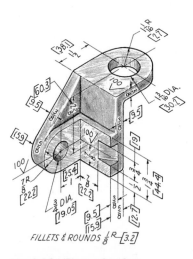

Fig. 8.84 □ Ejector bracket.
(Dimension values shown in [ ] are in millimeters.)

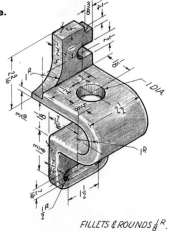

Fig. 8.86 □ Mounting bracket.

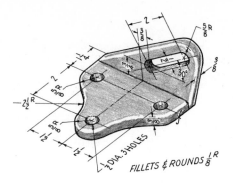

Fig. 8.87 □ Guide clip.

FILLETS & ROUNDS 1/8 R

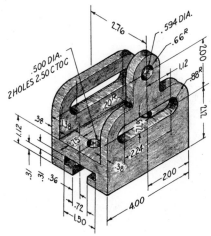

Fig. 8.88 □ Control guide.

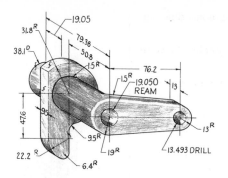

Fig. 8.89 □ Offset trip lever.
(Numerical values in millimeters.)

FILLETS & ROUNDS 3 R

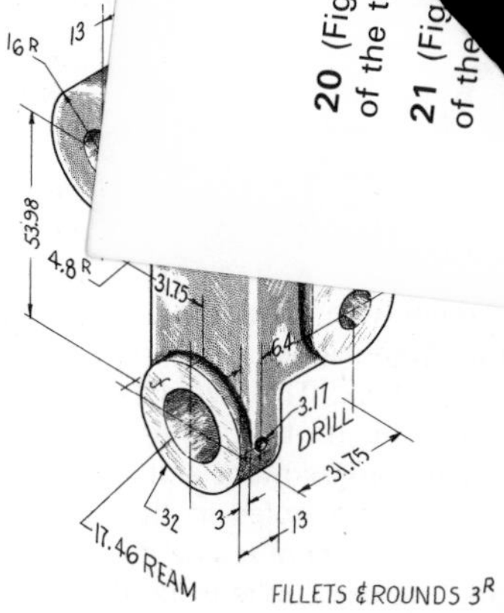

Fig. 8.90 □ Guide link.
(Numerical values in millimeters.)

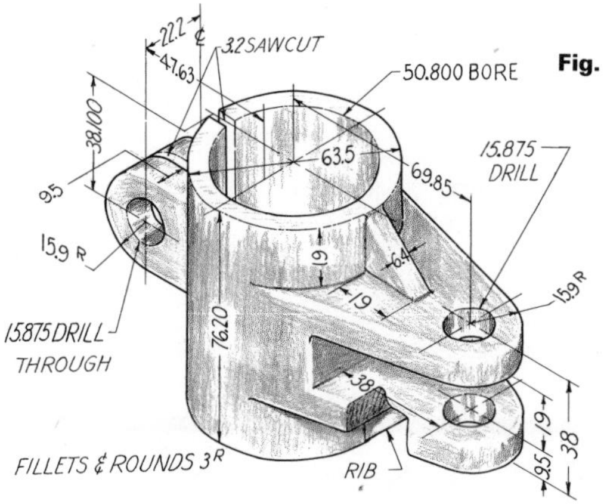

Fig. 8.91 □ Guide bracket.
(Numerical values in millimeters.)

FILLETS & ROUNDS 3 R

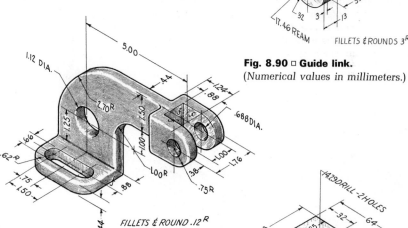

FILLETS & ROUND .12 R

Fig. 8.93 □ Arm bracket.

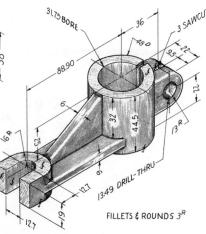

FILLETS & ROUNDS 3 R

Fig. 8.92 □ Shifter.
(Numerical values in millimeters.)

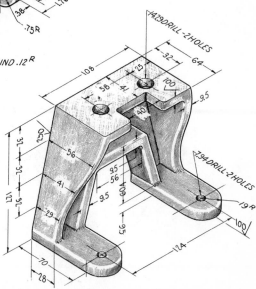

Fig. 8.94 □ Lathe leg.
(Numerical values in millimeters.)

). Make a complete orthographic drawing holder.

8.96). Make a complete orthographic drawing shaft bracket.

(Fig. 8.97). Make a three-view orthographic sketch of the motor bracket.

23 (Fig. 8.98). Make a complete three-view sketch of the tool rest and/or the tool rest bracket. The rectangular top surface of the tool rest is to be $1\frac{1}{8}$ in. above the center line of the hole for the $\frac{7}{16}$-in. bolt. The overall dimensions of the top are $1\frac{1}{4} \times 2\frac{1}{2}$ in. It is to be $\frac{1}{4}$ in. thick. The overall dimensions of the rectangular pad of the bracket are $1\frac{1}{4} \times 1\frac{7}{8}$ in. The center line of the adjustment slot is $\frac{9}{16}$ in. above the center line of the top holes in the rectangular pad and the distance from center line to center line of the slot is $1\frac{3}{8}$ in. The bracket is to be fastened to a housing with $\frac{1}{4}$-in. roundhead machine screws.

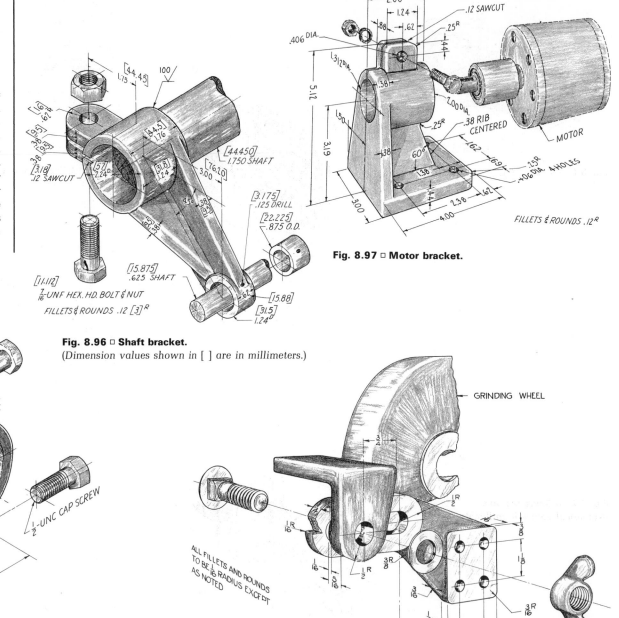

Fig. 8.96 □ Shaft bracket.
(Dimension values shown in [ ] are in millimeters.)

Fig. 8.97 □ Motor bracket.

Fig. 8.95 □ Tube holder.

Fig. 8.98 □ Tool rest and tool rest bracket.

Problems 11

Sectional views

The following problems were designed to emphasize the principles of sectioning. Those drawings that are prepared from the pictorials of objects may be dimensioned if the elementary principles of dimensioning (Chapter 14) are carefully studied.

24 (Fig. 8.99). Reproduce the two views of the hand wheel and change the right-side view to a full section.

25 (Fig. 8.100). Reproduce the circular front view of the pump cover and convert the right-side view to a full section.

26 (Fig. 8.101). Reproduce the top view of the control housing cover and convert the front view to a full section.

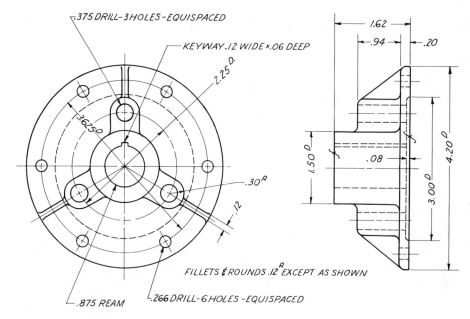

Fig. 8.100 □ **Pump cover.**

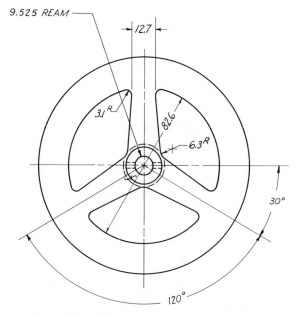

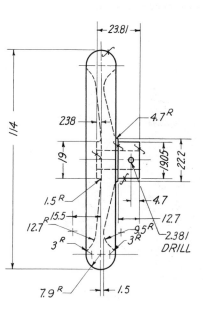

Fig. 8.99 □ **Hand wheel.**
(Numerical values in millimeters.)

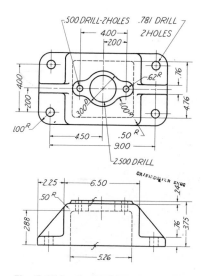

Fig. 8.101 □ **Control housing cover.**

27–32 (Figs. 8.102–8.107). These problems may be dimensioned, as are working drawings. For each object, the student should draw all the views necessary for a working drawing of the part. Good judgment should be exercised in deciding whether the sectional view should be a full section or a half section. After the student has made his decision, he should consult his class instructor.

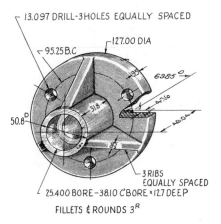

Fig. 8.102 □ Rod yoke.
(Numerical values in millimeters.)

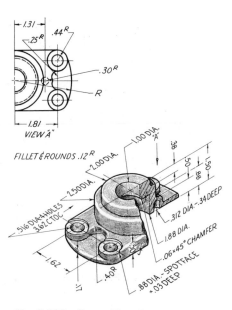

Fig. 8.103 □ Control housing cover.

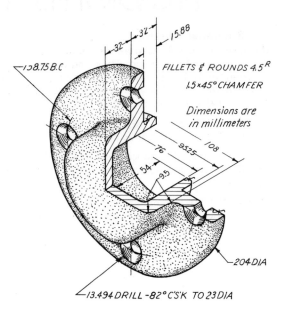

Fig. 8.104 □ Cover.
(Numerical values in millimeters.)

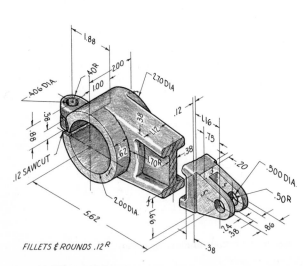

FILLETS & ROUNDS .12ᴿ

Fig. 8.105 □ Shifter link.

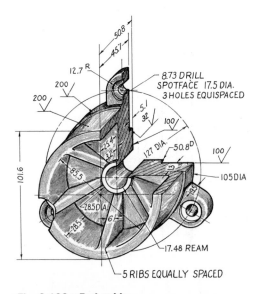

Fig. 8.106 □ End guide.
(Numerical values in millimeters.)

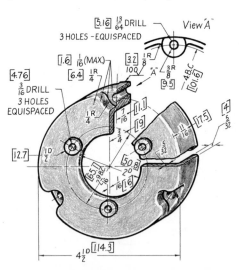

Fig. 8.107 □ Cover.
(Dimension values shown in [ ] are in millimeters.)

138

Problems III

Auxiliary views

The problems shown in Fig. 8.108 are designed to give the student practice in constructing auxiliary views of the inclined surfaces of simple objects formed mainly by straight lines. They will provide needed drill in projection if, for each of the objects in Fig. 8.108, an auxiliary is drawn showing the entire object. Complete drawings may be made of the objects shown in Figs. 8.109–8.114. If the views are to be dimensioned, the student should adhere to the rules of dimensioning given in Chapter 14 and should not take too seriously the locations for the dimensions on the pictorial representations.

33 (Fig. 8.108). Using instruments, reproduce the given views of an assigned object and draw an auxiliary view of its inclined surface.

34 (Fig. 8.109). Draw the necessary views of the anchor bracket. Make partial views for the top and end views.

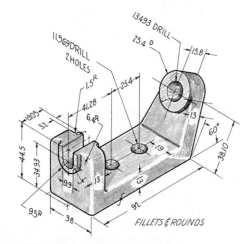

Fig. 8.109 □ **Anchor bracket.**
(Numerical values in millimeters.)

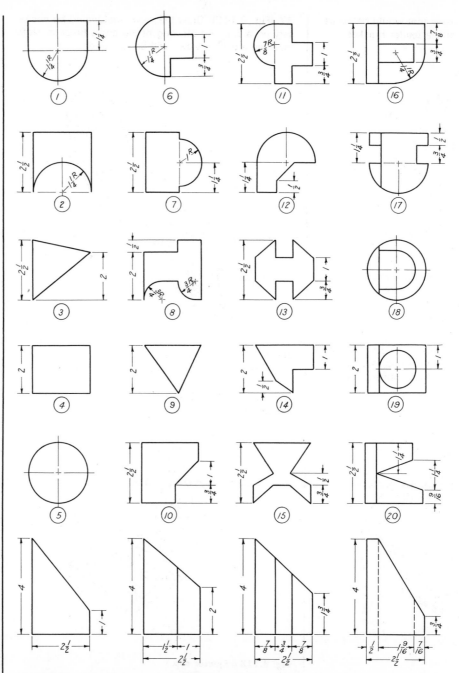

Fig. 8.108 □ Using instruments, reproduce the given views of an assigned object and draw an auxiliary view of its inclined surface.

35 (Fig. 8.110). Draw the views that would be necessary on a working drawing of the feeder bracket.

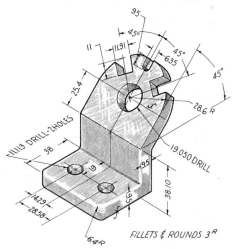

Fig. 8.110 □ Feeder bracket.
(*Numerical values in millimeters.*)

36 (Fig. 8.111). Draw the necessary views of the offset guide. It is suggested that partial views be used, except in the view where the inclined surface appears as a line.

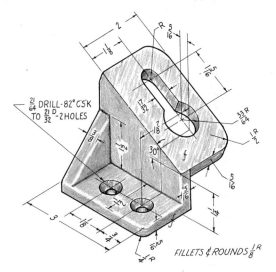

Fig. 8.111 □ Offset guide.

37 (Fig. 8.112). Draw the views that would be necessary on a working drawing of the angle bracket. Note that two auxiliary views will be required.

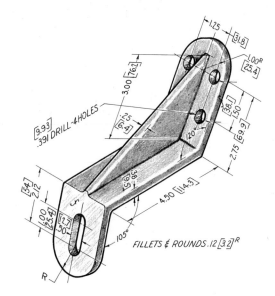

Fig. 8.112 □ Angle bracket.
(*Dimension values shown in [ ] are in millimeters.*)

38 (Fig. 8.113). Draw the necessary views of the ejector clip.

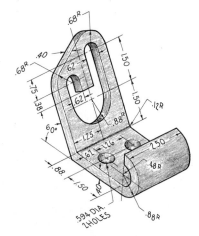

Fig. 8.113 □ Ejector clip.

39 Make a multiview drawing of the airplane engine mount shown in Fig. 8.114. The engine mount is formed of three pieces of steel plate welded to a piece of steel tubing. The completed drawing is to consist of four views. It is suggested that the front view be the view obtained by looking along and parallel to the axis of the tube. The remaining views that are needed are an auxiliary view showing only the inclined lug, a side view that should be complete with all hidden lines shown, and a partial top view with the inclined lug omitted.

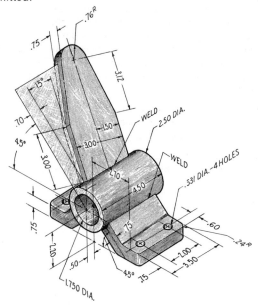

Fig. 8.114 □ Airplane engine mount.

40 (Fig. 8.115). Draw the views as given. Complete the top view.

41 (Fig. 8.116). Draw the views given and add the required primary and secondary auxiliary views.

42 (Fig. 8.117). Using instruments, draw a secondary auxiliary view that will show the true size and shape of the inclined surface of an assigned object. The drawing must also show the given principal views.

43 (Fig. 8.118). Draw the layout for the support anchor as given and then, using the double auxiliary view method, complete the views as required.

The plate and cylinder are to be welded. Since the faces of the plate show as oblique surfaces in the front and top views, double auxiliary views are necessary to show the thickness and the true shape.

Start the drawing with the auxiliary views that are arranged horizontally on the paper, then complete the principal views. The inclined face of the cylinder will show as an ellipse in top and front views; but do not show this in the auxiliary view that shows the true shape of the square plate.

How would you find the view that shows the true angle between the inclined face and the axis of the cylinder?

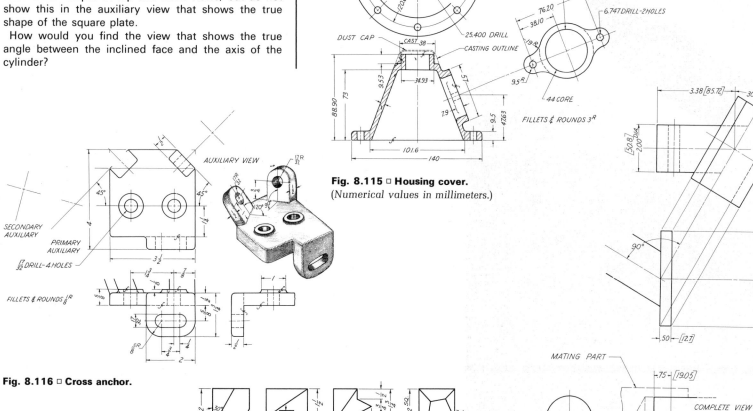

Fig. 8.115 □ **Housing cover.**
(*Numerical values in millimeters.*)

Fig. 8.116 □ **Cross anchor.**

Fig. 8.117 □ **Using instruments, draw a second auxiliary view that will show the true size and shape of the inclined surface of an assigned object.**
The drawing must also show the given principal views.

Fig. 8.118 □ **Support anchor.**
(*Dimension values shown in [ ] are in millimeters.*)

"Alcoa Seaprobe" at work.
The design of this recovery ship required the application of the principles of descriptive geometry in the development of creative ideas. (*Courtesy Aluminum Corporation of America.*)

Chapter 9

A □ Basic descriptive geometry

9.1 □ Introduction

On many occasions, problems arise in engineering design that may be solved quickly by applying the basic principles of orthographic projection.

If one thoroughly understands the solution for each of the problems presented, he should find it easy, at a later time, to analyze and solve almost any of the practical problems he may encounter.

It should be pointed out at the very beginning that to solve most types of problems one must apply the principles and methods used to solve a few basic problems, such as: (1) to find the true length of a line, (2) to find the point projection of a line, and (3) to find the true size and shape of a surface (Fig. 9.1). To find information such as the angle between surfaces, the angle between lines, or the clearance between members of a structure, one must use, in proper combination, the methods of solving these basic problems. Success in solving problems by projection depends largely on the complete understanding of the principles of projection, the ability to visualize space conditions, and the ability to analyze a given situation. Since the ability to analyze and to visualize are of utmost importance in engineering design, the student is urged to develop these abilities by resisting the temptation to memorize step procedures.

9.2 □ The projection of a point

Figure 9.2(a) shows the projection of point S on the three principal planes of projection and a supplementary plane A. The notation

Spatial geometry for design and analysis

used is as explained in Sec. 8.7. Point s^F is the view of point S on the frontal plane; s^H is the view of S on the horizontal plane; and s^P is its view on the profile plane. For convenience and ease in recognizing the projected view of a point on a supplementary plane, the supplementary planes are designated as A-planes and O-planes. A (auxiliary) planes are always perpendicular to one of the principal planes. Point s^A is the view of S on the A-plane. The view of S on an O (oblique) plane would be designated s^O. Additional O-planes are identified as O_1, O_2, O_3, etc., in the order that they follow the first O-plane.

Since it is necessary to represent on one plane (the working surface of our drawing paper) the views of point S that lie on mutually perpendicular planes of projection, the planes are assumed to be hinged so that they can be revolved, as shown in Fig. 9.2(b), until they are in a single plane, as in (c). The lines about which the planes of projection are hinged are called *reference lines*. A reference line is identified by the use of capital letters representing the adjacent planes, as FH, FP, FA, HA, AO, and so forth (see Fig. 9.2).

It is important to note in (c) that the projections s^F and s^H fall on a vertical line, s^F and s^P lie on a horizontal line, and s^H and s^A lie on a line perpendicular to the reference line HA. In each case this results from the fact that point S and its projections on adjacent planes lie in a plane perpendicular to the reference line for those planes [Fig. 9.2(a)]. This important principal of projection determines the location of views when the relationship of lines and planes form the problem.

9.3 □ The projection of a straight line

Capital letters are used for designating the end points of the actual line in space. In the projected views, these points are identified as shown in Fig. 9.3. The student should read Sec. 8.7, which presents the principles of multiview drawing. In particular, he should study the related illustration, which shows some typical line positions.

9.4 □ Projection of a plane in space

Theoretically, a plane is considered to be flat and unlimited in extent. A plane can be delineated graphically by: (1) two intersecting lines, (2) a line and a point not on the line, (3) two parallel lines, and lastly, (4) three points not on a straight line. For graphical purposes and to facilitate the solution of space problems as presented in this chapter, planes will be bounded and usually triangular. The space picture and multiview representation of a plane ABC are given in Fig. 9.4. The pictorial at the left in (a) shows the projected views on the principal planes of projection. In (b) the planes are shown being opened outward to be in the

Fig. 9.1 □ **The frame of this satellite could not have been designed without the application of descriptive geometry methods.** (*Courtesy TRW Systems Group.*)

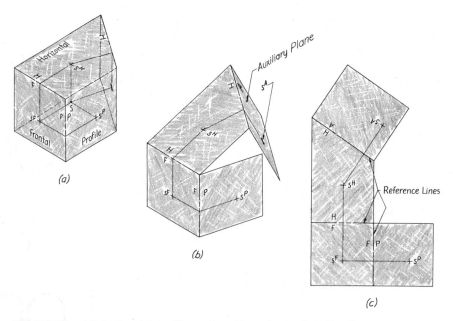

Fig. 9.2 □ **Frontal, horizontal, profile, and auxiliary views of a point S.**

plane of the paper, as in (c). It should be noted that plane ABC projects as a line (edge view) on the auxiliary plane. The three points A, B, and C of the plane are projected in the same manner as the single point S in Fig. 9.2 and are identified similarly (a^F, a^H, a^P, a^A, etc.).

9.5 □ Parallel lines

Any two lines in space must be either (1) parallel, (2) intersecting, or (3) nonintersecting and nonparallel (called *skew lines*). Figure 9.18 shows intersecting lines, while Fig. 9.17 shows skew lines. Parallel lines are shown in Fig. 9.5.

It might be stated as a rule of projection, with one exception, that when two lines are parallel their projections will be parallel in every view (Fig. 9.5). In other words the lines will appear to be parallel in every view in which both appear. This is true even though in specific views they may appear as points or their projections may coincide. In either case they are still parallel because both conditions indicate that the lines have the same direction. The exception that has been mentioned occurs when the F- and H-projections of two inclined profile lines are shown. For proof of parallelism a supplementary view should be drawn, which may or may not be the profile view.

The true or shortest distance between two parallel lines can be determined on the view that will show these lines as points. The true distance can be measured between the points (Fig. 9.5).

9.6 □ To determine the true length of an oblique line

In order to find the true length of an oblique line, it is necessary to select an auxiliary plane of projection that will be parallel to the line (Figs. 9.6 and 9.7).

Given: The F (frontal) view $a^F b^F$ and the H (top) view $a^H b^H$ of the oblique line AB (Fig. 9.6).

Solution: (1) Draw the reference line HA parallel to the projection $a^H b^H$. The A-plane for this reference line will be parallel to AB and perpendicular to the H-plane (see pictorial drawing). (2) Draw lines of projection from points a^H and b^H perpendicular to the reference line. (3) Transfer height measurements from the F-view to the A-view to locate a^A and b^A. In making this transfer of measurements, the students should attempt to visualize the space condition for the line and understand that, since the F-view and A-view both show height and because the planes of projection for these views are perpendicular to the H-plane, the perpendicular distance D_1 from the reference line HF to point a^F must be the same as the distance D_1 from the reference line HA to point a^A.

The projection $a^A b^A$ shows the true length of the line AB.

It was not necessary to use an auxiliary plane perpendicular to the H-plane to find the true length of line AB in Fig. 9.6. The

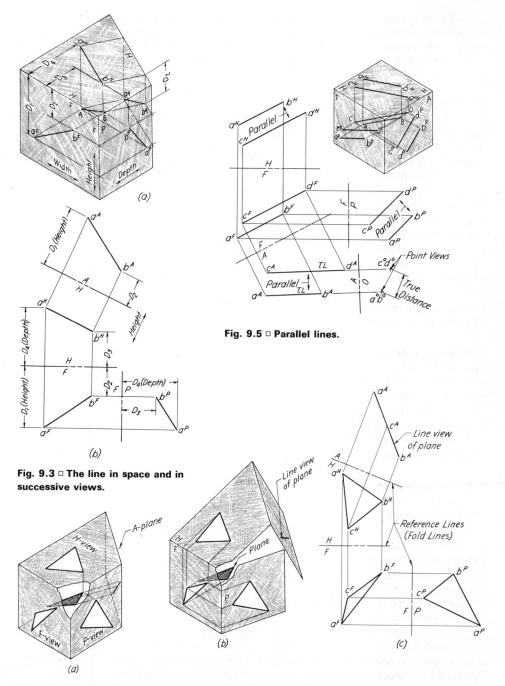

Fig. 9.5 □ Parallel lines.

Fig. 9.3 □ The line in space and in successive views.

Fig. 9.4 □ The plane in space and in successive views.

auxiliary plane could just as well have been perpendicular to either the *F*- or *P*-planes. Figure 9.7 shows the use of an auxiliary plane perpendicular to the frontal plane to find the true length of the line. In this case the auxiliary view has depth distances in common with the top view, as indicated.

9.7 □ Perpendicular lines

Lines that are perpendicular in space will have their projections perpendicular in any view that shows either or both of the lines in true length. A second rule of perpendicularity might be that, when a line is perpendicular to a plane, it will be perpendicular to every line in that plane. A careful study of Fig. 9.8 will verify these rules. For instance, it should be noted that the lines *AB* and *CD* lie in a plane that is outlined with broken lines and that $e^H f^H$ is perpendicular to $a^H b^H$ because $a^H b^H$ shows the true length of the line *AB*. In the *A*-view, we see that $e^A f^A$ is perpendicular to the line view of the plane and is therefore perpendicular to both *AB* and *CD*. The *O*-view shows the true shape (TSP) of the plane and the line *EF* as a point. This again verifies the fact that *EF* is perpendicular to the plane and to lines *AB* and *CD*. Otherwise line *EF* would not appear as a point. Note also that the *O*-view shows the true length (TL) of *AB* and *CD*.

9.8 □ Bearing of a line

The bearing of a line is the horizontal angle between the line and a north-south line. A bearing is given in degrees with respect to the meridian and is measured from 0° to 90° from either north (N) or south (S). The bearing reading indicates the quadrant in which the line is located by use of the letters N and E, S and E, S and W, or N and W, as N 48° E or S 54° 40′ W. The bearing of a line is measured in the *H*-view (Fig. 9.9).

9.9 □ Point view of a line (Fig. 9.10)

It was pointed out in the first section of this chapter that the solutions of many types of problems depend on an understanding of a few basic constructions. One of these basic constructions involves the finding of the view showing the point view or point projection of a line. For instance, this construction is followed when it is necessary to determine the dihedral angle between two planes, for the true size of the angle will appear in the view that shows the line common to the two planes as a point.

A line will show as a point on a projection plane that is perpendicular to the line. The observer's direction of sight must be along and parallel to the line. When a line appears in true length on one of the principal planes of projection, only an auxiliary view is needed to show the line as a point. However, in the case of an oblique line both an auxiliary and an oblique view are required,

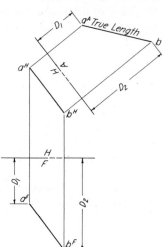

Fig. 9.6 □ To find the true length of an oblique line.

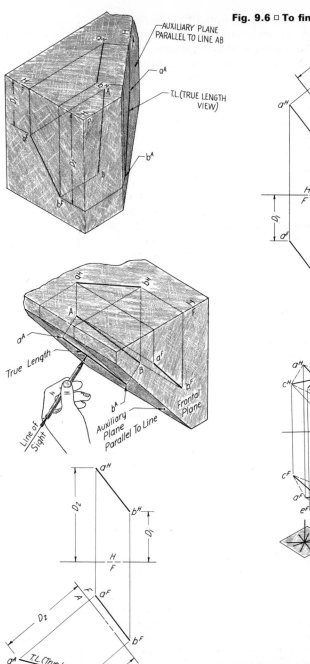

Fig. 9.7 □ To find the true length of a line.

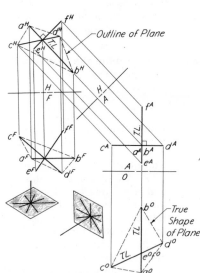

Fig. 9.8 □ Perpendicular lines.

146

for a point view must always follow a true-length view. In other words, the plane of projection for the view showing the line as a point must be adjacent to the plane for the true view and be perpendicular to it.

Given: The F-view $a^F b^F$ and the H-view $a^H b^H$ of the oblique line AB (Fig. 9.10).

Solution: (1) Draw the view showing the TL (true length) of AB. This is an auxiliary view drawn, as explained in Sec. 9.6. (2) Draw reference line AO perpendicular to the true-length projection $a^A b^A$. This reference line is for an O-plane that is perpendicular to the A-plane. (3) Draw a projection line from $a^A b^A$ and transfer the distance D_3 from the H-view to the O-view. It should be noted from the pictorial drawing that the distance D_3 is common to both of these views, and that points a^O and b^O coincide to give a point or end view of line AB.

9.10 □ To find the shortest distance from a point to a line

The shortest distance between a given point and a given straight line must be measured along a perpendicular drawn from the point to the line. Since lines that are perpendicular will have their projections show perpendicular in any view showing either or both lines in true length (Sec. 9.7), the perpendicular must be drawn in the view showing the given line in true length.

Given: The F- and H-views of the line AB and point C (Fig. 9.11).

Solution: (1) Draw the A(auxiliary) view showing the true-length view $a^A b^A$ of line AB and view c^A of point C. (2) Draw $c^A x^A$ perpendicular to $a^A b^A$. Line $c^A x^A$ is a view of the required perpendicular from point C to its juncture with line AB at point X. (3) Draw reference line AO parallel to $c^A x^A$. This reference line locates an O-plane, which will be parallel to the perpendicular CX and perpendicular to the A-plane. The O-view will show the true length of CX. Line CX does not show true length in any of the other views.

9.11 □ The principal lines of a plane (Fig. 9.12)

Those lines that are parallel to the principal planes of projection are the *principal lines of a plane*. A principal line may be either a horizontal line, a frontal line, or a profile line. Principal lines are true-length lines and one such line may be drawn in any plane to appear true length in any one of the principal views, as desired. This is an important principle that is the basis for the solution of many problems involving lines and planes.

9.12 □ To obtain the edge view of a plane

When a plane is vertical, an edge view of it will be seen from above and it will be represented by a line in the top view. Should

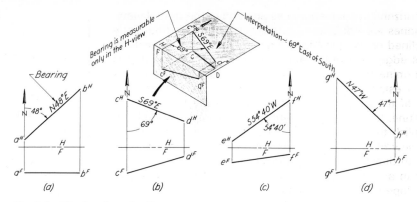

Fig. 9.9 □ The bearing of a line.

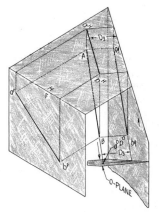

Fig. 9.10 □ The point view of a line.

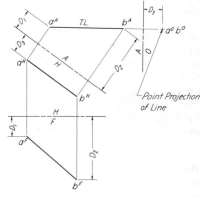

Auxiliary Plane Perpendicular to Horizontal Plane

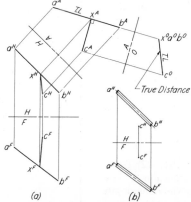

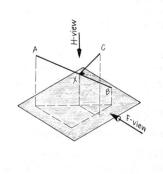

Fig. 9.11 □ To find the shortest distance from a point to a line.

147

a plane be horizontal, it will appear as an edge in the frontal view. However, planes are not always vertical or horizontal; frequently they are inclined or oblique to the principal planes of projection.

Finding the edge view of a plane is a basic construction that is used to determine the slope of a plane (dip), to determine clearance, and to establish perpendicularity. The method presented here is part of the construction used to obtain the true size and shape of a plane, to determine the angle between a line and plane, and to establish the location at which a line pierces a plane.

The edge view of an oblique plane can be obtained by viewing the plane with direction of sight parallel to it. The edge view will then appear in an auxiliary view. When the auxiliary view shows height, the slope of the plane is shown (Fig. 9.13).

Given: The F- and H-views of plane ABC.

Solution: (1) Draw the horizontal line AX in the plane. Because AX is parallel to the horizontal, a^Fx^F will be horizontal and must be drawn before the position of a^Hx^H can be established. (2) Draw reference line HA perpendicular to a^Hx^H. (3) Construct the A-view, which will show the plane ABC as a straight line. Since the F-view and A-view have height as a common dimension, the distances used in constructing the A-view were taken from the F-view. It should be noted by the reader that an edge view found by projecting from the front view will show the angle that the plane makes with the F-plane. Similarly, the edge view found by projecting from the side view will show the angle with the P-plane.

9.13 □ To find the true shape (TSP) of an oblique plane

Finding the true shape of a plane by projection is another of the basic constructions that the student must understand, for it is used to determine the solution of two of the problems that are to follow. In a way, the construction shown in Fig. 9.14 is a repetition of that shown in Fig. 8.74. However, repetition in the form of another presentation should help even those students who feel they understand the method for finding the true shape of an oblique surface of an object.

To see the true size and shape of an oblique plane an observer must view it with a line of sight perpendicular to it. To do this he must, as the first step in the construction, obtain an edge view of the plane. An O-view taken from the A-view will then show the true shape of the plane.

Given: The F- and H-views of plane ABC.

Solution: (1) Draw a frontal line CD in plane ABC. (2) Draw reference line FA perpendicular to c^Fd^F and construct the A-view showing the edge view of the plane. The auxiliary view has depth in common with the H-view. (3) Draw reference line AO parallel to the edge view in the A-view and construct the O-view, which

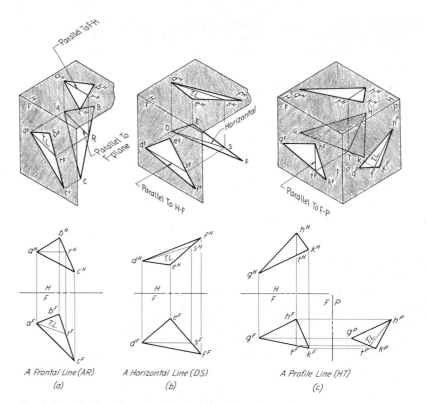

Fig. 9.12 □ The location of a principal line in a plane.

A Frontal Line (AR)
(a)

A Horizontal Line (DS)
(b)

A Profile Line (HT)
(c)

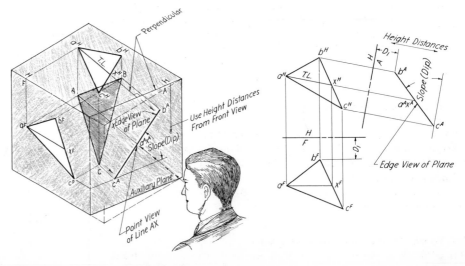

Fig. 9.13 □ To find the edge view of a plane.

148

will show the true size and shape of plane *ABC*. The needed distances from the reference line to points in the *O*-view are found in the *F*-view.

9.14 □ To find the piercing point of a line and a plane

Determining the location of the point where a line pierces a plane is another fundamental operation with which one must be familiar. A line, if it is not parallel to a plane, will intersect the plane at a point that is common to both. In a view showing the plane as an edge, the piercing point appears where the line intersects (cuts) the edge view. This method is known as the edge-view method to distinguish it from the cutting-plane method, which may be found in Sec. 9.24.

The simple cases occur when the plane appears as an edge in one of the principal views. The general case, for which we use an oblique plane, is given in Fig. 9.15.

Given: Plane *ABC* and line *ST*.

Solution: (1) Draw the horizontal line *BX* in *ABC*. (2) Draw reference line *HA* perpendicular to $b^H x^H$ and construct the *A*-view showing an edge view of the plane as line $a^A b^A c^A$ and the view of the line $s^A t^A$. Point p^A where the line cuts the edge view of the plane is the *A*-view of the piercing point. (3) Project point *P* back from the *A*-view first to the *H*-view and then to the *F*-view.

9.15 □ To determine the angle between a line and a given plane

The true angle between a given line and a given plane will be seen in the view that shows the plane as an edge (a line) and the line in true length. The solution shown in Fig. 9.16 is based on this premise. The solution as presented might be called the *edge-view method*.

Given: The *F*- and *H*-views of plane *ABC* and line *ST*.

Solution: (1) Draw the frontal line *BX* in the plane *ABC*. (2) Draw reference line *FA* perpendicular to $x^F b^F$ and construct the *A*-view. This view will show plane *ABC* as line $a^A b^A c^A$; however, since this view does not show *ST* in true length, the true angle is not shown. (3) Draw reference line *AO* parallel to the edge view of the plane and construct the *O*-view that will show line *ST* viewed obliquely and plane *ABC* in its true size and shape. (4) Draw reference line OO_1 parallel to $s^o t^o$ and construct the second oblique view.

Line $s^{O_1} t^{O_1}$ will show the true length of *ST* in this second oblique view. Plane *ABC* will be seen again as an edge (line), for it now appears on an adjacent view taken perpendicular to the view showing true shape (TSP). The required angle can now be meas-

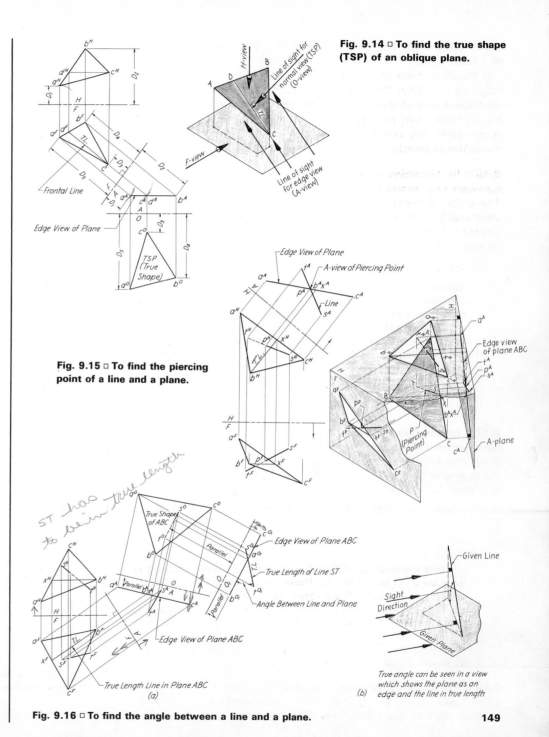

Fig. 9.14 □ To find the true shape (TSP) of an oblique plane.

Fig. 9.15 □ To find the piercing point of a line and a plane.

Fig. 9.16 □ To find the angle between a line and a plane.

149

ured between the true-length view of line ST and the edge view of plane ABC.

In the illustration of Fig. 9.16 three supplementary views were required to obtain the true angle. If the plane had appeared as an edge in one of the given views, only two supplementary views would have been needed; if it had appeared in true shape in a given view only one additional view, properly selected to show the plane as an edge and the line in true length, would be needed.

9.16 □ To determine the angle between two nonintersecting (skew) lines

The angle between two nonintersecting lines is measurable in a view that shows both lines in true length. The analysis and construction that might be followed to obtain the needed view is shown in Fig. 9.17.

Given: The two nonintersecting lines AB and CD.

Solution: (1) Draw the reference line HA parallel to $a^H b^H$ and construct the A-view that shows AB in true length and the view of the line CD. (2) Draw reference line AO perpendicular to $a^A b^A$ and draw the O-view in which the line AB will appear as a point $(a^o b^o)$. (3) Draw reference line OO_1, parallel to $c^o d^o$ and construct the O-view showing the true length of CD. Line AB must also show true length in this view since AB, showing as a point in the O-view, is parallel to the OO_1-plane. With both lines shown in true length in the O_1-view, the required angle may be measured in this view.

9.17 □ To determine the true angle between two intersecting oblique lines

Since, as previously stated, two intersecting lines establish a plane, the true angle between the intersecting lines may be seen in a true-shape view of a plane containing the lines (Fig. 9.14). In Fig. 9.18 line AC completes plane ABC containing the given lines AB and BC. It is necessary to find the true angle between AB and BC.

Solution: (1) Draw the frontal line XC. (2) Draw reference line FA perpendicular to $c^F x^F$ and construct the A-view showing an edge view of plane ABC. (3) Draw the reference line AO parallel to the edge view $a^A c^A b^A$ and construct the O-view, which will show the TSP of plane ABC. In this view it is desirable to show only the given lines. The true angle between AB and BC is shown by $a^o b^o c^o$. As a practical application this method might be used to determine the angle between two adjacent sections of bent rod, as shown in (b).

9.18 □ The distance between two parallel planes

When two planes are parallel their edge views will appear as parallel lines in the same view. The clearance or perpendicular

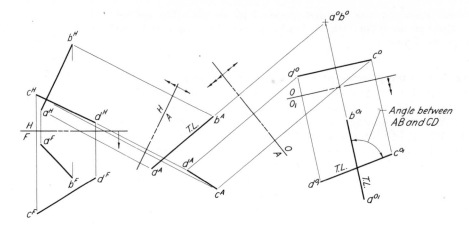

Fig. 9.17 □ To find the angle between two skew lines.

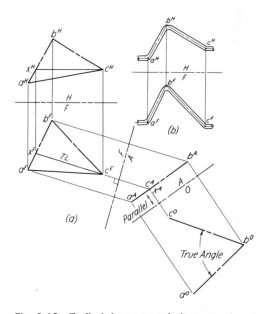

Fig. 9.18 □ To find the true angle between two intersecting lines.

distance between them can be measured in this view (Fig. 9.19). The existence of planes as parallel lines in a view is proof that they are parallel.

9.19 □ To find the dihedral angle between two planes

The angle between two planes is known as a *dihedral angle*. The true size of this angle between intersecting planes may be seen in a plane that is perpendicular to both. For this condition as set forth, the intersecting planes will appear as edges and the line of intersection of the two planes as a point. The true angle may be measured between the edge views of the planes (Fig. 9.20).

Given: The intersecting planes $ABCD$ and $CDEF$. The line of intersection is line CD, as shown in the pictorial drawing.

Solution: (1) Draw reference line FA parallel to $c^F d^F$ and construct the A-view. This view will show CD in true length ($c^A d^A$). (2) Draw reference line AO perpendicular to $c^A d^A$ and construct the adjacent O-view. Since this view was taken looking along line CD, points C and D are coincident and appear as a single point identified as $c^O d^O$. The intersecting planes show as edge views and the true angle between the given planes may be measured between these edge-view lines. When two planes are given that do not intersect, the dihedral angle may be found after the line of intersection has been determined.

9.20 □ To find the shortest distance between two skew lines

As was stated in Sec. 9.5, any two lines that are not parallel and do not intersect are called *skew lines*. The shortest distance between any such lines must be measured along one line and only one line that can be drawn perpendicular to both. This common perpendicular can be drawn in a view that is taken to show one line as a point. Its projection will be perpendicular to the view of the other line and will show in true length (Fig. 9.21).

Given: The F- and H-views of two skew lines AB and CD.

Solution: (1) Draw an A-view adjacent to the H-view to show line AB in true length ($a^A b^A$). Line CD should also be shown in this same view ($c^A d^A$). (2) Draw reference line AO perpendicular to $a^A b^A$ and draw the O-view in which line AB will appear as a point ($a^O b^O$). It is in this view that the exact location of the required perpendicular can be established. (3) Draw the line $e^O f^O$ through point $a^O b^O$ perpendicular to $c^O d^O$. The shortest distance between the skew lines now appears in true distance as the length of $e^O f^O$. Although the CD does not appear in true length in the O-view, $e^O f^O$ does, and hence $c^O d^O$ and $e^O f^O$ will appear perpendicular. (4) Complete the A-view by first locating point e^A on $c^A d^A$ and then draw $e^A f^A$ parallel to reference line AO. (5) Locate points E and F in the H- and F-views remembering that point E is located on line CD and point F on line AB.

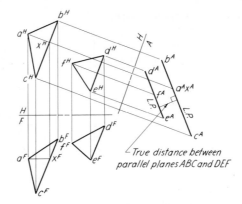

Fig. 9.19 □ Distance between parallel planes.

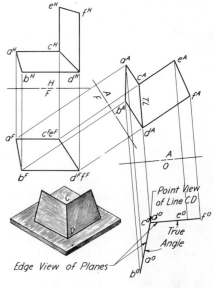

Fig. 9.20 □ To find the dihedral angle between two planes.

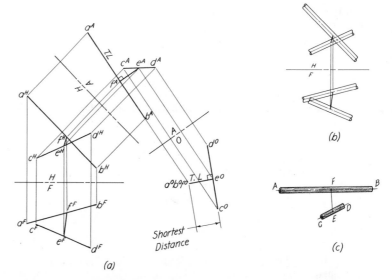

Fig. 9.21 □ To find the shortest distance between two skew lines.

In engineering design an engineer frequently has to locate and find the length of the shortest line between two skewed members in order to determine clearance or the length of a connecting member. In underground construction work one might use this method to locate a connecting tunnel.

In Fig. 9.21(b) and (c) we see the two rods for which the clearance distance was determined in (a). Lines AB and CD represent the center lines of the rods.

9.21 □ To find the true length of a line by revolution

In engineering design, it is frequently necessary to determine the true length of a line when constructing the development of a surface. The true lengths must be found of those lines that are not parallel to any coordinate plane and therefore appear foreshortened in all the principal views. (See Sec. 8.7, ''Projection of lines.'') The practical as well as theoretical procedure is to revolve any such oblique line into a position parallel to a coordinate plane such that its projection on that particular plane will be the same length as the line. In Fig. 9.22(a), this is illustrated by the edge AB on the pyramid. AB is oblique to the coordinate planes, and its projections are foreshortened. If this edge line is imagined to be revolved until it becomes parallel to the frontal plane, then the projection ab_r in the front view will be the same length as the true length of AB.

A practical application of this method is shown in Fig. 9.22(c). The true length of the edge AB in Fig. 9.22(a) would be found by revolving its top projection into the position ab_r, representing AB revolved parallel to the frontal plane, and then projecting the end point b_r down into its new position along a horizontal line through b. The horizontal line represents the horizontal plane of the base, in which the point B travels as the line AB is revolved.

Commercial draftsmen who are unfamiliar with the theory of coordinate planes find the true-length projection of a line by visualizing the line's revolution. They think of an edge as being revolved until it is in a plane perpendicular to the line of sight of an observer stationed an infinite distance away. The process corresponds to that used in drawing regular orthographic views (Sec. 8.4). Usually this method is more easily understood by a student.

Note in Fig. 9.22(a) and (b) that the true length of a line is equal to the hypotenuse of a right triangle whose altitude is equal to the difference in the elevation of the end points and whose base is equal to the top projection of the line. With this fact in mind, many draftsmen determine the true length of a line by constructing a true-length triangle similar to the one illustrated in Fig. 9.22(d).

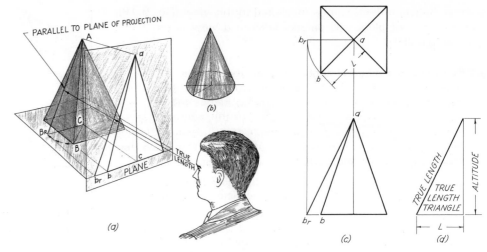

Fig. 9.22 □ True length of a line, revolution method.

B □ Intersections

9.22 □ Lines of intersection of geometric surfaces

The line of intersection of two surfaces is a line that is common to both. It may be considered the line that would contain the points in which the elements of one surface would pierce the other.

The line of intersection of two surfaces is found by determining a number of points common to both surfaces and drawing a line or lines through these points in correct order. The resulting line of intersection may be straight, curved, or straight and curved.

9.23 □ Determination of a piercing point
by inspection (Fig. 9.23)

It is easy to determine where a given line pierces a surface when the surface appears as an edge view (line) in one of the given views. For example, when the given line AB is extended as shown in (a), the F-view of the piercing point C is observed to be at c^F, where the frontal view of the line AB extended intersects the line view of the surface. With the position of c^F known, the H-view of point C can be quickly found by projecting upward to the H-view of AB extended.

In (b) the H-view (f^H) of the piercing point F is found first by extending $d^H e^H$ to intersect the edge view of the surface pierced by the line. By projecting downward, f^F is located on $d^F e^F$ extended.

In (c) the views of the piercing point K are found in the same manner as in (b), the only difference being that the edge view of the surface pierced by the line appears as a circle arc in the H-view instead of a straight line. It should be noted that a part of the line is invisible in the F-view because the piercing point is on the rear side of the cylinder.

The F- and H-views of the piercing point R in (d) may be found easily by projection after the P-view (r^P) of R has been once established by extending $p^P q^P$ to intersect the line view of the surface.

9.24 □ Determination of a piercing point
using a line-projecting plane

When a line pierces a given oblique plane and an edge view is not given, as in Fig. 9.24, a line-projecting plane (cutting plane) may be used to establish a line of intersection that will contain the piercing point. In the illustration, a vertical projecting plane was selected that would contain the given line RS and intersect the given plane ABC along line DE, as illustrated by the pictorial drawing.

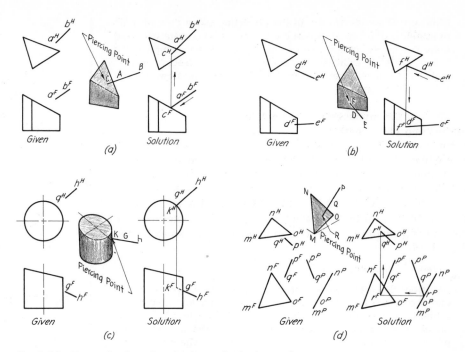

Fig. 9.23 □ Determination of a piercing point by inspection.

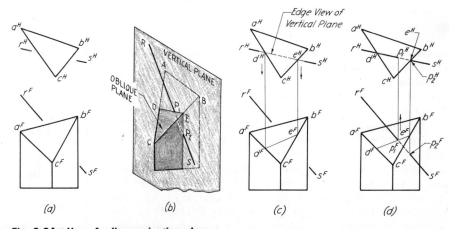

Fig. 9.24 □ Use of a line-projecting plane.

Solution: Draw the H-view of the projecting plane through $r^H s^H$ to establish $d^H e^H$, as shown in the H-view in (c). Locate $d^F e^F$ and draw the F-view of the line of intersection. Then, complete the line view $r^F s^F$ to establish $p_1{}^F$ at the point of intersection of $r^F s^F$ and $d^F e^F$. Finally, locate $p_1{}^H$ on $r^H s^H$ by projecting upward from $p_1{}^F$, as shown in (d).

9.25 □ To find the intersection of two planes, line-projecting plane method

The intersection of two oblique planes may be determined by finding where two of the lines of one plane pierce the other plane, as illustrated by the pictorial drawing in Fig. 9.25. The procedure that is illustrated employs line-projecting planes to find the piercing points of the lines XY and XZ and the oblique plane RST. Therefore, it might be said that the solution requires the determination of the piercing point of a line and an oblique plane, as explained in Sec. 9.24.

Given: The oblique planes RST and XYZ.

Solution: Since the (vertical) line-projecting plane $C_1 P_1$ is to contain the line XY of the plane XYZ, draw the line-view representation of this projecting plane to coincide with $x^H y^H$. Next, project the line of intersection AB between the line-projecting plane $C_1 P_1$ and plane RST from the top view, where it appears as $a^H b^H$ in the edge-view representation of $C_1 P_1$ to the front view. Then, since it is evident that the line AB is not parallel to XY, which lies in the projecting plane (see F-view), the line XY intersects AB. The location of this intersection at E is established first in the F-view, where the line $x^F y^F$ intersects $a^F b^F$ at e^F. The H-view of E, that is, e^H, is found by projecting upward from e^F in the F-view to the line view $x^H y^H$. The other end of the line of intersection between the two given planes at F is found by using the line-projecting plane $C_2 P_2$ and following the same procedure as for determining the location of point E.

9.26 □ To find the intersection of a cylinder and an oblique plane

There are two distinct and separate methods shown in Fig. 9.26 for finding the line of intersection of an oblique plane and a cylinder. Both methods appear together on the drawing at the left. The selected line method is illustrated pictorially in (b), while the cutting-plane method is shown in (c).

In the application of the selected line method, any line of the given plane, such as line BR, is drawn in the F- and H-views (a). It can be noted by observing, in the pictorial drawing in (b), that this particular line pierces the cylinder at points P_1 and P_2 to give two points on the line of intersection. On the multiview drawing, the locations of the H-views of points P_1 and P_2 can readily be recognized as being at the points labeled $p_1{}^H$ and $p_2{}^H$, where the

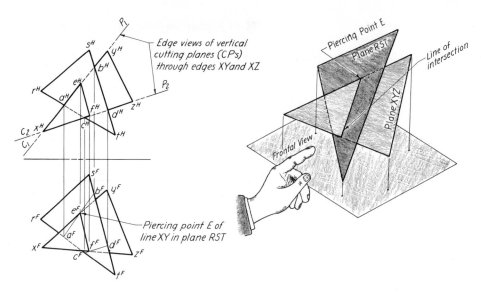

Fig. 9.25 □ To find the intersection line of two planes.

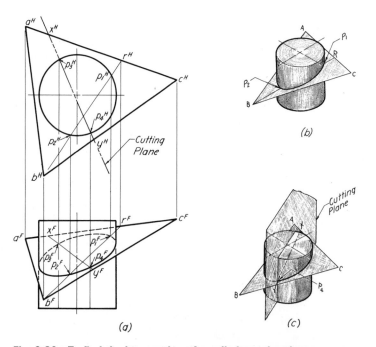

Fig. 9.26 □ To find the intersection of a cylinder and a plane.

line view $b^H r^H$ intersects the edge view of the surface of the cylinder. The F-views ($p_1{}^F$ and $p_2{}^F$) of points P_1 and P_2 were found by projecting downward from $p_1{}^H$ and $P_2{}^H$ to the line view $b^F r^F$. Additional points along the line of intersection, as needed, can be obtained by using other lines of the plane.

The line of intersection of this same plane and cylinder could have been almost as easily determined through the use of a series of line-projecting (cutting) planes passed parallel to the axis of the cylinder, as illustrated in (c). It should be noted that the vertical cutting plane shown cuts elements on the cylinder that intersect XY, the line of intersection of the cutting plane and the given plane, at points P_3 and P_4. Points P_3 [not visible in (c)] and P_4 are two points on the line of intersection, for they lie in both the cutting plane and the given plane and are on the surface of the cylinder. After the position of the cutting plane has been established in (a) by drawing the line representation in the H-view, x^H and y^H must be projected downward to the corresponding lines of the plane in the F-view. Line $x^F y^F$ as then drawn is the F-view of the line of intersection of the cutting plane and given plane. Finally, as the last step, the intersection elements that appear as points in the H-view at $p_3{}^H$ and $p_4{}^H$ must be drawn in the F-view. The F-views of points P_3 and P_4 ($p_3{}^F$ and $p_4{}^F$) are at the intersection of the F-views of the elements and the line $x^F y^F$.

A series of selected planes will give the points needed to complete the F-view of the intersection.

9.27 □ To find the intersection of a cone and an oblique plane

When the intersecting plane is oblique, as is true in Fig. 9.27, it is usually desirable to employ the cutting-plane method shown in (a) rather than to resort to the use of an additional view, as in (b). Since the given cone is a right cone, any one vertical cutting plane, passing through the apex O, will simultaneously cut straight lines on the conical surface and across the given oblique plane. The pictorial illustration shows that the cutting plane X-Y intersects the cone along the two straight-line elements O–2 and O–8 and the plane along the line RS. Points G and H, where the elements intersect the line RS, are points along the required line of intersection because both points lie in the given plane $ABCD$ and are on the surface of the cone. In (a), the line representation of the cutting plane XY in the H-view establishes the position of the H-view ($r^H s^H$) of the line RS. The H-views of elements O–2 and O–8 also lie in the edge view of the cutting plane XY. With this much known, the F-views of the two elements and line RS may be drawn. Points g^F and h^F are at the intersection of $r^F s^F$ and o^F–2 and o^F–8, respectively. A series of cutting planes passed similarly furnishes the points needed to complete the solution.

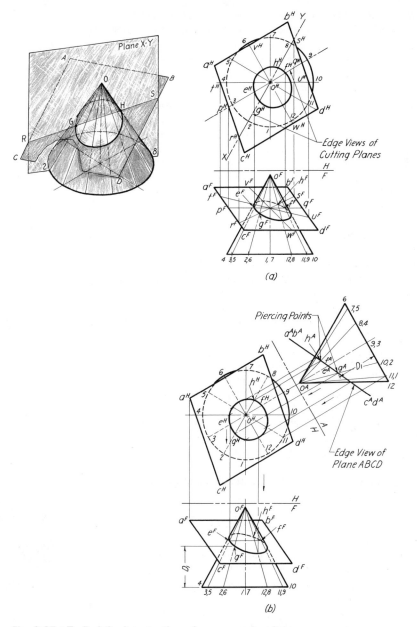

(a)

(b)

Fig. 9.27 □ To find the intersection of a cone and a plane.

At times one might prefer to determine the line of intersection through the use of a constructed auxiliary view that shows the given plane as an edge. In this case, when selected elements may be seen to intersect the line view of the plane in the auxiliary view, the solution becomes quite simple, because all that is required is to project the point of intersection of an element and the line view of the plane to the F- and H-views of the same element. For example, the A-view (o^A–8) of the element O–8 can be seen to intersect the edge line $a^A b^A c^A d^A$ at h^A, the A-view of point H. By projecting to line o^H–8, the H-view of H (h^H) may be easily established. The F-view of H (h^F) lies directly below h^H on o^F–8. Through the projected views of other points, located similarly, smooth curves may be drawn to form the F- and H-view representations, as shown.

9.28 □ To find the intersection of a sphere and an oblique plane

Horizontal cutting planes have been used to find the line of intersection of the sphere and oblique plane shown in Fig. 9.28. Two approaches to the solution have been given on the line drawing. The horizontal cutting planes, as selected, cut circles from the sphere and straight lines from the given oblique plane. For example, the cutting plane CP_3 in the F-view cuts the horizontal line 3–3 from plane $ABCD$ and a circle from the sphere. This circle appears as an edge in the F-view and shows in its true diameter. In the H-view it will show in true shape. The line and circle intersect at two points that also have been identified by the number 3, the number assigned to the cutting plane in which these two points lie. These points now located in the H-view are projected to the CP_3 line in the F-view. The curved line through points that have been determined by a series of planes, is a line common to both surfaces and is therefore the line of intersection.

This problem could also have been solved by using an auxiliary view showing the plane $ABCD$ as an edge. As before, the horizontal cutting planes will cut circles from the sphere and lines from the plane. However, in this case both the lines and the intersections show as points in the A-view. For each CP, these points must be projected from the auxiliary view to the corresponding circle in the H-view. The F-views of these points on the line of intersection may be found by projection and by using measurements taken from the A-view.

9.29 □ To find the intersection of two prisms

In Fig. 9.29, points A, C, and D, through which the edges of the horizontal prism pierce the vertical prism, are first found in the top view (a^H, c^H, and d^H) and are then projected downward to the corresponding edges in the front view. Point B, through which

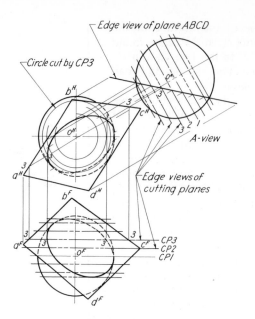

Fig. 9.28 □ To find the intersection of a sphere and an oblique plane.

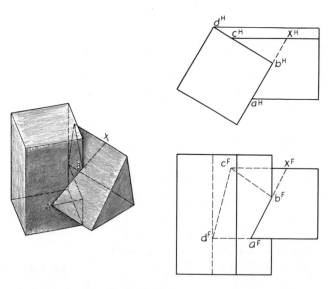

Fig. 9.29 □ Intersecting prisms.

the edge of the vertical prism pierces the near face of the triangular prism, cannot be found in this manner because the side view from which it could be projected to the front view is not shown. Its location, however, can be established in the front view without even drawing a partial side view, if some scheme like the one illustrated in the pictorial drawing is used. In this scheme, the intersection line AB, whose direction is shown in the top view as line a^Hb^H, is extended on the triangular face to point X on the top edge. Point x^H is projected to the corresponding edge in the front view and a light construction line is drawn between the points a^F and x^F. Since point B is located on line AX (see pictorial) at the point where the edge of the prism pierces the line, its location in the front view is at point b^F where the edge cuts the line a^Fx^F.

9.30 □ To determine the intersection of a prism and a pyramid using line-projecting planes

Frequently, it becomes necessary to draw the line of intersection between two geometric shapes so positioned that the piercing points of edges cannot be found by inspection if only the principal views are to be used. In this case, one must resort to the method discussed in Sec. 9.24 to determine where a line, such as the edge line GD of the prism shown in Fig. 9.30, pierces a surface. As illustrated by the pictorial drawing, a vertical plane passed through the edge DG of the prism, intersects the surface ABC of the pyramid along the MN that contains point D, the piercing point of DG. In (a), the H-view (m^Hn^H) of the line MN lies along d^Hg^H extended to m^H on the edge of the pyramid, because the H-view of the cutting plane appears as an edge that coincides with d^Hg^H. With the F-view of MN established by projecting downward from m^Hn^H in the H-view, the frontal view of the piercing point D is at d^F where the view of the edge line DG of the prism intersects m^Fn^F. The H-view of D is found by projecting upward from d^F. The two other piercing points, at E and F, are found in the same manner using two other line-projecting planes.

9.31 □ To find the intersection of two cylinders

If a series of elements are drawn on the surface of the small horizontal cylinder, as in Fig. 9.31, the points A, B, C, and D in which they intersect the vertical cylinder will be points on the line of intersection (see pictorial). These points, which are shown as a^H, b^H, c^H, and d^H in the top view, may be located in the front view by projecting them downward to the corresponding elements in the front view, where they are shown as points a^F, b^F, c^F, and d^F. The desired intersection is represented by a smooth curve drawn through these points.

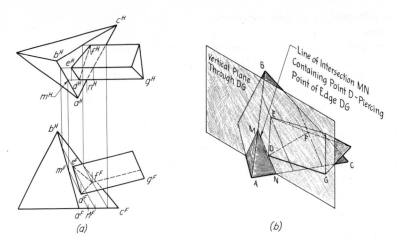

Fig. 9.30 □ Intersecting pyramid and prism.

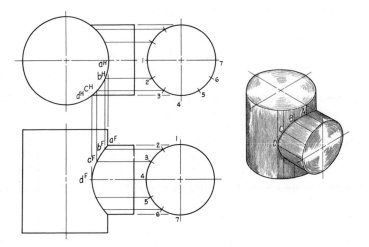

Fig. 9.31 □ Intersecting cylinders.

Spatial geometry for design and analysis □ **157**

9.32 □ To find the intersection of two cylinders oblique to each other

The first step in finding the line of intersection of two cylinders that are oblique to each other (Fig. 9.32) is to draw a revolved right section of the oblique cylinder directly on the front view of that cylinder. If the circumference of the right section is then divided into a number of equal divisions and elements are drawn through the division points, the points A, B, C, and D in which the elements intersect the surface of the vertical cylinder will be points on the line of intersection (see pictorial). In the case of the illustration shown, these points are found first in the top view and then are projected downward to the corresponding elements in the front view. The line of intersection in the front view is represented by a smooth curve drawn through these points.

9.33 □ To find the intersection of two cylinders using line-projecting (cutting) planes

The line of intersection of the two cylinders shown in Fig. 9.32 could have been determined through the use of a series of parallel line-projecting (cutting) planes passed parallel to their axes (Fig. 9.33). The related straight-line elements cut on the cylinders by any one cutting plane, such as C, intersect on the line of intersection of the cylinders. As many line-projecting planes as are needed to obtain a smooth curve should be used and they should be placed rather close together where a curve changes sharply.

9.34 □ To find the intersection of a cylinder and a cone

The intersection of a cylinder and a cone may be found by assuming a number of elements on the surface of the cone. The points at which these elements cut the cylinder are on the line of intersection (see Figs. 9.34 and 9.35). In selecting the elements, it is the usual practice to divide the circumference of the base into a number of equal parts and draw elements through the division points. To obtain needed points at locations where the intersection line will change suddenly in curvature, however, there should be additional elements.

In Fig. 9.34, the points at which the elements pierce the cylinder are first found in the top view and are then projected to the corresponding elements in the front view. A smooth curve through these points forms the figure of the intersection.

To find the intersection of the cone and cylinder combination shown in Fig. 9.35, line-projecting cutting planes were passed through the vertex O parallel to the axis of the cylinder to cut intersecting elements on both geometric forms. The partial auxiliary is needed to establish these planes, because it is only in a view showing the axis of the cylinder as a point that these planes and the surface of the cylinder will show as edge views. Each

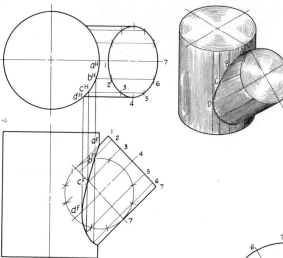

Fig. 9.32 □ Intersecting cylinders.

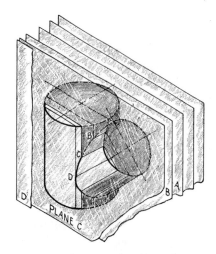

Fig. 9.33 □ To find the intersection of two cylinders using line-projecting planes.

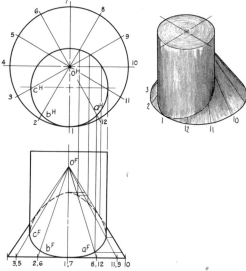

Fig. 9.34 □ Intersecting cylinder and cone.

cutting plane cuts one needed straight-line element from the conical surface and two straight-line elements from the cylindrical surface. Cutting plane 5, for example, cuts one element (numbered 5) from the near side of the cone and an upper and lower element from the surface of the cylinder. The intersection of these three elements, all numbered 5, establish the location of points A and B on the line of intersection.

An alternate method for finding the line of intersection of a cylinder and a right cone is illustrated in Fig. 9.36. Here horizontal cutting planes are passed through both geometric shapes in the region of their line of intersection. In each cutting plane, the circle cut on the surface of the cone will intersect elements cut on the cylinder at two points common to both surfaces (see pictorial). A curved line traced through a number of such points in different planes is a line common to both surfaces and is therefore the line of intersection.

9.35 □ To find the intersection of a prism and a cone

The complete line of intersection may be found by drawing elements on the surface of the cone (Fig. 9.37) to locate points on the intersection as explained in Sec. 9.34. To obtain an accurate curve, however, some thought must be given to the placing of these elements. For instance, although most of the elements may be equally spaced on the cone to facilitate the construction of its development, additional ones should be drawn through the critical points and in regions where the line of intersection changes sharply in curvature. The elements are drawn on the view that will reveal points on the intersection, then the determined points are projected to the corresponding elements in the other view or views. In this particular illustration a part of the line of intersection in the top view is a portion of the arc of a circle that would be cut by a horizontal plane containing the bottom surface of the prism.

C □ Vector geometry

9.36 □ Vector methods

In order to be successful in solving some types of problems that arise in design, a well-trained designer should have a working knowledge of ''vector geometry.'' The methods presented in this chapter should furnish the student with some background knowledge for solving force problems as they appear in the study of mechanics, strength of materials, and design. Through discreet use of the methods of vector geometry, as well as mathematical methods, it is possible to solve engineering problems quickly within a fully acceptable range of accuracy. Since any quantity having both magnitude and direction may be represented by a

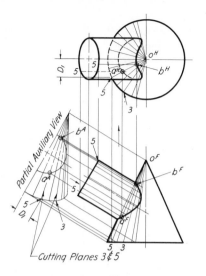

Fig. 9.35 □ To find the intersection of a cone and a cylinder.

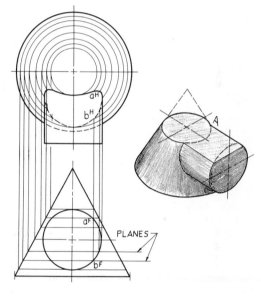

Fig. 9.36 □ Intersecting cylinder and cone.

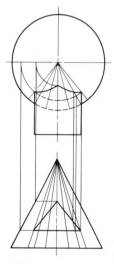

Fig. 9.37 □ Intersecting cone and prism.

fixed or rotating vector, vector operations are commonly used for problems in the design of frame structures, problems dealing with velocities in mechanisms, and for problems arising in the study of electrical properties. Because a student in a beginning course in engineering graphics should have basic principles rather than specialized cases presented to him, the methods given in this chapter for solving both two-dimensional and three-dimensional force problems deal mainly with static structures or, in other words, structures with forces acting so as to be in equilibrium. In a study of physics, graphical methods are useful for the composition and resolution of forces.

It is hoped that as a student progresses through his other undergraduate courses, he will desire to learn more about the use of vector methods for solving problems, and that he will become able to recognize the cases where he may have a choice between a graphical and an algebraic method. The graphical method is the better for many cases because it is much quicker and can be checked more easily.

An example of a vector addition is shown in Fig. 9.38. An airplane is flying north with a crosswind from the west. If the speed of the plane is 150 mph (miles per hour) and the wind is blowing toward the east at 60 mph, the plane will be flying NE (northeast) at 161.5 mph. Vectors can be used for problems of this type because forces acting on a body have both magnitude and direction.

9.37 □ A force

In our study of vector methods, a force may be defined as a cause which tends to produce motion in an object.

A force has four characteristics which determine it. First, a force has "magnitude." The value of this magnitude may be expressed in terms of some standard unit. It is usually given in pounds. Second, a force has a "line of action." This is the line along which the force acts. Third, a force has "direction." This is the direction in which it tends to move the object upon which it acts. Fourth, and last, a force has a "point of application." This is the place at which it acts upon the object, often assumed to be a point at the center of gravity.

9.38 □ A vector

A force can be represented graphically by a straight line segment with an arrowhead at one end. Such an arrow when used for this purpose is known as a "vector" (Fig. 9.39). The position of the body of the arrow represents the line of action of the force while the arrowhead points out the direction. The magnitude is represented to some selected scale by the overall length of the arrow itself.

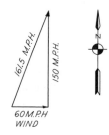

Fig. 9.38 □ A vector problem.

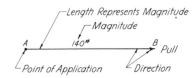

Fig. 9.39 □ A vector.

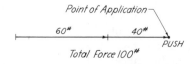

Fig. 9.40 □ A vector addition.

When a force acts in a two-dimensional plane, only one view of the vector is needed. However, if the force is in space, two views of the vector must be given.

9.39 □ Addition of vector forces—two forces

For a thorough understanding of the principles of vector addition, two simple examples will be considered first.

If one of two men who find it necessary to move a supply cabinet pushes on it with a 60 lb force while the other pushes in the same direction with a 40 lb force, the total force exerted to move the cabinet is 100 lb. The representation of two or more such forces in the manner shown in Fig. 9.40 amounts to a vector addition. Should these men be in a prankish mood and decide to push in opposite directions, as illustrated in Fig. 9.41, the cabinet might move provided the 20 lb resultant force were sufficient to overcome friction. The 20 lb resultant comes from a graphical addition.

Now let it be supposed that force A represented by F_A and force B represented by F_B in Fig. 9.42 act from a point P, the point of application. The resultant force on the body will not now be the sum of forces A and B, but instead will be the graphical addition of these forces as represented by the diagonal of a parallelogram having sides equal to the scaled length of the given forces. This single force R of 105.5 lb would produce an effect upon the body that would be equivalent to the combined forces F_A and F_B. The single force which could replace any given force system, is known as the *resultant* (R) for the force system.

Figure 9.42 shows that the resultant R divides the parallelogram into two equal triangles. Therefore, R could have been found just as well by constructing a single triangle as shown in Fig. 9.43 provided that the vector F_B' is drawn so that its tail-end touches the tip-end of F_A, and R is drawn with its arrow-end to the tip-end of F_B. Since either of the triangles shown in Fig. 9.42 could have been drawn to determine R, it should be obvious the resultant is the same regardless of the order in which the vectors are added. However, it is important that they be added tip-end to tail-end and that the vector arrows show the true direction for the action of the concurrent forces in the given system.

To find the resultant of two forces, which are applied as shown in Fig. 9.44, it is first necessary to move the vector arrows along their lines of action to the intersection point P before one can apply the parallelogram method.

The forces of a system whose lines of action all lie in one plane are called *coplanar forces*. Should the lines of action pass through a common point, the point of application, the forces are said to be *concurrent*. Figure 9.45 shows a system of forces that are both concurrent and coplanar.

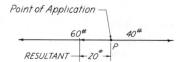

Fig. 9.41 □ Forces in opposite directions.

Fig. 9.42 □ Parallelogram of forces.

Fig. 9.43 □ A vector triangle.

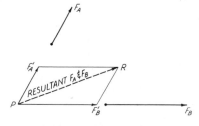

Fig. 9.44 □ Resultant of two forces.

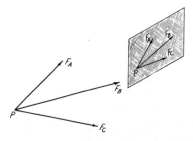

Fig. 9.45 □ Coplanar, concurrent forces.

9.40 □ Addition of vectors—three or more forces

The parallelogram method may be used to determine the resultant for a system of three or more forces that are concurrent and coplanar. In applying this method to three or more forces, it is necessary to draw a series of parallelograms, the number depending upon the number of vector quantities that are to be added graphically. For example, in Fig. 9.46 two parallelograms are required to determine the resultant R for the system. The resultant R_1 for forces F_A and F_B is determined by the first parallelogram to be drawn, and then R_1 is combined in turn with F_C by forming the second and larger parallelogram. By combining the forces in this way R becomes the resultant for the complete system.

Where a considerable number of vectors form a system, a somewhat less complicated diagram results, and less work is required when the triangle method is extended and applied to the formation of a vector diagram such as the one shown in Fig. 9.47. In this case the diagram is formed by three vector triangles, one adjacent to the other, and the resultant of forces F_A and F_B is combined with F_C to form the second triangle. Finally, by combining the resultant of the three forces F_A, F_B, and F_C with F_D, the vector R is obtained, which represents the magnitude and direction of the resultant of the four forces. In the construction F_B, F_C, and F_D in the diagram must be drawn so as to be parallel respectively to their lines of action in the system. However, the order in which they are placed in the diagram is optional as long as one vector joins another tip to tail.

9.41 □ Vector components

A *component* may be defined as one of two or more forces into which an original force may be resolved. The components, which together have the same action as the single force, are determined by a reversal of the process for vector addition, that is, the original force is resolved using the parallelogram method (see Fig. 9.48). The resolution of a plane vector into two components, horizontal and vertical, is illustrated in (a). In (b), the resolution of a force into components of specified direction is shown.

9.42 □ Forces in equilibrium

A body is said to be in equilibrium when the opposing forces acting upon it are in balance. In such a state the resultant of the force system will be zero. The concurrent and coplanar force system shown in Fig. 9.49 is in a state of equilibrium, for the vector triangle closes and each vector follows the other tip to tail.

An "equilibrant" is the force which will balance two or more forces and produce equilibrium. It is a force that would equal the resultant of the system but would necessarily have to act in an opposite direction.

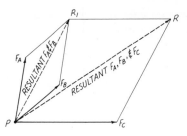

Fig. 9.46 □ Resultant of three or more forces with a common point of application (parallelogram method).

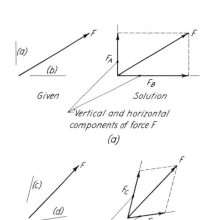

Fig. 9.48 □ Components.

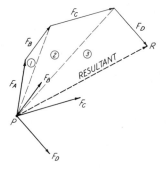

Fig. 9.47 □ Resultant of forces (polygon of forces).

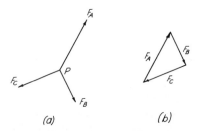

Fig. 9.49 □ Forces in equilibrium.

Figure 9.50 shows a weight supported by a short steel cable. The force to be determined is that needed to hold the weight in a state of equilibrium when it is swung from the position indicated by the broken lines into the position shown by solid lines.

This may be done by drawing a vector triangle with the forces in order from tip to tail. The 87.5 lb force vector represents the equilibrant, the force that will balance the 150 lb force and the 173 lb tension force now in the cable. The reader may wonder at the increase in the tension force in the cable from a 150 lb force when hanging straight down to a 173 lb force when the cable is at an angle of 30° with the vertical. It might help to realize that as the weight is swung outward towards a position where the cable will be horizontal, both the tension force and the equilibrant will increase. Theoretically it would require forces infinitely large to hold the system in equilibrium with the cable in a horizontal position.

In solving a force system graphically it is possible to determine two unknowns in a coplanar system.

Now suppose that it is desired to determine the forces acting in the members of a simple truss as shown in Fig. 9.51(a). To determine these forces graphically, one should isolate the joint supporting the weight and draw a diagram, known as a free-body diagram, to show the forces acting at the joint (b). Although the lines of this diagram may have any length, they must be parallel to the lines in the space diagram in (a). Since the boom will be in compression, a capital letter C has been placed along the line that represents the boom in the diagram. A letter T has been placed along the line for the cable because it will be in tension. Although the diagram may not have been essential in this particular case, such a diagram does play an important part in solving more complex systems.

In constructing the force polygon, it is necessary to start by drawing the vertical vector, for the load is the only force having a known magnitude and direction. After this vertical vector has been drawn to a length representing 1200 lb, using a selected scale, the force polygon (triangle) may be completed by drawing the remaining lines representing the unknown forces parallel to their known lines of action as shown in (a). The force polygon will close since the force system is in equilibrium.

The magnitude of the unknown forces in the members of the truss can now be determined by measuring the lines of the diagram using the same scale selected to lay out the length of the vertical vector. This method might be used to determine the forces acting in the members at any point in a truss.

9.43 □ Coplanar, noncurrent force systems

Forces in one plane having lines of action that do not pass through

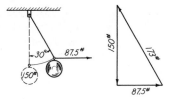

Fig. 9.50 □ **Determination of forces (graphically).**

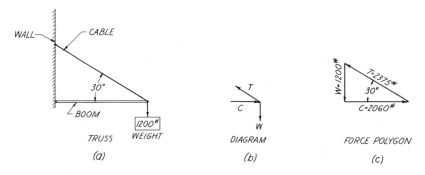

Fig. 9.51 □ **Determination of the forces at the joint of a simple truss.**

a common point are said to be *coplanar, nonconcurrent forces* (Fig. 9.52).

9.44 □ Two parallel forces

When two forces are parallel and act in the same direction, their resultant will have a line of action that is parallel to the lines of action of the given forces and it will be located between them. The magnitude of the resultant will be equal to the sum of the two forces [Fig. 9.53(a)], and it will act through a point that divides any perpendicular line joining the lines of action of the given forces inversely as the forces.

Should the two forces act in opposite directions, as shown in (b), the resultant will be located outside of them and will have the same direction as the greater force. Its magnitude will be equal to the difference between the two given forces. The proportion shown with the illustration in (b) may be used to determine the location of the point of application of the resultant. Those who prefer to determine graphically the location of the line of action for the resultant may use the method illustrated in Fig. 9.54. This method is based on well known principles of geometry.

With the two forces F_A and F_B given, any line 1–2 is drawn joining their lines of action. From this line two distances must be laid off along the lines of action of the given forces. If the given forces act in the same direction, then the distances are laid off in opposite directions from the line 1–2, (a). If they act in opposite directions, the distances must be laid off on the same side of line 1–2, (b). In Fig. 9.54(a), a length equal by scale to F_A was laid off from point 1 on the line of action of F_B. Then from point 2 a length equal to F_B was marked off in an opposite direction. These measurements located points 3 and 4, the end-points of the line intersecting line 1–2 at point O. Point O is on the line of action of the resultant R. In Fig. 9.54(b) this method has been applied to establish the location of the resultant for two forces acting in opposite directions.

9.45 □ Moment of a force

The "moment of a force" with respect to a point is the product of the force and the perpendicular distance from the given point to the line of action of the force. In the illustration, Fig. 9.55, the moment of the force F_A about point P is Mom. $= F_A \times d$. The perpendicular distance d is known as the lever arm of the force. Should the distance d be measured in inches and the force be given in pounds, the moment of the force will be in inch-pounds.

9.46 □ Force couples (Fig. 9.56)

Two equal forces that act in opposite directions are known as a "couple." A couple does not have a resultant, and no single force can counteract tendency to produce rotation. The measurement of this tendency is the moment of the couple that is the product of one of the forces and the perpendicular distance between them.

To prevent the rotation of a body that is acted upon by a couple, it is necessary to use two other forces that will form a second couple. The body acted upon by these couples will be in equilibrium if each couple tends to rotate the body in opposite directions and the moment of one couple is equal to the other.

9.47 □ String polygon—Bow's notation

A system for lettering space and force diagrams, known as "Bow's notation," is widely used by technical authors. Its use in this chapter will tend to simplify the discussions which follow.

In the space diagram, shown in Fig. 9.57(a), each space from the line of action of one force to the line of action of the next one is given a lower-case letter such as a, b, c, and d in alphabetical order. Thus the line of action for any particular force can be designated by the letters of the areas on each side of it. For example, in Fig. 9.59 the line of action for the 1080 lb force, acting downward on the beam, would be designated as line of action bc. On the force diagram, corresponding capital letters are used at the ends of the vectors. In Fig. 9.57(b), AB represents the magnitude of ab in the space diagram and BC represents the magnitude of bc.

To find the resultant of three or more parallel forces graphically, the "funicular" or "string polygon" is used. The magnitude and direction of the required resultant for the system shown in Fig. 9.57 are known. The magnitude, representing the algebraic sum of the given forces, appears as the heavy line AD of the force polygon. It is required to determine the location of its line of action. With the forces located in the space diagram and the force polygon drawn, the steps for the solution are as follows:

1 Assume a pole point O and draw the rays OA, OB, OC, and OD. Each of the triangles formed is regarded as a vector triangle with one side representing the resultant of the forces represented by the other two sides. For example: If we consider AB to be a resultant force, then OA and OB are two component forces that could replace AB. For the second vector triangle, OB and BC have OC as their resultant. OC, when combined with CD, will have OD as the resultant. OA and OD combine with AD, the final resultant of the system.

2 Draw directly on the space diagram the corresponding strings of the funicular polygon. The funicular polygon may be started at any selected point r along the line of action ab. The string ob will then be parallel to OB of the force polygon. From point

s, where *ob* intersects *bc*, draw *oc* parallel to *OC*. The line *oc* extended to *cd* establishes the location of point *t*. Line *od* drawn parallel to *OD* and line *oa* drawn parallel to *OA* intersect at point *p*. Point *p* is a point on the line of action of force *ad*, the resultant force *AD* for the given force system.

When one or more of a system of parallel forces are directed oppositely from the others, the magnitude and direction of the resultant will be equal to the algebraic sum of the original forces.

9.48 □ Coplanar, nonconcurrent, nonparallel forces

In further study of coplanar and nonconcurrent forces it might be supposed that it is necessary to determine the magnitude, direction, and line of action of the one force that will establish a state of equilibrium when combined with the given forces *AB*, *BC*, and *CD* of the force system shown in Fig. 9.58. The direction and line of action of the original forces are given in both the space diagram in (*a*) and the force polygon in (*b*). The magnitude and direction of the force that will produce equilibrium is represented by *DA*, the force needed to close the force polygon. With the force polygon completed, the next step is to assume a pole point *O* and draw the rays *OA*, *OB*, *OC*, and *OD*. Now *OA* and *OB* are component forces of *AB*, and *AB* might be replaced by these forces. To clarify this statement; each of the four triangles may be considered to be a vector triangle, and in the case of vector triangle *OAB*, *AB* can be regarded as the resultant for the other two forces *OA* and *OB*. It should be noted that component force *OB* of the vector triangle *OBC* must be equal and opposite in direction to component force *OB* of *OAB*.

All that remains to be done is to determine the line of action of the required force *DA* by drawing the string diagram as explained in Sec. 9.47, remembering that point *r* may be any point along the line of action of *ab*. The intersection point *p* for strings *oa* and *od* is a point along the line of action *da* of force *DA*. Although lines of action *ab*, *bc*, *cd*, and *da* were drawn to a length representing their exact magnitude in Fig. 9.58(*a*), they could have been drawn to a convenient length to allow for the construction of the string polygon, for these lines merely represent lines of action for the forces *AB*, *BC*, *CD*, and *DA*. The lines were presented in scaled length for illustrative purposes.

9.49 □ Equilibrium of three or more coplanar parallel forces

When a given system, consisting of three or more coplanar forces, is in equilibrium, both the force polygon and the funicular polygon must close. If the force polygon should close and the funicular polygon not close, the resultant of the given system will be found to be a force couple.

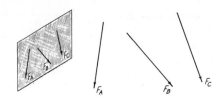

Fig. 9.52 □ Coplanar, nonconcurrent forces.

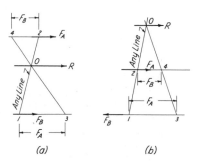

Fig. 9.54 □ Determination of the position of the resultant of parallel forces (graphical method).

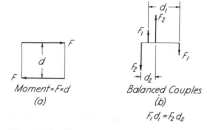

Moment = F×d
(*a*)

d_1 F_2

F_1

F_2 d_2

Balanced Couples
(*b*)

$F_1 d_1 = F_2 d_2$

Fig. 9.56 □ Force couples.

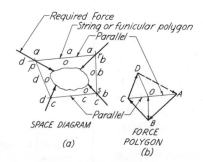

Required Force
String or funicular polygon
Parallel

SPACE DIAGRAM
(*a*)

FORCE POLYGON
(*b*)

Fig. 9.58 □ Funicular or string polygon.

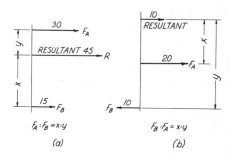

$F_A : F_B = x:y$ (*a*)

$F_B : F_A = x:y$ (*b*)

Fig. 9.53 □ Parallel forces.

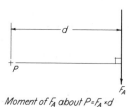

Moment of F_A about $P = F_A × d$

Fig. 9.55 □ Moment of a force.

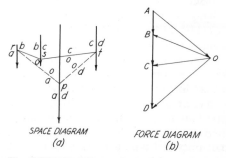

SPACE DIAGRAM
(*a*)

FORCE DIAGRAM
(*b*)

Fig. 9.57 □ Resultant of parallel forces—Bow's notation.

Two unknown forces of a parallel coplanar force system may be determined graphically by drawing the force and funicular polygons as shown in Fig. 9.59, since the forces are known to be in equilibrium, and all are vertical. Although one may be aware that the sum of the two reaction forces R_1 and R_2 is equal to the sum of forces AB, BC, CD, and DE, the magnitude of R_1 and R_2 as single forces is unknown. The location of point F in the force polygon, which is needed if one is to determine the magnitudes of R_1 and R_2, may be found by fulfilling the requirement that the funicular polygon be closed.

The funicular polygon is started at any convenient point along the known line of action of R_1, and successive strings are drawn parallel to corresponding rays of the force polygon. In area b the string will be parallel to OB, in area c the string will be parallel to OC and so on, until the string that is parallel to OE has been drawn. String of may then be added to close the funicular polygon. This closing line from point x to the starting point y determines the position of OF in the force polygon, for ray OF must be parallel to the string of. The magnitude of the reaction R_1 is represented to scale by vector FA, and R_2 is represented by the vector FE.

The graphical method for determining the values of wind load reactions for a roof truss having both ends fixed is shown in Fig. 9.60. The solution given is practically identical with the solution applied to the beam in Fig. 9.59.

9.50 □ Concurrent, noncoplanar force systems

Up to this point in our study of force systems, the student's attention has been directed solely to systems lying in one plane in order that the graphical methods dealing with the composition and resolution of forces could be presented in a clear and simple manner, free from the thinking needed for understanding force systems involving the third dimension.

In dealing with noncoplanar forces it is necessary to use at least two views to represent a structure in space. Although the methods as applied to coplanar force systems for solving problems may be extended to noncoplanar systems, the vector diagram for noncoplanar forces must have two views instead of one view as in the case of coplanar forces. If the student is to understand the discussions that are to follow he must grasp the idea that for the composition and resolution of noncoplanar forces he will work with two distinct and separate space representations, the space diagram for the given structure and the related vector diagram (force polygon). Figure 9.61 shows two views of a concurrent, noncoplanar force system not in equilibrium.

There are a few basic relationships that exist between a space diagram and its related vector diagram, which must be kept in mind when solving noncoplanar force problems. These rela-

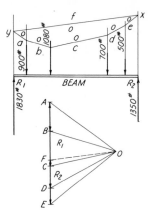

Fig. 9.59 □ To determine the reaction forces of a loaded beam.

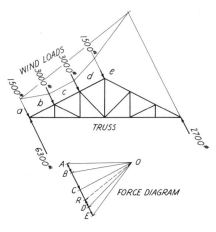

Fig. 9.60 □ The determination of the reactions for wind loads.

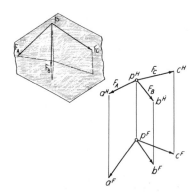

Fig. 9.61 □ Concurrent, noncoplanar forces.

tionships are: (1) in corresponding views (*H*-view and *H*-view, or *F*-view and *F*-view), each vector in the vector diagram will be parallel to its corresponding representation in the space diagram, (2) if a system of concurrent, noncoplanar forces is in equilibrium, the force polygon in space closes and the projection on each plane will close, and (3) the true magnitude of a force can be measured only in the vector diagram when it appears in true length or is made to do so.

9.51 □ Determination of the resultant of a force system of concurrent, noncoplanar forces

The parallelogram method for determining the resultant of concurrent forces as explained in Sec. 9.40 may be employed to find the resultant of the three forces *OA*, *OB*, and *OC* in Fig. 9.62. Any number of given concurrent, noncoplanar forces can be combined into their resultant by this method. In the illustration, forces *OA* and *OB* were combined into their resultant, which is the diagonal of the smaller parallelogram, then this resultant in turn was combined with the third force *OC* to obtain the final resultant *R* for the given system. Since the true magnitude of *R* can be scaled only in a view showing its true length, an auxiliary view was projected from the front view. The true length of *R* could also have been determined by revolution.

Since the single force needed to hold a force system in balance, known as the *equilibrant*, is equal to the resultant in magnitude but is opposite in direction, this method might be used to determine the equilibrant for a system of concurrent, noncoplanar forces.

In presenting this problem and the two problems that follow, it has been assumed that the student has read the previous sections of this chapter and that his knowledge of the principles of projection is sufficient for him to find the true length, having two views given, and to draw the view of a plane so that it will appear as an edge.

9.52 □ To find the three unknown forces of a simple load-bearing frame—special case

In dealing with the simple load-bearing frame in Fig. 9.63, it should be realized that this is a special case rather than a general one, for two of the truss members appear as a single line in the frontal view of the space diagram. This condition considerably simplifies the task of finding the unknown forces acting in the members, and it is this particular spatial situation that makes this a special case. However, it should be pointed out now that this condition must exist or be set up in a projected view when a vector solution is to be applied to any problem involving a system of concurrent, noncoplanar forces. More will be said about the necessity

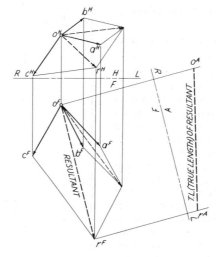

ORTHOGRAPHIC SOLUTION

Fig. 9.62 □ Determination of the resultant of concurrent, noncoplanar forces.

for having two unknown forces coincide in one of the views in the discussion in Sec. 9.54.

After the space diagram has been drawn to scale, the steps toward the final solution are as follows:

1   Draw a free-body diagram showing the joint at A, the joint at which the load is applied. The lines of this diagram may be of any length, but each line in it must be parallel respectively to a corresponding line in the view of the space diagram to which the free-body diagram is related. In this case, it is the horizontal (top) view. A modified form of Bow's notation was used for convenience in identifying the forces. The vertical load has been shown pulled to one side in order that this force can be made to fall within the range of the notation. This diagram is important to the solution of this problem, for it enables one to see and note the direction of all of the forces that the members exert upon the joint (note arrowheads). Capital letters were used on the free-body diagram to identify the spaces between the forces, rather than lower-case letters as is customary, so that lower-case letters could be used for the ends of the views of the vectors in the vector diagram.

2   Using a selected scale, start the two views of the vector diagram (b) by laying out vector RS representing the only known force, in this case the 1000 lb load. Since RS is a force acting in a vertical direction $r^F s^F$ will be in true length in the F-view and will appear as point $r^H s^H$ in the H-view.

3   Complete the H- and F-views of the vector diagram. Since each vector line in the top view must be parallel to a corresponding line in the top view of the space diagram, $s^H t^H$ must be drawn parallel to $a^H b^H$, $t^H u^H$ parallel to $a^H c^H$, and $u^H r^H$ parallel to $a^H d^H$. Since the forces acting at joint A are in equilibrium, the vector triangle will close and the vectors will appear tip to tail.

In the frontal view of the vector diagram, $s^F t^F$ will be parallel to $a^F b^F$, $t^F u^F$ will be parallel to $a^F c^F$, and $r^F u^F$ will be parallel to $d^F a^F$.

4   Determine the magnitude of the forces acting on joint A. Since vector RU shows its true length in the F-view, the true magnitude of the force represented may be determined by scaling $r^F u^F$ using the same scale used to lay out the length of $r^F s^F$. Although it is known that vectors ST and TU are equal in magnitude, it is necessary to find the true length representation of one or the other of these vectors by some approved method before scaling to determine the true value of the force.

An arrowhead may now be added to the line of action of each force in the free-body diagram to indicate the direction of the action. Since the free-body diagram was related to the top view of the space diagram, the arrowhead for each force will point in the same direction as does the arrowhead on the corresponding vector in the H-view of the vector diagram. These arrowheads show that the forces in members AB and AC are acting away from joint A and are therefore *tension forces*. The force in AD acts toward A and thus is a *compression force*.

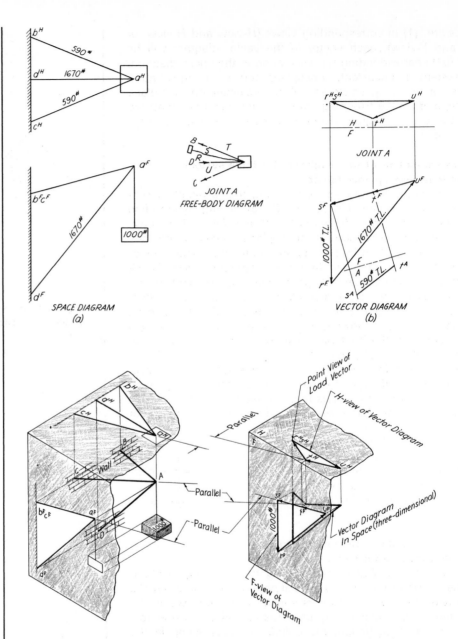

Fig. 9.63 □ **Solution of a concurrent, noncoplanar force system—special case.**

9.53 □ To find the three unknown forces
of a simple load-bearing frame—composition method (Fig. 9.64)

Since the forces in AB and AC lie in an inclined plane that appears as an edge in the F-view, they may be composed into a single force that will have the same effect in the force system as the forces it replaces. This replacement force along with the force in AD and the load now become the forces that would be acting in a simple load-bearing truss (See Fig. 9.51). After the vertical vector of this concurrent coplanar force system has been drawn to a selected scale, the triangular force polygon in (b) may be completed by drawing the remaining lines representing the forces in AD and the resultant R. These lines are drawn parallel to their known lines of action as shown in the F-view of the space diagram in (a). Finally, the true length view of R must be transferred to the auxiliary view in (a) and resolved into its component forces, acting in AB and AC, that collectively have the same action as the resultant R.

9.54 □ To find the three unknown forces
of a simple load-bearing truss—general case

For the general case shown in Fig. 9.65, the known force is in a vertical position as in the previous problem, but no two of the three unknown forces appear coincident in either of the two given views. For this reason, it is necessary at the very start to add a complete auxiliary view to the space diagram that will combine with the existing top view to give a point view of one member and a line view of two of the three unknown forces. To obtain this desired situation, one should start with the following steps, which will transform the general case into the special case with which one should now be familiar.

Step 1: Draw a true length line in the plane of two of the members. In Fig. 9.65(a) this line is DE, which appears in true length (TL) in the H-view.

Step 2: Draw the needed auxiliary view, taken so that DE will appear as a point (d^4e^4) and OB and OC will be coincident (line $o^4b^4c^4$). This construction involves finding the edge view of a plane (see Sec. 9.12). In this particular case, the auxiliary view has height in common with the frontal view.

Step 3: Draw the two views of the vector diagram by assuming the H-view and the A-view to be the given views of the special case. Proceed by the steps set forth for the special case in Sec. 9.52.

Step 4: Determine the magnitude of the forces and add arrowheads to the free-body diagram to show the direction of action of the forces acting on point O.

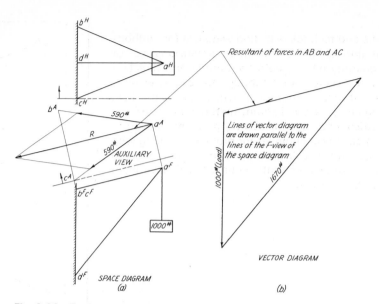

Fig. 9.64 □ To determine the three unknown forces of a simple load-bearing frame—composition method.

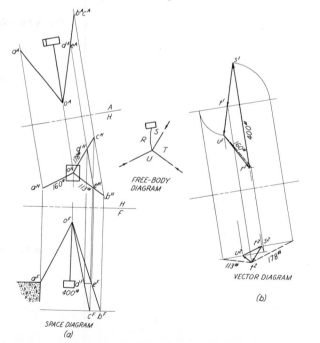

Fig. 9.65 □ Solution of a concurrent, noncoplanar force-system—general case.

9.55 □

In practice, engineers and technologists find wide use for methods that solve problems through the use of three-dimensional vector diagrams, for any quantity having both magnitude and direction may be represented by a vector. And, although the examples used in the chapter deal with static structures, which are in the field of the structural engineer, vector diagram methods are used frequently by the electrical engineer for solving problems arising in his field and by the mechanical engineer for problems dealing with bodies in motion. The student will without doubt encounter some of these methods in a textbook for a later course or will have them presented to him by his instructor.

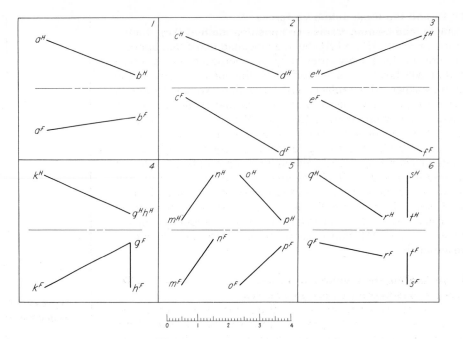

Fig. 9.66 □ True length of a line.

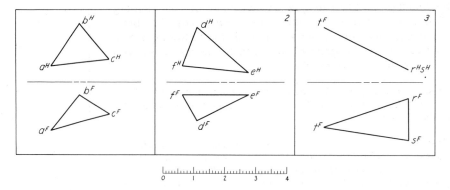

Fig. 9.67 □ True size and shape of a plane.

Problems

Descriptive geometry

Problems 1 through 5 have been selected and arranged to offer the student an opportunity to apply basic principles of descriptive geometry.

The problems can be reproduced to a suitable size by transferring the needed distances from the drawing to the scale that has been provided for each group of problems.

1 (Fig. 9.66). Reproduce the given views of the line or lines of a problem as assigned and determine the true length or lengths using the auxiliary view method explained in Sec. 9.6. A problem may be reproduced to a suitable size by transferring the needed distances from the drawing to the given scale to determine values. The distances, as they are determined, should be laid off on the drawing paper using a full-size scale.

2 (Fig. 9.67). Reproduce the given views of the plane of a problem as assigned and draw the view showing the true size and shape. Use the method explained in Sec. 9.13. Determine needed distances by transferring them from the drawing to the accompanying scale, by means of the dividers.

3 (Fig. 9.68). These problems are intended to give some needed practice in manipulating views to obtain certain relationships of points and lines. Determine needed distances by transferring them from the drawing to the accompanying scale, by means of the dividers.

    1 Determine the distance between points A and B.

    2 Draw the H- and F-views of a $\frac{1}{2}$-in. perpendicular erected from point N of the line MN.

    3 Draw the H-view of the $3\frac{3}{4}$-in. line ST.

    4 Draw the H- and F-views of a plane represented by an equilateral triangle and containing line AB as one of the edges. The added plane ABC is to be at an angle of 30° with plane $ABDE$.

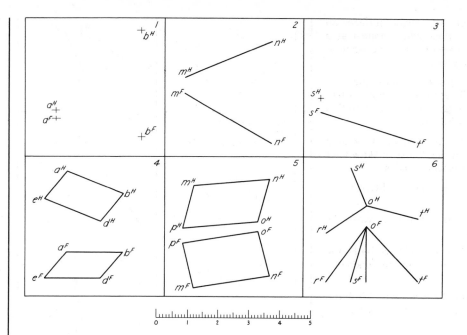

Fig. 9.68 □ **Relationships of points and lines.**

5 If the figure *MNOP* is a plane surface, an edge view of the surface would appear as a line. Draw such a view to determine whether or not *MNOP* is a plane.

6 A vertical pole with top *O* is held in place by three guy wires. Determine the slope in tangent value of the angle for the guy wire that has a bearing of N 23° W.

4 (Fig. 9.69). In this group of problems it is required to determine the shortest distance between skew lines and the angle formed by intersecting lines.

    1 Show proof that the plane *ABCD* is an oblique plane.

    2 Determine the shortest distance between the lines *MN* and *ST*.

    3 Through point *K* on line *GH* draw the *F*- and *H*-views of a line that will be perpendicular to line *EF*.

    4 Determine the angle between line *AB* and a line intersecting *AB* and *CD* at the level of point *E*.

    5 Through a point on line *MN* that is $1\frac{1}{4}$ in. from point *N*, draw the *F*- and *H*-views of a line that will be perpendicular to line *ST*.

    6 Erect a 1-in. perpendicular at point *K* in the plane *EFGH*. Connect the outer end point *L* of the perpendicular with *F*. Determine the angle between *LF* and *KF*.

5 (Fig. 9.70). These problems require that the student determine the angle between a line and a plane and the angle between two given planes.

    1 Determine the angle between the planes *MNQP* and *RST*.

    2 Determine the angle between the line *ST* and
    (a) The *H*-plane.
    (b) The *F*-plane.

    3 The line *EF* has a bearing of N 53° E. What angle does this line make with the *P*-plane?

    4 Draw the *F*- and *H*-views of a line through point *K* that forms an angle of 35° with plane *MNQR*.

    5 The top and front views of planes *ABC* and *RST* are partially drawn.

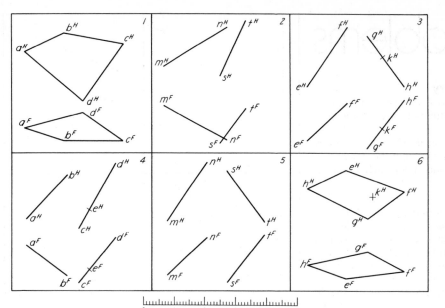

Fig. 9.69 □ **Shortest distance between skew lines and the angle formed by intersecting lines.**

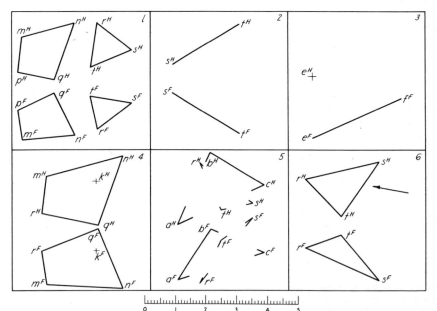

Fig. 9.70 □ **Angle between a line and a plane and the angle between two planes.**

(a) Complete the views including the line of intersection.

(b) Determine the angle between the line of intersection and the *H*-plane of projection.

6 Two views of a plane *RST* and the top view of an arrow are shown. The arrow, pointing downward and toward the left, is in a plane that forms an angle of 68° with plane *RST*. The arrow point is $\frac{1}{4}$ in. from the plane *RST*.

(a) Draw the front view of the arrow.

(b) Draw the top and front views of the line of intersection of the 68° plane and plane *RST*.

Intersections

6–7 (Figs. 9.71 and 9.72). Draw the line of intersection of the intersecting geometric shapes as assigned. Show the invisible portions of the lines of intersection as well as the visible. Consider that the interior is open.

8–9 (Figs. 9.73 and 9.74). Draw the line of intersection of the intersecting geometric shapes as assigned. It is suggested that the elements used to find points along the intersection be spaced 15° apart. Do not erase the construction lines. One shape does not pass through the other.

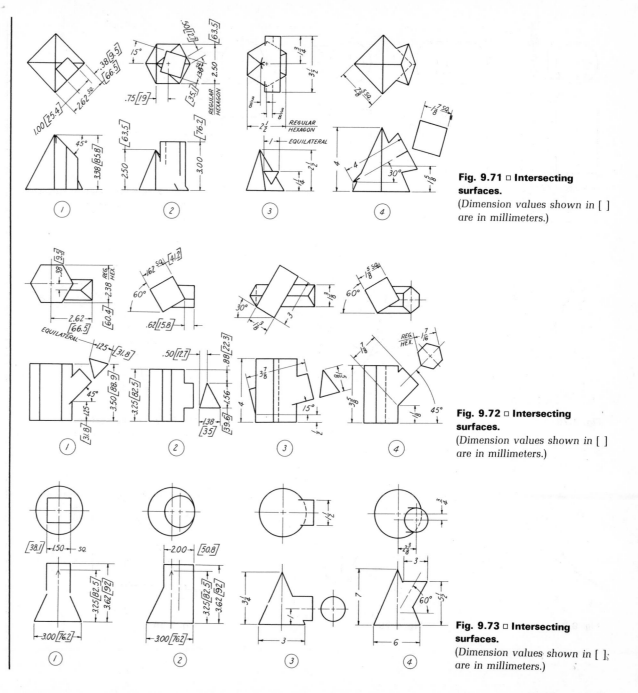

Fig. 9.71 □ **Intersecting surfaces.**
(*Dimension values shown in [ ] are in millimeters.*)

Fig. 9.72 □ **Intersecting surfaces.**
(*Dimension values shown in [ ] are in millimeters.*)

Fig. 9.73 □ **Intersecting surfaces.**
(*Dimension values shown in [ ] are in millimeters.*)

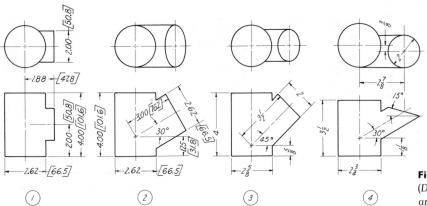

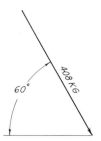

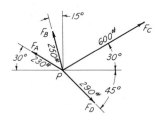

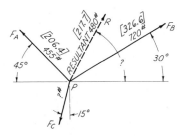

Fig. 9.74 □ Intersecting surfaces.
(*Dimension values shown in [ ]
are in millimeters.*)

Fig. 9.75

Fig. 9.76

Fig. 9.77 □ Force system.
(*Values shown in [ ] are in
kilograms.*)

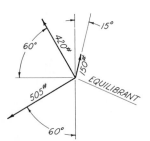

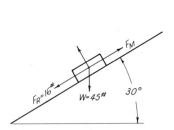

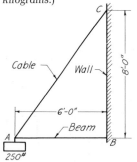

Fig. 9.78

Fig. 9.79

Fig. 9.80

Vector geometry

The following problems have been selected to emphasize the basic principles underlying vector geometry. By solving a limited number of the problems presented, the student should find that he has a working knowledge of some vector methods that are useful for solving problems in design that involve the determination of the magnitude of forces as well as their composition and resolution. The student is to select his own scale remembering that a drawing made to a large scale usually assures more accurate results.

1 (Fig. 9.75). A force of 408 kg acts downward at an angle of 60° with the horizontal. Determine the vertical and horizontal components of this force.

2 (Fig. 9.76). Determine the resultant force for the given coplanar, concurrent force system.

3 (Fig. 9.77). Determine the magnitude of the force F_C and the angle that the resultant force R makes with the horizontal for the given coplanar, concurrent force system.

4 (Fig. 9.78). Determine the magnitude and direction of the equilibrant for the given coplanar, concurrent force system.

5 (Fig. 9.79). A block weighing 45 lb is to be pulled up an inclined plane sloping at an angle of 30° with the horizontal. If the frictional resistance is 16 lb, what is the magnitude of the force F_M that is required to move the block uniformly up the plane?

6 (Fig. 9.80). A horizontal beam AB is hinged at B as shown. The end of the beam at A is connected by a cable to a hook in the wall at C. The load at A is 250 lb. Using the dimensions as given, determine the tension force in the cable and the reaction on the hinge at B. The weight of the beam is to be neglected.

7 (Fig. 9.81). A 272.2 kg load is supported by cables as shown. Determine the magnitude of the tension in the cables.

8 (Fig. 9.82). A ship that is being pulled through the entrance of a harbor is headed due east through a crosscurrent moving at 4 knots as shown. If the ship is moving at 12 knots, what is the speed of the tug boat?

9 (Fig. 9.83). Determine the magnitude of the reactions R_1 and R_2 of the beam with loads as shown.

10 (Fig. 9.84). Determine the magnitude of the reactions R_1 and R_2 of the beam.

11 (Fig. 9.85). Determine the magnitude of the reactions R_1 and R_2 for the roof truss shown. Each of the six panels is of the same length.

12 (Fig. 9.86). Determine the magnitude of the reactions R_1 and R_2 to the wind loads acting on the roof truss as shown.

13 (Fig. 9.87). A tripod with an 85 lb load is set up on a level floor as shown. Determine the stresses in the three legs due to the vertical load on the top.

14–15 (Figs. 9.88–9.89). Determine the stresses in the members of the space frame shown.

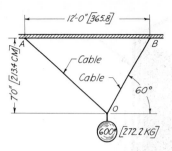

Fig. 9.81 □ Cable support system.
(*Values shown in [ ] are in the metric system.*)

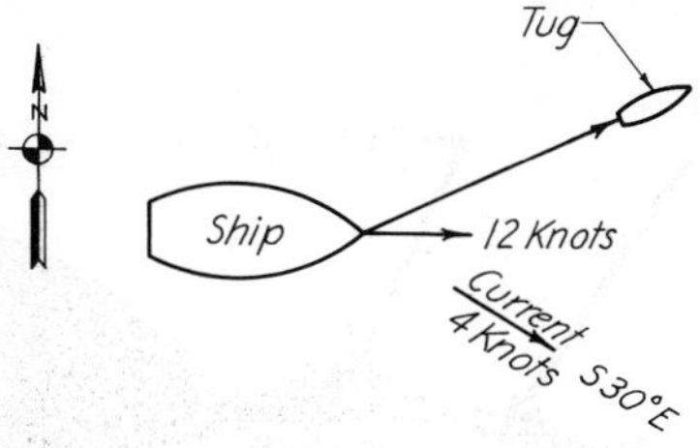

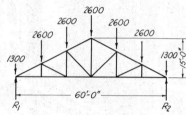

Fig. 9.82

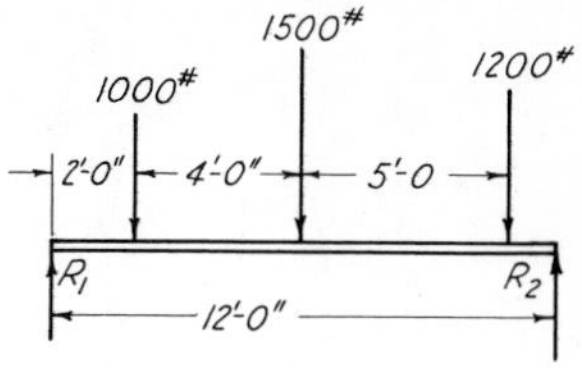

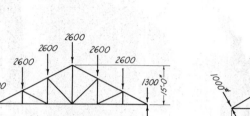

Fig. 9.83

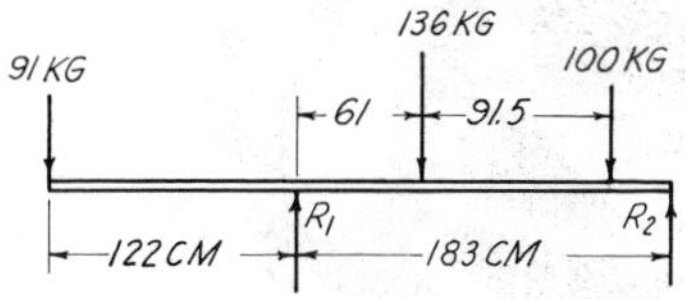

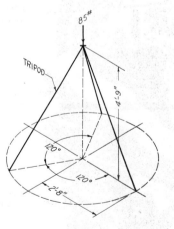

Fig. 9.84 □ Simple beam.
(*Values are in the metric system.*)

Fig. 9.85

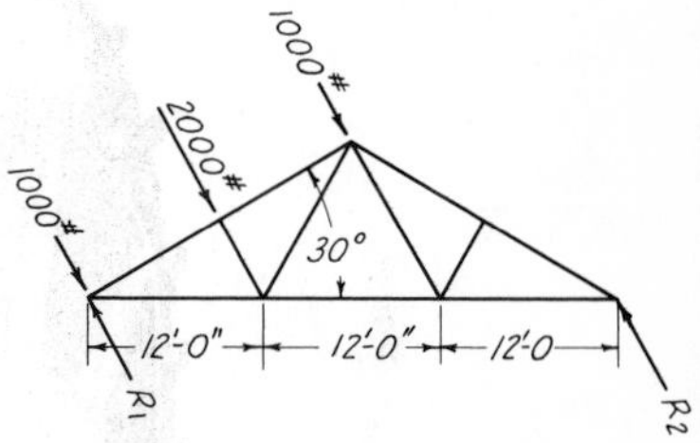

Fig. 9.86

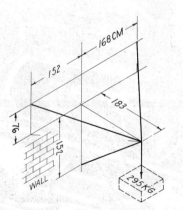

Fig. 9.87

Fig. 9.88 □ Space frame.
(*Values are in the metric system.*)

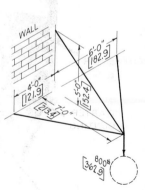

Fig. 9.89 □ Space frame.
(*Values shown in [ ] are in the metric system.*)

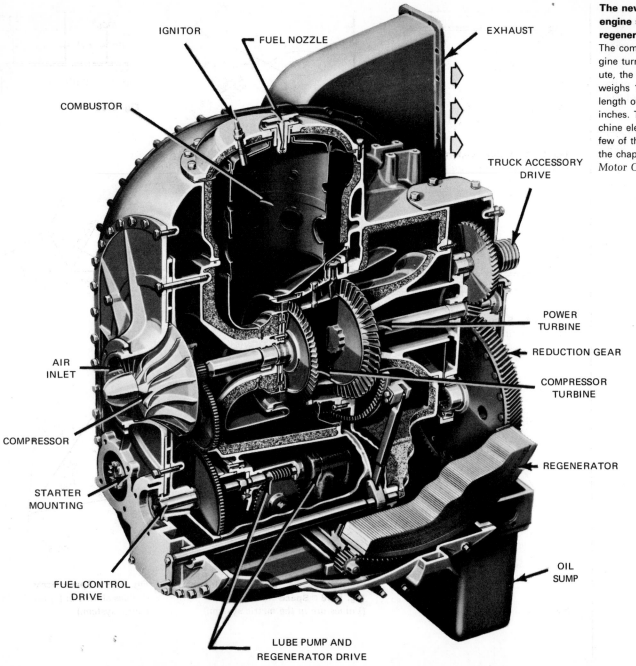

IGNITOR

FUEL NOZZLE

COMBUSTOR

EXHAUST

TRUCK ACCESSORY DRIVE

POWER TURBINE

REDUCTION GEAR

COMPRESSOR TURBINE

AIR INLET

COMPRESSOR

REGENERATOR

STARTER MOUNTING

OIL SUMP

FUEL CONTROL DRIVE

LUBE PUMP AND REGENERATOR DRIVE

The new research prototype gas turbine engine shown is a low pressure, regenerative power plant.
The compressor in the 375-horsepower engine turns at 37,500 revolutions per minute, the output shaft at only 3000 RPM. It weighs 1700 pounds installed, with a length of 40 inches and a height of 39 inches. The reader should note the machine elements that can be readily seen. A few of these elements will be discussed in the chapter that follows. (*Courtesy Ford Motor Company.*)

10.1 □ Gears

It is important for the designer of mechanical systems to know the general proportions and nomenclature pertaining to gearing. In Fig. 10.1 the nomenclature for bevel gears is shown. It will be noted that, in general, the definitions pertaining to gear-tooth parts can be represented in a right section of the gear.

The theory of gears is a part of the study of mechanisms. In working drawings of gears and toothed wheels it is necessary to draw at least one tooth of each gear. Some of the terms used in defining gear teeth are shown in Fig. 10.2.

Two systems of generating tooth curves are in general use, the *involute system* and the *cycloidal system*. The curve most commonly used for gear-tooth profiles is the involute of a circle.

An *involute* is the curve generated by a point on a straightedge as the straightedge is rolled on a cylinder. It also may be defined as the curve generated by a point on a taut string as the string is unwrapped from a cylinder. The circle from which the involute is developed is called the *base circle*.

A method of constructing an involute curve is shown in Fig. 10.3. Starting with point 0, on the base circle, divide the base circle into a convenient number of equal arcs of length 0–1, 1–2, 2–3, and so forth. (Where the lengths of the divisions on the base circle are not too great, the chord can be taken as the length of the arc.) Draw a tangent to the base circle at point 0, and divide this line to the left of 0 into equal parts of the same lengths as the arcs. Next, draw tangents to the circle from points 1, 2, 3, and so on. With the center of the base circle "*O*" as a pivot, draw

Geometry of machine elements: gears, cams, and linkages

concentric arcs from 1′, 2′, 3′, and so forth, until they intersect the tangent lines drawn from 1, 2, 3, and so forth. The intersection of the arcs and the tangents are points on the required involute curve, such as 1″, 2″, 3″, and so forth. The illustration in Fig. 10.3 shows the portion XY of the tooth outline as part of the involute curve.

The cycloidal system, as the name implies, has tooth curves of cycloidal form. A *cycloid* is the curve generated by a point on the circumference of a circle as the circle rolls on a straight line. If the circle rolls on the outside of another circle, the curve generated is called an *epicycloid*; if it rolls on the inside of another circle, the curve generated is called a *hypocycloid*. In Fig. 10.4, let R be the radius of the fixed circle and r be the radius of the rolling circle. Draw through a a circle arc, AB, concentric with the fixed circle. Lay off on the rolling circle a convenient number of divisions, such as 0–1, 1–2, 2–3, and so forth; then divide the fixed-circle circumference into divisions, of the same length, such as 0–1′, 1′–2′, 2′–3′, and so on. Through these points on the fixed circle, draw radii and extend them to intersect the arc AB, thus producing points a_1, a_2, a_3, and so on. These points will be the centers of the successive positions of the rolling circle. Draw the positions of the rolling circle, using the centers a_1, a_2, a_3, and so forth. Next draw, on the rolling circle with the center "O" of the fixed circle as the pivot point, concentric arcs through points 1, 2, 3, and so forth. The intersection of these arcs with the rolling circles about a_1, a_2, a_3, and so forth, determine points, such as 1″, 2″, 3″, and so forth, on the epicyclic curve. The illustration in Fig. 10.4 shows XY of the tooth outline as part of the epicyclic curve.

The hypocyclic curve construction is the same as that for the epicyclic curve. In the construction of the hypocyclic curve, if the rolling circle has a diameter equal to one-half of the diameter of the fixed circle, the hypocyclic curve thus generated will be a radial line of the fixed circle.

Gear terms

1 The addendum circle is drawn with its center at the center of the gear and bounds the ends of the teeth. See Fig. 10.2.

2 The dedendum circle, or root circle, is drawn with its center at the center of the gear and bounds the bottoms of the teeth. See Fig. 10.2.

3 The pitch circle is a right section of the equivalent cylinder the toothed gear may be considered to replace.

4 Pitch diameter is the diameter of the pitch circle.

5 The addendum is the radial distance from the pitch circle to the outer end of the tooth.

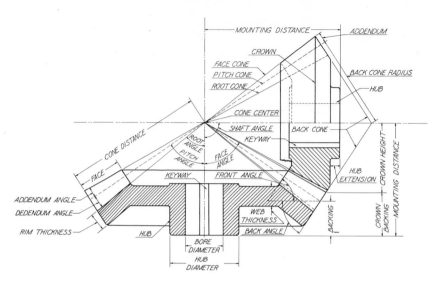

Fig. 10.1 □ **Bevel-gear nomenclature.**

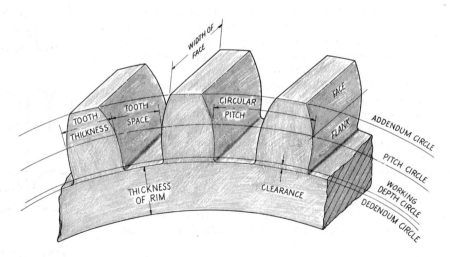

Fig. 10.2 □ **Spur-gear nomenclature.**

6 The dedendum is the radial distance from the pitch circle to the bottom of the tooth.

7 The clearance is the difference between the dedendum of one gear and the addendum of the mating gear.

8 The face of a tooth is that portion of the tooth surface lying outside the pitch circle.

9 The flank of a tooth is that portion of the tooth surface lying inside the pitch circle.

10 The thickness of a tooth is measured on the arc of the pitch circle. It is the length of an arc and not the length of a straight line.

11 The tooth space is the space between the teeth measured on the pitch circle.

12 Backlash is the difference between the tooth thickness of one gear and the tooth space on the mating gear, measured on the pitch circles.

13 The circular pitch of a gear is the distance between a point on one tooth and the corresponding point on the adjacent tooth, measured along the arc of the pitch circle. The circular pitches of two gears in mesh are equal.

14 The diametral pitch is the number of teeth per inch of pitch diameter. It is obtained by dividing the number of teeth by the pitch diameter.

15 The face of a gear is the width of its rim measured parallel to the axis. It should not be confused with the face of a tooth, for the two are entirely different.

16 The pitch point is on the line joining the centers of the two gears where the pitch circles touch.

17 The common tangent is the line tangent to the pitch circles at the pitch point.

18 The pressure angle is the angle between the line of action and the common tangent.

19 The line of action is a line drawn through the pitch point at an angle (equal to the pressure angle) to the common tangent.

20 The base circle is used in involute gearing to generate the involutes that form the tooth outlines. It is drawn from the center of each pair of mating gears tangent to the line of action.

21 When two gears mesh with each other, the larger is called the *gear* and the smaller the *pinion*.

It should be noted that *circular pitch* is a linear dimension expressed in inches, whereas *diametral pitch* is a ratio. There must be a whole number of teeth on the circumference of a gear. Thus it is necessary that the circumference of the pitch circle, divided by the circular pitch, be a whole number.

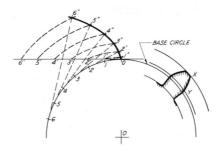

Fig. 10.3 □ Involute tooth.

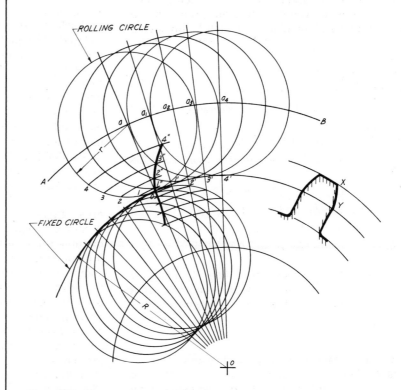

Fig. 10.4 □ Tooth curves—cycloidal form.

For circular pitch, let P' = circular pitch in inches, D = pitch diameter, and T = number of teeth. Then

$$TP' = \pi D, \qquad T = \frac{\pi D}{P'}, \qquad \cdot \; P' = \frac{\pi D}{T},$$

and

$$D = \frac{TP'}{\pi}$$

For diametral pitch, let P = diametral pitch, D = pitch diameter, and T = number of teeth. Then

$$T = PD, \qquad D = \frac{T}{P}, \qquad \text{and} \qquad P = \frac{T}{D}$$

The Brown and Sharpe $14\frac{1}{2}$-degree involute system has been adopted as one of the American standards and is commonly known as the $14\frac{1}{2}$-degree composite system. The tooth proportions of this system are given in terms of the diametral pitch P and circular pitch P'.

Pressure angle = $14\frac{1}{2}°$.

Addendum (inches) = 1/diametral pitch = $1/P$.

Dedendum (inches) = addendum plus clearance = $(1/P) + 0.05P'$.

Clearance = $0.05 \times$ circular pitch = $0.05P'$.

Whole depth of tooth = $2 \times$ addendum + clearance = $2 \times (1/P) + 0.05P'$.

Working depth of tooth = $2 \times$ addendum = $2 \times (1/P)$.

Thickness of tooth = circular pitch/2 = $P'/2$.

Width of tooth space = circular pitch/2 = $P'/2$.

Minimum radius of fillet = clearance = $0.05P'$.

In the above calculations the backlash is zero. Actually, however, it is common practice to provide backlash, and this is accomplished by using standard cutters and cutting the teeth slightly deeper than for standard teeth.

10.2 □ To lay out a pair of standard involute spur gears

The following facts are known regarding the laying out of a pair of standard spur gears: (1) number of teeth on each gear—large gear 24, small gear 16; (2) diametral pitch = 2; (3) pressure angle = $14\frac{1}{2}°$.

To draw a pair of spur gears, determine the pitch diameters as follows:

$$D = \frac{T}{P} = \frac{24}{2} = 12 \text{ in.} \quad \text{(for large gear)}$$

$$D = \frac{T}{P} = \frac{16}{2} = 8 \text{ in.} \quad \text{(for small gear)}$$

In Fig. 10.5, with radii O_1P and O_2P equal to 6 and 4 in., respectively, draw the pitch circles and, through P, draw the common tangent. Draw the line of action XY at an angle of $14\frac{1}{2}°$ to the common tangent. Drop perpendiculars from the centers O_1 and O_2, cutting the line of action at A and B, respectively. O_1A and O_2B are the radii of the base circles that can now be drawn.

From Sec. 10.1, determine the addendum and dedendum of the teeth, and draw in the respective addendum and dedendum circles.

Divide the pitch circle of the smaller gear into 16 equal parts and the pitch circle of the larger gear into 24 equal parts, which will give the circular pitch. Assuming that no allowance is made for backlash, bisect the circular pitch on each of the gears, which will give 32 equal divisions on the small gear and 48 equal divisions on the large gear.

At any point on the base circle of each gear, develop an involute (see Fig. 10.3) and draw in the curves between the base and addendum circles through alternate points on the pitch circles. This produces one side of all the teeth in each gear. The curve for the other side of the tooth is the reverse of the side just drawn. The part of the tooth between the base and dedendum circles is part of a radial line drawn from the base circles to the centers of the gears. The tooth is finished by putting in a small fillet between the working depth and dedendum circles.

10.3 □ Cams

A cam is a plate, cylinder, or any solid having a curved outline or curved groove that, by its oscillating or rotating motion, gives a predetermined motion to another piece, called the follower, in contact with it. The cam plays a very important part in the operation of many classes of machines. Cam mechanisms are commonly used to operate valves in automobiles and stationary and marine internal combustion engines. They also are used in automatic screw machines, clocks, locks, printing machinery, and in nearly all kinds of machinery that we generally regard as ''automatic machines.'' The applications of cams are practically unlimited, and their shapes or outlines are found in wide variety.

All cam mechanisms consist of at least three parts: (1) the cam, which has a contact surface either curved or straight; (2) the follower, whose motion is produced by contact with the cam surface;

and (3) the frame, which supports the cam and guides the follower.

The most common type of cam is the disc or plate cam. Here the cam takes the form of a revolving disc or plate, the circumference of the disc or plate forming the profile with which the follower makes contact. In Figs. 10.6 and 10.7, two simple examples of a disc cam and follower are shown. In Fig. 10.6, the cam is given a motion of rotation, thus causing the follower to rise and then return again to its initial position. In cams of this type it is necessary to use some external force, such as the spring, to keep the follower in contact with the cam at all times. Contact between the follower and the cam is made through a roller, which serves to reduce friction. It is sometimes necessary to use a flat-faced follower, instead of the roller type, an example of which is shown in Fig. 10.7. The follower face that comes in contact with the cam is usually provided with a hardened surface, to prevent excessive wear.

Another type of cam is one in which the follower is constrained to move in a definite path without the application of external forces. See Fig. 10.8. In this type, two contact surfaces of the follower bear on the cam at the same time, thus controlling the motion of the follower in two directions.

10.4 □ Design of a cam

The design of a cam outline is governed by the requirements with respect to the motion of the follower. In the layout of a cam, the initial position, displacement, and character of the motion of the follower are generally known. It is convenient to make first a graphical representation of the follower movement, a procedure that is called *making a displacement diagram*. This is a linear curve in which the length of the diagram represents the time for one revolution of the cam. The height of the diagram represents the total displacement of the follower; the length is made to any convenient length and is divided into equal time intervals, the total representing one rotation of the cam.

In Fig. 10.9 is shown a displacement diagram in which the follower rises 2 in. during 180° of rotation of the cam, then rests for 30° and returns to its initial position for the remainder of the cam revolution. Cam outlines should be designed to avoid sudden changes of motion at the beginning and end of the follower stroke. This can be accomplished by having a uniformly accelerated and decelerated motion at the beginning and end of the constant-velocity curve. The construction for uniformly accelerated motion is shown in Fig. 10.9. On a line, OX, making any convenient angle with OA, mark off any unit of length in this figure equal to Oa. The next point, b, is found by marking off, from O, 4 units of length. Point c is found by marking off 9 units of length. Next, project the intersection (point s) of time-unit 3 and the constant-

Fig. 10.5 □ To draw a pair of spur gears.

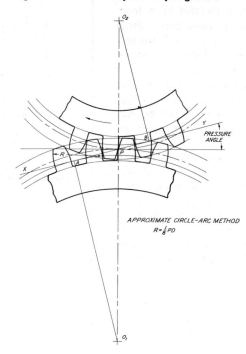

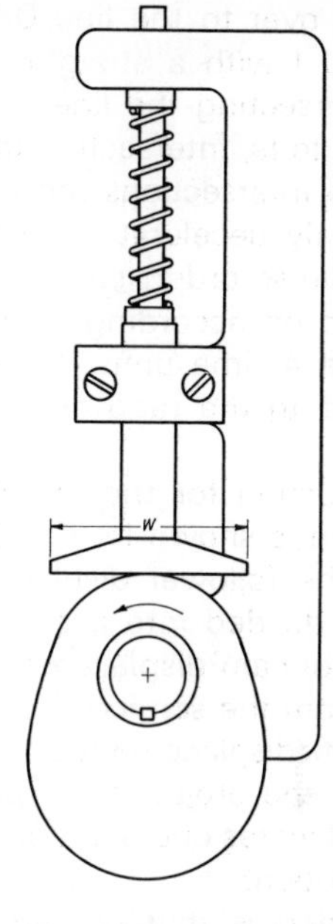

APPROXIMATE CIRCLE-ARC METHOD
$R = \frac{1}{8} PD$

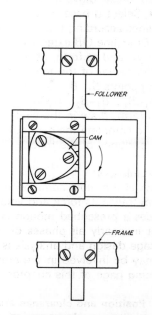

Fig. 10.7 □ Disc cam and follower.

Fig. 10.6 □ Disc cam and follower.

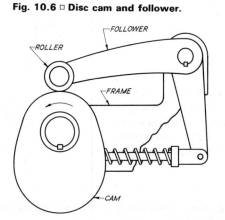

Fig. 10.8 □ Cam and follower.

Geometry of machine elements: gears, cams, and linkages □ **181**

velocity line over to the line OA, thus locating point t. Connect points c and t with a straight line and draw parallel lines from a and b intersecting the line OA. From these intersections draw lines parallel to ts, intersecting the time-unit lines 1 and 2, respectively. These intersections are points on the displacement curve. With uniformly decelerated motion, the series of points are laid off in the reverse order, such as 9–4–1. It will be noted that the units are laid off according to the square of the time unit. Thus, if there were 4 time units, the acceleration curve would be laid off according to the ratio of 1, 4, 9, 16, and the deceleration, 16, 9, 4, 1.

The construction for the displacement diagram for simple harmonic motion is shown in the same figure. A semicircle is drawn as shown, the follower displacement being used as a diameter, and is then divided into a convenient number of parts equal to the number of cam displacement units. Horizontal projection lines are drawn from the semicircle, and the intersections of these lines with the cam displacement lines are points on the displacement curve. Thus, the projection of point 15 on the semicircle to time-unit line 15 locates one point on the displacement curve for simple harmonic motion.

The next step is that of finding the cam profile necessary to produce these movements. The construction is shown in Fig. 10.10. Select a base circle of convenient size, and on it lay off radial lines according to the number of time units of cam displacement. Draw line OB extended to W, and on it lay off the distances y_1, y_2, y_3, and so forth, obtained from the displacement diagram, from the center of the roller shown in the starting position, thus locating points B_1, B_2, B_3, and so forth. With O as a center, draw arcs $B_1–B_1'$, $B_2–B_2'$, $B_3–B_3'$, and so forth, and at B_1', B_2', B_3', and so forth, draw in the circles representing the diameter of the roller. To complete the cam outline, draw a smooth curve tangent to the positions of the roller.

10.5 □ Linkages*
The link is the most common machine element. A link is a rigid bar that transmits force and velocity. An assemblage of links that produces a prescribed motion is called a linkage. Since linkages appear in nearly all phases of mechanical design, a knowledge of linkage design and analysis is necessary for all those designers who may be involved in the creation of useful mechanisms. See the facing page of this chapter.

10.6 □ Position and clearance analysis
In the preliminary design stage, a design layout man is frequently called upon to find the position of each link as the machine operates. When the layout has been completed, he must then check

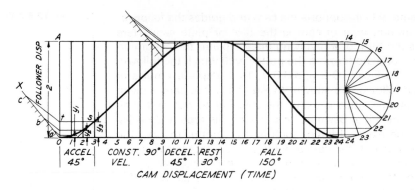

Fig. 10.9 □ A displacement diagram.

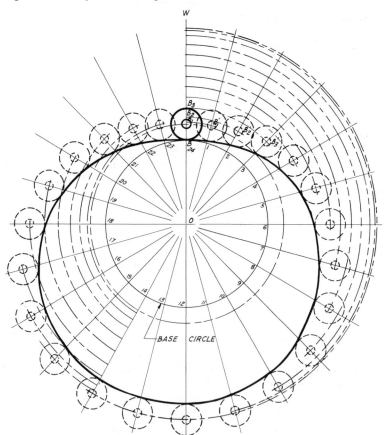

Fig. 10.10 □ Construction for cam profile.

to be sure that the linkage does not interfere with any other moving machine elements. Sufficient clearance must be left between rapidly moving machine parts to ensure that any deflection in the parts will still allow adequate clearance.

Mathematics, primarily geometry and trigonometry, can be used to calculate accurately both the position and clearances. On the other hand, graphical solutions provide sufficient accuracy with a minimum amount of effort and time. A graphical solution also allows the designer to visualize the problem easily and rapidly. Figure 10.11 shows how the layout man can analyze clearances between moving machine parts.

10.7 □ The 4-bar linkage*

The most common type of linkage found in machines is called a 4-bar linkage. This linkage is made up of four pin-connected links. Figure 10.12 shows a typical 4-bar linkage. Notice that link 1 is the distance between the two pin connections attached to the ground. The 4-bar linkage has one degree of freedom. This means that once the position of a link is defined with respect to another link the remaining links are also in a fixed position. For example, if the angle θ is defined in Fig. 10.12 then the angles β and Ψ can have only one value.

The 4-bar linkage is also important to the designer because many other linkages are analyzed in a similar manner. The slider crank and the scotch yoke, shown in Fig. 10.13, are examples of this type of mechanism.

10.8 □ Velocity analysis of a link*

A link does not have to be straight. It may be curved so that it does not interfere with another moving element of a mechanism. For analysis, however, we can replace a curved link by a straight link connecting the two ends. Figure 10.14 shows a curved link represented by a straight link. When the link is in motion, the velocity of points A and B will be different. The velocities of these two ends will be designated V_A and V_B respectively. Because velocity is a vector quantity, we may resolve each vector, V_A and V_B, into two components. Consider the components of V_A and V_B parallel to and perpendicular to the straight line connecting points A and B. In Fig. 10.14 the components parallel to the straight line connecting A and B are designated $V_A{}^t$ and $V_B{}^t$. The perpendicular components are designated $V_A{}^n$ and $V_B{}^n$.

Because the link is solid, $V_A{}^t$ and $V_B{}^t$ must be equal. If $V_A{}^t$ and $V_B{}^t$ were not equal, then the line joining A and B would have

*Prepared by Professor Wesley L. Baldwin, Mechanical Engineering Technology, Purdue University.

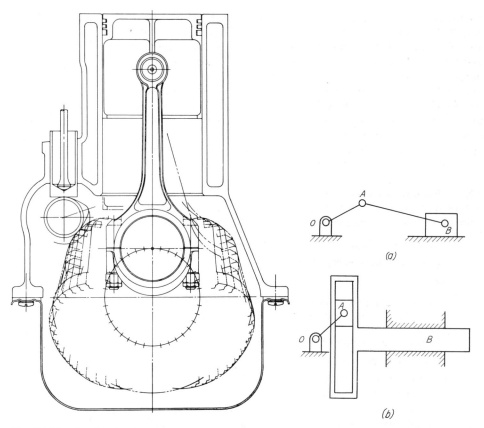

Fig. 10.11 □ Graphical solution for the clearances between moving machine elements. (*Courtesy General Motors Corporation.*)

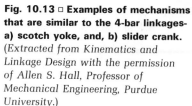

Fig. 10.13 □ Examples of mechanisms that are similar to the 4-bar linkages- a) scotch yoke, and, b) slider crank. (*Extracted from Kinematics and Linkage Design with the permission of Allen S. Hall, Professor of Mechanical Engineering, Purdue University.*)

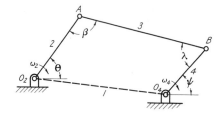

Fig. 10.12 □ Typical 4-bar linkage.

a changing length. The difference in velocity between points A and B must then be due entirely to the difference between $V_A{}^n$ and $V_B{}^n$. This difference between $V_A{}^n$ and $V_B{}^n$ is called the relative velocity between points A and B. The relative velocity is designated $V_{B/A}$, when $V_A{}^n$ is subtracted from $V_B{}^n$.

Since $V_A{}^t$ and $V_B{}^t$ are equal, $V_{B/A}$ is also equal to the difference between V_A and V_B. The difference between V_A and V_B, called the relative velocity $V_{B/A}$, lies in a direction perpendicular to the straight line joining A and B. The magnitude of $V_{B/A}$ is the same as the difference between $V_A{}^n$ and $V_B{}^n$. Figure 10.14 shows these relationships.

10.9 □ Velocity analysis of a 4-bar linkage*

A typical 4-bar linkage has link 2 driven at a constant angular velocity ω_2. To calculate the angular velocity, ω_4, of link 4:

1 Draw the linkage to scale in the desired position.

2 Calculate the velocity of point A.

$$V_A = \omega_2 \, (O_2A)$$

V_A is the velocity of point A and is perpendicular to (O_2A),
ω_2 is the angular velocity of link 2 measured in radians/second, and
(O_2A) is the length of link 2.

3 Construct vector V_A to scale.

4 From the tip of V_A construct a line perpendicular to link 3. This line lies in the direction of $V_{B/A}$.

5 From the tail of vector V_A construct a line perpendicular to (O_4B) to the line constructed in step 3. This line lies in the direction of V_B.

6 The line drawn in step 5 is the scaled velocity of point B.

7 Calculate ω_4.

$$\omega_4 = V_B/(O_4B)$$

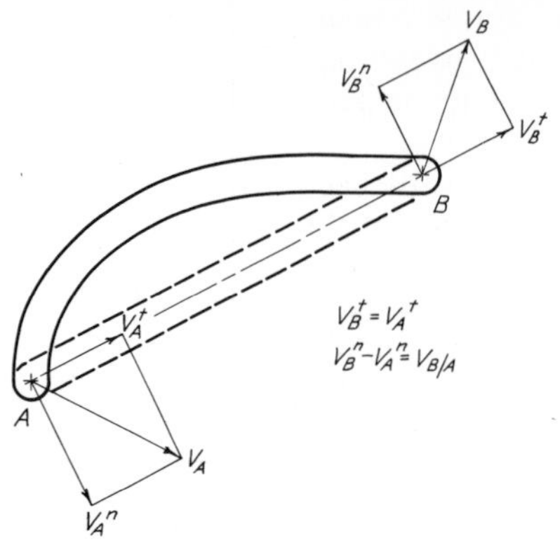

Fig. 10.14 □ A link in motion.

*Prepared by Professor Wesley L. Baldwin, Mechanical Engineering Technology, Purdue University.

Problems

Design geometry

The following exercises not only require the student to study and use certain common geometric constructions but also furnish additional practice in applying good line technique to the drawing of instrumental figures and practical designs. All work should be very accurately done. Tangent points should be indicated by a light, short dash across the line. Commonly used geometrical constructions may be found in Part C in the Appendix.

1 Construct an ellipse having a major diameter of $4\frac{1}{4}$ in. and a minor diameter of $2\frac{3}{4}$ in. Use the trammel method illustrated in Fig. C.27(Appendix).

2 Construct an ellipse having a major diameter of 4 in. and a minor diameter of $2\frac{3}{4}$ in. Use the concentric circle method illustrated in Fig. C.28. Find a sufficient number of points to obtain a smooth curve.

3 Construct a parabola with vertical axis. Make the focus $\frac{3}{4}$ in. from the directrix. Select a point on the curve and draw a line tangent to the parabola. Study Sec. C.35 and Fig. C.33.

4 Construct a hyperbola having a transverse axis of 1 in. and foci $1\frac{5}{8}$ in. apart. Study Sec. C.42 and Fig. C.39.

5 Construct the involute of an equilateral triangle with 1-in. sides. Study Secs. C.43 and C.45.

6 Construct the cycloid generated by a $1\frac{1}{2}$-in. circle. Study Sec. C.47 and Fig. C.41.

7 Construct the epicycloid generated by a $1\frac{1}{2}$-in. circle rolling on a 5-in. circle. Study Sec. C.48 and Fig. C.42.

8 Construct the hypocycloid generated by a $1\frac{1}{2}$-in. circle rolling on a $4\frac{1}{2}$-in. circle. Study Sec. C.49 and Fig. C.43.

9 Reconstruct the view of the wrench and hexagonal nut shown in Fig. 10.15. Mark all tangent points with short lines.

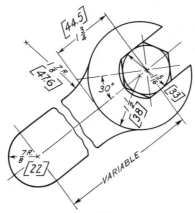

Dimensons in [ ] are in millimeters

Fig. 10.15 □ **Wrench.**

10 Construct the shape of the slotted guide shown in Fig. 10.16. Show all construction for locating centers, and mark points of tangency.

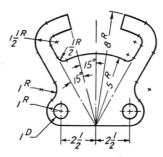

Fig. 10.16 □ **Slotted guide.**

11 Construct the adjustable Y-clamp shown in Fig. 10.17. Show all construction for locating centers and mark points of tangency.

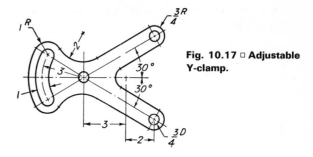

Fig. 10.17 □ **Adjustable Y-clamp.**

12 Reconstruct the end view of the dolly block shown in Fig. 10.18.

Fig. 10.18 □ **End view-dolly block.**

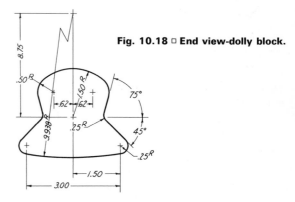

13 Reconstruct the view of the electrode shown in Fig. 10.19.

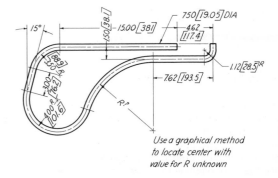

Use a graphical method to locate center with value for R unknown

Fig. 10.19 □ **Electrode.**
(Numerical values shown in [ ] are in millimeters.)

14 Reconstruct the view of the spline plate shown in Fig. 10.20.

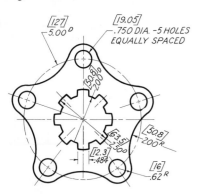

Fig. 10.20 □ Spline plate.
(Numerical values shown in [ ] are in millimeters.)

15 Reconstruct the view of the adjustment plate shown in Fig. 10.21.

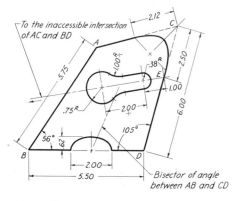

Fig. 10.21 □ Adjustment plate.

16 (Fig. 10.22). Part A is free to pivot about a shaft. If this part should be revolved in a counterclockwise direction as indicated by the arrows, it would contact surface C. Reproduce the drawing as given and show part A revolved until it is in contact with surface C. Use the symbolic line for showing an alternate position for part A in this new position. Show all geometric constructions clearly and do not erase construction lines.

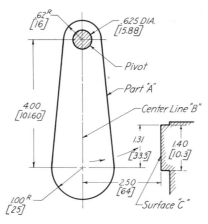

Dimensions in [ ] are in millimeters

Fig. 10.22 □ Geometric construction.

17 (Fig. 10.23). Part A revolves about shaft B in a clockwise direction from the position shown until surface C comes into contact with the cylindrical surface of the roller. Reproduce the drawing as given and show part A in its revolved position using the symbolic alternate position line for this new position. Show all geometric constructions clearly and do not erase construction lines.

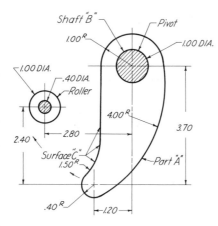

Fig. 10.23 □ Geometric construction.

18 (Fig. 10.24). The design of a counterweight system is such that point A (pivot point) of the counterweight and point C (end view of the axis of the shaft and roller) are in line horizontally. A point B on the counterweight is 4.12 from A on a line making an angle of 41° with AC through A. Points A and B are the centers of 1.60R and 1.00R arcs respectively.

Draw an arc of 3.20R tangent to the 1.00 and 1.60 arcs to form the lower portion of the outline of the counterweight. Complete the upper part of the outline by drawing a reverse curve tangent to the top of the 1.00 and 1.60 arcs. The radius of the curve tangent to the 1.60 arc is to be 43 percent of the total length of the two chords of the reverse curve. Using geometry, determine the location of the point of tangency of the counterweight and roller when the counterweight swings clockwise into contact with the outside surface of the stop-roller.

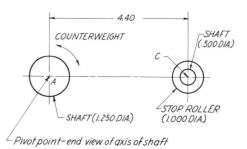

Fig. 10.24 □ Design of a counterweight system.

19 It is desired to know the angular displacement of the center line of the cam (Fig. 10.25) when the follower moves to a position .74 below the position shown. Show the cam and follower in their new positions in phantom outline. Use only an approved geometrical method to find the location of the centers of the arcs. A trial and error method is not acceptable. Show all construction and mark all points of tangency.

Determine the angle through which the center line of the cam has moved. Dimension the angle between the center lines.

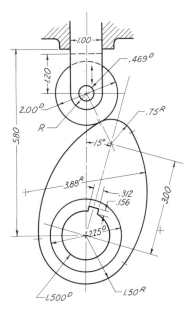

Fig. 10.25 □ Cam and follower.

	Gear					**Pinion**				
Prob. no.	Circular pitch (inches)	Diametral pitch	Pitch dia.	No. of teeth		Circular pitch (inches)	Diametral pitch	Pitch dia.	No. of teeth	Pressure angle
20	1.31			24		1.31			16	14½°
21		2		20			2		14	14½°
22		2.5	10				2.5	8		14½°
23	1.0			30		1.0			20	14½°
24	2.0			18		2.0			12	14½°
25		3	8				3	6		14½°

Gears and cams

Gear problems

20–25 Following the method shown in Sec. 10.2, lay out a pair of standard involute spur gears as assigned from the given table. The pinion is the driver.

Cam problems

Cam Data

Diameter of cam shaft . . . . $1\frac{1}{4}$ in.
Diameter of cam hub . . . . $2\frac{1}{4}$ in.
Diameter of roller . . . . 1 in.
Keyway . . . . $\frac{1}{4} \times \frac{1}{8}$ in.
Diameter of base circle . . . . $2\frac{3}{4}$ in.
Follower displacement . . . . 2 in.
Scale: full-size
Cam rotation: as noted

Determine points on the cam profiles at intervals of 15°.

26 Using the above data, design a plate cam to satisfy the following conditions: (a) a rise of 2 in. in 180°, with constant velocity, except for uniform acceleration for the first 30° and uniform deceleration for the last 45°; (b) rest 30°; and (c) return with simple harmonic motion. Use clockwise cam rotation.

27 Same as problem 26, except that the follower is of the flat-face type and is $2\frac{1}{2}$ in. wide.

28 Using the data for Problem 26, design a plate cam to satisfy the following conditions: (a) rise of 2 in. during 180°, the first 45° of which is uniformly accelerated motion, the next 60° being constant velocity, and the last 75° of rise being uniformly decelerated motion; (b) rest 15°; and (c) return to starting position with simple harmonic motion. Use counterclockwise cam rotation.

29 Same as Problem 28, except that the follower is of the flat-face type and is $2\frac{1}{2}$ in. wide.

30 Using the data for Problem 26, design a plate cam to satisfy the following conditions: (a) rise to 2 in. during 150°, by simple harmonic motion; (b) rest 30°; and (c) return to starting position during remainder of the revolution, with uniformly accelerated and decelerated motion, the value of the deceleration being twice that of the acceleration. Use clockwise cam rotation.

31 Using the data for Problem 26, except that the follower is to be of the flat-face type, $2\frac{1}{2}$ in. wide, design a plate cam to satisfy the following conditions: (a) rise of 2 in. with simple harmonic motion, in 120°; (b) rest 30°; (c) return in 150°, with constant velocity, except for uniform acceleration for the first 45° and uniform deceleration for the last 30° of fall; and (d) rest the balance of the revolution. Use counterclockwise cam rotation.

Linkage problems*

32 A 4-bar linkage has link 1 = 6 inches, link 2 = 2 inches, link 3 = 8 inches, link 4 = 5 inches. Draw the linkage in all positions for $\theta = 0°$ to 360° in 30° increments. Plot a graph of θ vs Ψ.

33 If link 2 in the problem above has a constant angular velocity of 10 rad/sec, find the angular velocity of link 4. Use the 12 positions indicated in problem 32. Plot θ vs ω_4.

*Prepared by Professor Wesley L. Baldwin, Mechanical Engineering Technology, Purdue University.

300 MPH Tracked-Air-Cushion-Vehicle (TACV).

The General Electric Company's Transportation Systems Division has been awarded a contract by the Department of Transportation's Office of High Speed Ground Transportation to make a preliminary design study for the TACV shown. This high-speed vehicle will "fly" at about 3/4-inch above and between a fixed guideway. The TACV is expected to receive community acceptability for intercity passenger service because of excellent ride quality and quietness of operation. Artist's and engineering renderings of this type are frequently needed to sell a design concept to public officials. Designers should be fully aware of the power of the picture. (*Courtesy General Electric Company.*)

Chapter 11

Pictorial presentation

11.1 □ Introduction

An orthographic drawing of two or more views describes an object accurately in form and size, but, since each of the views shows only two dimensions without any suggestion of depth, such a drawing can convey information only to those who are familiar with graphic representation. For this reason, multiview drawings are used mainly by engineers, draftsmen, contractors, and shopmen.

Frequently, however, engineers and draftsmen find they must use conventional picture drawings to convey specific information to persons who do not possess the trained imagination necessary to construct mentally an object from views. To make such drawings, several special schemes of one-plane pictorial drawing have been devised that combine the pictorial effect of perspective with the advantage of having the principal dimensions to scale. But pictorial drawings, in spite of certain advantages, have disadvantages that limit their use. A few of these are as follows:

1 Some drawings frequently have a distorted, unreal appearance that is disagreeable.

2 The time required for execution is, in many cases, greater than for an orthographic drawing.

3 They are difficult to dimension.

4 Some of the lines cannot be measured.

Even with these limitations, pictorial drawings are used extensively for technical publications, Patent Office records, piping diagrams, and furniture designs. Occasionally they are used, in one form or another, to supplement and clarify machine and structural details that would be difficult to visualize (Fig. 11.1).

11.2 □ Divisions of pictorial drawing

Single-plane pictorial drawings are classified in three general divisions: (1) axonometric projection, (2) oblique projection, and (3) perspective projection (Fig. 11.2).

Perspective methods produce the most realistic drawings, but the necessary construction is more difficult and tedious than the construction required for the conventional methods classified under the other two divisions. For this reason, engineers customarily use some form of either axonometric or oblique projection. Modified methods, which are not theoretically correct, are often used to produce desired effects.

A □ Axonometric projection

11.3 □ Divisions of axonometric projection

Theoretically, axonometric projection is a form of orthographic projection. The distinguishing difference is that only one plane is used instead of two or more, and the object is turned from its customary position so that three faces are displayed (Fig. 11.3). Since an object may be placed in a countless number of positions relative to the picture plane, an infinite number of views may be drawn, which will vary in general proportions, lengths of edges, and sizes of angles. For practical reasons, a few of these possible positions have been classified in such a manner as to give the recognized divisions of axonometric projection: (1) isometric, (2) dimetric, and (3) trimetric.

Isometric projection is the simplest of these, because the principal axes make equal angles with the plane of projection and the edges are therefore foreshortened equally.

11.4 □ Isometric projection

If the cube in Fig. 11.3 is revolved through an angle of 45° about an imaginary vertical axis, as shown in II, and then tilted forward until its body diagonal is perpendicular to the vertical plane, the edges will be foreshortened equally and the cube will be in the correct position to produce an isometric projection.

The three front edges, called isometric axes, make angles of approximately 35° 16′ with the vertical plane of projection, or picture plane. In this form of pictorial, the angles between the projections of these axes are 120°, and the projected lengths of the edges of an object, along and parallel to these axes, are approximately 81% of their true lengths. It should be observed that

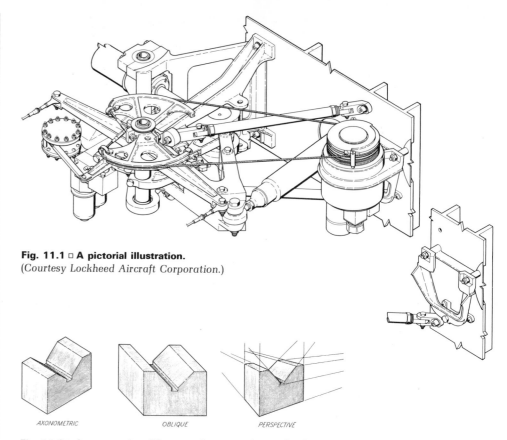

Fig. 11.1 □ A pictorial illustration.
(Courtesy Lockheed Aircraft Corporation.)

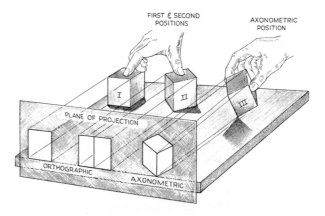

AXONOMETRIC   OBLIQUE   PERSPECTIVE

Fig. 11.2 □ Axonometric, oblique, and perspective projection.

FIRST & SECOND POSITIONS   AXONOMETRIC POSITION

PLANE OF PROJECTION

ORTHOGRAPHIC   AXONOMETRIC

Fig. 11.3 □ Theory of axonometric projection.

the 90° angles of the cube appear in the isometric projection as either 120° or 60°.

Now, if instead of turning and tilting the object in relation to a principal plane of projection, an auxiliary plane is used that will be perpendicular to the body diagonal, the view projected on the plane will be an axonometric projection. Since the auxiliary plane will be inclined to the principal planes on which the front, top, and side views would be projected, the auxiliary view, taken in a position perpendicular to the body diagonal, will be a secondary auxiliary view, as shown in Fig. 11.4.

11.5 □ Isometric drawing

Objects seldom are drawn in true isometric projection, the use of an isometric scale being inconvenient and impractical. Instead, a conventional method is used in which all foreshortening is ignored, and actual true lengths are laid off along isometric axes and isometric lines. To avoid confusion and to set this method apart from true isometric projection, it is called isometric drawing.

The isometric drawing of a figure is slightly larger (approximately 22½%) than the isometric projection, but, since the proportions are the same, the increased size does not affect the pictorial value of the representation (see Fig. 11.4). The use of a regular scale makes it possible for a draftsman to produce a satisfactory drawing with a minimum expenditure of time and effort.

In isometric drawing, lines that are parallel to the isometric axes are called *isometric lines*.

11.6 □ To make an isometric drawing of a rectangular object

The procedure followed in making an isometric drawing of a rectangular block is illustrated in Fig. 11.5. The three axes that establish the front edges, as shown in (b), should be drawn through point A so that one extends vertically downward and the other two upward to the right and left at an angle of 30° from the horizontal. Then the actual lengths of the edges may be set off, as shown in (c) and (d), and the remainder of the view completed by drawing lines parallel to the axes through the corners thus located, as in (e) and (f).

Hidden lines, unless absolutely necessary for clearness, always should be omitted on a pictorial representation.

11.7 □ Nonisometric lines

Those lines that are inclined and are not parallel to the isometric axes are called *nonisometric lines*. Since a line of this type does not appear in its true length and cannot be measured directly, its position and projected length must be established by locating its extremities. In Fig. 11.6, AB and CD, which represent the edges of the block, are nonisometric lines. The location of AB is estab-

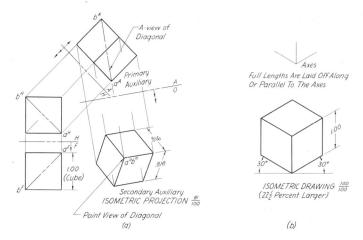

Fig. 11.4 □ Comparison of isometric projection and isometric drawing.

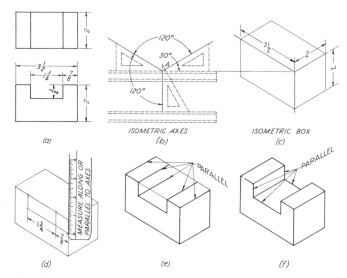

Fig. 11.5 □ Procedure for constructing an isometric drawing.

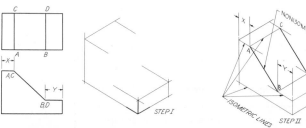

Fig. 11.6 □ Nonisometric lines.

lished in the pictorial view by locating points A and B. Point A is on the top edge, X distance from the left-side surface. Point B is on the upper edge of the base, Y distance from the right-side surface. All other lines coincide with or are parallel to the axes and, therefore, may be measured off with the scale.

The pictorial representation of an irregular solid containing a number of nonisometric lines may be conveniently constructed by the box method; that is, the object may be enclosed in a rectangular box so that both isometric and nonisometric lines may be located by points of contact with its surfaces and edges (see Fig. 11.7).

A study of Figs. 11.6 and 11.7 reveals the important fact that lines that are parallel on an object are parallel in the pictorial view, and, conversely, lines that are not parallel on the object are not parallel on the view. It is often possible to eliminate much tedious construction work by the practical application of the principle of parallel lines.

11.8 □ Coordinate construction method

When an object contains a number of inclined surfaces, such as the one shown in Fig. 11.8, the use of the coordinate construction method is desirable. In this method, the end points of the edges are located in relation to an assumed isometric base line located on an isometric reference plane. For example, the line RL is used as a base line from which measurements are made along isometric lines, as shown. The distances required to locate point A are taken directly from the orthographic views.

Irregular curved edges are most easily drawn in isometric by the offset method, which is a modification of the coordinate construction method (Fig. 11.9). The position of the curve can be readily established by plotted points located by measuring along isometric lines.

11.9 □ Angles in isometric drawing

Since angles specified in degrees do not appear in true size on an isometric drawing, angular measurements must be converted in some manner to linear measurements that can be laid off along isometric lines. Usually, one or two measurements taken from an orthographic view may be laid off along isometric lines on the pictorial drawing to locate an inclined edge that has been specified by an angular dimension. The scale used for the orthographic view must be the same as the one being used in preparing the pictorial drawing.

In Fig. 11.10(a), the position of the inclined line AB was established on the isometric drawing by using the distance X taken from the front view of the orthographic drawing. When an orthographic drawing has already been prepared to a different scale than the

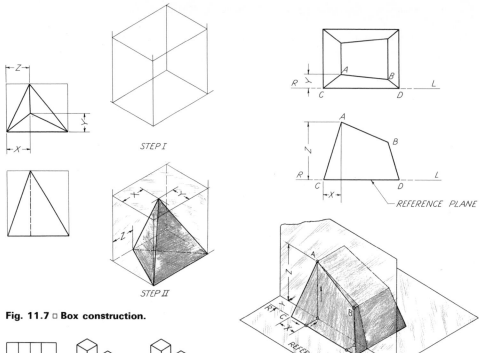

Fig. 11.7 □ Box construction.

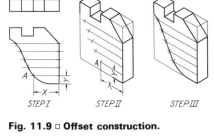

Fig. 11.9 □ Offset construction.

Fig. 11.8 □ Coordinate construction.

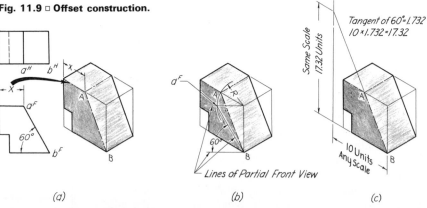

Fig. 11.10 □ Angles in isometric.

192

scale being used for the pictorial representation, one can draw a partial orthographic view and take off the needed dimensions. A practical application of this idea is shown in (*b*). By making the construction of a partial view at the place where the angle is to appear on the isometric drawing, the position of the required line can be obtained graphically.

If desired, the tangent method, as explained in Sec. C.7(Appendix), may be used, as shown in (*c*). In using this method, a length equal to 10 units (any scale) is laid off along an isometric line that is to form one side of the angle. Then, a distance equal to 10 times the tangent of the angle is set off along a second isometric line that represents the second leg of the right triangle in pictorial. A line drawn through the end points of these lines will be the required line at the specified angle.

11.10 □ Circle and circle arcs in isometric drawing

In isometric drawing, a circle appears as an ellipse. The tedious construction required for plotting an ellipse accurately (Fig. 11.11) often is avoided by using some approximate method of drawing. The representation thus obtained is accurate enough for most work, although the true ellipse, which is slightly narrower and longer, is more pleasing in shape. For an approximate construction, a four-center method is generally used.

To draw an ellipse representing a pictorial circle, a square is conceived to be circumscribed about the circle in the orthographic projection. When transferred to the isometric plane in the pictorial view, the square becomes a rhombus (isometric square) and the circle an ellipse tangent to the rhombus at the midpoints of its sides. If the ellipse is to be drawn by the four-center method (Fig. 11.12), the points of intersection of the perpendicular bisectors of the sides of the rhombus will be centers for the four arcs forming the approximate ellipse. The two intersections that lie on the corners of the rhombus are centers for the two large arcs, while the remaining intersections are centers for the two small arcs. Furthermore, the length along the perpendicular from the center of each arc to the point at which the arc is tangent to the rhombus (midpoint) will be the radius. All construction lines required by this method may be made with a T-square and a 30° × 60° triangle.

The amount of work may be still further shortened and the accuracy of the construction improved by following the procedure shown in Fig. 11.13. The steps in this method are as follows:

Step I: Draw the isometric center lines of the required circle.

Step II: Using a radius equal to the radius of the circle, strike arcs across the isometric center lines.

Step III–IV: Through each of these points of intersection erect a perpendicular to the other isometric center line.

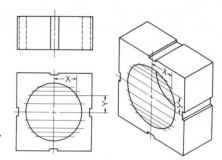

Fig. 11.11 □ To plot an isometric circle.

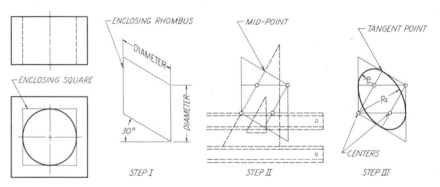

Fig. 11.12 □ Four-center approximation.

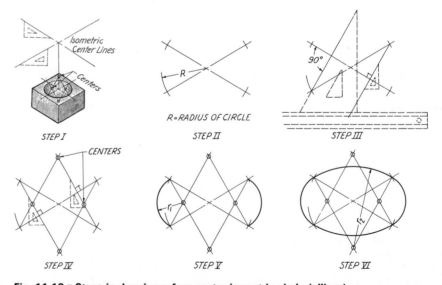

Fig. 11.13 □ Steps in drawing a four-center isometric circle (ellipse).

Steps V–VI: Using the intersection points of the perpendiculars as centers and lengths along the perpendiculars as radii, draw the four arcs that form the ellipse (Fig. 11.14).

A circle arc will appear in pictorial representation as a segment of an ellipse. Therefore, it may be drawn by using as much of the four-center method as is required to locate the needed centers (Fig. 11.15). For example, to draw a quarter circle, it is only necessary to lay off the true radius of the arc along isometric lines drawn through the center and to draw intersecting perpendiculars through these points.

To draw isometric concentric circles by the four-center method, a set of centers must be located for each circle (Fig. 11.16).

When several circles of the same diameter occur in parallel planes, the construction may be simplified. Figure 11.16 shows two views of an object and its corresponding isometric drawing. In Fig. 11.16, the centers for the ellipse representing the upper base of the large cylinder are found in the usual way, while the centers for the lower base are located by moving the centers for the upper base downward a distance equal to the height of the cylinder. By observing that portion of the object projecting to the right, it can be noted that corresponding centers lie along an isometric line parallel to the axis of the cylinder.

Circles and circle arcs in nonisometric planes may be plotted by using the offset or coordinate method. Sufficient points for establishing a curve must be located by transferring measurements from the orthographic views to isometric lines in the pictorial view. There is a rapid and easy way for drawing the cylindrical portion of the object shown in Fig. 11.17. The semi-circular arc must be plotted on the rear surface as the first step. Then, after this has been done, each point is brought forward to the inclined face. The offset distances (D_1, D_2, D_3, etc.) at each level are taken from the side view in (a).

The pictorial representation of a sphere is the envelope of all of the great circles that could be drawn on the surface. In isometric drawing, the great circles appear as ellipses and a circle is their envelope. In practice it is necessary, to draw only one ellipse, using the true radius of the sphere and the four-center method of construction. The diameter of the circle is the long diameter of the ellipse (Fig. 11.18).

11.11 □ Positions of isometric axes

It is sometimes desirable to place the principal isometric axes so that an object will be in position to reveal certain faces to a better advantage (Fig. 11.19).

The difference in direction should cause no confusion, since the angle between the axes and the procedure followed in constructing

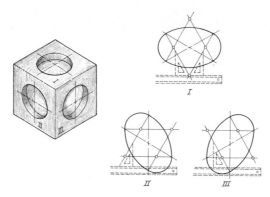

Fig. 11.14 □ Isometric circles.

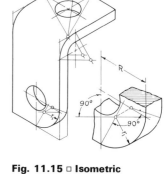

Fig. 11.15 □ Isometric circle arcs.

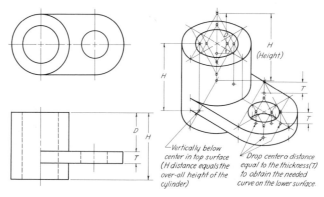

−Vertically below center in top surface (H distance equals the over-all height of the cylinder)

−Drop center a distance equal to the thickness(T) to obtain the needed curve on the lower surface.

Fig. 11.16 □ Isometric parallel circles.

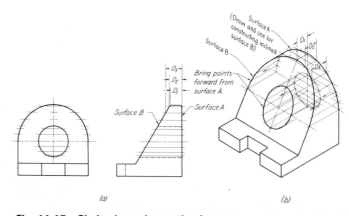

(a) (b)

Fig. 11.17 □ Circles in nonisometric planes.

the view are the same for any position. The choice of the direction may depend on the construction of the object, but usually this is determined by the position from which the object is ordinarily viewed.

Reversed axes (b) are used in architectural work to show a feature as it would be seen from a natural position below.

Sometimes long objects are drawn with the long axis horizontal, as shown in (c) and (d).

11.12 □ Isometric sectional views (Fig. 11.20)

Generally, an isometric sectional view is used for showing the inner construction of an object when there is a complicated interior to be explained or when it is desirable to emphasize features that would not appear in a usual outside view. Sectioning in isometric drawing is based on the same principles as sectioning in orthographic drawing. Isometric planes are used for cutting an object, and the general procedure followed in constructing the representation is the same as for an exterior view.

Figure 11.20 shows an isometric half section. It is easier, in this case, to outline the outside view of the object in full and then remove a front quarter with isometric planes.

Figure 11.21 illustrates a full section in isometric. The accepted procedure for constructing this form of sectional view is to draw the cut face and then add the portion that lies behind.

Section lines should be sloped at an angle that produces the best effect, but they should never be drawn parallel to object lines. Fig. 11.21 illustrates the slope that is correct for most drawings.

11.13 □ Dimetric projection

The view of an object that has been so placed that two of its axes make equal angles with the plane of projection is called a *dimetric projection*. The third axis may make either a smaller or larger angle. All of the edges along or parallel to the first two axes are foreshortened equally, while those parallel to the third axis are foreshortened a different amount. It might be said that dimetric projection, a division of axonometric projection, is like isometric projection in that the object must be placed to satisfy specific conditions. Similarly, a dimetric projection may be drawn by using the auxiliary view method. The secondary auxiliary view is the dimetric projection. The procedure is the same as for an isometric projection (Fig. 11.4), except that the line of sight is taken in the direction necessary to obtain the desired dimetric projection. Obviously, an infinite number of dimetric projections is possible.

In practical application, dimetric projection is sometimes modified so that regular scales can be used to lay off measurements to assumed ratios. This is called *dimetric drawing*. [Fig. 11.22(a)].

Fig. 11.18 □ Isometric drawing of a sphere.

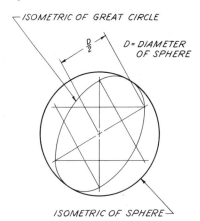

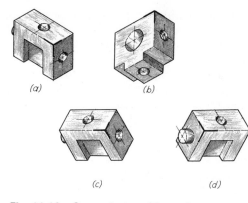

Fig. 11.19 □ Convenient positions of axes.

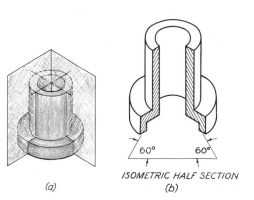

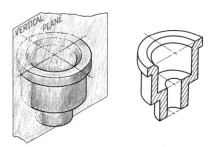

Fig. 11.20 □ Isometric half section.

Fig. 11.21 □ Isometric full section.

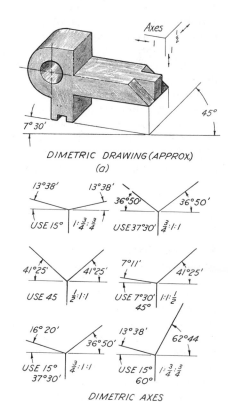

Fig. 11.22 □ Approximate dimetric drawing.

The angles and scales may be worked out* for any ratios, such as $1:1:\frac{1}{2}$ (full size : full size : half size); $1:1:\frac{3}{4}$ (full size : full size : three-fourths size). For example, the angles for the ratios $1:1:\frac{1}{2}$ are 7° 11' and 41° 25'. After the scales have been assumed and the angles computed, an enclosing box may be drawn in conformity to the angles and the view completed by following the general procedure used in isometric drawing, except that two scales must be used. The positions commonly used, along with the scale ratios and corresponding angles, are shown in Fig. 11.22(b). The first scale given in each ratio is for the vertical axis. Since two of the axes are foreshortened equally, while the third is foreshortened in different ratio, obviously two scales must be used. This is an effective method of representation.

11.14 □ Trimetric projection**
A trimetric projection of an object is the view obtained when each of the three axes makes a different angle with the plane of projection. As might be expected, a trimetric projection may be constructed by drawing successive auxiliary views. However, since there are an unlimited number of possible lines of sight that will produce unequal foreshortening in the directions of the three axes, considerable thought must be given to the selection of a position that will show the object pictorially to the best advantage in the second auxiliary view. Making this decision is not easy. This form of pictorial representation has been used to some extent by certain aircraft companies for the preparation of production illustrations.

B □ Oblique projection

11.15 □ Oblique projection
In oblique projection, the view is produced by using parallel projectors that make some angle other than 90° with the plane of projection. Generally, one face is placed parallel to the picture plane and the projection lines are taken at 45°. This gives a view that is pictorial in appearance, as it shows the front and one or more additional faces of an object. In Fig. 11.23, the orthographic and oblique projections of a cube are shown. When the angle is 45°, as in this illustration, the representation is sometimes called *cavalier projection*. It is generally known, however, as an *oblique projection* or an *oblique drawing*.

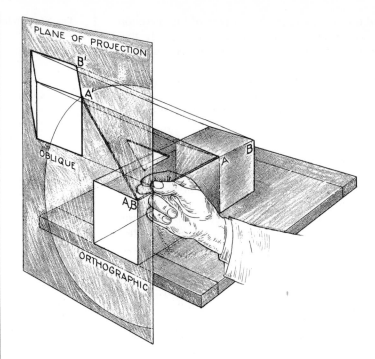

Fig. 11.23 □ Theory of oblique projection.

*Formula: $\cos \propto = -\sqrt{2s_1{}^2s_2{}^2 - s_2{}^4/2s_1s_2}$. In this formula, $\propto$ is one of the equal angles, s_1 is one of the equal scales, and s_2 is the third scale.
**Additional information on trimetric projection may be found in the author's text *Fundamentals of Engineering Drawing for Design, Communication, and Numerical Control*.

11.16 □ Principle of oblique projection

The theory of oblique projection can be explained by imagining a vertical plane of projection in front of a cube parallel to one of its faces (Fig. 11.23). When the projectors make an angle of 45° in any direction with the picture plane, the length of any oblique projection $A'B'$ of the edge AB is equal to the true length of AB. Note that the projectors could be parallel to any element of a 45° cone having its base in the plane of projection. With projectors at this particular angle (45°), the face parallel to the plane is projected in its true size and shape and the edges perpendicular to the picture plane are projected in their true length. If the projectors make a greater angle, the oblique projection will be shorter, while if the angle is less, the projection will be longer.

11.17 □ Oblique drawing

This form of drawing is based on three mutually perpendicular axes along which, or parallel to which, the necessary measurements are made for constructing the representation. Oblique drawing differs from isometric drawing principally in that two axes are always perpendicular to each other, while the third (receding axis) is at some convenient angle, such as 30°, 45°, or 60° with the horizontal (Fig. 11.25). It is somewhat more flexible and has the following advantages over isometric drawing: (1) circular or irregular outlines on the front face show in their true shape; (2) distortion can be reduced by foreshortening along the receding axis; and (3) a greater choice is permitted in the selection of the positions of the axes. A few of the various views that can be obtained by varying the inclination of the receding axis are illustrated in Fig. 11.24. Usually, the selection of the position is governed by the character of the object.

11.18 □ To make an oblique drawing

The procedure to be followed in constructing an oblique drawing of an adjustable guide is illustrated in Fig. 11.25. The three axes that establish the perpendicular edges in (b) are drawn through point O representing the front corner. OA and OB are perpendicular to each other and OC is at any desired angle (say, 30°) with the horizontal. After the width, height, and depth have been set off, the front face may be laid out in its true size and shape, as in (c), and the view can be completed by drawing lines parallel to the receding axis through the established corners. The circle and semicircle are shown parallel to the picture plane in order to avoid distortion and because, from the draftsman's standpoint, it is easier to draw a circle than to construct an ellipse.

In general, the procedure for constructing an oblique drawing is the same as for an isometric drawing.

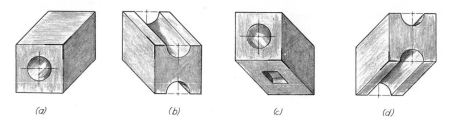

Fig. 11.24 □ Various positions of the receding axis.

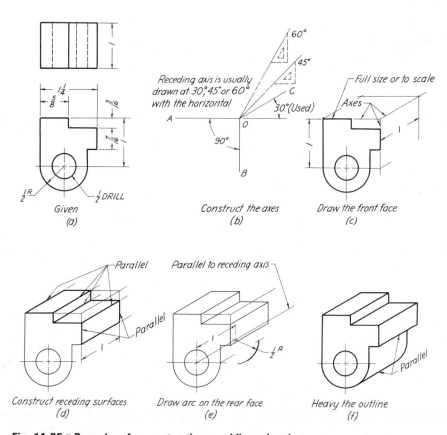

Given
(a)

Construct the axes
(b)

Draw the front face
(c)

Construct receding surfaces
(d)

Draw arc on the rear face
(e)

Heavy the outline
(f)

Fig. 11.25 □ Procedure for constructing an oblique drawing.

11.19 □ Rules for placing an object

Generally, the most irregular face, or the one containing the most circular outlines, should be placed parallel to the picture plane, in order to minimize distortion and simplify construction. By following this practice, all or most of the circles and circle arcs can be drawn with a compass, and the tedious construction that would be required to draw their elliptical representations in a receding plane is eliminated. In selecting the position of an object, two rules should be followed. The first is to place the face having the most irregular contour, or the most circular outlines, parallel to the picture plane. Note in Fig. 11.26 the advantage of following this rule.

When the longest face of an object is used as the front face, the pictorial view will be distorted to a lesser degree and, therefore, will have a more realistic and pleasing appearance. Hence, the second rule is to place the longest face parallel to the picture plane. Compare the views shown in Fig. 11.27 and note the greater distortion in (a) over (b).

If these two rules clash, the first should govern. It is more desirable to have the irregular face show its true shape than it is to lessen the distortion in the direction of the receding axis.

11.20 □ Angles, circles, and circle arcs in oblique

As previously stated, angles, circles, and irregular outlines on surfaces parallel to the plane of projection show in true size and shape. When located on receding faces, the construction methods used in isometric drawing may usually be applied. Figure 11.28 shows the method of drawing the elliptical representation of a circle on an oblique (receding) face. Note that the method is identical with that used for constructing isometric circles, except for the slight change in the position of the axes.

Circle arcs and circles on inclined planes must be plotted by using the offset or coordinate method (Fig. 11.29).

11.21 □ Reduction of measurements
in the direction of the receding axis

An oblique drawing often presents a distorted appearance that is unnatural and disagreeable to the eye. In some cases the view constructed by this scheme is so misleading in appearance that it is unsatisfactory for any practical purpose. As a matter of interest, the effect of distortion is due to the fact that the receding lines are parallel and do not appear to converge as the eye is accustomed to anticipating (Fig. 11.2).

The appearance of excessive thickness can be overcome somewhat by reducing the length of the receding lines. For practical

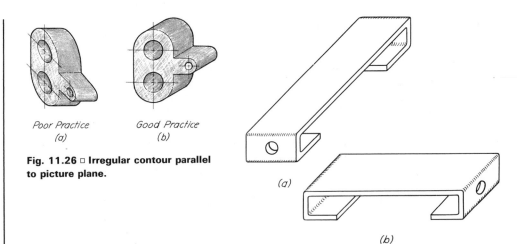

Poor Practice
(a)

Good Practice
(b)

Fig. 11.26 □ Irregular contour parallel to picture plane.

(a)

(b)

Fig. 11.27 □ Long axis parallel to picture plane.

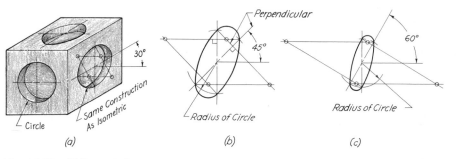

(a)   *(b)*   *(c)*

Fig. 11.28 □ Oblique circles.

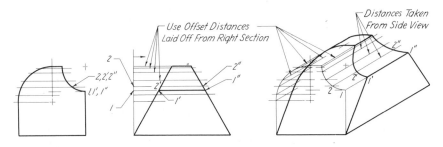

Fig. 11.29 □ Curved outlines on an inclined plane.

purposes, measurements usually are reduced one-half, but any scale of reduction may be arbitrarily adopted if the view obtained will be more realistic in appearance. When the receding lines are drawn one-half their actual length, the resulting pictorial view is called a *cabinet drawing*. Figure 11.30 shows an oblique drawing (*a*) and a cabinet drawing (*c*) of the same object, for the purpose of comparison.

11.22 □ Oblique sectional views

Oblique sectional views are drawn to show the interior construction of objects. The construction procedure is the same as for an isometric sectional view, except that oblique planes are used for cutting the object. An oblique half section is illustrated in Fig. 11.31.

11.23 □ Pictorial dimensioning

The dimensioning of isometric and other forms of pictorial working drawings is done in accordance with the following rules:

1 Draw extension and dimension lines (except those dimension lines applying to cylindrical features) parallel to the pictorial axes in the plane of the surface to which they apply (Fig. 11.32).

2 If possible, apply dimensions to visible surfaces.

3 Place dimensions on the object, if, by so doing, better appearance, added clearness, and easy readings result.

4 Notes may be lettered either in pictorial or as an ordinary drawings. When lettered as on ordinary drawings the difficulties encountered in forming pictorial letters are avoided (Fig. 11.32).

5 Make the figures of a dimension appear to be lying in the plane of the surface whose dimension it indicates, by using vertical figures drawn in pictorial (Fig. 11.32). (Note: Guide lines and slope lines are drawn parallel to the pictorial axes.)

11.24 □ Conventional treatment of pictorial drawings

When it is desirable for an isometric or an oblique drawing of a casting to present a somewhat more or less realistic appearance, it becomes necessary to represent the fillets and rounds on the unfinished surfaces. One method commonly used by draftsmen is shown in Figure 11.33(*a*). On the drawing in (*b*) all of the edges have been treated as if they were sharp. The conventional treatment for threads in pictorial is illustrated in (*b*) and (*c*).

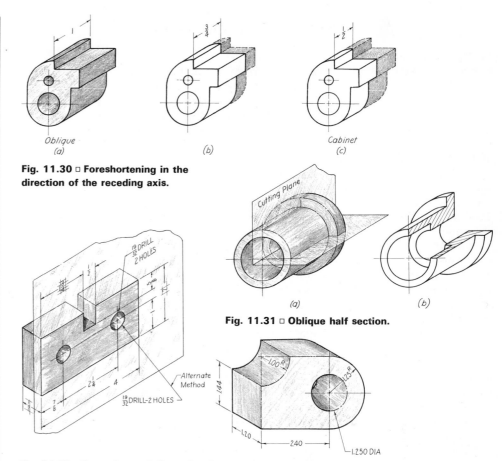

Fig. 11.30 □ Foreshortening in the direction of the receding axis.

Oblique (a)   (b)   Cabinet (c)

Fig. 11.31 □ Oblique half section.

(a)   (b)

Fig. 11.32 □ Extension and dimension lines in isometric (left); dimensions and notes in oblique (right).

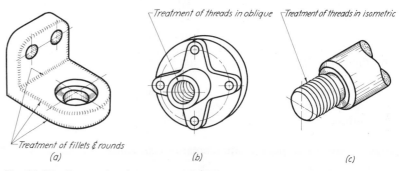

Fig. 11.33 □ Conventional treatment of fillets, rounds, and threads in pictorial.

199

C □ Perspective representation*

11.25 □ Perspective

In perspective projection an object is shown much as the human eye or camera would see it at a particular point. Actually, it is a geometric method by which a picture can be projected on a picture plane in much the same way as in photography. Perspective drawing differs from the methods previously discussed in that the projectors or visual rays intersect at a common point known as the *station point* (Fig. 11.36).

Since the perspective shows an object as it appears instead of showing its true shape and size, it is rarely used by engineers. It is more extensively employed by architects to show the appearance of proposed buildings, by artist-draftsmen for production illustrations, and by illustrators in preparing advertising drawings.

Figure 11.1 shows a type of production illustration that has been widely used in assembly departments as an aid to those persons who find it difficult to read an orthographic assembly. This form of presentation, which may show a mechanism both exploded and assembled, has made it possible for industrial concerns to employ semitrained personnel. Figure 11.44 shows a type of industrial drawing made in perspective that has proved useful in aircraft plants. Because of the growing importance of this type of drawing, and also because engineers frequently will find perspective desirable for other purposes, its elementary principles should be discussed logically in this text. Other books on the subject, some of which are listed in the bibliography, should be studied by architectural students and those interested in a more thorough discussion of the various methods.

The fundamental concepts of perspective can be explained best if the reader will imagine himself looking through a picture plane at a formal garden with a small pool flanked by lampposts, as shown in Fig. 11.34. The point of observation, at which the rays from the eye to the objects in the scene meet, is called the *station point*, and the plane on which the view is formed by the piercing points of the visual rays is known as the *picture plane* (*PP*). The piercing points reproduce the scene, the size which depends on the location of the picture plane.

It should be noted that objects of the same height intercept a greater distance on the picture plane when close to it than when farther away. For example, rays from the lamppost at 2 intercept a distance 1–2 on the picture plane, while the rays from the pole at 4, which actually is the same height, intercept the lesser dis-

*Additional information on perspective projection may be found in the author's text *Fundamentals of Engineering Drawing for Design, Communication, and Numerical Control.*

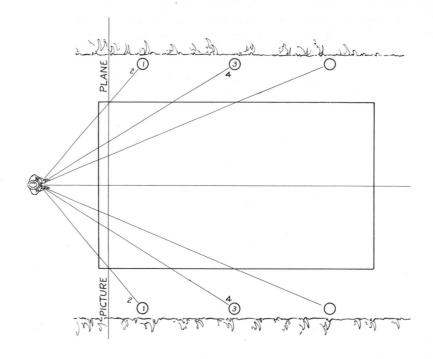

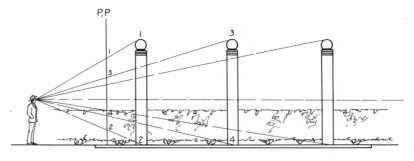

Fig. 11.34 □ The picture plane.

tance 3–4. From this fact it should be observed that the farther away an object is, the smaller it will appear, until a point is reached at which there will be no distance intercepted at all. This happens at the horizon.

Figure 11.35 shows the scene observed by the man in Fig. 11.34 as it would be formed on the picture plane. The posts farther from the picture plane diminish in height, as each one has a height on the picture plane equal to the distance it intercepts (Fig. 11.34). The lines of the pool and hedge converge to the center of vision or vanishing point, which is located directly in front of the observer on the horizon.

11.26 □ Perspective nomenclature

Figure 11.36 illustrates pictorially the accepted nomenclature of perspective drawing. The *horizon line* is the line of intersection of the horizontal plane through the observation point (eye of the observer) and the picture plane. The horizontal plane is known as the *plane of the horizon*. The *ground line* is the line of intersection of the ground plane and the picture plane. The *CV point* is the center of vision of the observer. It is located directly in front of the eye in the plane of the horizon on the horizon line.

11.27 □ Location of picture plane

The picture plane is usually placed between the object and the *SP* (station point). In parallel perspective (Sec. 11.32) it may be passed through a face of the object in order to show the true size and shape of the face.

11.28 □ Location of the station point

Care must be exercised in selecting the location for the station point, for its position has much to do with the appearance of the finished perspective drawing. A poor choice of position may result in a distorted perspective that will be decidedly displeasing to the eye.

In general, the station point should be offset slightly to one side and should be located above or below the exact center of the object. However, it must be remembered that the center of vision must be near the center of interest for the viewer.

One should always think of the station point as the viewing point, and its location should be where the object can be viewed to the best advantage. It is desirable that it be at a distance from the picture plane equal to at least twice the maximum dimension (width, height, or depth) of the object, for at such a distance, or greater, the entire object can be viewed naturally, as a whole, without turning the head.

A wide angle of view is to be avoided in the interest of good picturization. It has been determined that best results are obtained

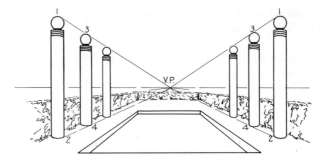

Fig. 11.35 □ **The picture (perspective).**

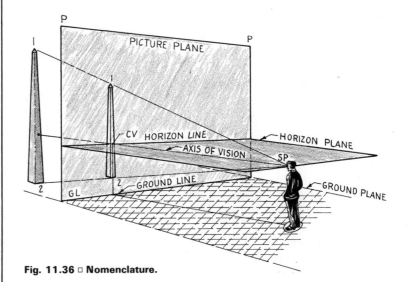

Fig. 11.36 □ **Nomenclature.**

when the visual rays from the station point (SP) to the object are kept within a cone having an angle of not more than 30° between diametrically opposing elements.

In locating an object in relation to the picture plane, it is advisable to place it so that both of the side faces do not make the same angle with the picture plane and thus will not be equally visible. It is common practice to choose angles of 30° and 60° for rectangular objects.

11.29 □ Position of the object in relation to the horizon

When making a perspective of a tall object, such as a building, the horizon usually is assumed to be at a height above the ground plane equal to the standing height of a man's eye, normally about 5 ft 6 in.

A small object may be placed either above or below the horizon (eye level), depending on the view desired. If an object is above the horizon, it will be seen looking up from below, as shown in Fig. 11.37. Should the object be below the horizon line, it will be seen from above.

11.30 □ Lines

The following facts should be recognized concerning the perspective of lines:

1 Parallel horizontal lines vanish at a single *VP* (vanishing point). Usually the *VP* is at the point where a line parallel to the system through the *SP* pierces the *PP* (picture plane).

2 A system of horizontal lines has its *VP* on the horizon.

3 Vertical lines, since they pierce the picture plane at infinity, will appear vertical in perspective.

4 When a line lies in the picture plane, it will show its true length because it will be its own perspective.

5 When a line lies behind the picture plane, its perspective will be shorter than the line.

11.31 □ Types of perspective

In general, there are two types of perspective: *parallel perspective* and *angular perspective*. In parallel perspective, one of the principal faces is parallel to the picture plane and is its own perspective. All vertical lines are vertical, and the receding horizontal lines converge to a single vanishing point. In angular perspective, the object is placed so that the principal faces are at an angle with the picture plane. The horizontal lines converge at two vanishing points.

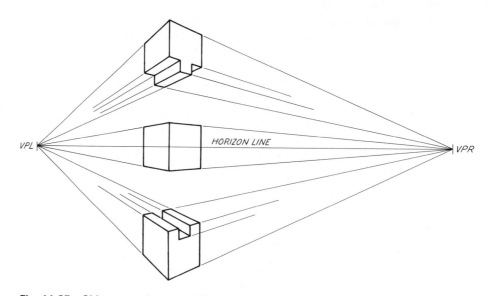

Fig. 11.37 □ Objects on, above, or below the horizon.

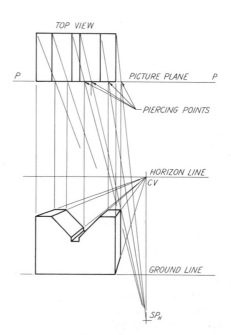

Fig. 11.38 □ Parallel perspective.

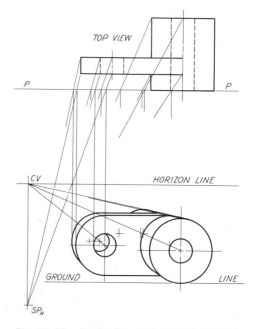

Fig. 11.39 □ Circles in parallel perspective.

11.32 □ Parallel perspective

Figure 11.38 shows the parallel perspective of a rectangular block. The PP line is the top view of the picture plane, SP_H is the top view of the station point, and CV is the center of vision. The receding horizontal lines vanish at CV. The front face, since it lies in the picture plane, is its own perspective and shows in its true size. The lines representing the edges back of the picture plane are found by projecting downward from the points at which the visual rays pierce the picture plane, as shown by the top views of the rays. Figure 11.39 shows a parallel perspective of a cylindrical machine part.

11.33 □ Angular perspective

Figure 11.40 shows pictorially the graphical method for the preparation of a two-point perspective drawing of a cube. To visualize the true layout on the surface of a sheet of drawing paper, it is necessary to revolve mentally the horizontal plane downward into the vertical or picture plane. On completion of Sec. 11.33, it is suggested that the reader turn back and endeavor to associate the development of the perspective in Fig. 11.41 with the pictorial presentation in Fig. 11.40. For a full understanding of the construction in Fig. 11.41, it is necessary to differentiate between the lines that belong to the horizontal plane and those that are on the vertical or picture plane. In addition, it must be fully realized that there is a top view for the perspective that is a line and that in this line view lie the points that must be projected downward to the perspective representation (front view).

Figure 11.41 shows an angular perspective of a block. The block has been placed so that one vertical edge lies in the picture plane. The other vertical edges are parallel to the plane, while all of the horizontal lines are inclined to it so that they vanish at the two vanishing points, VPL and VPR, respectively.

In constructing the perspective shown in this illustration, an orthographic top view was drawn in such a position that the visible vertical faces made angles of 30° and 60° with the picture plane. Next, the location of the observer was assumed and the horizon line was established. The vanishing points VPL and VPR were found by drawing a 30° line and a 60° line through the SP. Since these lines are parallel to the two systems of receding horizontal lines, each will establish a required vanishing point at its intersection with the picture plane. The vertical line located in the picture plane, which is its own perspective, was selected as a measuring line on which to project vertical measurements from the orthographic front view. The lines shown from these division points along this line to the vanishing points (VPL–VPR) established the direction of the receding horizontal edge lines in the

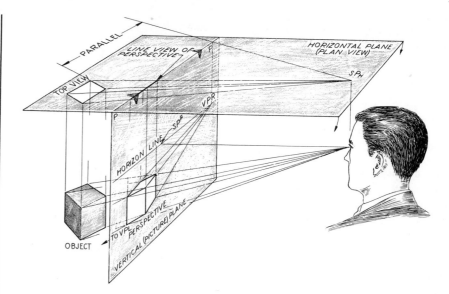

Fig. 11.40 □ Angular perspective.

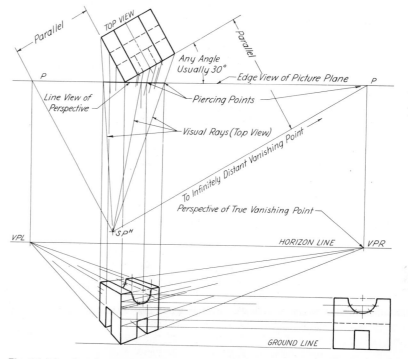

Fig. 11.41 □ Angular perspective.

perspective. The positions of the back edges were determined by projecting downward from the points at which the projectors from the station point (SP) to the corners of the object pierced the picture plane, as shown by the top view of the object and projectors.

11.34 □ Use of measuring lines

Whenever the vertical front edge of an object lies in the picture plane, it can be laid off full length in the perspective, because, theoretically, it will be in true length in the picture formed on the plane by the visual rays (see Fig. 11.41). Should the near vertical edge lie behind the picture plane, as is the case with the edge line AB in Fig. 11.42, the use of a measuring line becomes desirable. The measuring line $a'b'$ is the vertical edge AB moved forward to the picture plane, where it will appear in its true height. Some prefer to think of the vertical side as being extended to the picture plane so that the true height of the side is revealed. The length and position of AB is established in the perspective picture by first drawing vanishing lines from a' and b' to VPR; then the top view of the edge in the picture plane (point X) is projected downward to the front-view picturization. Points A and B must fall on the vanishing lines from a' and b' to VPR respectively.

A measuring line may be used to establish the "picture height" of any feature of an object. For example, in Fig. 11.42, the vertical measuring line through c' was used as needed to locate point C and the top line of the object in the perspective.

11.35 □ Circles in perspective

If a circle is on a surface that is inclined to the picture plane (PP), its perspective will resemble an ellipse. It is the usual practice to construct the representation within an enclosing square by finding selected points along the curve in the perspective, as shown in Fig. 11.43(a). Any points might be used, but it is recommended that points be located on 30° and 60° lines.

In (a), the perspective representation of the circle was found by using visual rays and parallel horizontal lines in combination. In starting the construction, the positions of several selected points, located on the circumference and lying on horizontal lines in the plane of the circle, were established in both views. After these lines had been drawn in perspective in the usual manner, the locations of the points along them were determined through the use of visual rays, as shown. Specifically, the position of a point in the perspective was found by projecting downward from the piercing point of the ray from the point and the picture plane (see line view) to the perspective view of the line on which the point must lie. In (b), the same method was applied to construct the perspective view of a circle in a horizontal plane. It should be noted

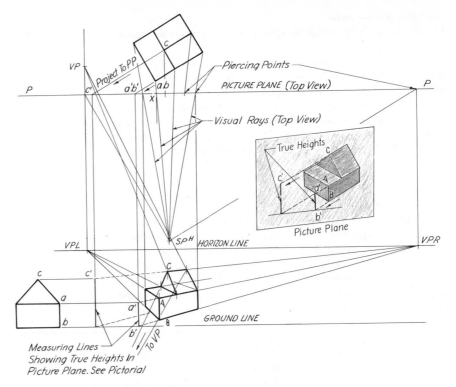

Fig. 11.42 □ Use of measuring lines.

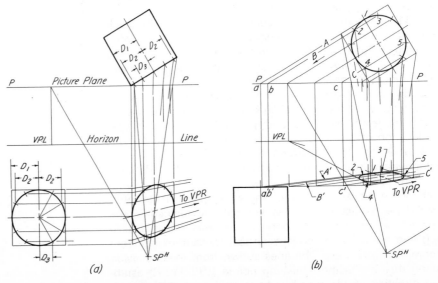

Fig. 11.43 □ Circles in perspective.

204

in this case that the horizontal lines, as established in the top view, were extended to cut the picture plane so that the true height of these lines at the plane could be used in the perspective view for locating the end points of the perspective representations (at the left).

D □ Industrial illustrations

11.36 □ Technical illustration drawings

The design and manufacturing procedures of present-day mass production require various types of pictorial illustrations to communicate ideas and concepts to large numbers of persons who are all working towards a common objective. These illustration drawings are of value at all stages of a project from the design phase, where they may be only pencil-shaded freehand sketches, through all of the stages of production that we may consider to include not only the assembly but final installation of the systems as well. They are used in operation and maintenance manuals to make complex and difficult tasks understandable to those persons who may be unable to interpret conventional drawings (Fig. 11.44). Pictorial illustrations range from simple types of line drawings that have already been discussed to artists' renderings that have the realism of photographs. An artist's rendering, such as the one shown in Fig. 11.45, is usually prepared to reinforce an oral or written report that is to be presented before a decision-making group that consists to some extent of persons who would otherwise be unable to understand construction details. Drawings of this type are depended on to sell a project.

Illustration drawings are used for many purposes in every field of engineering technology. They appear in advertising literature, operation and service manuals (Fig. 11.47), patent applications (Fig. 11.48), and textbooks (Fig. 11.49). Illustration drawings may be working drawings, assembly drawings, piping and wiring diagrams, and architectural and engineering renderings that are almost true-to-life. Typical examples of pictorial drawings that were prepared to facilitate assembly are shown in Figs. 11.1 and 5.8.

11.37 □ Design illustrations

Design illustrations are prepared to clarify conventional engineering design and production drawings and written specifications. These are used for the communication of ideas and concepts concerning the details of complicated designs. Properly prepared, drawings of this type reveal the relationship of the components of a system so clearly that the principles of operation of a unit can be understood by almost everyone, even by persons who may

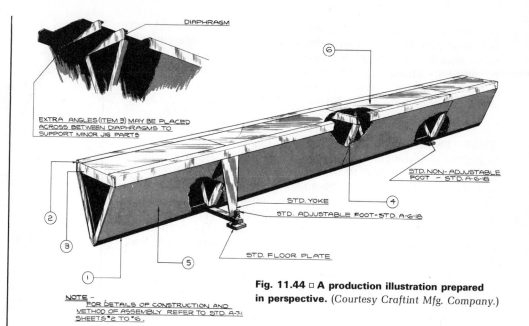

EXTRA ANGLES (ITEM 3) MAY BE PLACED ACROSS BETWEEN DIAPHRAGMS TO SUPPORT MINOR JIG PARTS

DIAPHRAGM

STD. NON-ADJUSTABLE FOOT - STD. A-6-18

STD. YOKE

STD. ADJUSTABLE FOOT-STD. A-6-18

STD. FLOOR PLATE

NOTE — FOR DETAILS OF CONSTRUCTION AND METHOD OF ASSEMBLY REFER TO STD. A-7-1 SHEETS *2 TO *6.

Fig. 11.44 □ A production illustration prepared in perspective. (*Courtesy Craftint Mfg. Company.*)

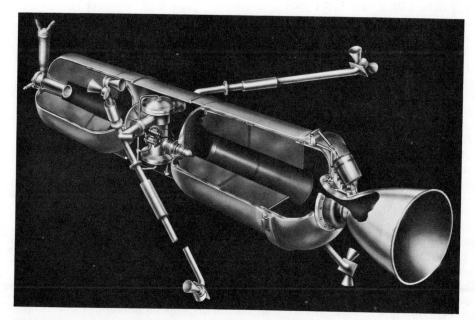

Fig. 11.45 □ An artist's concept of a rocket control system.
(*Courtesy Aeronutronic Division, Philco-Ford.*)

be relatively unfamiliar with graphic methods. It is common practice to prepare a series of such drawings to clarify complex details of construction, to indicate the function of closely related parts, and to reveal structural features. A pictorial of this type is shown in Fig. 11.46. It was used to supplement a technical paper presented at a Society of Automotive Engineers Congress.

11.38 □ Shading methods

A pictorial illustration can be improved and given a more realistic appearance by shading to produce the effect of surface texture. Shades and shadows may be added to give additional realism. The use of pencil shading is most common. Ink shading (see Fig. 11.49) produces clean illustrations of high quality that are well suited for reproductions in texts and brochures. The techniques of surface shading by means of ink lines will be discussed in Sec. 11.39.

Some of the other more basic methods of representing surface textures under light and shade involve the use of Rossboard, Craftint paper, Zip-a-tone overlay film, and the airbrush.

Most of the shaded pictorial drawings in this text were drawn on Rossboard. This popular drawing paper, with its rough plaster-type surface, is available in many textured patterns. Surface shading is done with a very soft pencil.

Craftint papers, single-tone or double-tone, have the pattern in the paper. Drawings are prepared on these papers in the usual manner and regular black waterproof drawing ink is used for the finished lines. Solid black areas are filled in with ink before the areas where shading is desired are brushed with a developer that brings out the surface pattern. These papers are available in many shading patterns. The drawing shown in Fig. 11.44 was prepared on Craftint paper.

Zip-a-tone clear cellulose overlay screens with printed shading patterns of dots and lines provide an easy method of surface shading suitable for high-quality printed reproductions (Fig. 11.47). The screen, backed with a clear adhesive, is applied as a sheet or partial sheet to the areas to be shaded. Unwanted portions are removed with a special cutting needle or razor blade, before the screen is rubbed down firmly to complete the bond between screen and paper.

A high degree of realism can be achieved using a small spray gun that in the language of the artist is known as an *airbrush*. This delicate instrument sprays a fine mist of diluted ink over the surface of the drawing to produce variations of tone. A capable artist can produce a representation that has the realism of a photograph. The pictorial illustration in Fig. 11.45 is an excellent example of airbrush rendering by a commercial illustrator. The airbrush

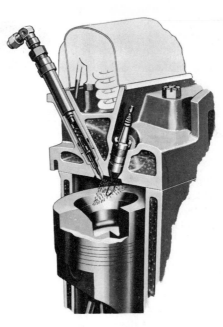

Fig. 11.46 □ A new combustion process developed by the Ford Motor Company is shown here in a cutaway view.
This process was developed to meet federal antipollution automobile requirements for 1976. A spray of gasoline enters from the left into the specially shaped combustion chamber. Caught in swirling air, the spray is mixed and then ignited by the long spark plug electrodes. (*Courtesy Ford Motor Company.*)

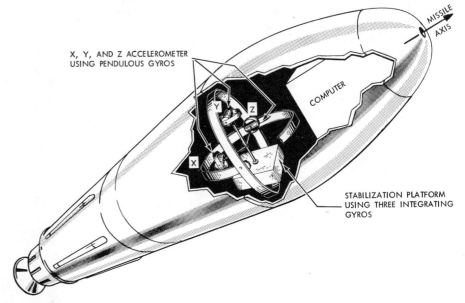

X, Y, AND Z ACCELEROMETER USING PENDULOUS GYROS

COMPUTER

MISSILE AXIS

STABILIZATION PLATFORM USING THREE INTEGRATING GYROS

Fig. 11.47 □ This pictorial representation shows the principle elements of the inertial guidance system in a missile vehicle.
(*Courtesy General Motors Corporation.*)

is sometimes used to retouch photographs to improve their appearance for reproduction.

11.39 □ Surface shading by means of lines (Fig. 11.48)

Line shading is a conventional method of representing, by ruled lines, the varying degrees of illumination on the surfaces of an object. It is a means of giving clearer definition to the shapes of objects and a finished appearance to certain types of drawings. In practice, line shading is used on Patent Office drawings, display drawings, and on some illustrations prepared for publications. It is never used on ordinary drawings, and for this reason few draftsmen ever gain the experience necessary to enable them to employ it effectively.

In shading surfaces, the bright areas are left white and the dark areas are represented by parallel shade lines (Fig. 11.49). Varying degrees of shade may be represented in one of the following ways:

1 By varying the weight of the lines while keeping the spacing uniform, as in Fig. 11.49.

2 By using uniform straight lines and varying the spacing.

3 By varying both the weight of the lines and the spacing.

The rays of light are assumed to be parallel and coming from the left, over the shoulder of the draftsman.

Although good line shading requires much practice and some artistic sense, a skillful person should not avoid shading the surfaces of an object simply because he never before has attempted to do so. After careful study, he should be able to produce fairly satisfactory results. Often the shading of a view makes it possible to eliminate another view that otherwise would be necessary.

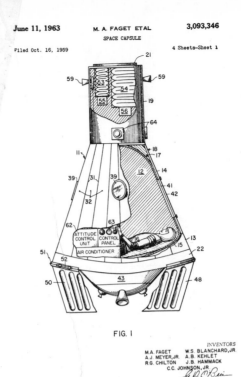

Fig. 11.48 □ A patent drawing with surface shading.

Fig. 11.49 □ A line-shaded pictorial drawing.
(Courtesy Socony-Vacuum Oil Company.)

Problems

The student will find that a preliminary sketch will facilitate the preparation of isometric and oblique drawings of the problems of this chapter. On such a sketch he may plan the procedure of construction. Since many engineers and designers frequently find it necessary to prepare pictorial sketches during discussions with untrained persons who cannot read orthographic views, it is recommended that some problems be sketched freehand on either plain or pictorial grid paper.

1–8 (Figs. 11.50–11.57). Prepare instrumental isometric drawings or freehand sketches of the objects as assigned.

9–10 (Figs. 11.50–11.51). Prepare instrumental oblique drawings or freehand sketches of the objects as assigned.

11 (Fig. 11.55). Make an oblique drawing of the locomotive driver nut.

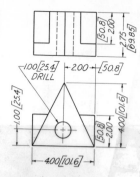

Fig. 11.50 □ A-block.
(*Dimension values shown in [ ] are in millimeters.*)

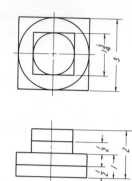

Fig. 11.51

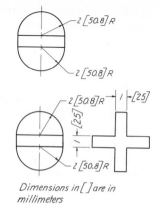

Dimensions in [ ] are in millimeters

Fig. 11.52

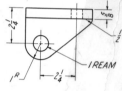
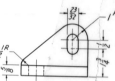

Fig. 11.53

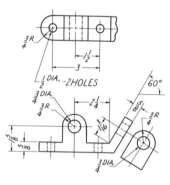

Fig. 11.54

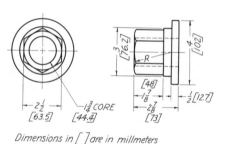

Dimensions in [ ] are in millimeters

Fig. 11.55 □ Locomotive driver nut.

12 (Fig. 11.56). Make an oblique drawing of the adjustment cone.

13 (Fig. 11.57). Make an oblique drawing of the feeder guide.

14 (Fig. 11.58). Make an isometric drawing of the hinge bracket.

15 (Fig. 11.59). Make an isometric drawing of the alignment bracket.

16 (Fig. 11.60). Make an isometric drawing of the stop block.

17–19 (Figs. 11.50–11.52). Make an angular perspective drawing as assigned.

20–22 (Figs. 11.56–11.58). Make a parallel perspective drawing as assigned.

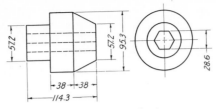

Dimensions are in millimeters

Fig. 11.56 □ Adjustment cone.

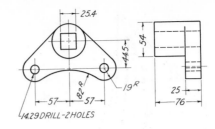

Fig. 11.57 □ Feeder guide.
(Dimension values are in millimeters.)

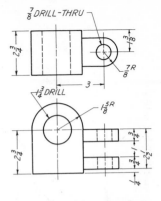

Fig. 11.58 □ Hinge bracket.

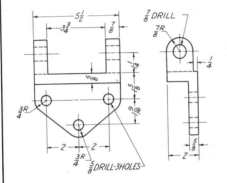

Fig. 11.59 □ Alignment bracket.

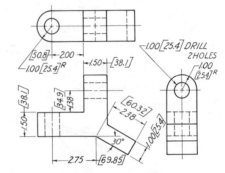

Fig. 11.60 □ Stop block.
(Dimension values shown in [ ] are in millimeters.)

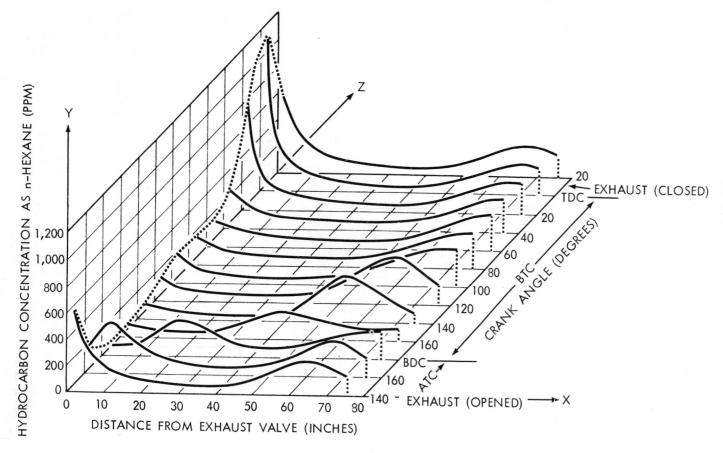

A three-dimensional graph.
The curves at left show the hydrocarbon concentration in the exhaust pipe as functions of crank angle and distance from the exhaust valve. Distance from the exhaust valve is plotted along the X-axis; engine crank angle, beginning with the time the exhaust valve opens, along the Z-axis; and hydrocarbon concentration is plotted along the Y-axis. Three-dimensional charts are used to show the relationship among three variables. Their construction is based upon pictorial projection. *Reprinted from the General Motors Engineering Journal with permission. (Courtesy General Motors Corp.)*

Chapter 12

Graphs and charts for communication, design, and analysis

A □ Engineering graphs and charts

12.1 □ Introduction

A properly designed graphical representation will convey correlated data and facts to an average individual more rapidly and effectively than will a verbal, written, or tabulated description, because a visual impression is easily comprehended and requires less mental effort than would be necessary to ascertain the facts from complex tables and reports (Figs. 12.1 and 12.2). It is because of this that diverse kinds of graphs and charts have been developed to present scientific, statistical, and technical information. Note how quickly the relationship presented by the line graph in Fig. 12.4 can be interpreted.

Engineers, even though they are concerned mainly with technical graphs, should be familiar also with the popular forms, for every industrial concern frequently must prepare popular types of graphs in order to strengthen their relationship with the public.

As much drafting skill is required in the execution of a graph as in making any other type of technical drawing. Good appearance is important and can be achieved only with the help of good lettering and smooth, uniform, and properly contrasted lines.

12.2 □ Classification of charts, graphs, and diagrams

Graphs, charts, and diagrams may be divided into two classes in accordance with their use and then further subdivided according to type. When classified according to use, the two divisions are, first, those used for strictly scientific and technical purposes and, second, those used for the purpose of popular appeal. The classification according to type is as follows:

1 Rectilinear charts

2 Semilogarithmic charts

3 Logarithmic charts

4 Barographs and area and volume charts

5 Percentage charts

6 Polar charts

7 Trilinear charts

8 Alignment charts (nomographs)

9 Pictorial charts

12.3 □ Quantitative and qualitative charts and graphs

In general, charts and diagrams are used for one of two purposes, either to read values or to present a comparative picture relationship between variables. If a chart or graph is prepared for reading values, it is called a *quantitative* graph; if prepared for presenting a comparative relationship, it is called *qualitative*. Obviously, some charts serve both purposes and cannot be classified strictly as either type. One of these purposes, however, must be predominant. Since a number of features in the preparation depend on the predominant purpose, such purpose must be determined before attempting to construct a graph.

12.4 □ Ordinary rectangular coordinate graphs

Most engineering graphs prepared for laboratory and office use are drawn on ruled, rectangular graph paper and are plotted in the first quadrant (upper-right-hand), with the intersection of the X-(horizontal) axis and Y-(vertical) axis at the lower left used as the zero point or origin of coordinates. The paper is ruled with equispaced horizontal and vertical lines, forming small rectangles. The type most commonly used for chart work in experimental engineering is $8\frac{1}{2} \times 11$ in. and is ruled to form $\frac{1}{20}$-in. squares [Fig. 12.3(a)], every fifth line being heavy. Another type of paper frequently used, which is suitable for most laboratory reports in technical schools, has rulings that form 1-mm and 1-cm squares [Fig. 12.3(b)]. Other rulings run $\frac{1}{10}$, $\frac{1}{8}$, or $\frac{1}{4}$ in. apart. Ordinarily the ruled lines are spaced well apart on charts prepared for reproduction in popular and technical literature (Fig. 12.2). The principal advantage of having greater spacing between the lines is that large squares or rectangles tend to make the graph easier to read. Ready-printed graph papers are available with various rulings in several colors.

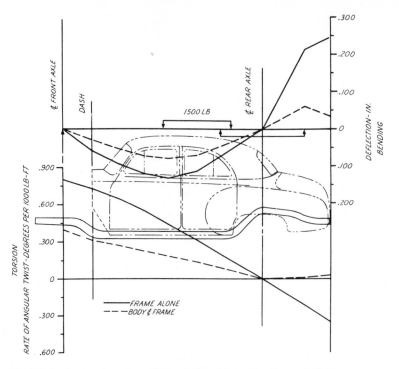

Fig. 12.1 □ An engineering graph reprinted from the General Motors Engineering Journal.

(*Courtesy General Motors Corporation.*)

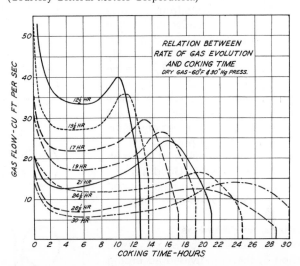

Fig. 12.2 □ An engineering graph prepared for publication.

(*Courtesy Blast Furnace and Steel Plant Magazine.*)

Ordinary coordinate line graphs are used extensively because they are easily constructed and easily read. The known relationship between the variables is expressed by one or more continuous lines, which may be straight, broken, or curved.

The graph in Fig. 12.4 shows the approximate barometric pressure at different heights above sea level.

A graphical representation may be drawn easily and correctly if, after the required data have been assembled, careful consideration is given to the principles of curve drawing discussed in the following sections.

12.5 □ The determination of the variables for ordinate and abscissa

The independent variable, the quantity arbitrarily varied during the experiment, usually is chosen for the abscissa (Fig. 12.5). Certain kinds of experimental data, however, such as a stress–strain diagram (Fig. 12.6), are plotted with the independent variable along the ordinate.

12.6 □ The selection of suitable scales

The American Society of Mechanical Engineers, in a standard for engineering and scientific graphs, recommends:*

  (a) Very careful consideration should be given to the choice of scales since this has a controlling influence on the slope of the curve. The slope of the curve, as a whole and also at intermediate points, provides a visual impression of the degree of change in the dependent variable for a given increment in the independent variable. Creating the right impression of the relationship to be shown by a line graph is, therefore, probably controlled more critically by the relative stretching of the vertical and horizontal scales than by any other feature involved in the design of the graph.
  (b) The range of scales should be chosen to insure effective and efficient use of the coordinate area in attaining the objective of the chart.
  (c) The zero line should be included, if visual comparison of plotted magnitudes is desired.

(If the chart is quantitative, the intersection of the axes need not be at the origin of coordinates. If it is qualitative, however, both the ordinate and abscissa generally should have zero value at the intersection of the axes, as in Fig. 12.2.)

  (d) For arithmetic scales, the scale numbers shown on the graph and space between coordinate rulings should preferably correspond to 1, 2, or 5 units of measurement, multiplied or divided by 1, 10, 100, etc.

*These statements were abstracted from the *American Standard for Engineering and Scientific Graphs for Publication.*

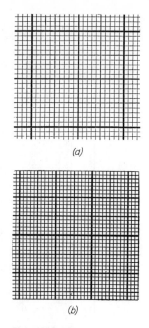

Fig. 12.3 □ Types of graph paper.

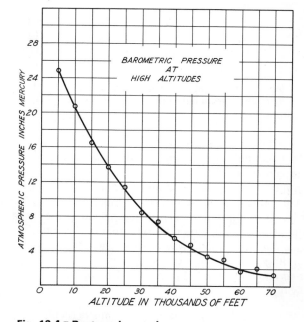

Fig. 12.4 □ Rectangular graph.

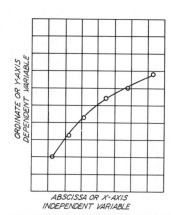

Fig. 12.5 □ Independent and dependent variables.

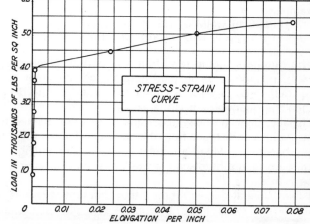

Fig. 12.6 □ Stress-strain diagram.

(Other units could be used except for the fact that they create situations wherein it becomes difficult to interpolate values. For example, one square should equal one of the following.)

0.01	0.1	1	10	100	etc.
0.02	0.2	2	20	200	etc.
0.04	0.4	4	40	400	etc.
0.05	0.5	5	50	500	etc.
etc.	etc.	etc.	etc.	etc.	etc.

(e) The horizontal (independent variable) scale values should usually increase from left to right and the vertical (dependent variable) from bottom to top.

12.7 □ Locating the axes and marking the values of the variables

On graphs prepared for laboratory reports and not for publication, the axes should be located 1 in. or more inside the border of the coordinate ruling (Fig. 12.7). When selecting the scale units and locating the axes, it should be remembered that the abscissa may be taken either the long way or short way of the coordinate paper, depending on the range of the scales.

Concerning the numbers, the ASME standard recommends the following:

The use of many digits in scale numbers should be avoided. This can usually be accomplished by a suitable designation in the scale caption.

Examples:
PRESSURE, MM OF HG $\times 10^{-5}$;
RESISTANCE, THOUSANDS OF OHMS.

The numbers should read from the bottom when possible (Fig. 12.7). For the sake of good appearance, they never should be crowded. Always place a cipher to the left of the decimal point when the quantity is less than one.

Usually, only the heavy coordinate lines are marked to indicate their values or distance from the origin, and, even then, the values may be shown only at a regular selected interval (Fig. 12.7). The numbers should be placed to the left of the Y-axis and just below the X-axis.

When several curves representing different variables are to appear on the same graph, a separate axis generally is required for each variable (Fig. 12.8). In this case, a corresponding description should be given along each axis. The axes should be grouped at the left or at the bottom of the graph, unless it is desirable to place some at the right or along the top.

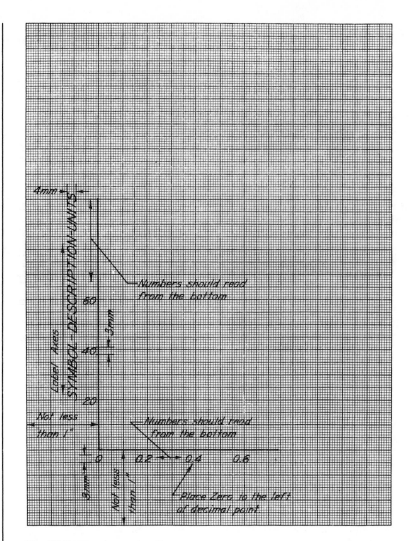

Fig. 12.7 □ Graph construction.

12.8 □ Indicating plotted points representing the data

If the data represent a set of experimental observations, the plotted points of a single-curve graph should be marked by small circles approximately 0.1 in. in diameter (Fig. 12.9). The following practice is recommended: open circles, filled-in circles, and partially filled-in circles (○ ● ◖), rather than crosses, squares, and triangles, should be used to differentiate observed points of several curves on a graph. Filled-in symbols may be made smaller than those not filled in.

Mathematical curves are frequently drawn without distinguishing marks at computed positions.

12.9 □ Drawing a curve

Since most physical phenomena are continuous, curves on engineering graphs usually represent an average of plotted points (Fig. 12.10). Discontinuous data should be plotted with a broken line, as shown in Fig. 12.11.

It is preferable to represent curves by solid lines. If more than one curve appears on a graph, differentiation may be secured by varied types of lines; but the most important curve should be represented by a solid one. A very fine line should be used for a quantitative curve if values are to be read accurately. A heavy line ($\frac{1}{40}$-in. width) is recommended for a qualitative curve. It should be observed in Figs. 12.10 and 12.11 that the curve line does not pass through open circles.

For ordinary qualitative graphs (Fig. 12.8), the ASME standard proposes:

(a) When more than one curve is presented on a graph, relative emphasis or differentiation of the curves may be secured by using different types of line, i.e., solid, dashed, dotted, etc. A solid line is recommended for the most important curve.

(b) When more than one curve is presented on a graph, each should bear a suitable designation.

(c) Curves should, if practicable, be designated by brief labels placed close to the curves (horizontally or along the curves) rather than by letters, numbers or other devices requiring a key.

12.10 □ The labeling of the scales

Each scale caption should give a description of the variable represented and the unit of measurement. The captions on engineering graphs frequently contain an added identifying symbol, such as

N–EFFICIENCY–PERCENT
P–OUTPUT–HP

All lettering should be readable from the bottom and right side of the graph (not the left side). When space is limited, standard abbreviations should be used, particularly for designating the unit

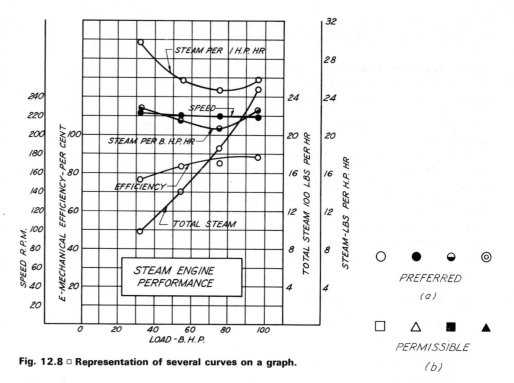

Fig. 12.8 □ Representation of several curves on a graph.

Fig. 12.9 □ Identification symbols.

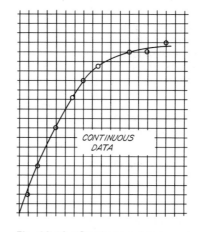

Fig. 12.10 □ Continuous curve.

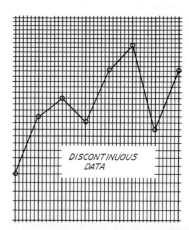

Fig. 12.11 □ Discontinuous data.

of measurement. To avoid confusing the reader, the engineer should use only recognized word contractions.

12.11 □ Titles, legends, notes, and so on

The title of a graph should be clear, concise, complete, and symmetrical. It should give the name of the curve, the source of the data, the date, and other important information (Fig. 12.12). It should be so placed that it gives a balanced effect to the completed drawing (Fig. 12.2). In addition to the title, a wiring diagram, pictorial diagram, formula, or explanatory note is often necessary to give a clear picture of the nature of the experiment. For example, if there is any great irregularity in the plotted points or if there is a condition that may have affected the values as shown by the data, a note of explanation should be given. A legend or key is sometimes included to explain a set of curves in greater detail.

In commercial practice, alcohol is often used to clear a rectangular area of coordinate lines in order that the title may be printed in an open space.

12.12 □ Procedure for making a graphical representation in ink

1 Select the type of coordinate paper.

2 Determine the variables for ordinate and abscissa.

3 Determine the scale units.

4 Locate the axes and mark the scale values in pencil.

5 Plot the points representing the data. [Many draftsmen ink the symbol (○ ◑) indicating the points at this stage.]

6 Draw the curve. If the curve is to strike an average among the plotted points, a trial curve should be drawn in pencil. If the curve consists of a broken line, as is the case with discontinuous data, the curve need not be drawn until the graph is traced in ink.

7 Label the axes directly in ink.

8 Letter the title, notes, and so on. The title should be lettered on a trial sheet that can be used as a guide for lettering directly in ink on the graph.

9 Check the work and complete the diagram by tracing the curve in ink.

12.13 □ Logarithmic graphs

Logarithmic coordinate graphs are constructed on prepared paper on which the parallel horizontal and parallel vertical rulings are spaced proportional to the logarithms of numbers (Fig. 12.13). This type of graph has two principal advantages over the ordinary coordinate type. First, the error in plotting or reading values is a constant percentage, and, second, an algebraic equation of the

STRESS - STRAIN DIAGRAM
FOR
COMPRESSION
IN
CAST IRON

Fig. 12.12 □ A title.

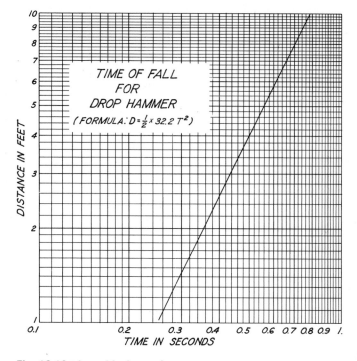

TIME OF FALL
FOR
DROP HAMMER
(FORMULA: $D = \frac{1}{2} \times 32.2\, T^2$)

Fig. 12.13 □ Logarithmic graph.

form $y = ax^b$ appears as a straight line if x has a value other than zero. The exponent b may be either plus or minus.

The equation for a falling body, $D = \frac{1}{2}gt^2$, is represented in Figs. 12.13 and 12.14. A practical application of interest to engineers is in the design of drop hammers. In this equation, based on uniform accelerated motion, t represents time in seconds and D the distance traveled in t seconds by a freely falling body with no initial velocity. Observe that the plotted points form a parabolic curve on ordinary coordinate graph paper and a straight line on logarithmic paper. To draw the line on the graph in Fig. 12.13, it is necessary to calculate and locate only two points, while in Fig. 12.14, several points must be plotted to establish the location of the corresponding curved-line representation. The line on Fig. 12.13 has a slope of 2:1, because the exponent of t is 2. Therefore, the line could be drawn by utilizing one point and the slope, instead of plotting two points and joining them with a straight line.

Log paper is available with rulings in one or more cycles for any range of values to be plotted. Part-cycle and split-cycle papers may also be purchased.

12.14 □ Semilogarithmic graphs

Semilogarithmic paper has ruled lines that are spaced to a uniform scale in one direction and to a logarithmic scale in the other direction (Fig. 12.15). Charts drawn on this form of paper are used extensively in scientific studies, because functions having values in the form of geometric progressions are represented by straight lines. In any case, the main reason for the use of semilogarithmic paper is that the slope of the resulting curve indicates rate of change rather than amount of change, the opposite being true in the case of curves on ordinary coordinate graph paper. Persons who are interested may determine the rate of increase or decrease at any point by measuring the slope. A straight line indicates a constant rate of change. In commercial work this form of paper is generally called "ratio paper," and the charts are known as "rate-of-change charts."

As previously stated, the choice of a type of graph paper depends on the information to be revealed. Curves drawn on uniform coordinate graph paper to illustrate the percentage of expansion or contraction of sales, and so on, present a misleading picture. The same data plotted on semilogarithmic paper would reveal the true rate of change to the business management. For this reason, semilogarithmic paper should be used whenever percentage of change rather than quantity change is to be shown. In scientific work, when the value of one variable increases in a geometric progression and the other in an arithmetic progression, this form is valuable.

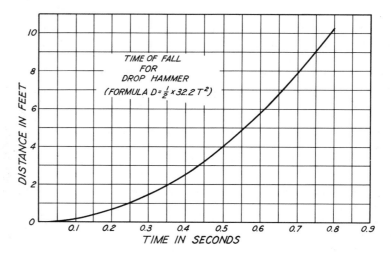

Fig. 12.14 □ Coordinate graph.

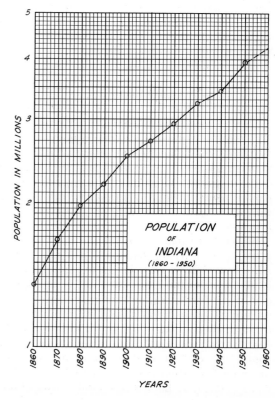

Fig. 12.15 □ Semilogarithmic chart.

Graphs and charts for communication, design, and analysis □ **217**

12.15 □ Bar charts

Bar charts or barographs are used principally in popular literature covering economic and industrial surveys. They are a simple diagrammatic form giving a pictorial summary of statistical data and can be easily understood by the average person. Logarithmic and uniform coordinate graphs are less suited for this purpose, because few people know the procedure for reading curves or understand their picture qualities.

Whenever values or quantities are illustrated, as in Fig. 12.16, by consecutive heavy bars whose lengths are proportional to the amounts they represent, the resulting representation is called a *bar chart*.

The bars on this type of diagram may be drawn either horizontally or vertically, but all should start at the same zero line. Their lengths should be to some fixed scale, the division values of which may be given in the margin along the bottom or left side of the graph. When it is necessary to give the exact values represented, the figures should be placed along each bar in a direction parallel to it. To place the values at the end gives the illusion of increasing the length of the bars. Usually, the names of the items are lettered to the left of the vertical starting line on a horizontal chart and below the starting line on a vertical chart.

12.16 □ Area (Percentage) charts

An area diagram can be used profitably when it is desirable to present pictorially a comparison of related quantities in percentage. This form of representation illustrates the relative magnitudes of the component divisions of a total of the distribution of income, the composition of the population, and so on. Two common types of the various forms of area diagrams used in informative literature are illustrated in Figs. 12.17 and 12.18. Percentages, when represented by sectors of a circle or subdivisions of a bar, are easy to interpolate.

The pie chart (Fig. 12.17) is the most popular form of area diagram, as well as the easiest to construct. The area of the circle represents 100% and the sectors represent percentages of the total. To make the chart effective, a description of each quantity and its corresponding percentage should be lettered in its individual sector. All lettering should be completed before the areas are cross-hatched or colored if this is to be done. The percentage bar chart shown in Fig. 12.18 fulfills the same purpose as the pie chart. The overall area of the bar represents 100%. Note that each percentage division is cross-hatched in a different direction. The descriptions may be placed on either side of the bar; the percentages should be on the bar or at the side.

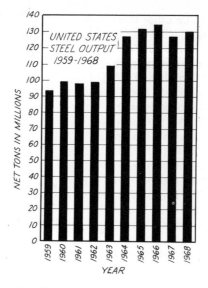

Fig. 12.16 □ A bar chart.

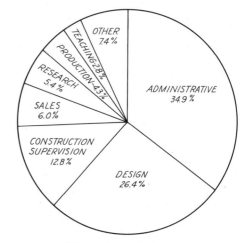

PROFESSIONAL ENGINEERS
THE KIND OF WORK THEY DO

Fig. 12.17 □ A pie chart.

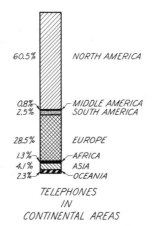

Fig. 12.18 □ Percentage bar chart.

12.17 □ Polar charts

Certain types of technical data can be more easily plotted and better represented on polar coordinate paper. Polar charts are drawn by self-recording instruments; polar diagrams and plotted polar curves representing various kinds of scientific data are very common. Polar curves are used to represent the intensity of diffused light, intensity of heat, and so on. The polar chart in Fig. 12.19 gives, in terms of candlepower, the intensity of light in two planes.

12.18 □ Trilinear charts

Trilinear charts are used principally in the study of the properties of chemical compounds, mixtures, solutions, and alloys (Fig. 12.20). Basically this is a 100% chart the use of which, owing to its geometric form, is limited to the investigation of that which is composed of three constituents or variables. Its use depends on the geometric principle that the sum of the three perpendiculars from any point is equal to the altitude. If the altitude represents 100%, the perpendiculars will represent the percentages of the three variables composing the whole.

The ruling can be accomplished conveniently by dividing any two sides of the triangle into the number of equal-percentage divisions desired and drawing through these points lines parallel to the sides of the triangle.

B □ Empirical equations

12.19 □ Empirical equations

In all phases of engineering work considerable experimentation is done with physical quantities, and the engineer (or engineering technologist) in his work is usually the person most concerned with the behavior of quantities in relation to one another. Often it is known that the subject of the experiment obeys some physical law which can be expressed by a mathematical equation, but the exact equation is unknown. Then, the person performing the experiment is faced with the task of finding an equation to fit the data that has been obtained. The three articles that follow discuss means of arriving at an equation from a graphical study of the data. An equation determined in this manner is an *empirical equation*. Since the unknown law may be quite complex, a single empirical equation to fit the whole range of data may not exist. However, in the majority of such cases a series of the various empirical equations with limited coverage (parameters) can be found.

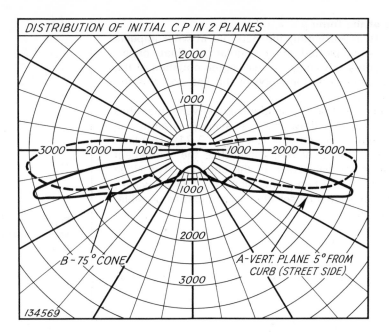

Fig. 12.19 □ Polar chart.
(*Courtesy General Electric Company.*)

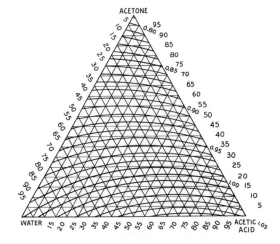

Fig. 12.20 □ Trilinear chart.
(*Courtesy American Chemical Society.*)

12.20 □ Equations of the form: $y = a + bx$

If the plotted points, representing the data, lie in what appears to be a straight line when plotted on rectangular coordinate paper, the equation of the data is a linear or first degree equation of the form: $y = a + bx$, where b is the slope of the line and a is the y-intercept $(x = 0)$.

After it has been decided that the relationship between quantities is linear, the next step is to draw the best average straight line that will be representative of the data (see Fig. 12.21). This line, extended if necessary to the y-axis, establishes the y-intercept $(x = 0)$ which will be the value of a in the equation. The slope can be determined from any two points along the line. However, for accuracy the points should be selected as far apart as possible. The value of b is expressed as

$$b = \frac{y_2 - y_1}{x_2 - x_1}$$

Thus, for the line in Fig. 12.21, having selected two points as illustrated,

$$b = -\frac{3 - 1}{6 - 3} = -\frac{2}{3} \text{ (slope)}$$

With the slope now known and the value of a having been read directly on the y-axis as 5, the equation of the line can be written as

$$y = 5 - \frac{2}{3}x$$

If it is not reasonable to include $x = 0$ in the plot of the data, then a pair of simultaneous equations are set up using two points on the line:

$$y_1 = a + bx_1$$
$$y_2 = a + bx_2$$

This system can then be solved for a and b.

12.21 □ Equations of the form: $y = ax^b$

If the data points seem to lie in a straight line on logarithmic graph paper, the equation of the data is a power equation of the form: $y = ax^b$. Power equations of this type, where one quantity varies directly as some power of another, appear as either parabolic or hyperbolic curves when plotted on rectangular coordinate paper, depending upon whether the exponent is positive or negative. For positive values of b (except for unity) the curves are parabolic; negative values produce hyperbolic curves.

When the equation is placed in logarithmic form and rewritten

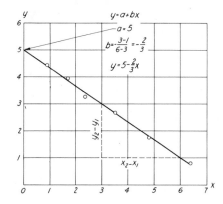

Fig. 12.21 □ **Determination of empirical equations of the form** $y = a + bx$ **(first degree).**

as: $\log y = \log a + b \log x$, it is now in the same form as the equation of a straight line. In this case, when $x = 1$ then $y = a$ because $\log 1 = 0$.

Hence, after the data have been plotted (y vs. x) on logarithmic paper as in Fig. 12.22, a representative (average) straight line should be drawn. The line extended to the y-axis establishes a at the y intercept ($x = 1$). The value of a can be read directly, since the spacing is logarithmic. As before, two points along the line are used in the equation that follows to determine b:

$$b = \frac{\log y_2 - \log y_1}{\log x_2 - \log x_1}$$

For the two points selected along the line in Fig. 12.22,

$$b = \frac{1.477 - 0.699}{1.772 - 0.398} = 0.57$$

Thus, the equation is $y = 3x^{0.57}$.

If $x = 1$ is not included in the plot of the data, then one may resort to a pair of simultaneous equations

$$\log y_1 = \log a + b \log x_1$$
$$\log y_2 = \log a + b \log x_2$$

In solving for a (also b) the solution gives the value for $\log a$. The real value of a (and b) must be found in log tables.

12.22 □ Equations of the form: $y = ab^x$

If the data points form a nearly straight line on semilogarithmic graph paper (Fig. 12.23), the equation of the data is an exponential equation of the form: $y = ab^x$. This equation may be re-written as: $\log y = \log a + x \log b$, which once again is in the same form as the equation of a straight line. In this case, when $x = 0$ then $y = a$.

With the data for the exponential equation plotted directly on semilogarithmic graph paper, one must again draw the most representative line along the path of the plotted points. This line extended to the y-axis ($x = 0$) determines the value of a. Again, two points must be selected along the line and their coordinates substituted in the equation

$$\log b = \frac{\log y_2 - \log y_1}{x_2 - x_1}$$

Then, with the logarithm of the value of b known, the real value of b as needed for the equation can be determined from log tables. For the two points selected for the line in Fig. 12.23

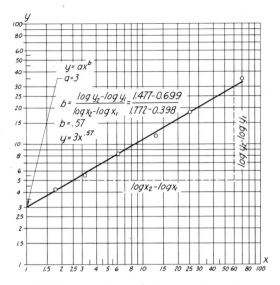

Fig. 12.22 □ Determination of empirical equations of the form $y = ax^b$ (power).

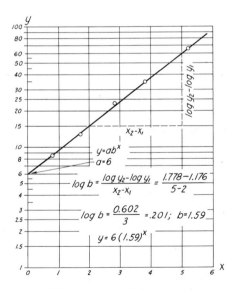

Fig. 12.23 □ Equations of the form $y = ab^x$.

$$\log b = \frac{1.778 - 1.176}{5 - 2} = \frac{0.602}{3} = .201$$

Hence,

$$b = 1.59$$

With the value of a read on the y-axis as 6, the formula for the data may be written as: $y = 6\,(1.59)^x$.

If $x = 0$ is not included in the plot of the data, then one must solve the simultaneous equations

$$\log y_1 = \log a + x_1 \log b$$
$$\log y_2 = \log a + x_2 \log b$$

Then, since the solution gives both a and b in terms of logarithms, log tables must be used to determine the real values of a and b.

When natural logarithms are available, the equation $y = ab^x$ may be changed to $y = Ae^{mx}$. This latter equation, in turn, may be written as: $\ln y = \ln A + mx$, which is still the equation of a straight line. The value of A can be determined as before at $x = 0$, and

$$m = \frac{\ln y_2 - \ln y_1}{x_2 - x_1}$$

where the value of m is found directly.

When $x = 0$ is not included, the equations to be used are

$$\ln y_1 = \ln A + mx_1$$
$$\ln y_2 = \ln A + mx_2$$

C □ Graphical calculus

12.23 □ Graphical calculus

In solving engineering problems, it is frequently desirable and often necessary to present a graphical analysis of empirical data. Even though it is often possible to make an evaluation through the use of analytical calculus, a graphical representation is more meaningful because it is pictorial in character. Graphical integration and differentiation are particularly desirable for problems for which only a set of values are known, or for curves that have been produced mechanically, as in the case of steam engine indicator diagrams, or if the results cannot be determined by the analytical methods of calculus.

The following sections are devoted to the graphical rules and methods for determining derived curves. Discussion of the inter-

Fig. 12.24 □ Illustration of the principle of integration.

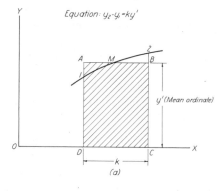

(b)

Equation: $y_2 - y_1 = ky'$

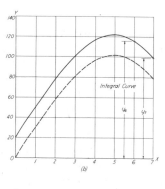

y' (Mean ordinate)

(a)

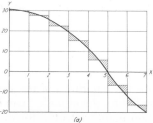

(a)

Fig. 12.26 □ The integration of a curve.

Fig. 12.25 □ The integration of a curve.

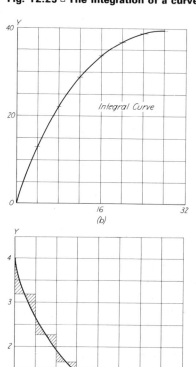

Integral Curve

(b)

(a)

pretation of results has been intentionally omitted since interpretation is not usually graphical and, therefore, not within the scope of this text.

12.24 □ Graphical integration

In deriving curves of a higher order, the principle is applied that the area bounded by two successive ordinates, the curve, and the axis is equal to the difference in magnitude of the corresponding ordinates of the integral curve. Figure 12.24 illustrates this principle of graphical integration. In (a) an increment of a curve is shown enlarged. The area under the curve will be approximately equal to the area of the shaded rectangle $ABCD$ when the line AB is drawn so that the area $AM1$ above the curve is approximately equal to the area $MB2$ below the curve. With a little practice one will find it easy to establish a line such as AB quite accurately by eye if a strip of celluloid or a triangle is used through which the curve can be seen.

By applying the principle of graphical integration to a series of increments, an integral curve may be drawn as shown in Fig. 12.25. At this point, it should be recognized that since the difference between successive ordinates represents increase in area, the difference between the final ordinate and the initial ordinate represents the total area between these ordinates that is bounded by the curve and the X-axis.

The scale selected for the Y-axis of the integral curve need not be the same as the scale for the given curve.

Portions of a lower-order curve that are above the X-axis are considered to be positive whereas areas below with negative ordinates are recognized as negative [see Fig. 12.26(a)]. Since the negative area between any two ordinates on the lower-order curve represents only the difference in the length of the corresponding ordinates on the integral curve, the length of y_7 is less than the length of y_6 by an amount equal to the negative area. Also, because areas represent only differences in length of successive ordinates of the integral curve, the initial point on the integral curve might have any value and still fulfill its purpose. For example, either integral curve shown in Fig. 12.26(b) is a satisfactory solution for the curve in (a).

Figure 12.27 shows the derived curves for a falling drop hammer. It is common practice, when drawing related curves, to place them in descending order as shown, that is, the lower-order curve is placed below.

In (a) the straight line represents a uniform acceleration of 32.2 ft per sec per sec, which is the acceleration for a freely falling body. The initial velocity is 0. The units along the X-axis represent time in seconds and the units along the Y-axis represent acceleration in feet per second per second.

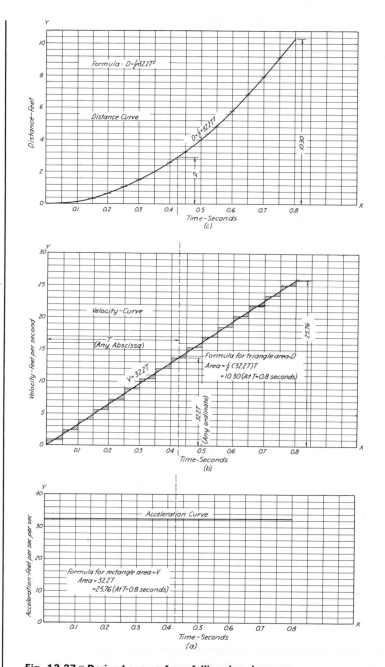

Fig. 12.27 □ Derived curves for a falling drop hammer.

Since the acceleration is uniform, the velocity will be a straight line of constant slope, (b). The length of the last ordinate is equal to the total area under the acceleration curve, namely 25.76 ft per sec (fps).

The distance curve, which is obtained by integrating the velocity time curve, is shown in (c). The length of the ordinate at any interval point is equal to the total area below the velocity curve between the origin and the point $[D = A = \frac{1}{2} (32.2T)T]$. See Figs. 12.24, 12.25 and 12.26.

12.25 □ To integrate a curve by the ray polygon method

An integral curve may be drawn by a purely graphical process known as the ray polygon method.

This method of integrating the area under a curve is illustrated in Fig. 12.28.

Divide the X-axis into intervals and draw ordinates at the division points. Then, select the pole point P at some convenient location that will make the distance d equal to any number of the full units assigned to the X-axis. The selection of the number of units for the distance d determines the length of the scale along a Y-axis for the integral curve. To establish relationship between y_1 and y_0 the following equation based on similar right triangles can be written:

$$y_1 : k = y_0 : d$$
$$y_1 \cdot d = k \cdot y_0$$
$$y_1 = \frac{k \cdot y_0}{d} = \frac{k}{d} \cdot y_0$$

Determine the mean ordinate for each strip and transfer its height to the Y-axis as length OA, OB, OC, and so forth. Draw rays from P to points A, B, C, D, E, and F.

To construct the integral curve, start at O and draw a line parallel to PA cutting the first vertical through R at R'. Through R' draw a line parallel to PB until it cuts the second vertical through S at S'. Repeat this procedure to obtain points T', U', and so on. Points R', S', T', U', V', and W' are points on the required integral curve. In Fig. 12.28 the integral curve is constructed on the same coordinate axes as the lower-order curve.

12.26 □ Pole method applied to the construction of an integral curve

The first of the several examples that have been selected to illustrate the use of the so-called pole method for the construction of an integral graph, has a sloping straight line, plotted on rectangular coordinates, as the given curve of lower order (see Fig. 12.29). It should be noted in this case that the integral graph has been

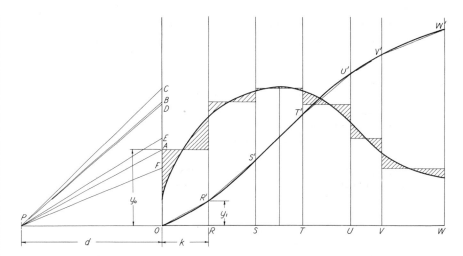

Fig. 12.28 □ Use of ray polygon.

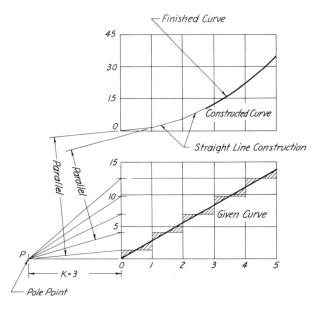

Fig. 12.29 □ Pole method.

placed directly above the lower graph and in projection with it so that the same intervals may be used for both. As shown, ordinates were first drawn to divide the chart area into vertical strips. With this done the horizontal lines were drawn next, one for each of the strips at mean-ordinate height, to form the five rectangular areas that are assumed to approximate the area under the curve (sloping straight line). Then, these mean-ordinate heights must be projected horizontally to the y-axis. With the pole distance k determined and the pole point P located along the x-axis extended, ray lines are drawn from P to the points on the y-axis representing the heights of the mean ordinates. The manner in which the distance k affects the steepness of the integral curve, as illustrated in Fig. 12.31, will be discussed later. The first point on the integral curve was found by drawing a line through the origin parallel to the ray for the corresponding interval below, in this case the lowest ray. The second segment is then drawn parallel to the ray for the second interval through the end of the first segment. Other intervals are treated similarly and then the smooth curve, that is, the integral curve, is drawn through the points obtained. Since the finished curve has in reality been drawn through the end-points of chords by this method, most persons identify this construction procedure by calling it the ''chordal method.''

In Fig. 12.30, the so-called chordal method using a pole has been applied to the integration of an irregular curve. The intervals between the ordinates need not be equal as shown. In fact, the number and spacing of the ordinates will usually be determined by the shape of the curve and one soon discovers that it is wise to space ordinates closer together where there is a sharp change or a reversal of curvature and farther apart where the curve is flat. For this illustration, an example was purposely selected that would have a portion of the curve below the x-axis. This area below represents negative area that must be subtracted from the positive area above the x-axis. When a pole is used, this subtraction is taken care of graphically by the negative slope of the rays for the intervals at the location where the curve lies below the x-axis.

Ordinarily, the modulus of the ordinate (y-axis) scale of the integral curve must be smaller than the modulus of the given (lower-order) curve if the graph for the integral curve is to be of reasonable size. When the height of the derived curve is somewhat near the same as the height of the given curve, the general over-all appearance of the graphs will be satisfactory and the scales will ordinarily be large enough to be read with reasonable accuracy. At the very beginning when it becomes necessary to determine the maximum value to be read on the ordinate scale for the integral curve, one must first remember that the value will be equal numerically to the value for the total area under the given curve. This value may be found either by adding the areas of the individual rectangles

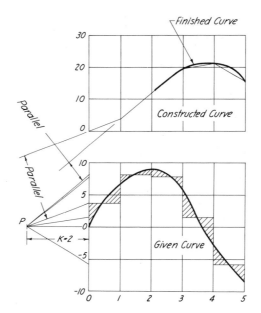

Fig. 12.30 □ Use of pole point.

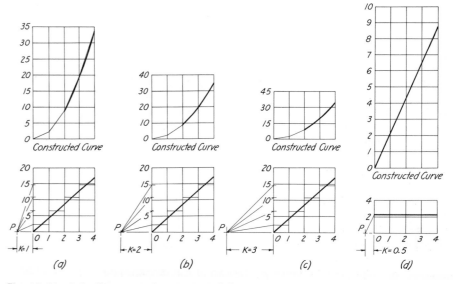

Fig. 12.31 □ Pole distance and steepness of the curve.

or by calculating the area of a single large rectangle that is estimated to be equal to the total area under the curve. Once this maximum value is known it is relatively easy to select a value for k that will give good proportion to the integral curve and at the same time permit one to make accurate readings.

When the pole distance used is equal to the length of one abscissa unit as shown in Fig. 12.31(a), the same distance represents the same number of units on both scales. In (b), where a pole distance of 2 has been used, ten units are represented on the scale of the constructed curve by the same distance that represents 5 units on the scale of the curve below. As can be easily seen, a three-unit pole distance reduces the height of the constructed curve still more (c). When the pole distance is one-half unit, the constructed graph becomes very high, too high to be appropriate in the case shown in (d). However, there may be conditions encountered now and then when the use of a pole distance of one-half unit is desirable. An area graph is shown in Fig. 12.32, where a five-unit pole distance was used to construct the work (integral) curve.

An interesting application of graphical integration is shown in Fig. 12.33. Let it be supposed that an irregularly shaped plot of land, lying along a lake shore and a road, is to be divided into three equal lots. The area to be divided was first plotted (feet against feet) on the lower graph. Its boundary lines in addition to the shore line at the north, are the x and y axes, representing property lines, and the right-of-way line of S.R. 60. It was decided that the division lines were to run due north and south, perpendicular to the south property line. With the lower curve drawn, the integral curve was constructed by the pole-and-ray method as shown and the total area was found to be 180,000 sq ft, an area that, in this particular case, can be divided evenly into three lots, each having 60,000 sq ft. Horizontal lines, drawn through points representing 60,000 and 120,000 values on the scale of the area graph to the area curve and thence downward to the shore line, located the end-points of the division lines for the lots.

12.27 □ Graphical differentiation

Curves of a lower order are derived through the application of the principle that the ordinate at any point on the derived curve is equal to the slope of a tangent line at the corresponding point on the given curve. The slope of a curve at a point is the tangent of the angle with the X-axis formed by the tangent to the curve at the point. For all practical purposes, when constructing a derivative curve, the slope may be taken as the rise of the tangent line parallel to the Y-axis in one unit of distance along the X-axis, or the slope of the tangent equals y_1/k as shown in Fig. 12.34.

Figure 12.34 illustrates the application of this principle of graphi-

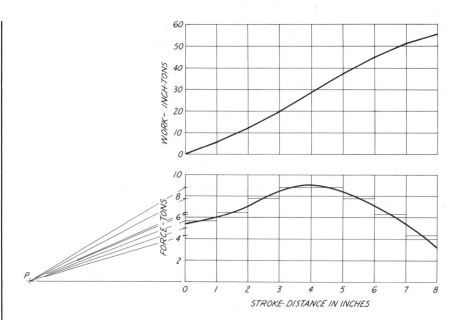

Fig. 12.32 □ Example of graphical integration.

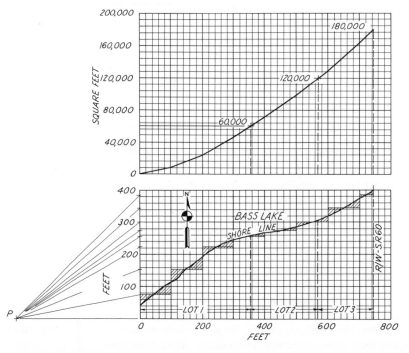

Fig. 12.33 □ Example of graphical integration.

cal differentiation. The length of the ordinate y_1' at point A' on the derived curve is equal to the slope y_1/k at point A on the given curve as shown in (a). When the slope is zero as at point C, the length of the ordinate is zero and point C' lies on the X-axis for the derived curve. When the slope is negative, as shown at D, the ordinate is negative and lies below the X-axis.

The graph shown in Fig. 12.35 is composed of segments of straight lines. Since the slope is constant for the interval 0–1, the derivative curve in the interval is a horizontal line. In the interval 1–2, the slope is also constant but of a lesser magnitude. Thus, the derivative curve is composed of straight line segments as shown in (b).

At this point in the discussion of graphical calculus it becomes possible to determine the relationship between the principles of integration and differentiation and to show that one is derived from the other.

From inspection of the graphs shown in Fig. 12.35, equations may be formulated as follows:

By the principle of differentiation

$$y_2' = \frac{y_2 - y_1}{k_2}$$

in the interval 1–2 where

$$\frac{y_2 - y_1}{k_2}$$

represents the slope of AB. The area under the curve in (b) in the interval 1–2 is equal to

$$y_2' \cdot k_2 = \frac{y_2 - y_1}{k_2} \cdot k_2 = y_2 - y_1 \text{ (integral curve)}$$

and

$$y_2' = \frac{y_2 - y_1}{k_2} \text{ (differential curve)}$$

In constructing a derivative curve, the determination of the tangent lines is often difficult because the direction of a tangent at a particular point is usually not well defined by the curvature of the graph. Two related schemes that may be used for constructing tangents are shown in Fig. 12.36(a) and (b). In (a) the tangent is drawn parallel to a chord of the curve, the arc of which is assumed to approximate the arc of a parabola. A sufficiently accurate location for the point of tangency T_1 may be determined by drawing a line from the mid-point of the chord to the arc parallel to an assumed direction for the diameter of the parabola. When

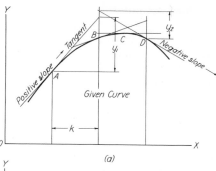

Fig. 12.34 □ Illustration of the principle of differentiation.

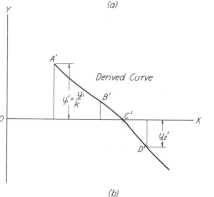

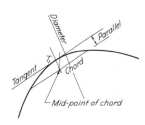

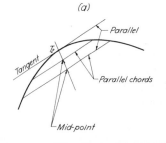

(a)

(b)

Fig. 12.36 □ Construction of a tangent line.

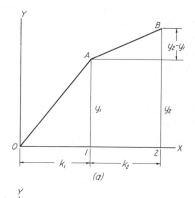

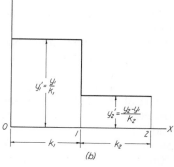

Fig. 12.35 □ Graphical differentiation.

working with small segments of the curve, one may assume the diameter to be either horizontal or vertical.

A more accurate construction is shown in (b) where the tangent is drawn parallel to two parallel chords. The point of tangency T_2 is determined by connecting the mid-points of the chords and extending this line to the curve. This line determines the direction of the diameter.

The construction in (b) using two chords to establish a tangent is applicable to any curve that may be approximated by a portion of a circle, ellipse, parabola, or hyperbola.

Since a tangent is assumed to be parallel to a chord, it is common practice to use chords instead of tangents for constructing a derivative curve, as shown in Fig. 12.37(a). The slope is plotted on an ordinate located midway in the corresponding interval of the derived curve.

A derivative curve can also be drawn using the ray polygon method as explained in Sec. 12.25 in reverse (see Fig. 12.38). The lines PA, PB, and PC of the ray polygon are drawn parallel to the tangents at points T_1, T_2, and T_3. Point Q is found by drawing a line horizontally from point A to the ordinate through the point of contact of the tangent parallel to PA. Points R and S are found similarly.

12.28 □ Drawing a derivative curve by the pole method

Before attempting to construct a derivative curve by the pole method, the beginner must have a clear understanding of the "slope law" as it applies to the differentiation, and he must be able to think of the derived curve being sought as the curve of slopes (slope locus) for the given curve. If deemed necessary, the first part of the discussion given in the previous section should be reviewed before continuing. In making a start, the given curve should first be analyzed and certain facts noted concerning the slope at specific points along the curve. For example, in Fig. 12.39, it should be recognized that the slope for the first and second intervals is constant and positive, a fact indicated by the single straight line sloping upward across both strips. In the third interval (2 to 3), the curve has a constant zero slope; while again, for the fourth and fifth intervals, the slope is constant and positive. Finally, the curve for the last three intervals has a constant negative slope, because the straight line, this time downward, indicates decreasing values.

A pole distance must now be selected, remembering that the distance in abscissa units determines the ordinate scale ratio for both curves. Then, with the location of the pole point P established, rays are drawn, parallel to the straight lines of the given curve until they intersect the Y-axis for the derived curve. These

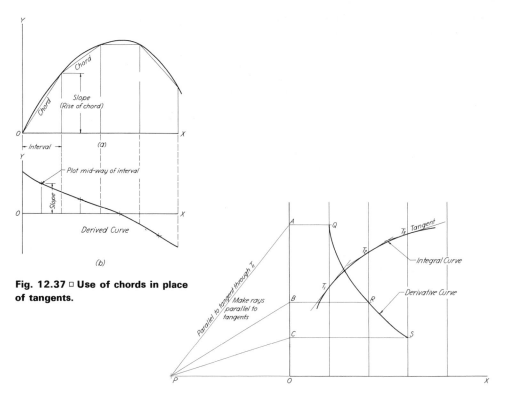

Fig. 12.37 □ Use of chords in place of tangents.

Fig. 12.38 □ Ray polygon method.

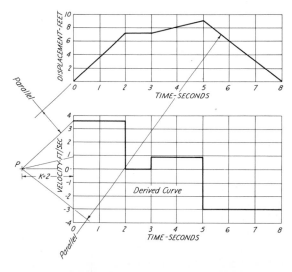

Fig. 12.39 □ Derivative curve constructed by the pole method.

228

points of intersection, projected horizontally, determine the derived curve that is composed of straight lines in this particular case.

In Fig. 12.40, the derived curve (slope locus) was obtained from the given (upper) curve by using tangents. Through the pole point P, a ray was drawn parallel to each of the several tangent lines. Then, the intersection point of each ray with the Y-axis was projected horizontally across to the corresponding ordinate as shown. The drawing of a smooth curve through these points completed the graph.

Chords were used instead of tangent lines in Fig. 12.41. The procedure that is followed here may be thought of as being the reverse of the chordal method illustrated in Fig. 12.29. As the initial step for the construction, chords of the given curve were drawn as shown. In selecting these chords, an effort was made in every case to see that the curve would be nearly symmetrical about a perpendicular bisector of the subtended chord. Also, as is appropriate and desirable, shorter chords were used where the curvature is sharper. After the pole point P had been located and the axes established for the derivative curve, rays were drawn through P, parallel to the chords, to an intersection with the Y-axis. Next, the tangent point (such as T_1) was located for each of the subtended arcs of the given curve, using the method shown at the right in (b). In doing this it was assumed that the chords would be parallel to tangents at points, such as T_1, where the perpendicular bisector of the chord would cut the subtended arc. These tangent points, when projected downward to corresponding horizontal lines drawn from the points along the Y-axis, at the rays, determine the derivative curve. For example, in (b) point U_1 in interval A is found by projecting T_1 downward to a horizontal line drawn through the point of intersection of the Y-axis and the related ray through P. The related ray in this case is the one that is parallel to the chord of the arc of the given curve in interval A. The desired derivative curve was completed by drawing a smooth curve through U_1 and the several other points shown, all of which were located in the same manner.

When using the chordal method to determine a derivative curve, some people frequently prefer to use equally spaced ordinates and then confine each chord to a single interval. Although this procedure proves to be quite satisfactory in many cases, there are times when desired results cannot be obtained for there are either no points or too few points given to draw a smooth curve at critical locations. In such cases, some extra construction will be needed. However, the easiest and best solution would have resulted from a careful choice of chords of the given curve at the start.

Figure 12.42 shows the differentiation of a curve using chords instead of tangents. The given distance-time curve was plotted from data obtained for a passenger train leaving a small station

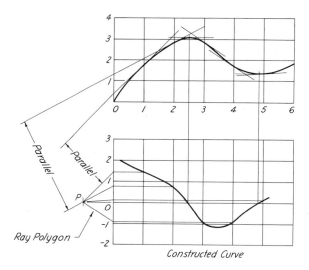

Fig. 12.40 □ **Use of tangents.**

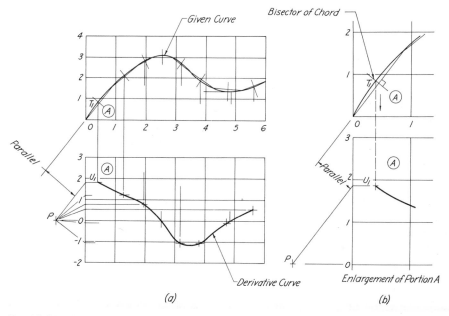

Fig. 12.41 □ **Chordal method.**

Graphs and charts for communication, design, and analysis □ **229**

near a large city. The velocity and acceleration curves reveal that the train moves with a constant acceleration for approximately 100 sec until it reaches a velocity of 55 mph. From this point it travels with a constant velocity toward its destination.

D □ Nomography: alignment charts

12.29 □ Alignment charts (nomographs)

The purpose of alignment charts is to eliminate many of the laborious calculations necessary to solve formulas containing three or more variables. Such a chart is often complicated and difficult to construct, but if it can be used repeatedly, the labor involved in making it will be justified. In the commercial field, these charts appear in varied forms, which may be very simple or very complicated (Fig. 12.43).

Briefly stated, the simplest form of alignment chart consists of a set of three or more inclined or vertical scales so spaced and graduated as to represent graphically the variables in a formula. The scales may be divided into logarithmic units or some other types of functions, depending upon the form of equation. As illustrated in Fig. 12.44, the unknown value may be found by aligning a straight-edge to the points representing known values on two of the scales. With a scale or triangle so placed, the numerical value representing the solution of the equation can be read on the third scale at the point of intersection.

Since alignment charts in varied forms are being used more and more by engineers and technologists, it is desirable that students studying in the fields dealing with the sciences have some knowledge of the fundamental principles underlying their construction. However, in any brief treatment, directed toward a beginner, it is impossible to explain fully the mathematics involved in the construction of the many and varied types (Fig. 12.43). Therefore, our attention here must be directed toward an understanding of a few of the less complicated straight-line forms with the hope that the student will gather sufficient knowledge to construct simple charts for familiar equations (Fig. 12.51).

12.30 □ Forms of alignment charts

Examples of some of the forms that alignment charts may have are shown in outline in Fig. 12.43. Examples of forms of proportional charts are illustrated in (a), (b), and (c). Miscellaneous forms are shown in (d), (e), and (f). One may obtain information needed for the construction of proportional type charts, concurrent scale charts, four-variable N-charts, and charts having a curved scale from any of the several books on nomography that are listed in the bibliography of this text.

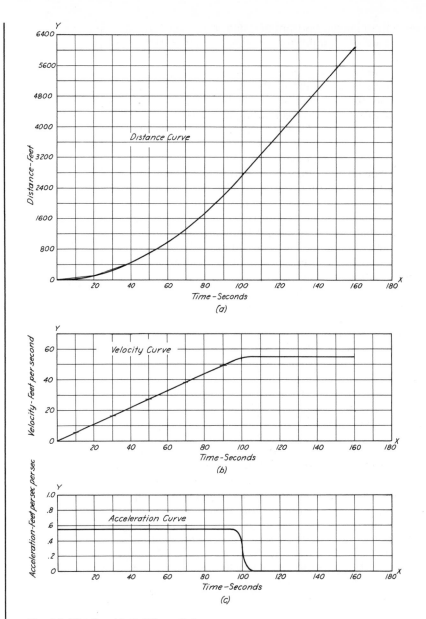

Fig. 12.42 □ **Graphical differentiation.**

12.31 □ Construction of simple alignment charts

In this limited study, the explanation for the constructions will be based on the principles of plane geometry. The two forms to be considered for formulas are the parallel-scale chart and the Z-chart, also called an N-chart (Fig. 12.44).

Without giving thought at this time to the geometry underlying the construction of alignment charts and to the selection of scales, the methods that might be used to construct simple charts for graphical addition and subtraction and graphical multiplication and division might well be considered [see Fig. 12.44(a) and (b)].

A parallel-scale alignment chart of the type shown in (a), prepared for the purpose of making additions and subtractions, could be constructed as follows:

Step I: Draw three vertical straight lines spaced an equal distance apart.

Step II: Draw a horizontal base line. This line will align and establish the origins (0) of the three scales.

Step III: Using an engineer's decimal scale, mark off a series of equal lengths on scales S_A and S_B. Start at the base line in each case. Mark the values of the graduations upward on both scales starting with 0 at the base line.

Step IV: Mark off on the S_C scale a series of lengths that are half as long as those on scales S_A and S_B. Number the graduation marks starting at the base line.

In using this chart to add two numbers, say 2 and 4, one may align the ruling edge of a triangle through 2 on the S_A scale and 4 on the S_B scale, then, read their sum at the point where the edge of the triangle crosses the S_C scale (see line X). To subtract one number from another, say 8 from 14, the edge of the triangle should be placed so as to pass through 8 on scale S_A and 14 on S_C. The difference, read on the S_B scale, will be 6 as shown by line Y.

If logarithmic scales are used for this form of chart as in (b) instead of natural scales as in (a), a chart for multiplication and division results, for the log of the product of two numbers is equal to the sum of the logs of the factors. Thus, by a method of addition a product can be obtained.

Necessary information for the construction of logarithmic scales is given in Sec. 12.33.

A Z-chart (also called an N-chart), which will give the product of two numbers, is shown in (c). For example, if line X is assumed to represent the edge of a triangle so placed as to pass through 5 on the S_A scale and 3 on the S_B scale, it can be seen that $3 \times 5 = 15$.

A simple Z-chart that has been prepared solely for straight multi-

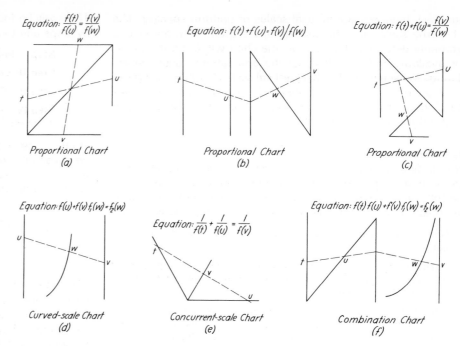

Equation: $\frac{f(t)}{f(u)} = \frac{f(v)}{f(w)}$

Proportional Chart
(a)

Equation: $f(t) + f(u) = f(v)/f(w)$

Proportional Chart
(b)

Equation: $f(t) + f(u) = \frac{f(v)}{f(w)}$

Proportional Chart
(c)

Equation: $f(u) + f(v) \, f_1(w) = f_2(w)$

Curved-scale Chart
(d)

Equation: $\frac{1}{f(t)} + \frac{1}{f(u)} = \frac{1}{f(v)}$

Concurrent-scale Chart
(e)

Equation: $f(t) \, f(u) + f(v) \, f_1(w) = f_2(w)$

Combination Chart
(f)

Fig. 12.43 □ Some common chart forms.

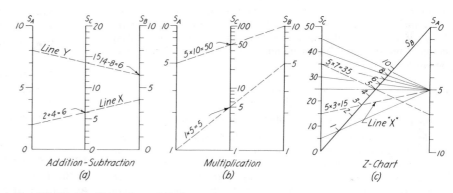

Addition-Subtraction
(a)

Multiplication
(b)

Z-Chart
(c)

Fig. 12.44 □ Parallel-scale and Z-charts.

plication will have outside vertical scales of uniform spacing. The length of the scales and the spacing of the graduation marks can be arbitrarily determined as long as the chart will fit on the paper that is available and provided that the graduations cover the desired range in each case. The two vertical scales begin with 0 (zero) value at opposite ends of the diagonal scale so that the values of the graduations read in increasing magnitude upward on one scale and downward on the other.

12.32 □ Definitions

Before starting a discussion on the construction of scales, it is necessary that the student have an understanding of the meaning of the following terms and expressions that are commonly used when constructing alignment charts for solving equations:

Constant. A quantity whose value remains unchanged in an equation.

Variable. A quantity capable of taking values in an equation. A variable is designated by some letter, usually one of the latter letters of the alphabet.

Function of a Variable. A mathematical expression for a combination of terms containing a variable, usually expressed in abbreviated form as $f(x)$ which is understood to mean "function of x." An equation usually contains several functions of different variables, such as $f(r) + f(s) = f(t)$; or $f(u) \cdot f(v) = f(w)$.

Functional Modulus. A proportionality multiplier that is used to bring a range of values of a particular function within a selected length for a scale. For instance, with the upper and lower limits of a function known and a definite length L chosen for the scale, the value of the functional modulus (m) can be found by dividing L by the amount of the difference between the upper and lower limits of the function. The scale equation for determining m may be written

$$m = \frac{L}{f(u_2) - f(u_1)}$$

where $f(u_2)$ and $f(u_1)$ are the upper and lower limits.

Scale. A graduated line that may be either straight or curved. When the graduation marks are equally spaced, that is, when the distance between marks is the same for equal increments of the variable as the variable increases in magnitude, the scale is known as a uniform scale. When the lengths to the graduation marks are laid off to correspond to scale values of the function of a variable, the scale is called a functional scale.

12.33 □ Construction of a functional scale

Let it be supposed that it is necessary to construct a functional scale, 5 in. in length, for $f(u) = u^2/2$ with u to range from 0 to

10. It will be found desirable to make the necessary computations by steps and to record the scale data in tabular form (Fig. 12.45).

Step I: Record the values of u in the table.

Step II: Compute the values of the function.

Step III: Determine the functional modulus m.

Step IV: Multiply the recorded values of the function by the functional modulus.

In this case and in many other cases, the functional modulus may be chosen by inspection. For this problem, the over-all length (L) of the scale will be 5 in. when $m = 0.10$. If the scale equation is used to determine the functional modulus then

$$m = \frac{L}{f(u_2) - f(u_1)} = \frac{5}{(10^2/2) - 0}$$
$$= \frac{5}{50} = 0.10$$

All that now remains to be done to construct the scale is to lay off the computed distances along the line for the scale and mark the values at the corresponding interval points (Fig. 12.46).

Although a logarithmic scale might be constructed in this same manner, much time can be saved by using the graphic method shown in Fig. 12.47 for subdividing the scale between its endpoints. To apply this method to a scale that has already been laid off to a predetermined length with the end-points of the range (say 1 to 10) marked, the steps of the construction are as follows:

Step I: Draw a light construction line through point 1 making any convenient angle with the scale line.

Step II: Using a printed log scale, mark off points on the auxiliary line.

Step III: Draw a line through the 10 point on the construction line and the 10 point on the scale, then, through the remaining points draw lines parallel to this line through the 10's. These will divide the scale in proportion to the logarithms of numbers from 1 to 10.

12.34 □ Parallel-scale charts for equations of the form $f(t) + f(u) = f(v)$

An alignment chart that is designed for solving an equation that can be set up to take this form will have three parallel functional scales that may be either uniform or logarithmic depending upon the equation. More information will be presented later concerning parallel-scale charts with logarithmic scales. In this section, attention will be directed to charts having scales with uniform spacing.

Before one can start constructing the chart, he must determine by calculation certain necessary information. First, he must determine how the scales are to be graduated, and second, he must calculate the ratio for the scale spacing.

To be competent to design parallel-scale charts, one must have a full understanding of the geometric basis for their construction. The explanation to follow is associated with the line layout in Fig. 12.48. Three parallel scales S_A, S_B, and S_C are shown with the origins t_0, u_0, and v_0 on line AB. Line (1) and line (2) are drawn parallel to AB through points v and u respectively on the isopleth. By similar triangles (shown shaded);

$$\frac{L_t - L_v}{a} = \frac{L_v - L_u}{b}$$

Now if

$$m_t = \frac{L_t}{f(t) - f(t_0)}$$

then

$$L_t = m_t[f(t) - f(t_0)]$$

When the function of t_0 is zero, the equation becomes

$$L_t = m_t f(t)$$

Similarly, when v_0 is zero,

$$L_v = m_v f(v)$$

and, when u_0 is zero,

$$L_u = m_u f(u)$$

Substituting these values,

$$\frac{m_t f(t) - m_v f(v)}{a} = \frac{m_v f(v) - m_u f(u)}{b}$$

Collecting terms,

$$m_t f(t) + \left(\frac{a}{b}\right) m_u f(u) = \left(\frac{a}{b}\right) m_v f(v) + m_v f(v)$$

$$= m_v \left(1 + \frac{a}{b}\right) f(v)$$

But $f(t) + f(u) = f(v)$ only if the coefficient of the three terms are equal, therefore,

$$m_t = \left(\frac{a}{b}\right) m_u = \left(1 + \frac{a}{b}\right) m_v$$

u	0	1	2	3	4	5	6	7	8	9	10
u^2	0	1	4	9	16	25	36	49	64	81	100
$u^2/2$	0	0.5	2.0	4.5	8.0	12.5	18.0	24.5	32.0	40.5	50.0
$m(u^2/2)$	0	0.05″	0.2″	0.45″	0.80″	1.25″	1.80″	2.45″	3.20″	4.05″	5.00″

Fig. 12.45 □ Table.

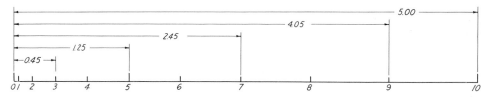

Fig. 12.46 □ Functional scale for $u^2/_2$.

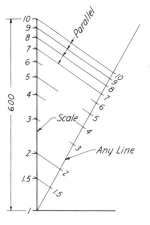

Fig. 12.47 □ Graduating a log scale.

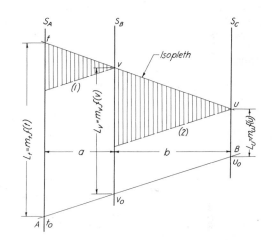

Fig. 12.48 □ Geometric basis for construction of parallel-scale alignment charts.

Graphs and charts for communication, design, and analysis □ **233**

and

$$\frac{m_t}{m_u} = \frac{a}{b} \qquad \text{Eq. (1)}$$

Now since

$$m_t = \left(1 + \frac{a}{b}\right)m_v$$

and

$$\frac{a}{b} = \frac{m_t}{m_u}$$

Then,

$$m_t = \left(1 + \frac{m_t}{m_u}\right)m_v$$

$$m_v = \frac{m_t}{\left(1 + \dfrac{m_t}{m_u}\right)}$$

and finally,

$$m_v = \frac{m_t m_u}{m_t + m_u} \qquad \text{Eq. (2)}$$

Now suppose that it is desired to construct a chart having the form of $t + u = v$ and that t is to have a range from 0 to 10 and u from 0 to 20 (Fig. 12.49). It has been determined that the scale lengths should be 6 in.

Then

$$m_t = \frac{6}{10.0} = \frac{6}{10} = 0.6;$$

$$m_u = \frac{6}{20.0} = \frac{6}{20} = 0.3$$

and,

$$m_v = \frac{m_t m_u}{m_t + m_u} = \frac{0.6 \times 0.3}{0.6 + 0.3}$$

$$= \frac{0.18}{0.90} = 0.20 \qquad \text{Eq. (2)}$$

To determine the ratio of the scale spacing:

$$\frac{m_t}{m_u} = \frac{a}{b} = \frac{0.6}{0.3} = \frac{2}{1} \qquad \text{Eq. (1)}$$

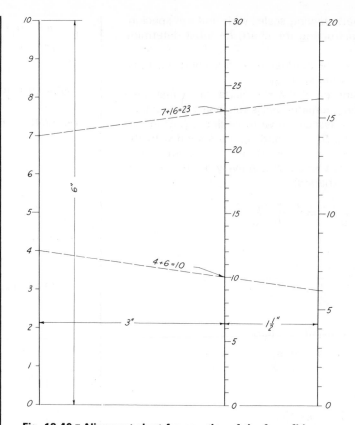

Fig. 12.49 □ Alignment chart for equation of the form $f(t) + f(u) = f(v)$.

For convenience, distance a between scales can be made 3 in. Distance b must then be $1\frac{1}{2}$ in. to satisfy the proportion of 2 to 1. The total width of the chart will be $4\frac{1}{2}$ in.

Since in this particular case the modulus for each of the scales is in full tenths and the scales are to be uniformly divided, an engineer's scale may be used to mark off the scales of the chart. With other conditions, it would be necessary to prepare either a table such as the one shown in Fig. 12.45 or to divide the line between its end-points using a geometric method.

12.35 □ Parallel-scale charts for equations of the form $f(t) \cdot f(u) = f(v)$

Equations of this form may be rewritten so as to take the form $f(t) + f(u) = f(v)$ by using logarithms for both sides of the equation. For example, suppose that it is desirable to prepare a chart for $M = Wl$, a formula commonly used by designers and engineers. In this equation, M is the maximum bending moment at the point of support of a cantilever beam; W is the concentrated load; and l is the distance from the point of support to the load.

Given:

$$M = Wl$$

Rewritten:

$$\log M = \log W + \log l$$

Another example would be the formula for determining the discharge of trapezoidal weirs.

Given:

$$Q = 3.367Lh^{3/2}$$

Rewritten:

$$\log Q = \log 3.367 + \log L + 1.5 \log h$$

In this last equation, Q is discharge in cubic feet per second; L is the length of the crest in feet (width of weir); and h is the observed head (depth of water).

For the purpose of our discussion, the formula $P = I^2 R$ will be used where P is power in watts; I is current in amperes; and R is resistance in ohms (Fig. 12.50). It has been determined that I must vary from 1 to 10 amp and R from 1 to 10 ohms. A chart of this type might be used to determine the power loss in inductive windings. The length of the scales is to be 5 in.

The steps for the construction are as follows:

Step I: Write the equation in standard form.

$$\log P = 2 \log I + \log R$$

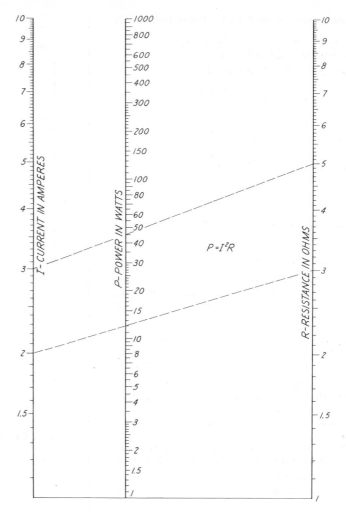

Fig. 12.50 □ Alignment chart for the equation $P = I^2R$.

Step II: Determine the moduli m_I and m_R for the outside scales.

$$m_I = \frac{5}{2 \log 10 - 2 \log 1} = \frac{5}{2} = 2.5$$

$$L_I = 2.5\,(2 \log I) = 5 \log I$$

$$m_R = \frac{5}{\log 10 - \log 1} = \frac{5}{1} = 5$$

$$L_R = 5 \log R$$

Step III: Determine m_P and L_P for the P-scale.

$$m_P = \frac{2.5 \times 5}{2.5 + 5} = \frac{12.5}{7.5} = \frac{5}{3}$$

$$L_P = \frac{5}{3} \log P$$

Step IV: Determine the ratio for the spacing of the scales.

$$\frac{m_I}{M_R} = \frac{2.5}{5} = \frac{1}{2} \text{ (ratio)}$$

Step V: Draw three vertical lines 1 in. and 2 in. apart and add a horizontal base line. By using these selected values, the ratio for spacing will be maintained and the chart will have good proportion.

Step VI: Graduate the scales for I and R using the method shown in Fig. 12.47. In this particular case, both logarithmic scales will be alike and will range from 1 on the base line upward to 10.

Step VII: Graduate the P-scale. By substituting values in the equation, it will be found that P will range from 1 w to 1000 w; therefore, the scale will be a three-cycle logarithmic scale, and the graphic method may be used for locating the graduation marks.

Figure 12.51 shows another parallel-scale alignment chart. This particular one could be used to determine the volume of a cylindrical tank when the diameter and height are known.

The formula for the volume of a cylinder, as given in the illustration, may be rewritten as

$$\log V = \log (\pi/4) + 2 \log D + \log H$$

Suppose that it has been decided, as in this case, that both the diameter (D) and the height (H) are to vary from 1 ft to 10 ft, and that the length of the two outside scales is to be 6 in. For the moment, the constant term $\log (\pi/4)$ can be ignored, for it can be accounted for at a later time by shifting the V-scale for the volume upward until it is in the position for furnishing a correct reading for a particular value as computed using the formula.

The moduli can be determined as previously explained.

$$m_D = \frac{6}{2 \log 10 - 2 \log 1} = 3$$

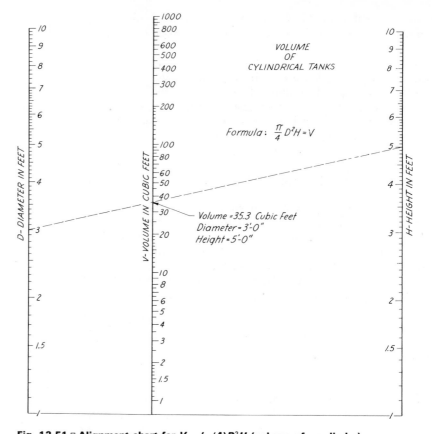

Fig. 12.51 □ **Alignment chart for $V = (\pi/4)D^2H$ (volume of a cylinder).**

and

$$L_D = 3(2 \log D) = 6 \log D$$

$$m_H = \frac{6}{\log 10 - \log 1} = 6$$

and

$$L_H = 6 \log H$$

$$m_V = \frac{m_D \times m_H}{m_D + m_H} = \frac{3 \times 6}{3 + 6} = \frac{18}{9} = 2$$

and

$$L_V = 2(\log V)$$

Scale spacing ratio:

$$\frac{m_D}{m_H} = \frac{3}{6}$$

For convenience, the scales can be spaced at 2 in. and 4 in., giving a chart that is a square in form.

The scales for D and H may be graduated by using the graphical method illustrated in Fig. 12.47. The scale for V will be a three-cycle scale with a 6-in. range of length between the 1 and 1000 values. Since the value of V at the base line must result from the substitution in the equation of the values of 1 and 1 for the other two scales, the resulting value of 0.7854 cu ft is the volume at the base line. The most convenient procedure to follow in graduating the V-scale is to start with the 0.7854 value at the base line. By so doing, the constant term, log $\pi/4$, is taken into account. Should one desire to determine the distance to be laid off along the V-scale from the base line to the 1, it will be found to be equal to two times the difference between the logs of the numbers in this particular case, for the value of m_v is 2.

12.36 □ Three-scale alignment chart—simplified (graphical) construction

After the student has acquired a thorough understanding of the theory of alignment charts, which includes a full knowledge of the geometric basis underlying their construction, he will be in a position to simplify his construction work and make it purely graphical through the use of tie-lines. Of course, this can only be done after the chart form has been identified, the equation converted, and the ranges for the variables (as needed) have been decided upon. The use of tie-lines to locate check points on the third scale of a three-scale alignment chart is shown in Fig. 12.52.

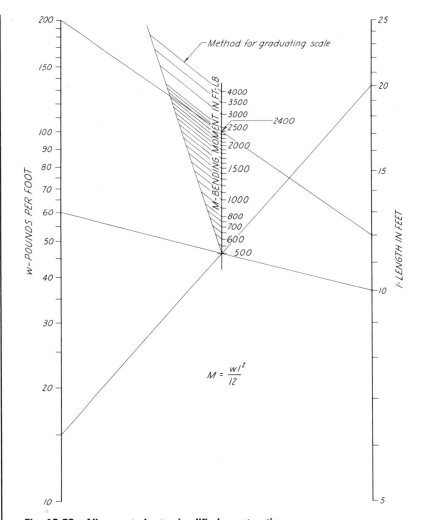

Fig. 12.52 □ Alignment chart—simplified construction.

Given:

$$M = \frac{wl^2}{12},$$

the equation for the bending moment (maximum) of a beam, having a uniformly distributed load and fixed at both ends. In this equation, w is the load in pounds per foot of length and l is the span in feet.

It was decided that the alignment chart would fulfill most needs if w had limits from 10 to 200 lb per ft and l had limits from 5 to 25 ft.

Equation rewritten:

$$\log w + 2 \log l = \log M + \log 12$$

The steps for the construction of the chart, with only minor calculations that may be performed mentally, are as follows:

1 Draw two vertical lines (say 5 in. long) at any convenient distance apart.

2 Graduate the w-scale (left-hand line) from the lower point 10 to the uppermost point 200. (In this case, use was made of the method shown in Fig. 12.47 to locate the scale values between 10 and 200.)

3 Graduate the l-scale with the lower point as 5 and the point at the top of the scale as 25. (Scale values located as in Step 2.)

4 Using a value of 10 for l and 500 for M yields $w = 60$ for the tie-line from the 60 value on the w-scale to the 10 value on the l-scale. Again, with $l = 20$ and $M = 500$, w is found to be 15 for the second tie-line through $M = 500$. The intersection of these two tie-lines is a check point on the third vertical scale, a point that establishes not only the position of the scale line but the 500 value as well.

5 Locate a second check point on the M-scale by drawing a third tie line. In this case it was found to be convenient to use $w = 200$ and $l = 12$ to yield $M = 2400$.

6 Graduate the M-scale as shown (read Sec. 12.33).

12.37 □ Graphical construction of a Z-chart

A simplified graphical construction may be employed to construct a Z-chart (N-chart) such as the one shown in Fig. 12.53, for the equation $V = 2.467Dd^2$, which gives the volume of a torus. The given equation was considered to be so arranged that the vertical V- and D-scales would be uniformly graduated. The d-scale on the diagonal representing a variable to a power, is nonuniformly graduated. For this form of chart, one should recall that the vertical scales begin at zero and run in opposite directions (see Sec.

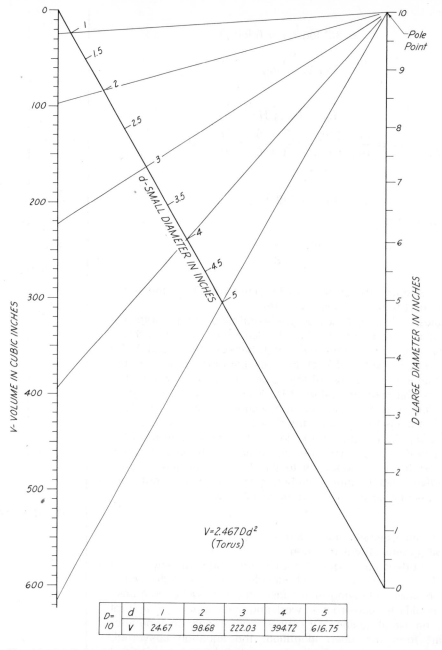

Fig. 12.53 □ Chart for equation $V = 2.467Dd^2$.

D=10	d	1	2	3	4	5
	V	24.67	98.68	222.03	394.72	616.75

12.31). The diagonal connecting these two zero points completes the usual N-shape.

The steps in making the construction are as follows:

Step I: Draw the scale lines for the variables V and D at any convenient distance apart, in this case 3.50 in. Then, lay off the D-scale to some selected length, say 6 in., and locate the zero point of V-scale opposite the 10 of the D-scale.

Step II: Graduate the V-scale uniformly, being certain that the range is sufficient to make possible a full reading when using the maximum values of the other scales.

Step III: Graduate the D-scale for the selected range of values, in this case from 0 to 10 in.

Step IV: Graduate the d-scale by selecting one convenient value of D and substituting it repeatedly in the original equation, each time with a different value of d which is to be marked on the diagonal scale. In so doing, values of V are found that establish the ends of tie-lines from the pole point (10) selected on the D-scale. For example, when the values 10 and 1 are substituted for D and d, respectively, in the given equation, V is found to be 24.67 cu in. With this known, a tie-line drawn from the pole to 24.67 on the V-scale, will locate $d = 1$ on the diagonal scale. The volumes for a range of d from 1 to 5 in. are shown in the table below the illustration. These values were used to draw the tie-lines shown.

12.38 □ Four-variable relationship—
parallel-scale alignment chart

A four-variable parallel-scale alignment chart may be constructed for an equation of the form $f(t) + f(u) + f(v) = f(w)$ when the equation has been rewritten as follows:

$$f(t) + f(u) = f(k)$$
$$f(k) + f(v) = f(w)$$

When rewritten in this form, the four-variable chart can then be constructed as two three-scale nomographs with the k-scale common to both. The k-scale, which serves as a pivot line in using the chart, need not be graduated but it must be remembered that it does have a modulus and does represent a function in both equations.

In Fig. 12.54, a four-variable alignment chart is shown for the addition of numbers. For this particular construction, it was arbitrarily decided that t should have a range from 0 to 5, u from 0 to 10, and v from 0 to 20. The procedure for the construction of this chart is as follows:

Step I: Draw the vertical lines (stems) for the t- and u- scales at any convenient distance apart and, by means of the engineer's scale or graphically, graduate t from 0 to 5 and u from 0 to 10.

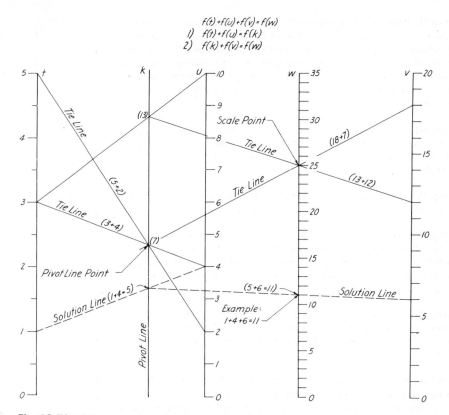

$f(t) + f(u) + f(v) = f(w)$
1) $f(t) + f(u) = f(k)$
2) $f(k) + f(v) = f(w)$

Fig. 12.54 □ Alignment chart for the equation $f(t) + f(u) + f(v) = f(w)$.

Step II: Draw the two intersecting tie-lines between the t and u scales that determine the position of the pivot line k at their point of intersection (7). Intersecting tie-lines could have been drawn from other graduation marks along the t and u scales as long as the sum of the values at the ends of one of the selected tie-lines equals the sum of the values at the ends of the other tie-line intersecting it.

Step III: Draw a third tie-line to locate another check point along the pivot line k. In this case, the tie-line was drawn from 3 on the t-scale to 10 on the u-scale to locate the mark for 13 on the k-scale.

Step IV: Decide upon the best location for the v-scale, then draw the stem line and graduate it uniformly from 0 to 20.

Step V: Using the 7 and 13 value marks that have now been established on the k-scale, draw the two tie-lines ($18 + 7 = 25$ and $13 + 12 = 25$) that will determine the location of the w-scale at their point of intersection as well as the position of the 25 graduation mark.

Step VI: Draw the w-scale and graduate it from 0 to 35, the 35 value representing the largest sum obtainable from the use of the t-, u-, and v-scales as they have been graduated for this particular chart.

The procedure for finding the sum of the numbers 1, 4, and 6 is illustrated by the broken lines extending between the scales. First, a straightedge can be laid across the t- and u-scales through the 1 and 4 values, as shown, to determine the position on the pivot line (k) that the graduation mark would have that represents their sum, in this case 5. Then the straightedge should be shifted so that its edge will pass through the 5 position on the pivot line and the graduation mark for the 6 value on the v-scale. The sum of the three given numbers can then be read as 11 on the w-scale.

The four-scale alignment chart, shown in Fig. 12.55, for the computation of simple interest, may be easily constructed once the equation $I = PRT$ has first been written in the logarithmic form of

$$\log P + \log R + \log T = \log I,$$

and then rewritten as

$$\log P + \log R = \log k \qquad (1)$$

and

$$\log k + \log T = \log I \qquad (2)$$

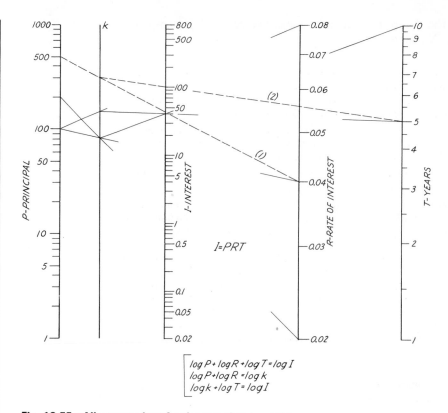

Fig. 12.55 □ Alignment chart for the equation *I* = *PRT*.

Graphs and charts for communication, design, and analysis □ **240**

The equation as first rewritten is in the form of $f(t) + f(u) + f(v) = f(w)$ and as arranged in (1) and (2) in the form of

$$f(t) + f(u) = f(k)$$
$$f(k) + f(v) = f(w)$$

Now, since equations (1) and (2) have the same form as those for the previous problem, the alignment chart for the equation $I = PRT$ is constructed similarly except that logarithmic scales must be used in this case in place of natural (uniformly divided) scales. As before, tie-lines were used to position the k and I scales and to locate needed check points. The procedure for reading the chart is illustrated by the broken lines, numbered (1) and (2).

The reader is urged to follow through the construction of the alignment chart for the equation $I = PRT$ mentally, step by step. At the start, however, he must recognize that the k-scale belongs first to the PRk chart and then to the kTI three-scale chart that together make the complete chart for the four variables P, R, T, and I.

12.39 □ Use of alignment charts in industry
Alignment charts are used widely by designers in industry. An example of one of the many alignment charts used by engineers working for one of the world's largest industrial organizations can be seen in Fig. 12.56. Its use has been explained on the chart.

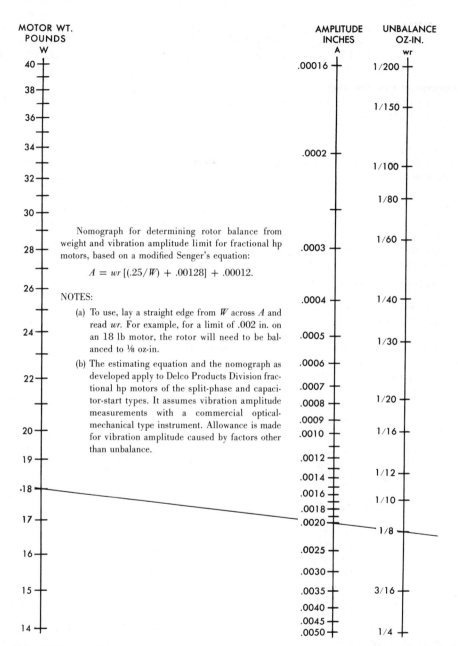

Nomograph for determining rotor balance from weight and vibration amplitude limit for fractional hp motors, based on a modified Senger's equation:

$$A = wr \, [(.25/W) + .00128] + .00012.$$

NOTES:

(a) To use, lay a straight edge from W across A and read wr. For example, for a limit of .002 in. on an 18 lb motor, the rotor will need to be balanced to ⅛ oz-in.

(b) The estimating equation and the nomograph as developed apply to Delco Products Division fractional hp motors of the split-phase and capacitor-start types. It assumes vibration amplitude measurements with a commercial optical-mechanical type instrument. Allowance is made for vibration amplitude caused by factors other than unbalance.

Fig. 12.56 □ An alignment chart.
(*Courtesy General Motors Corporation.*)

Problems

Technical graphs and charts

The following problems have been designed to emphasize the fundamental principles underlying the preparation and use of technical graphs and charts.

1 Determine the values for the following equations, as assigned, and plot the curve in each case for quantitative purposes.

Parabola	$Y = 4x^2$, x from 0 to 5
Ellipse	$Y^2 = 100 - 2x^2$
Sines	$Y = \sin x$, x from 0° to 360°
Cosines	$Y = \cos x$, x from 0° to 360°
Logarithms	$Y = \log x$, x from 1 to 10
Reciprocals	$Y = 1/x$, x from 1 to 10

2 In a hydraulics laboratory, the construction of a quantitative curve that would give the weight of water contained in tubes of various diameters and lengths was desired. This was accomplished by filling tubes of known diameters with water to a depth of 1 ft and observing the weight of water thus added. The water was kept at a temperature for maximum density and the following data were obtained:

D = Diameter of tube in inches	W = Weight of 1-ft column of water
2	1.362
2½	2.128
3	3.064
3½	4.171
4	5.448
4½	6.895
5	8.512
5½	10.299
6	12.257
6½	14.385
7	16.683
7½	19.152
8	21.790

On a sheet of graph paper [Fig. 12.3(b)], plot the above data. Place the axes 3 cm in from the edges. Letter the title in any convenient open space.

3 Owing to the uncontrollable factors, such as lack of absolute uniformity of material or test procedure, repeated tests of samples of material do not give identical results. Also, it has been observed in many practical situations that

a Large departures from the average seldom occur.

b Small variations from average occur quite often.

c The variations are equally likely to be above average and below average.

The foregoing statements are borne out by the accompanying data showing the results of 4,000 measurements of tensile strength of malleable iron.

On a sheet of coordinate graph paper [Fig. 12.3(a) or (b)] prepare a graph showing frequency of occurrence of various strength values as ordinates and tensile strength as abscissa. Draw a smooth symmetrical curve approximating the given data.

Range of tensile strength values in pounds per square inch	Number of observations
Under 45,000	0
45,000–45,999	1
46,000–46,999	2
47,000–47,999	3
48,000–48,999	6
49,000–49,999	20
50,000–50,999	232
51,000–51,999	376
52,000–52,999	590
53,000–53,999	740
54,000–54,999	771
55,000–55,999	604
56,000–56,999	383
57,000–57,999	184
58,000–58,999	60
59,000–59,999	20
Over 60,000	8
	4000

4 On a sheet of paper of the type shown in Fig. 12.3(b), plot a curve to represent the data given below. (Note: For *stress-strain diagrams*, although the load is the independent variable, it is plotted as ordinate,

contrary to the general rule as given in Sec. 12.5. Figure 12.6 shows a similar chart. In performing tests of this nature, some load is imposed before any readings of elongation are taken.)

It is suggested that the label along the abscissa be marked "Strain, 0.00001 in. per in."; then fewer figures will be required along the axis.

Stress, pounds per square inch	Strain, inches per inch
3,000	0.0001
5,000	0.0002
10,000	0.00035
15,000	0.00054
20,000	0.00070
25,000	0.00090
30,000	0.00106
32,000	0.00112
33,000	0.00130
34,000	0.00140

Graphical calculus

The following problems have been designed to emphasize the fundamental principles underlying graphical integration and differentiation and to offer the student an opportunity to acquire a working knowledge of the methods that are commonly used by professional engineers.

5 A beam 10 ft long is uniformly loaded at 20 lb per ft as shown in Fig. 12.57. Plot distance in feet along the X-axis. Draw (1) the integral curve to show the shearing force, and (2) the second integral curve to show the bending moment.

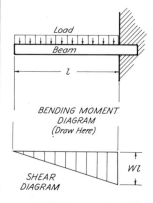

Fig. 12.57

6 A beam 15 ft long and supported at both ends is loaded uniformly at 18 lb per foot as shown in Fig. 12.58. Plot distance in feet along the X-axis and load along the Y-axis. Draw (1) the integral curve to show the shearing force, and (2) the second integral curve to show the bending moment.

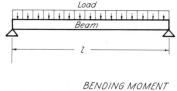

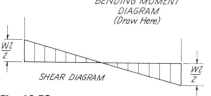

Fig. 12.58

7 Plot the points given in the table and draw a smooth curve. Construct the derivative curve. Write the equation of the derivative curve.
Equation of given curve $y = \frac{1}{2}x^2 + 10$.

x	0	1	2	3	4	5	6	7	8	9	10
y	10	10.5	12	14.5	18	22.5	28	34.5	42	50.5	60

8 Construct the distance-time, velocity-time, and acceleration-time curves for an automobile moving as follows: time = 10 sec, acceleration = 5 ft per sec per sec throughout the interval, initial velocity = 0.

9 The passenger train for which derived curves are shown in Fig. 12.42 is brought to a stop with a constant negative acceleration of 1.0 ft per sec per sec. Before applying the brakes the train was traveling with a constant velocity of 55 ft per sec. Construct the curves showing acceleration-time, velocity-time, and distance-time relationships.

Alignment charts

The following problems have been designed to emphasize the fundamental principles underlying the preparation and use of alignment charts.

10–20 Prepare an alignment chart for the given equation. Chart scales should have a sufficient number of division marks to enable the user to obtain some reasonably accurate results (readings). In each problem the range for two of the variables has been given. The range of the third variable must make possible the use of the full range of each of the other two variables.

10 Construct an alignment chart for the multiplication of numbers from 1 to 100 (read Sec. 12.31 and study Fig. 12.44).

11 Construct an alignment chart of the form $t + u = v$. Let t vary from 0 to 10 and u from 0 to 15 (read Sec. 12.34 and study Fig. 12.49).

12 Construct an alignment chart for determining the area of a triangle. The student is to determine for himself the range for each of the scales.

13 Construct an alignment chart for determining the volume of a cylinder. The diameter is to range from 1 to 5 in. and the height from 1 to 10 in. The volume is to be in cubic inches (read Sec. 12.35).

14 Make an alignment chart for determining the volume of a paraboloid, $V = 1/8\pi ab^2$, where a is the length (measured along the axis) and b is the diameter of the base. Let a vary from 1 to 20 in. and b from 1 to 10 in.

15 Make an alignment chart for the discharge of trapezoid weirs, formula $Q = 3.367Lh^{3/2}$. Q is discharge in cubic feet per second; L is length of crest in feet (width of weir); and, h is the head (depth of water). Let L vary from 1 to 10 ft and h from 0.5 to 2 ft (read Secs. 12.35 and 12.36).

16 Make an alignment chart for the formula $P = I^2R$ as explained in Sec. 12.35. Let I vary from 1 to 20 amp and R from 1 to 10 ohms.

17 Make an alignment chart for the formula $R = E/I$ where E is the electromotive force in volts; I is current in amperes; and, R is resistance in ohms. Let R vary from 1 to 20 ohms and I from 1 to 100 amp (read Secs. 12.35 and 12.36).

18 Make an alignment chart for the formula $I = bd^3/36$ where I is the moment of inertia of a triangular section; b is the length of the base in inches; and d is the depth of the section (altitude of triangle). Let b vary from 1 to 10 in. and d from 1 to 10 in. (read Secs. 12.35 and 12.36).

19 Make an alignment chart for the formula $M = wl^2/8$ where M is the bending moment in foot-pounds, w is the load in pounds per foot, and l the length of span in feet. Let w vary from 10 to 200 lb per ft and l from 5 to 25 ft (see Fig. 12.52).

20 Make an alignment chart (Z-chart) for the equation for the volume of a right circular cylinder, $V = \pi r^2 h/144$, where V is the volume in cubic feet, r is the radius of the base in inches and h is the height in feet. Let r vary from 0 to 20 in. and h from 0 to 20 ft.

Part 4

Computer aided design

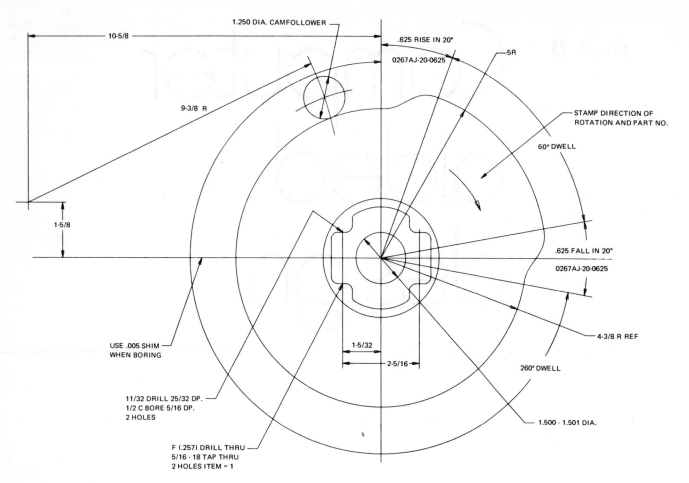

10-5/8

1.250 DIA. CAMFOLLOWER

.625 RISE IN 20°

0267AJ-20-0625

5R

9-3/8 R

STAMP DIRECTION OF
ROTATION AND PART NO.

60° DWELL

1-5/8

.625 FALL IN 20°

0267AJ-20-0625

4-3/8 R REF

USE .005 SHIM
WHEN BORING

1-5/32

2-5/16

260° DWELL

11/32 DRILL 25/32 DP.
1/2 C BORE 5/16 DP.
2 HOLES

1.500 - 1.501 DIA.

F (.257) DRILL THRU
5/16 - 18 TAP THRU
2 HOLES ITEM = 1

The Westinghouse Electric Corporation uses a CalComp plotter on-line with an IBM 360 computer to aid in the design of special purpose cams.
The layout drawing shown was prepared using a computer program developed at the Lamp Division of this company. It is reported that the typical cam design now requires approximately five minutes card punching time, two minutes computer time, and five minutes or less plotting time for a total of twelve minutes. Manual methods for the same task usually required twenty-five to thirty hours. (*Courtesy Westinghouse Electric Corporation.*)

Chapter 13

A □ The computer and computer graphics

13.1 □ The computer (Fig. 13.1)

A computer is not an electronic brain, as many would have us believe. Rather, it is an electronic calculating machine working in a manner that would be considered inefficient if it were done by a human being, since essentially the machine only adds and subtracts. And yet, by doing what it has been directed to do, with greater than human speed, the computer is capable of masterful performances that are carried out with technical precision. At times, it seems almost to perform miracles—but only almost.

A computer differs from what one ordinarily considers a calculator in that it has components for an internal memory storage system for instructions. It can be said that a computer can calculate, store, compare, correct itself, and then calculate some more *as programmed*.

13.2 □ Use of the computer (Fig. 18.1)

Computers must be instructed in considerable detail. However, the computer program represents more than just detailed instructions. It includes the problem definition, analysis, and flow charting that are a part of the initial preparation (Fig. 13.2). When the program is written in actual machine coding, the programmer must first analyze the problem in terms of operations that the computer can perform and then write the program supplying tables, formulas, codes, and so forth, as needed for the specific application. To do this, the programmer must understand the computer in detail.

Computer-aided design and automated drafting

Since this method of programming is often difficult, even for a qualified person, and at times is found to be impractical, several recognized programming languages have been developed that permit the programmer to give instructions to the computer by using statements and symbols, each statement representing many machine language instructions (Fig. 13.3). Use of these high-level languages, such as FORTRAN (FORmula TRANslation), eliminates much of the painstaking detail of computer programming. Furthermore, since FORTRAN statements closely resemble mathematical terminology, a working knowledge of the language can be acquired with about a week of instruction. With a program called a compiler, the computer translates FORTRAN programs into more detailed instructions for its own operation. Other programming languages are COBOL (COmmon Business Oriented Language), and ALGOL (ALGOrithmic Language).

Instructions, fed to the computer by punched cards, paper tape, or magnetic tape (Fig. 13.4), constitute the so-called program.

13.3 □ Language of the computer (input)

Punched cards, punched tapes, and magnetic tapes (Fig. 13.4) with information recorded as magnetized spots (called *bits*) are the conventional means for making contact with an electronic computer in the only language that the computer recognizes, the language of electrical impulses. The information supplied by these patterns of holes or magnetic spots are translated by a "sensing element" or "feeler" into the on–off binary language of the computer. Up-to-date punched-tape and punched-card sensing elements use photoelectric cells that transmit current when light reaches them. Paper tapes are punched by an input tape-punching device.

On magnetic tape, data are recorded in continuous parallel tracks. Usually, there are either six or eight data tracks and one checking track, depending on the format. Clocking mechanisms and changes in magnetic polarity indicate separate bits in the tracks. (A bit is a binary digit, either 0 or 1.) In the tape drive, a magnetic read–write head simultaneously reads a bit at a time from each of the tracks. The six or eight bits read represent a complete character. (Eight bits, called a *byte*, can represent two decimal digits or one character.) Using binary-coded decimal coding, the letters of the alphabet as well as decimal numbers may be recorded on magnetic tape. The BCD format also provides for punctuation marks and permits the use of special characters.

Punched cards, such as the one shown in Fig. 13.4, are stacked into a "card reader," which senses the cards and translates the information furnished by the perforations into the pulse language of the computer.

To permit more efficient use of a computer, the pulses punched into the paper may be transferred first to a magnetic tape and

Fig. 13.1 □ A computer system.
Shown is the IBM System/370 Model 155 with a variety of input and output units. The keyboard/printer operator console, attached to the Model 155's central processing unit (center), can be duplicated in a second key operational area of an installation to increase the efficiency of computer room personnel. While the central processor performs arithmetic and logical operations at speeds measured in billionths of a second, it can draw upon vast amounts of data stored in magnetic disk units such as the IBM 2314, at right. The unit at right foreground represents the new IBM 3211 printer that operates at 2,000 lines a minute. (*Courtesy International Business Machines Corporation.*)

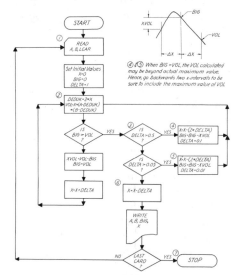

Fig. 13.2 □ Flow chart for maximum volume problem (see Fig. 13.3).
A flow chart presents in graphical form the logic of a programming task facing the programmer.

then from it to the computer. Magnetic tapes can deliver magnetized pulses at the rate of 180,000 characters (or 360,000 decimal digits) per second, a stretch of time in which a human can say only "one and two" at the usual word rate of a speaking voice.

13.4 □ Computer storages

Every computer must have a memory system; that is, it must have repositories for the storage of information. Input and output storages receive data needed for calculation and note the results before these results are typed out by the high-speed printer. There are also intermediate (short) storages that receive results that the computer will need for continued calculation. A special component for intermediate storage, the accumulator, receives totals to be returned immediately to the matrix.

The three categories of computer storage are (1) the main memory; (2) auxiliary storage, which includes direct-access devices such as magnetic disc units and magnetic drums; and (3) bulk storage, involving the use of punched cards and magnetic tape. (See Fig. 13.1.)

13.5 □ Computer output

Up to this point, we have focused our attention on how information is fed into electronic computers by means of an input unit and to some extent on how the computer digests data and makes the necessary calculations. All of this action would quite naturally be useless, however, if the computer could not express the results of the calculations (Fig. 13.3). The apparatus needed for this purpose is called an *output unit*. Several types of units are available to satisfy different requirements.

One such unit is the tape punch having perforating pins that are actuated by the electrical impulses of the computer. A tape-punch unit is capable of punching 50 or more characters per second in the paper tape. Punched tapes are deciphered by a printer that converts every character into either a letter or a number. Output card punches are also available. Punched-card information may be converted to printed copy through the computer or by using special auxiliary equipment.

When the results obtained represent an intermediate step and there is no need to read the results at the time, the pulses issued by the computer are picked up on magnetic tape to be held in reserve, say, for use a month later. When it becomes necessary to learn what is on a magnetic tape, it may be run through a complicated electronic device that sends the magnetic impulses to a high-speed output printer. This high-speed printing device is almost a miracle in itself because it can print a full line at a time (up to 132 characters) at a rate of 1,100 lines a minute.

```
$JOB        NCS,FE,BE464/HAMMOND,TIME=002,PAGES=010
C A PROGRAM TO DETERMINE THE LARGEST VOLUME THAT CAN BE FORMED FROM A
C RECTANGULAR SHEET OF DIMENSIONS A BY B BY CUTTING SQUARES OF SIZE X FROM THE
C FOUR CORNERS OF THE SHEET AND THEN FOLDING UP THE RECTANGLES SO FORMED. THE
C FORMAT STATEMENTS LIMIT THE VALUES OF A AND B TO 100.00 IF THE DECIMAL POINT
C PUNCHED ON THE DATA CARD, AND 1000.00 IF THE DECIMAL POINT IS NOT REPEAT NOT
C PUNCHED.  IF LARGER VALUES ARE DESIRED THE FORMAT STATEMENTS MUST BE CHANGED.
C THE VALUE OF X WILL BE DETERMINED TO THE NEAREST 0.01 UNIT, AND THE VOLUME TO
C THE NEAREST 0.0001 UNIT.  DOUBLE PRECISION MUST BE USED BECAUSE OF THE LARGE
C VALUES AND THE SMALL DIFFERENCES IN THE VALUES.
1        DOUBLE PRECISION X,BIG,DELTA,DEDUK,VOL,XVOL
2      1 READ (1,100)A,B,LCAR
3    100 FORMAT(2F6.2,I3)
C SET INITIAL VALUE FOR SIDE OF SQUARE TO BE REMOVED.
4        X=0.0D0
C SET INITIAL VALUE OF VOLUME IN BIG. BIG WILL ALWAYS CONTAIN THE LAST VOLUME.
5        BIG=0.0D0
C SET FIRST VALUE FOR INCREMENTING THE SIZE OF THE SQUARE TO BE REMOVED.
6        DELTA=1.0D0
C DETERMINE THE AMOUNT TO BE DEDUCTED FROM THE SIDES OF THE BASIC RECTANGLE.
7      2 DEDUK=2.0*X
C DETERMINE THE VOLUME AFTER FOLDING UP THE SIDES.
8        VOL=X*(A-DEDUK)*(B-DEDUK)
C LOOK FOR A DECREASE IN VOLUME.
9        IF(BIG.GT.VOL)GO TO 3
C DETERMINE THE CHANGE IN VOLUME.
10       XVOL=VOL-BIG
C PUT VOLUME JUST CALCULATED INTO BIG.
11       BIG=VOL
C INCREMENT THE VALUE OF THE SIDE OF THE SQUARE.
12       X=X+DELTA
C REPEAT CALCULATIONS.
13       GO TO 2
C RUN THROUGH AN 'IF' FILTER TO DETERMINE THE VALUE OF THE NEXT INCREMENT TO BE
C USED FOR THE SIDE OF THE SQUARE.
14     3 IF(DELTA.GT.0.5)GO TO 4
15       IF(DELTA.GT.0.05)GO TO 5
16       GO TO 6
C RESET THE VALUE OF THE SIDE OF THE SQUARE BACK TWO INCREMENTS TO INSURE THAT
C IT IS LESS THAN THAT NECESSARY TO PRODUCE THE LARGEST VOLUME.
17     4 X=X-(2.0*DELTA)
C RESET BIG TO CORRESPOND TO THE X JUST SET. BIG MUST BE TRUNCATED TO REMOVE
C ERROR CAUSING RESIDUALS FOR THE NEXT TEST OF (BIG.GT.VOL).
18       IBIG=BIG-XVOL
19       BIG=IBIG
C SET NEXT VALUE OF INCREMENT.
20       DELTA=1.0D-1
C REPEAT CALCULATIONS.
21       GO TO 2
22     5 X=X-(2.0*DELTA)
23       IBIG=BIG-XVOL
24       BIG=IBIG
C SET NEXT VALUE OF INCREMENT.
25       DELTA=1.0D-2
C  REPEAT CALCULATIONS.
26       GO to 2
C THE VOLUME IS NOW CALCULATED TO THE DESIRED ACCURACY. SINCE THE LAST VOLUME
C SHOWED A DECREASE, BIG IS THE MAXIMUM VOLUME AND THEREFORE X MUST BE REDUCED
C BY ONE INCREMENT TO CORRESPOND.
27     6 X=X-DELTA
28       WRITE(3,200)A,B,BIG,X
29   200 FORMAT(1X,'IF A=',F7.2,' AND B=',F7.2,'. THE VOLUME=',F12.5,' WHEN
       1 X=',F6.2/)
C CHECK TO SEE IF LAST DATA CARD HAS BEEN READ.
30       IF(LCAR)1,1,7
31     7 STOP
32       END
```

```
$DATA
IF A=  25.00 AND B=  25.00, THE VOLUME=  1157.40685 WHEN X=  4.17
IF A=  25.00 AND B=  50.00, THE VOLUME=  3007.03181 WHEN X=  5.28
IF A=  25.00 AND B=  75.00, THE VOLUME=  4930.70458 WHEN X=  5.64
IF A=  25.00 AND B= 100.00, THE VOLUME=  6870.46676 WHEN X=  5.81
IF A=  50.00 AND B=  50.00, THE VOLUME=  9259.25815 WHEN X=  8.33
IF A=  50.00 AND B=  75.00, THE VOLUME= 16504.77956 WHEN X=  9.81
IF A=  50.00 AND B= 100.00, THE VOLUME= 24056.25877 WHEN X= 10.57
IF A=  75.00 AND B=  75.00, THE VOLUME= 31250.00000 WHEN X= 12.50
IF A=  75.00 AND B= 100.00, THE VOLUME= 47379.72378 WHEN X= 14.14
IF A= 100.00 AND B= 100.00, THE VOLUME= 74074.07185 WHEN X= 16.67
```

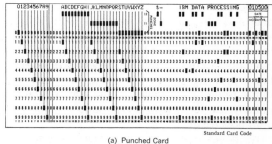

(a) Flat

(b) Folded

Fig. 13.3 □ Computer print-out of maximum volume problem (Fig. 13.2).
The program was written by Robert H. Hammond, Director of the Freshman Engineering Division-North Carolina State University.

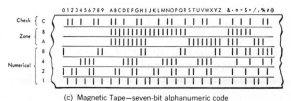

(a) Punched Card

Standard Card Code

(b) Paper Tape—eight channel code

(c) Magnetic Tape—seven-bit alphanumeric code

Fig. 13.4 □ Data recording media—punched card, papertape, and seven-track magnetic tape.
(Courtesy International Business Machines Corporation.)

Information may be transferred, without changing the code, from magnetic tapes to punched tapes or cards and, as also might be expected, from punched cards to magnetic tapes. This is accomplished by devices known as *converters*.

13.6 □ Communicating with a computer

Improvement in programming is gradually eliminating the formidable barrier that has existed between the computer and its prospective users. In the years ahead computers will become commonplace and they will have capabilities that will make it possible for them to be used by almost anyone. The day when each of us can consult with his own computer lies just ahead. With this thought in mind, it will be shown how man and computer can work together as partners, with man in control of the creative aspect of the design process. To instruct the computer, the user needs little more than a working knowledge of basic graphics, since the programming barrier is fading away to a great extent in this area.

13.7 □ Computer graphics

In addition to making extensive use of conventional computer techniques for engineering and scientific analysis, the engineering team can, with the aid of one of several graphic data processing systems, bring a large-scale digital computer to bear on those problems that can best be analyzed and finally formalized by using and modifying graphical representation. This great step forward in automated graphics makes it possible for an engineer or designer to do at least a part of his creative work on a graphic input–output cathode-ray console that is connected directly to the computer (see Fig. 13.5). The means for man–machine communication is not in the future; it is available now and several practical systems are in daily use for design and production work in the aerospace and automotive fields.

The direct man–computer interaction that this system affords leads to quick answers for even small problems and permits concepts to be evaluated and tested and then accepted or rejected. This relationship in which the computer has become man's immediate partner in creative design became possible only when a way was found to make the computer correctly interpret drawings and, if desired, make a drawing after the intentions of its human partner had been made known. Properly used, these systems can relieve both the engineer and the man at the drawing board of endless calculations and of much of the tedious work connected with making layout and detail drawings. Such systems bring together the talents and creativity of the user and the power and speed of the digital computer in conjunction with digitized plotters or CRT consoles with appropriate supporting equipment.

In general, the term computer-aided design implies that the com-

Fig. 13.5 □ Man-machine communication.
"At the General Motors Research Laboratories, using the new GM DAC-I system (Design Augmented by Computers), a research engineer checks out a computer program that allows him to modify a design 'drawing.' A touch of the electric 'pencil' to the tube face signals the computer to begin an assigned task, in this case, 'Line Deletion,' where indicated. The man may also instruct the computer using the keyboard at right, the card reader below the keyboard, or the program control buttons below the screen. Hundreds of special computer programs, written by GM computer research programmers, are needed to carry out these studies in man–machine communications."

puter assists the designer in analyzing and modifying previously created design within new design parameters that he, the designer, has established. Almost any designer or draftsman can use graphic data processing effectively and can analyze, modify, and distribute design information even though he may have only a limited knowledge of computer programming. This is true because the designer at the console and the computer are in two-way communication while using the designer's own graphic language of lines, symbols, pictures, and words.

The computer system described thus does not replace the man, nor does it eliminate the requirement that a designer have a working knowledge of engineering graphics. The graphic language, so fundamental to the design process, has now become a computer language. As such, it has proved to be a reliable means for a conversational type of man–machine communication.

The GM DAC–1 (Fig. 13.5) system that will be discussed in Secs. 13.8–13.10 is no longer in general use. However, full descriptions of its operations and capabilities have been given here in this text in order to point up the potential of a complete system of this type. Ordinarily, it is much more practical for an industrial organization to assemble a needed graphic system using units now available on the market than to develop a special "one-of-kind" system that may not justify the cost and may soon be out-dated. In a sense, it can be said that DAC–1 was developed as somewhat of a laboratory experiment to determine the potential of a total system of computer graphics in the area of design. As such, it represented a great leap forward at that time.

13.8 □ The main elements of the DAC–1 digital graphic input–output system

The main elements of the GM graphic data processing system shown in Fig. 13.6 are as follows:

1 A large-scale digital computer system (with supporting software) that is capable of storing vast amounts of information and retrieving specifically desired information rapidly.

2 One or more graphic consoles with viewing screens.

3 A film recorder (35 mm) for "hard copy" of graphic information that has been computer generated.

4 A 35 mm film scanner for the input of data in graphical form. This unit of the system converts images recorded on microfilm to digital form for further processing.

13.9 □ The DAC–1 display console

It is at the display console that the man–machine conversation takes place. The principal feature of the console is the cathode-ray-tube (CRT) screen on which the computer-programmed graphic

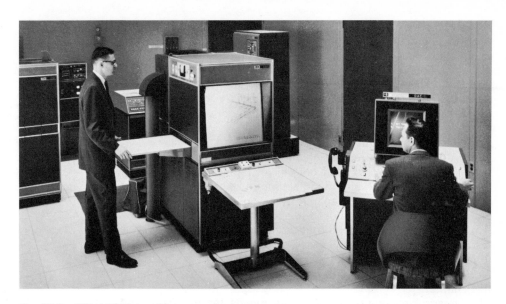

Fig. 13.6 □ GM DAC–1 graphic processing components.
(Courtesy of General Motors Corporation.)

and alphanumeric information is displayed (Fig. 13.5). Close at hand on the console are the keyboards and the light pencil that provide convenient means of entering and modifying rapidly displayed computer-programmed graphic and alphanumeric information. At the console unit the user has direct access to data (previously stored in the system's main or auxiliary storage units in digital form) that he may study and then modify and redisplay.

13.10 □ GM DAC–1 image processing unit

The DAC–1 computerized design and drafting system provides for both the input and output of data in a graphical form. A direct output for engineering and manufacturing use can be in the form of hard copy, a drawing prepared by an on-line plotter, or a programmed tape that can be used to actuate an automated production machine. At this point in our discussion our attention will be directed to the special image processor that enables the computer to read and generate drawings (Fig. 13.7). The use of tapes with plotters adaptable for off-line operation will be discussed later in this chapter.

The image processing unit provides means for the input and output of data in graphical form. Primarily, this component contains a CRT photo-recorder, projectors, a CRT photo-scanner, an input camera, and the rapid film processing equipment needed to develop 35 mm film. When the film is developed within the recorder, the image can be viewed on a rear projection screen that is located for easy viewing by the user (Fig. 13.6).

After an input copy has been inserted into the image processor on a tray and then photographed and developed, the 35-mm negative is positioned so that the CRT photo-scanner can "read it." The film scanner converts the image furnished by the 35-mm negative (light lines on dark background) directly to digital data that is sent either to the computer or to memory storage. Basically, the near miracle is accomplished via an electronic beam that, when it determines whether each of many addressable points is above or below a set level of light intensity, notes the result in computer (digital) language. The digital information thus developed may now be either modified or stored permanently.

From the high-speed recorder of the image processor, on 35 mm film, come the permanent records of drawings, charts, and parts lists that are needed by designers and draftsmen. The film, exposed to a high-resolution CRT and developed within the recorder, can be viewed at development on a rear projection screen, as stated previously. Finally, conventional working copies (on paper) may be readily obtained using modern viewer–copier equipment and recorder-produced exposures mounted in aperture cards. It was the development of image processing units, such as the DAC–1 described, that extended the capability of computers to

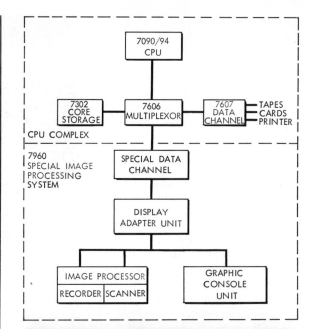

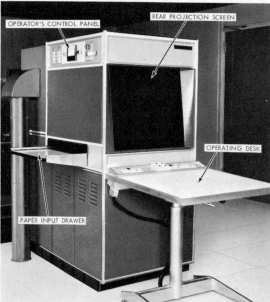

Fig. 13.7 □ **The DAC–1 image processing unit.**
(*Courtesy General Motors Corporation.*)

the point where they can accept, interpret, analyze, relate, and finally produce needed output in graphic form.

13.11 □ The IBM 2250 display unit (Fig. 13.8)

Images are generated on the CRT at the console by computer-programmed positioning and deflection of the CRT's electronic beam to the more than one million possible points (grid format of 1024 by 1024 addressable points) of the 12 in. × 12 in. effective display area. The beam is deflected to each point on the screen that is addressed by the program, either intensifying it or not intensifying it, as directed. The image as we see it on the face of the CRT (screen) is produced by the beam hitting a phosphor coating causing the coating to glow for just an instant. Operating under this condition, the information normally displayed fades rapidly (within a fraction of a second); therefore, the display must be regenerated approximately 30–40 times every second for a flicker-free representation of high resolution. A buffer is used to regenerate the display independently of the computer.

Alphanumeric symbols may be formed by synthesizing the characters from a series of individually programmed dots or line segments, either horizontal or vertical. To improve computer efficiency, an optional character generator can be used for computer-independent formation of all alphanumeric symbols.

The position-indicating light pencil, when used in conjunction with the keyboards and programs in storage, provides the means whereby the image displayed can be altered (Fig. 13.5). It may be used at will to create, remove (erase), enlarge, or rearrange any part of a displayed image that, as a whole, may consist of points, line segments of any length and orientation, simple and complex geometric shapes, and appropriate graphical symbols and statements. The light pencil (receiving device), having a photocell element that senses a spot of light directly under its point, signals the system. In turn, based on where the cathode-ray beam was scanning at that instant, the system sets a register, which indicates the position of the light pencil on the face of the display. When the control system feeds this positional information to the computer, the computer in turn relates the identified portion of the image to the equivalent digital representation stored in memory.

After a picture has once been generated, either the man with his light pencil or the computer may direct the other to reconsider a portion of the image. The computer may react, as programmed, to any portion of the display by placing an X at the point in the display that the man should consider. Thus, the machine suggests while the man directs. Each reacts to the other.

The programmed function keyboard (Fig. 13.8, bottom left) consists of program control keys, status indicator lights, and sensing switches for use with interchangeable (plastic) descriptive over-

Fig. 13.8 □ An example of a CRT display.

The IBM 2250, an advanced graphics display unit, permits users to exchange visual information with IBM's lowest-cost computing system, the desk-sized IBM 1130 (right). This particular unit was developed especially for engineers and designers who need fingertip access to their computers. The display enables them to work with charts, diagrams, drawings, or printed letters and numbers directly on the face of the television-like screen. With the electronic ''light pen'' the designer can revise images, add or delete lines or change dimensions. He also can change information and images on the screen through the 1130 and 2250 keyboards (note programmed function and alphanumeric keyboards). A civil engineer, for example could use the system to help design a building or a bridge, or an electrical engineer could use it to diagram new circuits. As the designer works he can call on the 1130 to perform calculations and update information previously stored in the system. If the user wants a more powerful computer, he can link his combined 1130/2250 system to a larger IBM System/360 through standard communication lines. (*Courtesy of International Business Machines Corporation.*)

lays. An overlay identifies a series of program functions contained in the computer memory system. When a key is depressed, the computer receives a signal to act on the image as directed by the program subroutine associated with the key selected. For example, the signal given might direct the computer to either enlarge, reduce, or delete a portion of the programmed image displayed. Any of these actions can be performed by the computer in a fraction of a second. To complete the same task manually might take minutes or even hours.

With the alphanumeric typewriter-type keyboard shown in Fig. 13.8, the user can compose statements consisting of letters, numbers, or symbols and can perform editing functions. The message, as composed, is displayed on the screen for verification. It may then be transferred on demand to the computer's main memory. The statement or message composed could be a parts list, a dimension value, an electrical value for a circuit component, a descriptive note, or a drawing title.

Through the use of graphic data processing as described, the designer can create a complete digital description of a design within the system.

13.12 □ Design programs

It has been stated in Section 13.7 that the designer can use a computer-aided design system even though he may know little about computer programming. This statement can be true only if it is possible for him to select from a library of previously created design programs those programs suitable for the design functions required for the specific job that faces him. Since it is not to be expected that he will be able to do this on all occasions, the designer or draftsman is fortunate if he has at least a basic understanding of how programs are prepared.

Programs and related design data are stored on tapes or disc files as digital descriptions and operational routines. Sets of words and numbers may represent standard graphic items, such as dimensions, part descriptions, and even formulas. In time, when graphic data processing systems have wider use and have been improved, more standard computer programs will become available and the tedious task of writing and developing programs may, to a great extent, be relegated to the past.

13.13 □ How a graphic data processing system is used

At present, how do we use an existing graphic data processing system?

First, the designer writes out statements that give a short description of his problem. In the case of auto-body design, the problem is very often a descriptive geometry problem requiring drawings and sketches as guides. These statements are then key-

punched on cards and entered into the computer memory. With this much accomplished, the designer, at the display console, is ready to command the computer to perform the functions requested. In executing the program, the computer checks each statement and displays an error comment if any error is detected. When and if this happens, the designer inspects the statement, makes the needed corrections, and instructs the computer to proceed.

Once the program has been executed, the computer should respond by displaying a simple line drawing, as shown in Fig. 13.5. In inspecting the solution to determine whether or not it is acceptable, the engineer using DAC–1 may enlarge any portion and view it from any angle. If not satisfied, he may then modify the design by entering other statements, by changing statement parameters, or by simply adding or deleting lines. When he is satisfied, he can, as has been explained in Sec. 13.10, direct the computer to produce either permanent copy or control tapes for automatic drafting machines or machine tools. Since the program is general, it may now be placed in the library of design programs for use at a later time for a specific job.

13.14 □ Design of electronic circuits

Using a graphic data processing system, a designer has almost unlimited flexibility in designing an electronic circuit because so little time is needed to rearrange and analyze components. For example, if a designer should wish to see what would happen if a component were changed (or eliminated) in an electronic circuit that has been displayed on the screen along with the waveform that circuit would produce, he can get an immediate answer. By entering the new value and working with his light pencil and the program in storage, he can see on the screen almost immediately the new waveform that would result from the change. At the same time, the old circuit has been stored in the computer's memory and may be recalled if needed.

13.15 □ Achievements of graphic input–output systems

Computers with graphic input–output systems are capable of performing marvelous feats. For example, either a single feature or a related group of features in a design display may be rotated, repositioned, enlarged, or reduced in size at will. When special attention must be given to an individual feature requiring development, that feature may be shown enlarged. This may be done with the rest of the graphic image out of view, that is, not showing on the screen. If need be, a standard element, such as a rivet or a cap screw that has been drawn and stored away in computer memory, may be added by pushing an appropriate button. Once the task of adding necessary details has been completed, the feature can again be brought back to scale and the complete image shown as before.

The first program to provide a graphical link with the computer was developed at Massachusetts Institute of Technology's Lincoln Laboratory some years ago. The first program, known as SKETCHPAD, proved beyond any doubt that a computer system could be developed that would interact with a creative designer.

It was found that a program could be written for an animated movie showing a small automobile moving along a winding road. Other programs were developed to show parts in motion, which, for example, made it possible to show linkages in action. SKETCHPAD proved to be an excellent draftsman, since, as straight lines, circles and curves were all drawn freehand, irregularities could be removed at a push of a button.

Programs exist for the creation of exotic shapes (airfoils, hydrofoils, etc.). A program, prepared at MIT under the supervision of Professor Coons, shows that by varying chords a shape can be changed accordingly and then reevaluated. In a motion picture showing actual use of this program, the shape displayed was rotated in alternate directions and tilted at different angles to permit the operator to view it from several directions, even from the rear.

In computer-aided design, new developments occur with increasing frequency and the "state of the art" advances steadily despite the fact that new achievements come only at great expense. As of today, we are moving away to some extent from gibberish geometric languages and English-like statements in using sketches to talk to the computer. In the near future, voice control may be added to the SKETCHPAD concept, and drawings and rough sketches coupled with the voice will be accepted methods for input. When this point has been reached, automated drafting and design will be an actuality and a computer graphic system will not be as inconvenient to use as it is now.

B □ Automated drafting

13.16 □ Digital drafting systems
Present models of automated drafting machines now on the market are digitally controlled. That is, the input (instructions) to the control system of the drafting machine is entered in digital form on punch cards, punch tape, or magnetic tape. The information supplied may be in either straight binary or binary-coded decimal form, the latter format being the most widely accepted.

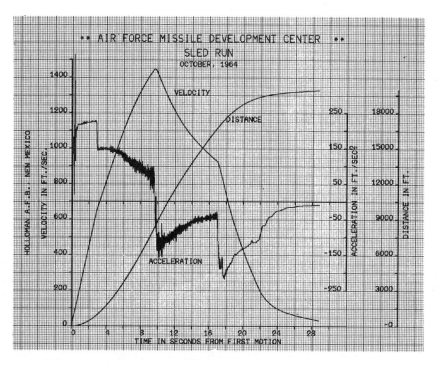

Fig. 13.9 □ An example of computer-plotter output.
Plotters make rough sledding easier at Holloman. Shown above is the plot of dynamic time history for a rocket sled run at Holloman Air Force Base high speed track. To cope with growing volume of computed data, the Air Force Missile Development Center at Holloman turns out a large percentage of its scientific and test data in this form. A general plot program and two CalComp Model 565 plotters are used to produce fully annotated linear or logarithmic charts, ready for direct reproduction. The pair of 565's is operated on-line with a CDC 3600 160A computer system. A time share program is used for simultaneous fullspeed operation of the plotters, three line printers, a card reader, a card punch, and a data link. (*Courtesy California Computer Products, Inc.*)

A few of the many fine automated drafting machines in use today are The Gerber Plotting System, The Orthomat (Universal Drafting Machine Corp.), The Coradomat, The CalComp System (California Computer Products—Fig. 13.9), and The IBM 1620 Drafting System with the 1627 Plotter.

The all-digital design approach of high-speed $X-Y$ plotting and drafting machines provides for flexible internal programming so that the plotter will execute on command a circle, a dashed line, a center line, and so forth. The Gerber Plotting System produces high-quality displays of computer-generated digital information. The Orthomat can automatically translate complex formulas into reliable line drawings.

When one is thinking of drafting machines, automated drafting may be defined as being the drawing of lines (ink or pencil) or the formation of scribe (groove) marks on a working surface with little or no human intervention (see Fig. 13.10). The medium used for the drawing may be either paper, plastic, film, metal, or any material with a scribe coat.

Of most interest to the user of an all-digital electronic plotter are the control system and the drafting equipment itself. Basically, the control system consists of a punch card or tape reader (punch or magnetic) and a translator. The translator converts input information received from the reader into a series of command signals to the servomotors of the drawing head. The drawing head responds as directed, checks its position, and then indicates that the command, as given, has been performed. Informing the control system when a command has been accomplished is made possible by a so-called feedback device, a mechanical–electrical system designed to provide mechanical error correction.

Depending on the model, a drafting machine can have either a horizontal, vertical, or tilting table. Although the horizontal table requires more floor space than a vertical table and cannot be viewed as readily, there are other advantages to its use that cause it to be favored for very accurate work. A tilting table can be positioned horizontally or it can be adjusted to be almost vertical.

A drawing head has a turret (revolving) mechanism that may have four, six, or eight stations, depending again on the model. The stations accommodate the pens, pencils, or scribes that can be selected and used under program control. Ordinarily Rapidograph-type pens are used; however, ball-point-type pens serve better for drawing at high speed. For wide lines felt marker pens have been used and found to be satisfactory.

A rotary scriber is primarily used for printed circuitry. The turret has the same basic station arrangement as the turret for the normal drafting machine. However, instead of pens at each of the positions there are small end-mill-type tools that cover a range of sizes.

The two types of control systems are the incremental system and

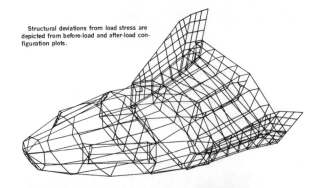

Structural deviations from load stress are depicted from before-load and after-load configuration plots.

Fig. 13.10 □ Analysis load configuration of the prime maneuverable re-entry vehicle.

The data processing center of this company is presently utilizing a plotting system to study load configurations on the prime maneuverable re-entry vehicle. A mathematical structure is shown being plotted in what is considered an unloaded configuration (see plot above). Then, theoretical load configurations are applied through the computer to different parts of the structure. New plotting tapes are generated and replots are made of the new load structure. By visually comparing before- and after-load plots, company engineers are able to determine deviation and stress of the structure due to loading. (*Courtesy Martin Marietta Corporation.*)

the absolute system. At the present time, the incremental with its delta motion is the most widely used. This system requires the ΔX and ΔY values for the distance from the "where-now" location to the next position. The somewhat different approach provided by the absolute system requires that the X and Y coordinates of the next point be related to the very first, that is, the starting point (set point or zero point).

In addition to the $X-Y$ coordinate information that is needed to meet position requirements, there are control commands associated with the drawing of a line segment, such as: move to origin, move and plot, pen up and move, pen down and move, rotate turret, draw dotted line, pause and change pen, and so forth.

Circles and curves must be approximated with straight lines. For these to be drawn accurately, acceleration and deceleration blocks and feed numbers must be calculated. These data can best be calculated by a computer.

The drafting machine (plotter) shown in Fig. 13.10 is suitable for a wide range of applications from contour map generation to automatic drafting and graphical verification of machine-tool tapes. A four-station drawing head is provided, with each station accommodating a standard holder for either an inking or a scribing device. The carriages are positioned by means of digital servo systems. The effective working area is 60 in. on both the X and Y axes for all styli (marking devices).

All control functions, operator functions and input–output devices are contained in the separate control unit, which includes a general-purpose computer. The computer includes a 4 K core memory and high-speed arithmetic unit.

The computer continuously monitors the position of the stylus carriage (marking device carrier) and controls the drive system as required by operator control and input–output devices. The standard input–output device is an ASR–33 Teletypewriter with paper tape reader and punch (magnetic tape input optional). Additional controls have been provided for manual entry. The digital servo control signals from the computer are processed within the control unit to form suitable outputs to the servomotors.

Performance capabilities

Straight-line interpolation is performed digitally. The format may be either coordinate, delta, or incremental.

Stored in prime form in core memory are 64 alphanumeric characters and symbols. These characters, having a prime height of .064 in., may be drawn by any of the styli, scaled up to 1.024 in., and rotated to eight angular orientations. With the X and Y coordinates of the first character of a line specified, subsequent charac-

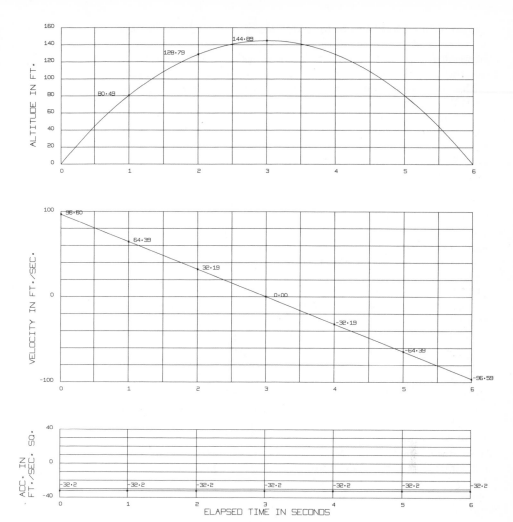

Fig. 13.11 □ **Acceleration, velocity and distance curves.**
These plots were done on a digital plotter connected on-line with an IBM 1130 computer.

ters are spaced to the right to form lines of up to 120 characters per line.

Accuracy relative to reference point is .002 in. over the entire working area for both static and dynamic operation; repeatability is .001 in. The encoder resolution is .00025 in. for the X-axis and .000375 in. for the Y-axis.

A few of the optional features available to users are (1) additional characters and symbols, up to 255 in number; (2) interpolation of circular, elliptical, or other conic sections; (3) routing head in lieu of a stylus carriage; (4) photographic head suitable for generation of printed-circuit work in lieu of stylus carriage; (5) digitizing head for operation with CRT monitor in lieu of stylus carriage (see Sec. 13.22); and (6) a drive system for paper advance.

At the present time, there are several recognized computer programming systems in general use. Although all of them no doubt meet the needs for which they were created, only six will be mentioned here. They are APT (Automatically Programmed Tools), which is particularly suited for use in engineering education; AUTO-PROMPT, developed by IBM; ALLADIN, developed by the Allison Division of General Motors; AUTODRAFT, which represents the joint efforts of IBM and North American Aviation, Inc.; UDRAFT–8, developed by the Universal Drafting Machine Corporation; and TRIDM, developed and tested in the classroom at North Carolina State University. Both APT and AUTOPROMPT were developed primarily for N/C (numerically controlled) machine tools. ALLADIN, AUTO-DRAFT, and TRIDM are drafting language systems.

In preparing a program, it is necessary for the draftsman to describe the shapes that are to be drawn. A description may be either in digital form or an equation given in the vocabulary of the system being used. The program that is prepared using these descriptions becomes, as one might expect, a complete sequence of coded instructions that must be followed by the drafting machine. This encoded information is then keypunched and fed to the computer. The computer in turn performs the calculations necessary to digitize it for either punched-tape or magnetic tape output.

A plot of acceleration, velocity, and distance curves is shown in Fig. 13.11.

13.17 □ TRIDM

TRIDM is an advanced version of a computer-graphics language originally developed for small computers such as the IBM 1130 by Dr. Bertram Herzog of the University of Michigan. The improvements effected at North Carolina State University in his DRAWL 1130 largely in terms of data structure, subroutine simplification, and input–output options, resulted in the version known as TRIDM. At this time both Dr. Herzog and North Carolina State

```
**** NCSU'S TRIDM FOR OS/360/370          VER 3  MOD 03      REV  1 JUNE 72   BY LLR ****

$ SIZE0         12.00    10.00     0.00     SIZE CARD FOR PRINTER-PLOT
$ OBJSOSB        0.00     0.00     0.00     THIS IS A SUPPORT BLOCK-LOCATED AT ORIGI
OBJS   1 HAS    44 POINTS
$ AXISO          2.00     3.00     0.00     SET AXES FOR ORTHOGRAPHIC VIEWS
$ DRAW2SB        0.00     0.00     0.00     TOP VIEW (VIEW FROM +Y)
DRAW   1
X-AXIS = 2.00     Y-AXIS = 3.00     N= 0
THE DRAWING HAS BEEN COMPLETED.
$ DRAW3SB        0.00     0.00     0.00     FRONT VIEW(VIEW FROM -Z)
DRAW   1
X-AXIS = 2.00     Y-AXIS = 3.00     N= 0
THE DRAWING HAS BEEN COMPLETED.
$ DRAW1SB        0.00     0.00     0.00     RIGHT SIDE (VIEW FROM +X)
DRAW   1
X-AXIS = 2.00     Y-AXIS = 3.00     N= 0
THE DRAWING HAS BEEN COMPLETED.
$ AXISO          3.50     4.00     0.00     SET AXIS FOR 3-VIEW ORTHOGRAPHIC
$ SHIFOF1       -3.00    -3.50     0.00     FRONT VIEW LOCATION
MATRIX  -1
$ TRANOB1        0.00     0.00     0.00     MOVE OBJECT FOR FRONT VIEWING
TRANS  -1 = MATRIX  -1(   1).
$ COMBOC1        0.00     0.00     0.00     TOP VIEW SETUP
$ ROTA1        -90.00     0.00     0.00     ROTATE OBJECT--TOP VIEWING
$ SHIFO         -3.00     0.50     0.00     SHIFT OBJECT -- TOP VIEWING
$ ECOMO          0.00     0.00     0.00
MATRIX  -17
$ TRANOT1        0.00     0.00     0.00     TOP VIEW LOCATION
TRANS  -4 = MATRIX  -17(   1).
$ COMBOC2        0.00     0.00     0.00     RIGHT SIDE VIEW SETUP
$ ROTA2         90.00     0.00     0.00     ROTATE OBJECT--RIGHT SIDE VIEW
$ SHIFO          0.50    -3.50     0.00     SHIFT OBJECT --RIGHT SIDE VIEW
$ ECOMO          0.00     0.00     0.00
MATRIX  -33
$ TRANOR1        0.00     0.00     0.00     RIGHT SIDE VIEW LOCATION
TRANS  -7 = MATRIX  -33(   1).
$ ASMBOA1        0.00     0.00     0.00     LOCK ALL VIEWS INTO COMMON PICTURE
ASMB  -10 HAS    3 PARTS
$ DRAWOA1        0.00     0.00     0.00     TAKE 3-VIEW ORTHOGRAPHIC VIEW
DRAW -10
X-AXIS = 3.50     Y-AXIS = 4.00     N= 0
THE DRAWING HAS BEEN COMPLETED.
$ COMBOC3        0.00     0.00     0.00     BEGIN ORIENTATION FOR PERSPECTIVE
$ SHIFO         -2.50    -1.50     0.00     MOVE TO LEFT OF Y-AXIS AND CENTER ON X
$ ROTA2         50.00     0.00     0.00     ROTATE +50 ABOUT Y
$ ROTA1        -22.00     0.00     0.00     ROTATE -22 ABOUT X
$ PERS3          5.00     0.00     0.00     PERSPECTIVE FROM -5
$ECOMO           0.00     0.00     0.00     PERSPECTIVE SETUP
$TRANOP1         0.00     0.00     0.00
TRANS -14 = MATRIX  -49(   1).
$ DRAWOP1        0.00     0.00     0.00     PERSPECTIVE VIEW
DRAW -14
X-AXIS = 3.50     Y-AXIS = 4.00     N= 0
THE DRAWING HAS BEEN COMPLETED.
$ SHIFOS1        2.50     2.00     0.00     PLACE PERSPECTIVE IN UPPER RIGHT QUADRAN
$ TRANOP2        0.00     0.00     0.00     PLACE PERSPECTIVE IN UPPER RIGHT QUADRAN
TRANS -17 = MATRIX  -65(  -14).
$ ASMBOA2        0.00     0.00     0.00     LOCK ALL VIEWS INTO COMMON PICTURE
ASMB  -20 HAS    4 PARTS
$ DRAWOA2        0.00     0.00     0.00     DRAW ORTHOGRAPHIC & PERSPECTIVE COMBINED
DRAW -20
X-AXIS = 3.50     Y-AXIS = 4.00     N= 0
THE DRAWING HAS BEEN COMPLETED.
$ SCALOS2        0.50     0.00     0.00     REDUCE SCALE
MATRIX  -81
$ TRANOA3        0.00     0.00     0.00     REDUCE SCALE
TRANS -25 = MATRIX  -81(  -20).
$ DRAWOA3        0.00     0.00     0.00     REDUCED SCALE-ORTHOGRAPHIC/PERSPECTIVE
DRAW -25
X-AXIS = 3.50     Y-AXIS = 4.00     N= 0
THE DRAWING HAS BEEN COMPLETED.
```

Fig. 13.12 □ The print-out of the program statements needed for preparing the orthographic and perspective drawings shown in Fig. 13.13. Note that the top line reads NCSU'S TRIDM.

University have waived rights to TRIDM. Dr. Herzog has retained rights to DRAWL 70 and DRAWL 71 for larger computers, while North Carolina State retains all rights to TRIDMN, a later, more efficient, and more versatile version of TRIDM.

For a computer-graphics language to meet the requirements of engineering education, it must be natural and must be applicable to a wide range of design and information display problems. TRIDM, utilizing command keywords that have been developed through classroom testing and are natural for the designer, offers an unusual set of options to informations users. TRIDM has been used in the design of a geodetic dome structure and a space shuttle vehicle as well as for information displays relating to studies such as highway routing alternatives, the mapping of drug reactions within the human body, and the evaluation of educational programs. A person closely identified with this work is Professor Byard Houck in the field of engineering graphics at North Carolina State University. Professor Houck prepared the program given in Fig. 13.12 and supplied the plotter-prepared drawings in Fig. 13.13.

The student will find it worthwhile to read through the program and relate the given comment statements to the plotter-prepared drawings. Note that the top view was drawn first, and then the front and right side views follow in order. Next, well down in the program there is a statement BEGIN ORIENTATION FOR PERSPECTIVE. This is the point at which the drawing of the perspective started. Finally, after the drawing shown had been completed another drawing was made in response to the program statement REDUCE SCALE. A half-size representation was drawn with the views and the perspective.*

13.18 □ IBM 1130 data presentation system

The IBM 1130 Data Presentation System is a computer program. The basic 1130 system for DPS consists of a 1627 plotter and an IBM 1442 card read punch. The system can be used by anyone, even by persons who may have no previous computer knowledge. If the user desires a graph, chart, or whatever, he or she merely feeds the computer a limited number of command phrases that describe the representation wanted and the format of the files where the data is stored. The computer provides the chart, graph,

*TRIDM was given to all persons who attended the Computer-graphics Summer School that was held at the time of the Annual Meeting of the American Society for Engineering Education in Lubbock, Texas, in June 1972. Another approach to computer-graphics was presented at this same meeting by Professor Clarence E. Hall of Louisiana State University. Those who may be interested in Professor Hall's method of presenting computer-graphics to engineering students will find his method presented in detail in Vol. 36, No. 2 (Spring 1972) issue of *Engineering Design Graphics*, published by the Division of Engineering Design Graphics of ASEE.

Fig. 13.13 □ **The drawings prepared by a plotter following the instructions given by the program shown in Fig. 13.12.**

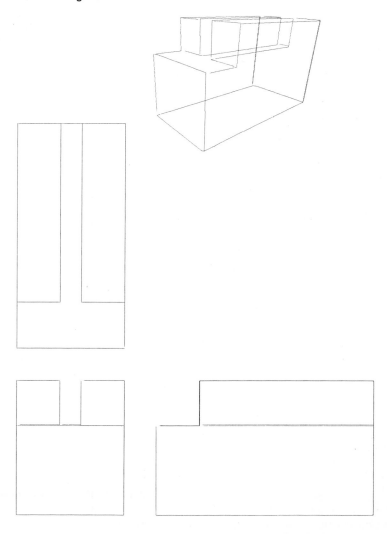

etc., that is needed. The graphic representation requested is drawn by the computer's 1627 plotter (Fig. 13.14). The 1627 plotter receives numerical data (plotting coordinates) from the computer and converts the information into graphic form. DPS processes any data that can be presented graphically and which has been recorded on punched cards or in the 1130 disk file. The Data Presentation System is designed in three modules. The most advanced level is called the "graphic report generator." This is a fully developed user-oriented system that requires no additional user program to construct a graph illustrating data in an 1130 file. A few of the characteristics that illustrate the versatility of the graphic report generator are: 1) as many as eight scales can be drawn in any location, 2) information may be selected from several data sets for the same plotted display, 3) scales may be linear, logarithmic, or polar, 4) scales are automatically set to keep all data on the plot, 5) data may be plotted as point plots, line plots, and logarithmic plots. Circles and spirals may also be generated. As can be observed in Fig. 13.15, the system is capable of constructing straight-line figures or perspective views of geometric shapes. From tabular results calculated by the computer, the plotter can be made to trace the shape of a cam being designed. The laborious work involved in structural detailing can be handled by the 1130 computer and the 1627 plotter. For example, in one design organization the system is used to calculate the dimensions for a steel member and then the 1627 produces the needed drawing with the required dimensions appearing along the view. The time needed to produce a detail sheet is reduced as much as 40 percent by this method thus making additional man hours available for technical design. Figure 13.16 shows a modified standard drawing prepared by the 1627 plotter.

13.19 □ UDRAFT–8

The Universal Drafting Machine Corporation now has available a software program for automatic drafting that can be used by persons having no knowledge of computer programming. The easily understood language, known as UDRAFT–8, can be used to produce tapes for drawings to be made on Universal Drafting Machine Corporation's Orthomat Systems. A feature of UDRAFT–8 is that it can be implemented on the drafting system computer thus permitting the system to prepare and edit input tapes. By using simple statements, combined alphanumeric capability, macro library, and other needed features, the designer or draftsman can prepare complex drawings with minimal effort and at a great saving in time.

13.20 □ CRT plotting systems (Fig. 13.18)

The CRT microfilm printer–plotter has undergone gradual improvement since the first system was placed in use in 1960. During

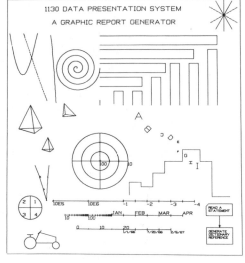

Fig. 13.14 □ The IBM 1130 Data Presentation System.
The 1627 plotter may be observed to the left of the man, at about the level of the chair-seat. (*Courtesy International Business Machines Corporation.*)

Fig. 13.15 □ Examples of graphic output–1130 Data Presentation System.
(*Courtesy International Business Machines Corporation.*)

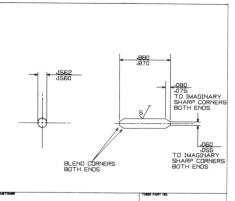

Fig. 13.16 □ A modified standard.
(*Courtesy International Business Machines Corporation.*)

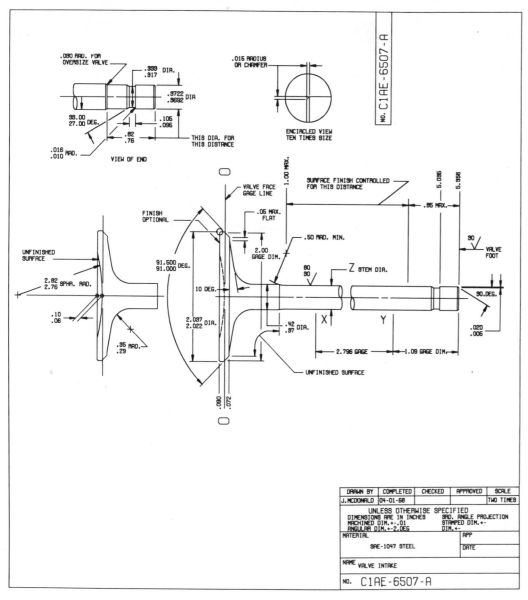

Fig. 13.17 □ **A detail drawing produced by a plotter.**
(*Courtesy Ford Motor Company.*)

Fig. 13.18 □ **A microfilm plotter.**
(*Courtesy Computer Industries, Inc.*)

the first years, the cost of CRT plotters was so high that they could be found only at aerospace and government research centers. However, extensive research and development work has led to the recent announcement of several economical systems with the flexibility needed for high-speed computer output, data analysis, automatic drafting, and circuit board design. Use of CRT printer–plotters is expected to rise rapidly. CRT plotters should soon be appearing in increasing numbers at medium-size computer centers and in the design rooms of the larger industrial organizations.

CRT microfilm printer–plotters have been discussed under automated drafting as such because they can be operated off-line. CRT plotters could have been included just as well with that portion of this chapter covering computer-aided design.

In the case of the model shown in Fig. 13.18, input data, decoded and displayed on a high-resolution cathode-ray tube, is photographed using a high-speed 35 mm magazine camera. This film is fed from a continuous roll that may be up to 400 ft. in length.

Long continuous charts can be produced using program-controlled precision film advance that permits the butting of consecutive data frames. The so-called "thru-put" is 120 full printing format frames per minute.

The average plotting time for a typical plot is less than 1 sec. Should a "quick-look" hard copy be needed, the auxiliary hard copy camera can be used to obtain copy directly on 9-in. paper within 4 sec. This is accomplished by an integrally mounted processor that is relatively inexpensive.

When output is to have a repetitive format, it is possible to superimpose a form on the CRT display. To do this saves the computer time needed to generate the form each time.

13.21 □ Digitizing

The computer can be made to "see" drawings, sketches, and photographs but it can be made to do so only through reading either a deck of cards or a roll of paper or magnetic tape produced by a digitizer (Fig. 13.19). What it "sees" as input must be in digital form and a drawing of any complexity must be received by the computer as a series of digits, that is, pairs of coordinate points. In the early days digitizing was a difficult and time-consuming task because it was necessary to "break-up" a drawing into straight line segments, the "break-up" theory being based on the fact that two points define a straight line. The readings for pairs of digits for straight line segments were then manually established or directly read off of graph paper. All readings were carefully recorded to be subsequently punched into a deck of cards that would be read into the computer via a card reader.

Manual digitizers soon followed. These in their simplest form

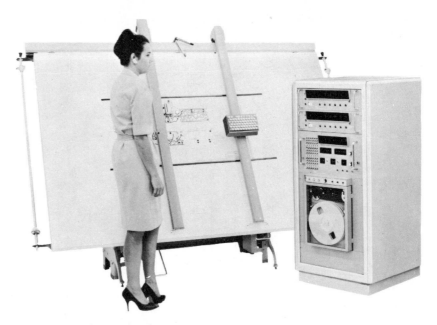

Fig. 13.19 □ **Digitizing a printed circuit.**
Output on magnetic tape. (*Courtesy Computer Industries, Inc.*)

resemble a drafting board with a movable rail that is equipped with a cursor, mounted so as to be free to move in the X and Y directions. With the drawing mounted on the board, the operator first positions the cursor with the cross-hairs centered over the end point of a line segment and then pushes a button to punch the coordinate values of the point onto a card or paper tape. Digitizers of this type utilize cable driven, rotary, optical encoders. These encoders send out electrical pulses that are translated into coordinate measurements by a control unit.

At the present time there are a wide variety of digitizers on the market with optional features. Of importance to designers in the electronic field is the grid recognition capability of some of these models since electronic (printed circuit) artwork generation is usually done so that features are dimensioned to integer multiples of some grid spacing, say 0.05 in.

Even though manual digitizing is satisfactory in every way for drawings composed mainly of straight lines and discrete points, manual digitizers require considerable painstaking effort when circular arcs and other mathematically defined curves are being prepared for the computer through digitizing. For this reason, computer programs have been developed that permit curves and repetitive patterns to be described in easily prepared coded statements that can be punched into cards to be read and interpreted by the computer in accordance with programmed instructions.

While manual digitizing coupled with part programming by means of statements works well for most types of graphics, limitations are encountered when the computer must be made to ''see'' some exotic curve that cannot be mathematically defined. The digitizing of such curves is discussed in Sec. 13.22. Since manual digitizing would be too slow and inaccurate for the complex curves that are so common in body design, the digitizing procedure is mechanized and a television camera and monitor screen are used for magnification of the image of the line to be digitized, as shown in Fig. 13.21.

New forms of automatic digitizers are now available. For example, one digitizer has a special head that locks onto a line being digitized and follows along it automatically while sending out digital information as directed by the program of the mini-computer control unit. Another type of automatic digitizer uses a ''scanner.'' In this case, the graphic representation is mounted around a drum that spins while a photosensitive receiver travels across the spinning drum in an axial direction while converting the light level differences being received into data for computer processing.

The graf/pen shown in Fig. 13.20 uses hypersonic impulses generated at the point of the stylus as its means of measurement. The times required for the sound wave to reach two linear microphone sensors are converted into X and Y coordinates. The sensors

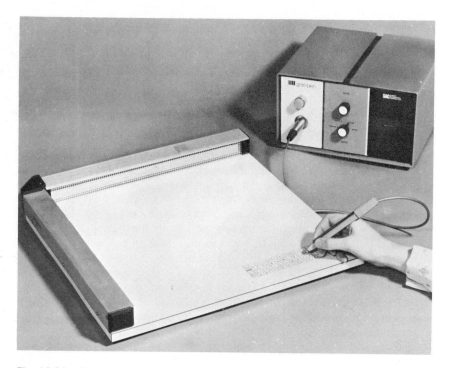

Fig. 13.20 □ Graf/pen sonic digitizer—gp3.
(*Courtesy SAC—Science Accessories Corporation.*)

detect each sonic impulse and convert the impulse to electrical signals for the control unit. This widely used digitizer may be used wherever data in graphical form must be entered into a digital system. In using the graf/pen the operator merely traces the line. Graf/pen units are available that provide digital readings in either the English or the metric system.

13.22 □ Use of the computer in automobile-body design

The Ford Motor Company has recently adopted a numerically controlled die-making process that eliminates much of the time-consuming drafting work that lies between the stylist's clay model and the body dies needed for a new car. Traditionally, many man-hours were spent preparing drawings from the clay model, making templates from these drawings, and finally producing "tracer masters" from a carved-wood model. It was from these masters that a set of body dies were cut by large milling machines. The new system bypasses much of the drafting work and nearly all of the handwork by body craftsmen.

For this numerically controlled die-making process a contour digitizing system is used, which includes a coordinatograph with magnetic reading heads, a television camera, and a lens system mounted on the reading arm. The image displayed on the screen at the operator's console is magnified 15 times for increased optical accuracy. See Fig. 13.21.

Basically, the system translates drawings of body contours into computer language. To accomplish this the operator takes "readouts" at predetermined layout points. With a television camera positioned at a point, the operator pushes the read-out button and the coordinates of the point, shown on the screen, are punched on a card. These subject data on punched cards (converted to magnetic tape) are then incorporated into the Ford Computer Program and output tapes are generated for verification and for numerically controlled drafting (Fig. 13.22). Orthographic and perspective drawings are prepared as required for verification and checks are made on the computer-generated cutter path. On completion of these steps, the data are transferred to punched paper tape. This final tape is prepared for a numerically controlled, multiaxis machine that cuts the needed body dies automatically.

13.23 □ Artwork generators

Artwork generators, which are now widely used in the electronics field, can be said to be high-precision automatic drafting machines. These machines are equipped with a photo-exposure head. They produce finished graphic drawings utilizing a beam of light focussed on a photosensitive emulsion surface. The servo commands that direct the movement of the light beam are fed to the generator control unit by means of punched paper tape or magnetic tape.

Fig. 13.21 □ Contour Data Reader.
(*Courtesy Ford Motor Company.*)

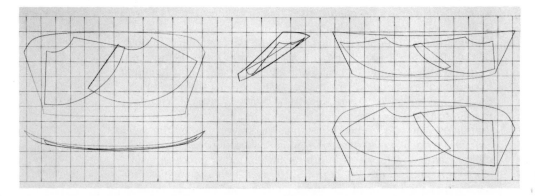

Fig. 13.22 □ Contour digitizer computer generated drawings.
(*Courtesy Ford Motor Company.*)

If desired, the artwork generator (drafting machine) may be operated directly on-line with the computer. However, most artwork generators are operated off-line.

13.24 □ Applications of plotting and digitizing systems

Computer-operated drafting machines are widely used for the preparation of perspective and axonometric drawings and integrated-circuit-board layouts. They do mapping, perform lofting, and plot impeller, airscrew, and wing sections. By utilizing their now recognized capabilities, drafting systems can prepare a perspective of an aerodynamic configuration composed of station lines or plot a series of compound surfaces and almost immediately present a perspective of these same surfaces for inspection and evaluation. This has been demonstrated using AUTOPROMPT as a drafting language.

To recreate the maneuvers of aircraft, animated motion pictures have been made with the cockpit and terrain features both included to provide orientation and to relate most of what a pilot would see from his position in the plane. Of course, a higher degree of realism could have been obtained by making a motion picture from the plane doing the maneuver, but only at a much higher cost. The automated drawing was done by a Mosely plotter and a little final rendering was added to bring about some realism. The creation of these highly complex three-dimensional scenes for a rough movie-type sequence has been largely the work of several outstanding engineering "graphicians" at The Boeing Company. The results of their work appear in a book entitled *Computer Graphics in Communication*.* The author is William A. Fetter, who was a member of a small group of persons assigned to the project. Some of the work of these men included maneuver simulation of a bank and turn and then a dive and pullout, pilot views of landing on a CVA–19 under simulated heavy seas, and pilot views of landing at an airport.

Three-dimensional graphs with scales and rotated perspective views have been prepared using an Orthomat plotter. Properly programmed, other plotters can perform the same tasks.

Now coming to the attention of people in industry is the application of plotting systems to the graphic interpretation of machining routines that have been programmed by a computer. The graphic interpretation provides a quick check and eliminates the need for running tests of routines on a numerically controlled machine tool. As stated elsewhere in this chapter, drafting machines can also produce drawings to verify cutter-tool paths specified by machine control tapes. For quality control, a drafting ma-

*William A. Fetter, *Computer Graphics in Communication* (New York: McGraw-Hill, Inc., 1966).

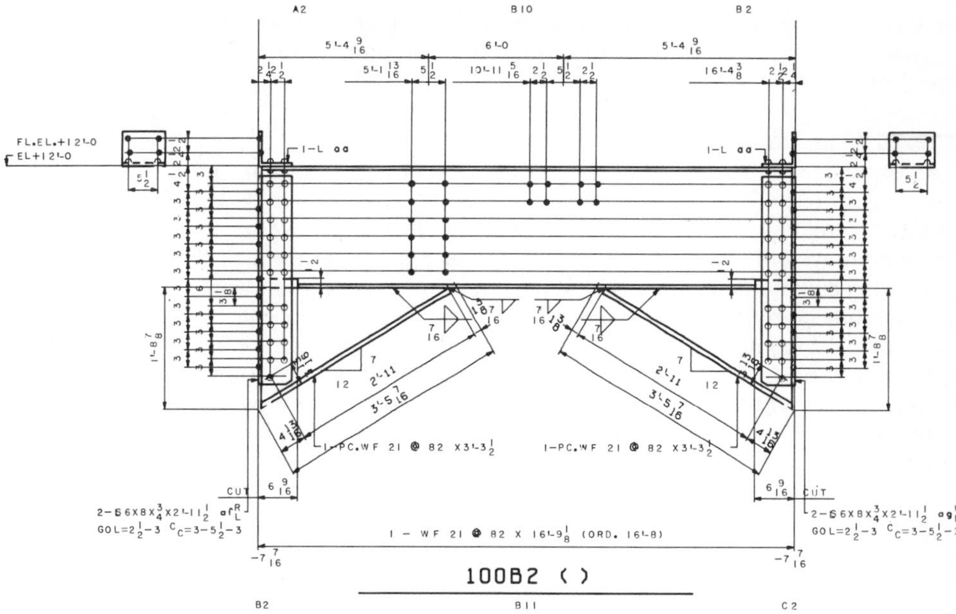

Fig. 13.23 □ A structural drawing prepared on a plotter.
This drawing was made using the CONSTRUCTS System of automated drafting. (*Courtesy Meiscon Division, Control Data Corporation.*)

chine can be used as an aid in establishing checkpoints on inspection tapes and in making scribed optical comparator charts and boards.

13.25 □ The future

At some point in the future it may be economically feasible to develop a system that will computer-generate detail drawings by software techniques from a design layout. Plans are now being made for systems that will permit a designer to prepare his design layout in collaboration with a suitably programmed computer. With the layout completed and checked, the next logical step will be to have the computer generate the needed detail and assembly drawings, parts list, and machine tool tapes with little, if any, human intervention. Only when we have reached the point where this can be accomplished can we say that we have real "automatic drafting."

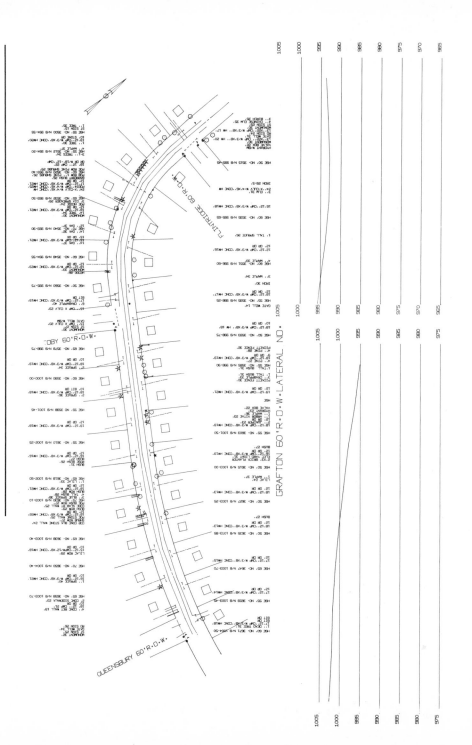

Fig. 13.24 □ A highway route plan and profile.

Program systems have been developed to produce civil engineering plans for highway, sewer, and drainage construction. The route plan and profile sheet shown was produced on a CalComp 702 in less than twenty-five minutes. All of the lines and lettered information were generated, except the title block at the bottom. The keypunch operator interprets the field notes and punches the required information into cards. This information is then fed into the computer to produce a magnetic tape which contains the needed drafting instructions. Johnson & Anderson, Inc. started to develop their system in late 1967. Their initial success encouraged further work towards the complete system that they now use to make their drawing at much less cost than by manual methods. To improve the total system further, they modified their method of taking field notes and changed the graphic read-out procedure. Increased efficiency is expected in the near future. (*Courtesy Johnson & Anderson, Inc., Consulting Engineers, Pontiac, Michigan.*)

Fig. 13.25 □ A plot of a subdivision.
The computer program designated SAMPS (Subdivision and Mapping System) plots a complete subdivision map with bearings, distances, and other information ordinarily given on a map of this type. As developed by PMT Associates of Sacramento, California, the program can also be used for plotting control networks, for surveying jobs and for primary control of aerial photographs. The program provides for choice of plotting scale, rotation of plotting axes, plotting of lines and points, plotting of a north arrow, and annotation of distances and bearings. The subdivision plot shown below was drawn on a CalComp Model 702 flatbed plotter operating on-line with an IBM 1130 computer. (*Courtesy of PMT Associates-Engineers, Land Surveyors, Planners.*)

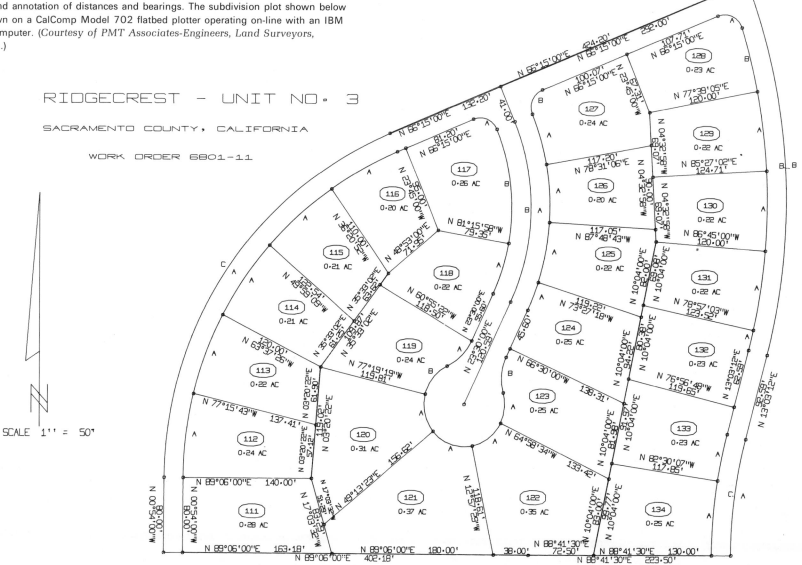

Part 5

Design for production

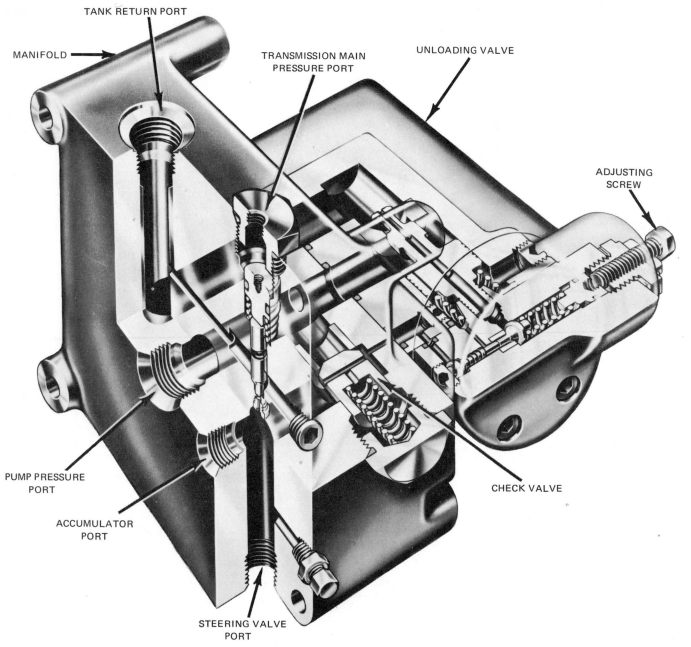

TANK RETURN PORT

MANIFOLD

TRANSMISSION MAIN
PRESSURE PORT

UNLOADING VALVE

ADJUSTING
SCREW

PUMP PRESSURE
PORT

ACCUMULATOR
PORT

STEERING VALVE
PORT

CHECK VALVE

**The manifold and unloading control
valve shown is part of the hydraulic
steering and braking circuits.**
It maintains the hydraulic system at a constant pressure. Units of this type require
the use of springs and fasteners. (*Courtesy
General Motors Corporation.*)

Chapter 14

Graphic standards

A □ Screw threads, fasteners, springs, bearings, and welds

Screw threads

14.1 □ Introduction

In the commercial field, where the practical application of engineering drawing takes the form of working drawings, knowledge of screw threads and fasteners is important. There is always the necessity for assembling parts either with permanent fastenings such as rivets, or with bolts, screws, and so forth, which may be removed easily.

Design engineers, detailers, and draftsmen must be completely familiar with the common types of threads and fastenings, as well as with their use and correct methods of representation, because of the frequency of their occurrence in structures and machines. Information concerning special types of fasteners may be obtained from manufacturers' catalogues.

A young technologist in training should study Fig. 14.1 to acquaint himself with the terms commonly associated with screw threads.

14.2 □ Threads

The principal uses of threads are (1) for fastening, (2) for adjusting, and (3) for transmitting power. To satisfy most of the requirements of the engineering profession, the different forms of threads shown in Fig. 14.2 are used.

The (ANSI) Unified thread form (Fig. 14.2), now recognized as the standard thread form in the United States, Great Britain and Canada, is a modification of the American National (N) thread form in use by American industry since 1935 (still in use to a limited extent). Since the thread forms are essentially the same, the new *Unified* thread form has found ready acceptance and it can be said to have superseded the old (N) form.

The sharp V thread is used to some extent where adjustment and holding power are essential.

For the transmission of power and motion, the modified square, Acme, and Brown and Sharpe worm threads have been adopted. The modified square thread, which is now rarely used, transmits power parallel to its axis. A still further modification of the square thread is the stronger Acme, which is easier to cut and more readily disengages split nuts (as lead screws on lathes). The Brown and Sharpe worm thread, with similar proportions but with longer teeth, is used for transmitting power to a worm wheel.

The knuckle thread, commonly found on incandescent lamps, plugs, and so on, can be cast or rolled.

The Whitworth and buttress threads are not often encountered by the average engineer. The former, which fulfills the same purpose as the American Standard thread, is used in England but is also frequently found in this country. The buttress or breech-block thread, which is designed to take pressure in one direction, is used for breech mechanisms of large guns and for airplane propeller hubs. The thread form has not been standardized and appears in different modified forms.

The Dardelet thread is self-locking in assembly.

14.3 □ American–British unified thread

The Unified Thread Standard came into existence after the representatives of the United States, Great Britain, and Canada signed a unification agreement on Nov. 18, 1948, in Washington, D.C. The Unified thread is a general-purpose thread for screws, bolts, nuts, and other threaded parts (ANSI B1.1–1960).

14.4 □ Multiple-start threads (Fig. 14.3)

Whenever a quick advance is desired, as on fountain pens, valves, and so on, two or more threads are cut side by side. Two threads form a double (2-START) thread; three, a triple (3-START) thread; and so on. A thread that is not otherwise designated is understood to be a single-start thread. All threads included in ANSI B1.1–1960 are single-start unless specifically identified as being multiple-start.

In drawing a single or an odd-number multiple-start thread, a

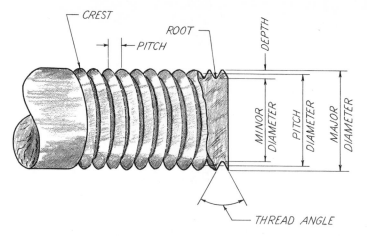

Fig. 14.1 □ Screw thread nomenclature.

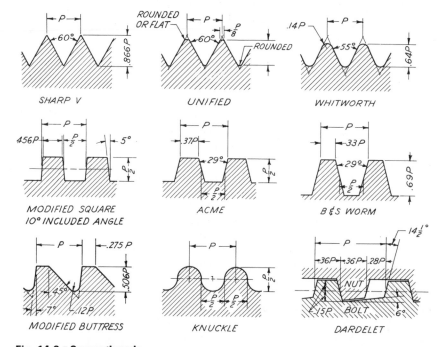

Fig. 14.2 □ Screw threads.

crest is always diametrically opposite a root; in a double or other even-number multiple-start thread, a crest is opposite a crest and a root opposite a root.

14.5 □ Right-hand and left-hand threads

A right-hand thread advances into a threaded hole when turned clockwise; a left-hand thread advances when turned counterclockwise. They can be easily distinguished by the thread slant. A right-hand thread on a horizontal shank always slants upward to the left (\) and a left-hand, upward to the right (/). A thread is always considered to be right-hand if it is not otherwise specified. A left-hand thread is always marked LH on a drawing.

14.6 □ Pitch

The pitch of a thread is the distance from any point on a thread to the corresponding point on the adjacent thread, measured parallel to the axis, as shown in Fig. 14.1.

14.7 □ Lead

The lead of a screw may be defined as the distance advanced parallel to the axis when the screw is turned one revolution (Fig. 14.3). For a single thread, the lead is equal to the pitch; for a double (2-START) thread, the lead is twice the pitch; for a triple (3-START) thread, the lead is three times the pitch, and so on.

14.8 □ Detailed screw-thread representation

The true representation of screw threads by helical curves, requiring unnecessary time and laborious drafting, is rarely used. The detailed representation, closely approximating the actual appearance, is preferred in commercial practice, for it is much easier to represent the helices with slanting lines and the truncated roots and crests with sharp ''V's'' (Fig. 14.4). Since detailed rendering is also time consuming, its use is justified only in those few cases where appearance and permanency are important factors and when it is necessary to avoid the possibility that confusion might result from the use of one of the symbolic methods. The preparation of a detailed representation is a task that belongs primarily to a draftsman, the engineer being concerned only with specifying that this form be used.

14.9 □ American national standard conventional thread symbols (Fig. 14.5)

To save valuable time and expense in the preparation of drawings, the American National Standards Institute has adopted the ''schematic'' and ''simplified'' series of thread symbols to represent threads having a diameter of 1 in. or less.

The root of the thread for the simplified representation of an

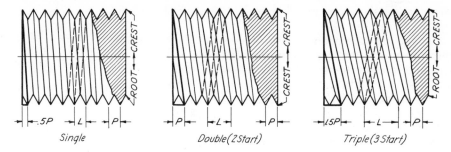

Fig. 14.3 □ Single-start and multiple-start threads.

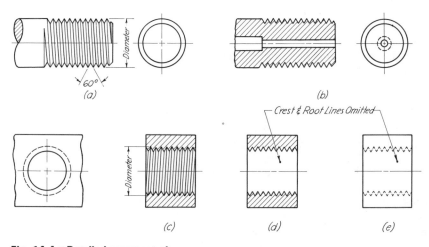

Fig. 14.4 □ Detailed representation.

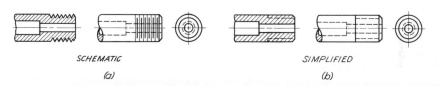

Fig. 14.5 □ External thread representation.

external thread is shown by invisible lines drawn parallel to the axis [Fig. 14.6(a)].

The schematic representation consists of alternate long and short lines perpendicular to the axis. Although these lines, representing the crests and roots of the thread, are not spaced to actual pitch, their spacing should indicate noticeable differences in the number of threads per inch of different threads on the same working drawing or group of drawings (Fig. 14.7). The root lines are made heavier than the crest lines (Fig. 14.8).

Before a hole can be tapped (threaded), it must be drilled to permit the tap to enter. See Table 9 in Appendix E for tap drill sizes for standard threads. Since the last of the thread cut is not well formed or usable, the hole must be shown drilled and tapped deeper than the screw will enter (Fig. 14.10). To show the threaded portion extending to the bottom of the drilled hole indicates the use of a bottoming tap to cut full threads at the bottom. This is an extra and expensive operation not justified except in cases where the depth of the hole and the distance the screw must enter are limited.

Figure 14.9 shows a simplified method of representation for square threads.

14.10 □ Threads in section

The detailed representation of threads in section, which is used for large diameters only, is shown in Fig. 14.4.

The simplified and schematic representations for threads of small diameter are shown in Figs. 14.6 and 14.7.

A sectioned assembly drawing is shown in Fig. 14.10. When assembled pieces are both sectioned, the detailed representation is used, and the thread form is drawn.

14.11 □ Unified and American screw-thread series

The Unified and American screw-thread series, as given in ANSI B1.1–1960 consists of six series and a selection of special threads that cover special combinations of diameter and pitch. Each series differs from the other by the number of threads per inch for a specific diameter (see Tables 9 and 10 in Appendix E).

The coarse-thread series (UNC and NC) is designated UNC for sizes above $\frac{1}{4}$ in. in diameter. This series is recommended for general industrial use.

The fine-thread series (UNF and NF), designated UNF for sizes above $\frac{1}{4}$ in., was prepared for use when a fine thread is required and for general use in the automotive and aircraft fields.

The extra-fine-thread series (UNEF and NEF) is used for automotive and aircraft work when a maximum number of threads is required for a given length.

The 8-thread series (8N) is a uniform pitch series for large diame-

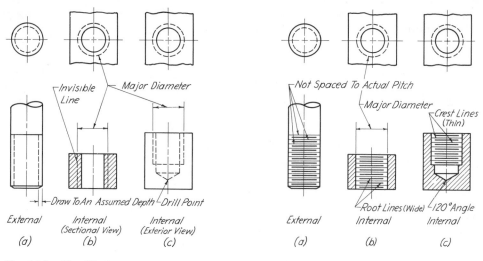

Fig. 14.6 □ **Simplified representation.**

External
(a)

Internal (Sectional View)
(b)

Internal (Exterior View)
(c)

Invisible Line   *Major Diameter*   *Draw To An Assumed Depth*   *Drill Point*

Fig. 14.7 □ **Schematic representation.**

External
(a)

Internal
(b)

Internal
(c)

Not Spaced To Actual Pitch   *Major Diameter*   *Crest Lines (Thin)*   *Root Lines (Wide)*   *120°Angle*

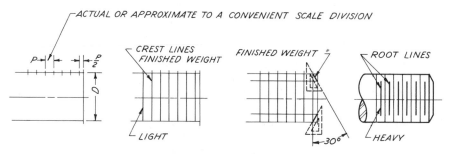

Fig. 14.8 □ **Drawing conventional threads—schematic representation.**

ACTUAL OR APPROXIMATE TO A CONVENIENT SCALE DIVISION

CREST LINES FINISHED WEIGHT   *FINISHED WEIGHT*   *ROOT LINES*   *LIGHT*   *30°*   *HEAVY*

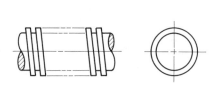

Fig. 14.9 □ **Simplified representation of a square thread.**

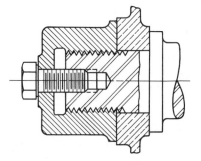

Fig. 14.10 □ **Threads in section.** (*ANSI Y14.6–1957.*)

ters. It is sometimes used in place of the coarse-thread series for diameters greater than 1 in. This series was originally intended for high-pressure joints.

The 12-thread series (12UN or 12N) is a uniform pitch series intended for use with large diameters requiring threads of medium-fine pitch. This series is used as a continuation of the fine-thread series for diameters greater than $1\frac{1}{2}$ in.

The 16-thread series (16UN or 16N) is a uniform pitch series for large diameters requiring a fine-pitch thread. This series is used as a continuation of the extra-fine-thread series for diameters greater than 2 in.

14.12 □ Unified and American screw-thread classes

Classes of thread are determined by the amounts of tolerance and allowance specified. Under the unified system, classes 1A, 2A, and 3A apply only to external threads; classes 1B, 2B, and 3B apply to internal threads. Classes 2 and 3 from the former American Standard (ASA) have been retained without change in the new Unified and American Thread Standard for use in the United States only, but they are not among the unified classes even though the thread forms are identical. These classes are used with the American thread series (NC, NF, and N series), which covers sizes from size 0 (.060) to 6 in.

Class 1A and class 1B replace class 1 of the old American Standard.

Class 2A and class 2B were adopted as the recognized standards for screws, bolts, and nuts.

Class 3A and class 3B invoke new classes of tolerance. These classes along with class 2A and class 2B have largely replaced class 2 and class 3 for most applications. Class 2 and class 3 are defined in the former standard ANSI(ASA) B1.1–1935 as follows:

> *Class 2 Fit.* Represents a high quality of commercial thread product and is recommended for the great bulk of interchangeable screw-thread work.

> *Class 3 Fit.* Represents an exceptionally high quality of commercially threaded product and is recommended only in cases where the high cost of precision tools and continual checking is warranted.

14.13 □ Identification symbols for unified screw threads

Threads are specified under the unified system by giving the diameter, number of threads per inch, initial letters (UNC, UNF, etc.), and class of thread (1A, 2A, and 3A; or 1B, 2B, and 3B) (see Fig. 14.11).

Unified and American National threads are specified on drawings, in specifications, and in stock lists by thread information given as

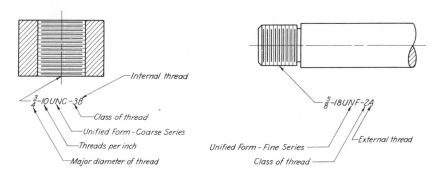

Fig. 14.11 □ Unified thread identification symbols.

shown in Fig. 14.12. A multiple-start thread is designated by specifying in sequence the nominal size, pitch, and lead.

14.14 □ Square threads

Square threads can be completely specified by a note. The nominal diameter is given first, followed by the number of threads per inch and the type of thread (see Fig. 14.12).

14.15 □ ANSI standard (NPT) pipe thread

The pipe taper thread, illustrated in Fig. 14.13, is similar to the Unified thread and has the same thread angle; but it is tapered $\frac{1}{16}$ in. per in., to insure a tight joint at a fitting. The crest is flattened and the root is filled in so that the depth of the thread is $0.80P$. The number of threads per inch for any given nominal diameter can be obtained from Table 28 in Appendix E.

Straight pipe thread, having the same number of threads per inch as taper thread, is in use for pressure-tight joints for couplings, for pressure-tight joints for grease and oil fittings, and for hose couplings and nipples. This thread may also be used for free-fitting mechanical joints. Usually a taper external thread is used with a straight internal thread, as pipe material is sufficiently ductile for an adjustment of the threads.

The NPT threads are for general use. When taper pipe threads are used and the joints are made up wrench tight with a lubricant or sealer, they will be pressure tight.

14.16 □ Specification of pipe threads

In specifying pipe threads, the ANSI recommends that the note be formulated using symbolic letters as illustrated in Fig. 14.14. For example, the specification for a 1 in. standard pipe thread should read:

$$1''-\text{NPT}$$

The letters NPT, following the nominal diameter, indicate that the thread is ANSI (American National) Standard (N), pipe (P), taper (T) thread.

Continuing with the same scheme of using letters, the specification for a 1 in. straight pipe thread would read:

$$1''-\text{NPS [American National (N)—pipe (P)—Straight (S)]}$$

The form of note given in Fig. 14.14 (b) reading 1″ AM STD PIPE THD is quite commonly used in practice.

Identification symbols and dimensions of American National pipe threads are given in the ANSI Standard for Pipe Threads (ANSI B2.1–1960).

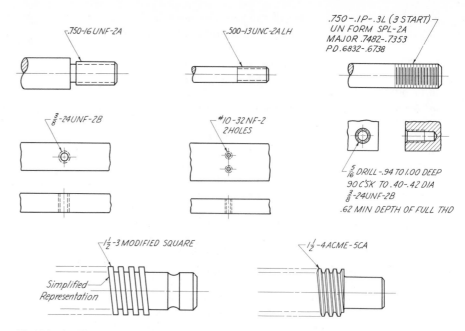

Fig. 14.12 □ Thread identification symbols.

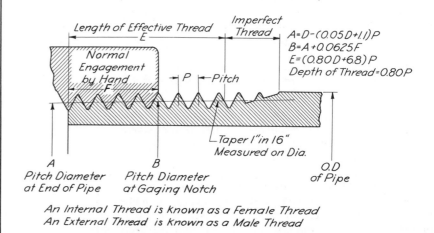

An Internal Thread is known as a Female Thread
An External Thread is known as a Male Thread

Fig. 14.13 □ Standard pipe thread.

Fasteners

14.17 □ American national standard bolts and nuts (Fig. 14.15)

Commercial producers of bolts and fasteners manufacture their products in accordance with the standard specifications given in the American National Standard entitled *Square and Hexagon Bolts and Nuts* (Revised 1965).* See Table 13 in Appendix E.

The ANSI has approved the specifications for three series of bolts and nuts:

1 *Regular Series.* The regular series was adopted for general use.

2 *Heavy Series.* Heavy boltheads and nuts are designed to satisfy the special commercial need for greater bearing surface.

3 *Light-Series Nuts.* Light nuts are used under conditions requiring a substantial savings in weight and material. They are usually supplied with a fine thread.

The amount of machining is the basis for further classification of hexagonal bolts and nuts in both the regular and heavy series as unfinished and semi-finished.

The chamfer angle on the tops of heads and nuts is 30° on hexagons and 25° on squares, but both are drawn at 30° on bolts greater than 1 in. in diameter.

Bolts are specified in parts lists and elsewhere by giving the diameter, number of threads per inch, series, class of thread, length, finish, and type of head.

Example:

$\frac{1}{2}$–13 UNC–2A × $1\frac{3}{4}$ SEMIFIN HEX HD BOLT

Frequently it is advantageous and practical to abbreviate the specification thus:

Example:

$\frac{1}{2}$ × $1\frac{3}{4}$ UNC SEMIFIN HEX HD BOLT

The engineer and experienced draftsman wisely resort to some form of template for drawing the views of a bolthead or nut (see Fig. A.5).

14.18 □ Studs

Studs, or stud bolts, which are threaded on both ends, as shown in Fig. 14.16, are used where bolts would be impractical and for parts that must be removed frequently (cylinder heads, steam chest covers, pumps, and so on). They are first screwed permanently into the tapped holes in one part before the removable member with its corresponding clearance holes is placed in posi-

*ANSI B18.2–1965.

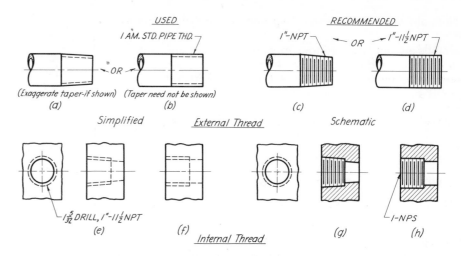

Fig. 14.14 □ Representation and specification of pipe threads.

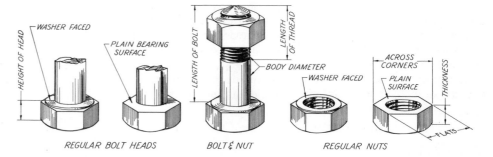

Fig. 14.15 □ American Standard bolts and nuts.

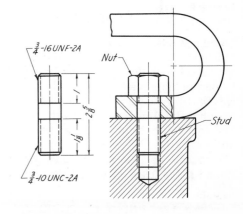

Fig. 14.16 □ Stud bolt.

tion. Nuts are used on the projecting ends to hold the parts together.

Since studs are not standard they must be produced from specifications given on a detail drawing. In dimensioning a stud, the length of thread must be given for both the stud end and nut end along with an overall dimension. The thread information is given by note.

In a bill of material, studs may be specified as follows:

Example:
$\frac{1}{2}$–13 UNC–2A × $2\frac{3}{4}$ STUD

It is good practice to abbreviate the specification thus:

Example:
$\frac{1}{2}$ × $2\frac{3}{4}$ STUD

14.19 □ Cap screws (Fig. 14.17)

Cap screws are similar to machine screws. They are available in four standard heads, usually in finished form. When parts are assembled, the cap screws pass through clear holes in one member and screw into threaded holes in the other. Hexagonal cap screws have a washer face $\frac{1}{64}$ in. thick with a diameter equal to the distance across flats. All cap screws 1 in. or less in length are threaded very nearly to the head.

Cap screws are specified by giving the diameter, number of threads per inch, series, class of thread, length, and type of head.

Example:
$\frac{5}{8}$–11 UNC–2A × 2 FIL HD CAP SC

It is good practice to abbreviate the specification thus:

Example:
$\frac{5}{8}$ × 2 UNC FIL HD CAP SC

14.20 □ Machine screws

Machine screws, which fulfill the same purpose as cap screws, are used chiefly for small work having thin sections (Fig. 14.18). Under the approved American National Standard they range from No. 0 (.060 in. diam.) to $\frac{3}{4}$ in. (.750 in. diam.) and are available in either the American Standard Coarse or Fine-Threaded Series. The four forms of heads shown in Fig. 14.19 have been standardized.

To specify machine screws, give the diameter, threads per inch, thread series, class of thread, length, and type of head.

Example:
No. 12–24 NC–3 × $\frac{3}{4}$ FIL HD MACH SC

It is good practice to abbreviate by omitting the thread series and class of fit.

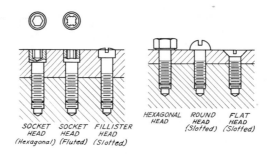

Fig. 14.17 □ Cap screws.

SOCKET HEAD (Hexagonal) SOCKET HEAD (Fluted) FILLISTER HEAD (Slotted)

HEXAGONAL HEAD ROUND HEAD (Slotted) FLAT HEAD (Slotted)

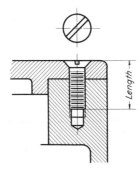

Fig. 14.18 □ Use of a machine screw.

Length

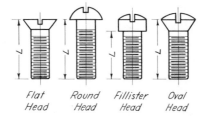

Flat Head Round Head Fillister Head Oval Head

Fig. 14.19 □ Types of machine screws.

Example:

No. 12–24 × ¾ FIL HD MACH SC

14.21 □ Fastener engagement

Unless a fastening carries a constant and appreciable fatigue stress, the usual practice is to have it enter a distance related to its nominal diameter (Fig. 14.10). If the depth of the hole is not limited, it should be drilled to a depth of 1 diameter beyond the end of the fastener to permit tapping to a distance of $\frac{1}{2}$ diameter below the fastener.

For fastenings and other general-purpose applications, the engagement length should be equal to the nominal diameter (D) of the thread when both components are of steel. For steel external threads in cast iron, brass, or bronze, the engagement length should be $1\frac{1}{2}D$. When assembled into aluminum, zinc, or plastic the engagement should be $2D$.

14.22 □ Set screws

Set screws are used principally to prevent rotary motion between two parts, such as that which tends to occur in the case of a rotating member mounted on a shaft. A set screw is screwed through one part until the point presses firmly against the other part (Fig. 14.20).

The several forms of safety heads shown in Fig. 14.21 are available in combination with any of the points. Headless set screws comply with safety codes and should be used on all revolving parts. The many serious injuries that have been caused by the projecting heads of square-head set screws have led to legislation prohibiting their use in some states [Fig. 14.20(c)].

Set screws are specified by giving the diameter, number of threads per inch, series, class of thread, length, type of head, and type of point.

Example:

$\frac{1}{4}$–20 UNC–2A × $\frac{1}{2}$ SLOTTED CONE PT SET SC

The preferred abbreviated form gives the diameter, number of threads per inch, length, type of head, and type of point.

Example:

$\frac{1}{4}$–20 × $\frac{1}{2}$ HEX SOCKET CONE PT SET SC

14.23 □ Keys

Keys are used in the assembling of machine parts to secure them against relative motion, generally rotary, as is the case between shafts, cranks, wheels, and so on. When the relative forces are not great, a round key, saddle key, or flat key is used (Fig. 14.22). For heavier duty, rectangular keys are more suitable (Fig. 14.23).

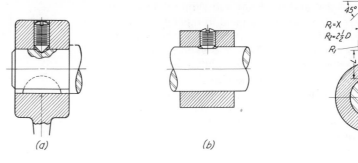

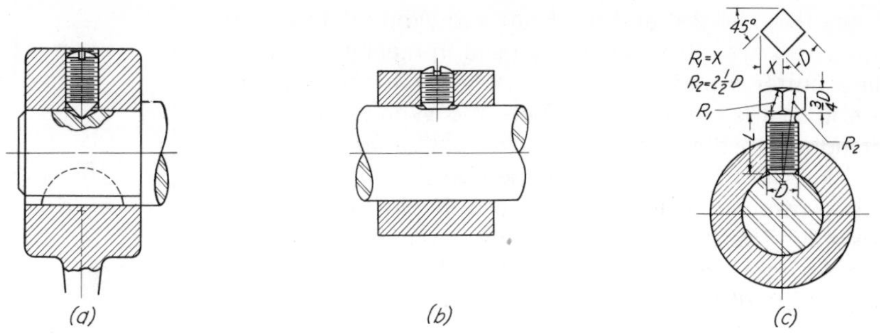

Fig. 14.20 □ Use of set screws.

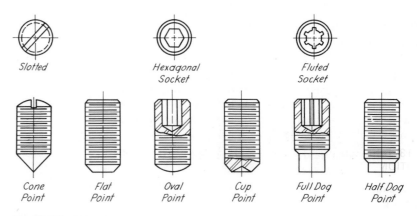

Fig. 14.21 □ Set screws.

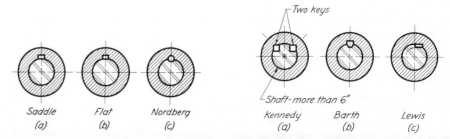

Fig. 14.22 □ Light duty keys.

Fig. 14.23 □ Heavy duty keys.

The square key (Fig. 14.24) and the Pratt and Whitney key (Fig. 14.25) are the two keys most frequently used in machine design. A plain milling cutter is used to cut the keyway for the square key, and an end mill is used for the Pratt and Whitney keyway. Both keys fit tightly in the shaft and in the part mounted on it.

The gib-head key (Fig. 14.26) is designed so that the head remains far enough from the hub to allow a drift pin to be driven to remove the key. The hub side of the key is tapered $\frac{1}{8}$ in. per ft to ensure a fit tight enough to prevent both axial and rotary motion. For this type of key, the keyway must be cut to one end of the shaft.

14.24 □ Woodruff keys

A Woodruff key is a flat segmental disc with either a flat or a round bottom (Fig. 14.27). It is always specified by a number, the last two digits of which indicate the nominal diameter in eighths of an inch, while the digits preceding the last two give the nominal width in thirty-seconds of an inch.

14.25 □ Taper pins

A taper pin is commonly used for fastening collars and pulleys to shafts, as illustrated in Fig. 14.28. The hole for the pin is drilled and reamed with the parts assembled. When a taper pin is to be used, the drawing callout should read as follows:

#3 (.213) DRILL AND REAM FOR #4 TAPER PIN WITH PC #6 IN POSITION

Drill sizes and exact dimensions for taper pins are given in Table 27 in Appendix E.

14.26 □ Locking devices

A few of the many types of locking devices that prevent nuts from becoming loose under vibration are shown in Figs. 14.29 and 14.30.

Figure 14.29 shows six forms of patented spring washers. The ones shown in (D), (E), and (F) have internal and external teeth.

In common use is the castellated nut with a spring cotter pin that passes through the shaft and the slots in the top [Fig. 14.30(a)]. This type is used extensively in automotive and aeronautical work.

Figure 14.30(b) shows a regular nut that is prevented from loosening by an American National Standard jam nut.

In Fig. 14.30(c) the use of two jam nuts is illustrated.

A regular nut with a spring-lock washer is shown in Fig. 14.30(d). The reaction provided by the lock washer tends to prevent the nut from turning.

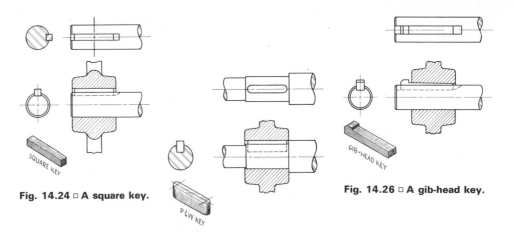

Fig. 14.24 □ A square key.

Fig. 14.25 □ A Pratt and Whitney key.

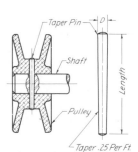

Fig. 14.26 □ A gib-head key.

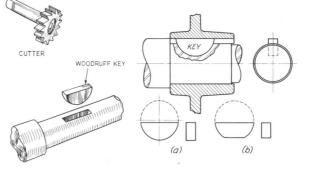

Fig. 14.27 □ A Woodruff key.

Fig. 14.28 □ Use of a taper pin.

Fig. 14.29 □ Special lock washers.

A regular nut with a spring cotter pin through the shaft, to prevent the nut from backing off, is shown in Fig. 14.30(e).

Special devices for locking nuts are illustrated in Fig. 14.30 (f) and (g). A set screw may be held in position with a jam nut, as in (h).

14.27 □ Rivets

Rivets are permanent fasteners used chiefly for connecting members in such structures as buildings and bridges and for assembling steel sheets and plates for tanks, boilers, and ships. They are cylindrical rods of wrought iron or soft steel, with one head formed when manufactured. A head is formed on the other end after the rivet has been put in place through the drilled or punched holes of the mating parts. A hole for a rivet is generally drilled, punched, or punched and reamed $\frac{1}{16}$ in. larger than the diameter of the shank of the rivet. Small rivets, less than $\frac{1}{2}$ in. in diameter, may be driven cold, but the larger sizes are driven hot. For specialized types of engineering work, rivets are manufactured of chrome–iron, aluminum, brass, copper, and so on. Standard dimensions for small rivets are given in Table 22 in Appendix E.

The type of rivets and their treatment are indicated on drawings by the American National Standard conventional symbols.

14.28 □ Riveted joints

Joints on boilers, tanks, and so on, are classified as either lap joints or butt joints. Lap joints are generally used for seams around the circumference (Fig. 14.31). Butt joints are used for longitudinal seams, except on small tanks where the pressure is to be less than 100 lb per sq in. (Fig. 14.32).

Springs

14.29 □ Springs

In production work, a spring is largely a matter of mathematical calculation rather than drawing, and it is usually purchased from a spring manufacturer, with the understanding that it will fulfill specified conditions. For experimental work, and when only one is needed, it may be formed by winding oil-tempered spring wire or music wire around a cylindrical bar. As it is wound, the wire follows the helical path of the screw thread. For this reason the steps in the layout of the representation for a spring are similar to the screw thread.

Single-line symbols for the representation of springs are shown in Fig. 14.33.

When making a detail working drawing of a spring, it should be shown to its free length. On either an assembly or detail drawing,

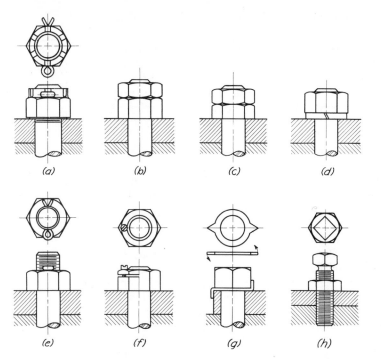

(a)   (b)   (c)   (d)

(e)   (f)   (g)   (h)

Fig. 14.30 □ Locking schemes.

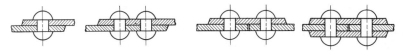

Fig. 14.31 □ Lap joints.   **Fig. 14.32 □ Butt joints.**

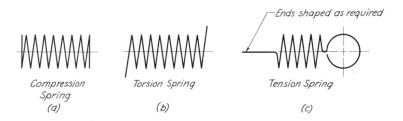

Compression Spring
(a)

Torsion Spring
(b)

Ends shaped as required

Tension Spring
(c)

Fig. 14.33 □ Single line representation of springs.
(*ANSI Z14.1–1946.*)

a fairly accurate representation, neatly drawn, will satisfy all requirements.

It is common practice in industry to rely on a printed spring drawing, accompanied by a filled-in printed form, to convey the necessary information for the production of a needed spring by a reliable manufacturer. The best procedure is to give the spring characteristics along with a list of necessary dimensions and then depend on an experienced spring designer, at the plant where the spring will be produced, to finalize the design.

The method of representing and dimensioning a compression spring is shown in Fig. 14.34. The spring is represented by a rectangle with diagonals (printed or drawn). Pertinent information is added as dimensions and notes. Either an ID (inside diameter) or an OD (outside diameter) dimension is given, depending on how the spring is to be used.

A method of representing and dimensioning an extension spring is shown in Fig. 14.35. Although drawings of this type may be printed forms, it is often necessary to prepare a drawing showing the ends and a few coils, since extension springs may have any one of a wide variety of types of ends. Needed information is presented in tabular form, as shown, with or without a complete spring design. When no printed form is available, a drawing similar to the one shown in Fig. 14.35 must be prepared.

A torsion spring offers resistance to a torque load. The extended ends form the torque arms, which are rotated about the axis of the spring. One method of representing and dimensioning torsion springs is shown in Fig. 14.36. A printed form may be used when there is sufficient uniformity in product requirements to warrant the preparation of a printed form or several printed forms. When a printed form is not available, a drawing similar to the one shown must be prepared.

The term *flat spring* includes all springs made of a strip material. One method of representing and dimensioning flat springs is shown in Fig. 14.37.

Bearings

14.30 □ Bearings

A draftsman ordinarily is never called on to make a detail drawing of a ball or roller bearing, because bearings of these two types are precision-made units that are purchased from reliable manufacturers. All draftsmen working on machine drawings, however, should be familiar with the various types commonly used and should be able to represent them correctly on an assembly drawing (Fig. 14.38). An engineer will find it necessary to determine shaft-mounting fits and housing-mounting fits from a manufacturer's

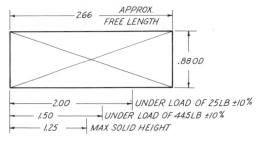

MATERIAL: .105 HARD DRAWN SPRING STEEL WIRE
12 COILS 10 ACTIVE
CLOSED ENDS GROUND
FINISH: PLAIN

Fig. 14.34 □ Drawing showing a compression spring.

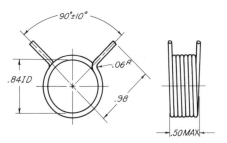

MATERIAL: .059 MUSIC WIRE
6¾ COILS - RIGHT HAND - NO INITIAL TENSION
TORQUE : 2.5 INCH LB AT 155° DEFLECTION SPRING MUST
DEFLECT 180° WITHOUT PERMANENT SET AND
MUST OPERATE FREELY ON .750 DIA SHAFT
FINISH: CADMIUM PLATE

Fig. 14.36 □ The representation and dimensioning of a torsion spring.

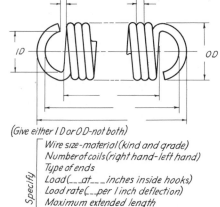

(Give either I D or O D-not both)

Specify:
Wire size-material (kind and grade)
Number of coils (right hand - left hand)
Type of ends
Load (___ at ___ inches inside hooks)
Load rate (___ per 1 inch deflection)
Maximum extended length
Finish, etc.

Fig. 14.35 □ Extension spring drawing.

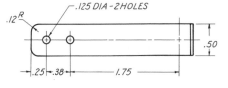

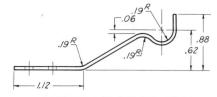

MATERIAL: .049 × .50 SPRING STEEL
HEAT TREAT: 44-48C ROCKWELL
FINISH: BLACK OXIDE AND OIL

Fig. 14.37 □ A drawing showing a flat spring.

handbook, in order to place the correct limits on shafts and housings. Figure 14.39 shows two types of ball bearings. A roller bearing is shown in Fig. 14.40.

Ball bearings may be designed for loads either perpendicular or parallel to the shaft. In the former, they are known as radial bearings and in the latter, as thrust bearings. Other types, designated by various names, are made to take both radial and thrust loads, either light or heavy. In most designs, bearings are forced to take both radial and thrust loads. Ball bearings are designated by a letter and code number, the last number of which represents the bearing bore. They may be extra light, light, medium, or heavy and still have the same bore number. That is, bearings of different capacities and different outer diameters can fit shafts having the same nominal size.

Roller bearings are designed for both radial and thrust loads. The bearing consists of tapered rollers that roll between an inner and an outer race. The rollers are enclosed in a retainer (cage) that keeps them properly spaced.

Welds

14.31 □ Welding processes

For convenience, the various welding processes used in commercial production may be classified into three types: pressure processes, nonpressure processes, and casting processes. The nonpressure processes are arc welding and gas welding. Metallic arc welding is the joining of two pieces of metal through the use of a sustained arc formed between the work and a metal rod held in a holder. The intense heat melts the metal of the work and at the same time heats the end of the electrode, causing small globules to form and cross the arc to the weld. In gas welding, the heat is produced by a burning mixture of two gases, which ordinarily are oxygen and acetylene. The weld is formed by melting a filler rod with the torch flame, along the line of contact, after the metal of the work has been preheated to a molten state. This method is essentially a puddling process, in that the weld is produced by a small moving molten pool that is maintained by the flame constantly directed upon it. Resistance welding is a pressure process, the fusion being made through heat and mechanical pressure. The work is heated by a strong electrical current that passes through it until the fusion temperature is reached; then pressure is applied to create the weld.

The forms of resistance welding are: projection welding, seam welding, spot welding, and flash welding. In spot welding, the parts are overlapped and welds are made at successive single spots. A seam weld is similar to a spot weld, except that a continu-

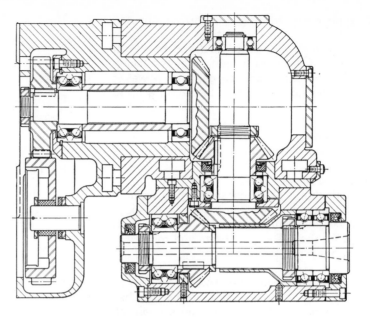

Fig. 14.38 □ **Ball bearings—typical mountings.**
(*Courtesy New Departure Bearing Company.*)

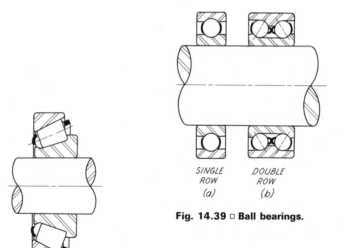

SINGLE ROW (a)   DOUBLE ROW (b)

Fig. 14.39 □ **Ball bearings.**

Fig. 14.40 □ **Roller bearing.**

ous weld is produced. In projection welding, one part is embossed and welds are made at the successive projections. In making a flash weld, the two pieces to be joined are held end to end in jaws and act as electrodes. At the right instant, after the facing metal has been heated by the arc across the gap, the power is shut off and the two ends are forced together to cool in a fused position.

14.32 □ Types of welded joints
Figure 14.41 shows the ordinary types of welded joints.

14.33 □ Working drawings of welded parts
Figure 14.42 shows a part that is to be constructed by welding rolled shapes. It should be noted that the joint is completely specified through the use of a welding symbol. A careful study will show that the drawing, except for the absence of fillets and rounds, is very much like a casting drawing.

14.34 □ Arrow-side and other-side welds (Fig. 14.43)
''The use of the words 'far side' and 'near side' in the past has led to confusion, because when joints are shown in section, all welds are equally distant from the reader, and the words 'near' and 'far' are meaningless. In the present system the joint is the basis of reference. Any joint whose welding is indicated by a symbol will always have an 'arrow-side' and an 'other-side.' ''*

14.35 □ Welding symbols
An enlarged drawing of the approved welding symbol is shown in Fig. 14.43, along with explanatory notes that indicate the proper locations of the marks and size dimensions necessary for a complete description of a weld.

The arrow is the basic portion of the symbol. It points toward the joint where the required weld is to be made, as in Fig. 14.42.

If the weld is on the arrow-side, the symbol indicating the type of weld is placed below or to the right of the base line, depending upon whether that line is horizontal or vertical (Fig. 14.43). If the weld is located on the other-side, the symbol should be above or to the left.

To indicate that a weld is to be made all around a connection, as is necessary when a piece of tubing must be welded to a plate, a weld all-around symbol, a circle, is placed as shown in Fig. 14.43.

The size of a weld is given along the base of the arrow, at the side of the symbol, as shown in Fig. 14.42. If the welds on the arrow-side and the other-side of a lap joint are the same size, only

*Extracted from ANSI Z32.2.1–1949.

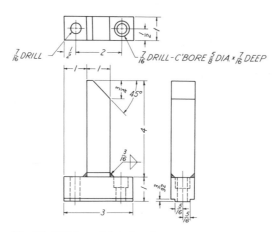

Fig. 14.41 □ Types of welds.

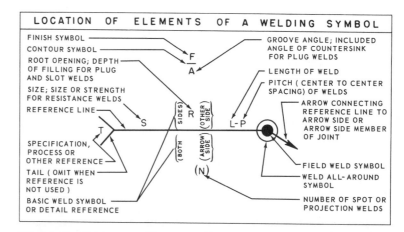

Fig. 14.42 □ A welding drawing.
(*Courtesy Lincoln Electric Company.*)

Fig. 14.43 □ The basic welding symbol.
(*Based on ANSI Y32.3–1969.*)

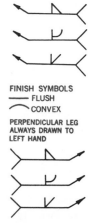

284

one dimension should be given. If they are not the same size, each dimension should be placed beside its associated symbol.

14.36 □ Gas and arc welding symbols

In order to satisfy the need for a standard group of symbols that could be understood in all manufacturing plants, the American Welding Society recommended in 1940 a set of conventional symbols so designed that each symbol resembled in a general way the type of weld it represented. Figure 14.44 shows a condensed table of symbols taken from ANSI Y32.3–1969. The symbols shown are the same as those first proposed by the American Welding Society.

14.37 □ Resistance welding

Figure 14.45 shows the symbols for the four principal types of resistance welding. The method of specifying resistance welds differs from the methods used for arc and gas welds. In the former, the strength of a weld is given in units instead of size, and the symbols do not show the form of the weld. The strength of spot and projection welds is given in units of pounds per weld. The strength for seam welds is given in units of pounds per linear inch.

B □ Fundamentals and techniques of dimensioning

14.38 □ Introduction

A detail drawing, in addition to giving the shape of a part, must furnish information such as the distances between surfaces, locations of holes, kind of finish, type of material, number required, and so forth. The expression of this information on a drawing by the use of lines, symbols, figures, and notes is known as *dimensioning*.

Intelligent dimensioning requires engineering judgment and a thorough knowledge of the practices of patternmaking, forging, and machining.

14.39 □ Theory of dimensioning

Any part may be dimensioned easily and systematically by dividing it into simple geometric solids. Even complicated parts, when analyzed, usually are found to be composed principally of cylinders and prisms and, frequently, frustums of pyramids and cones. The dimensioning of an object may be accomplished by dimensioning each elemental form to indicate its size and relative location from a center line, base line, or finished surface. A machine drawing

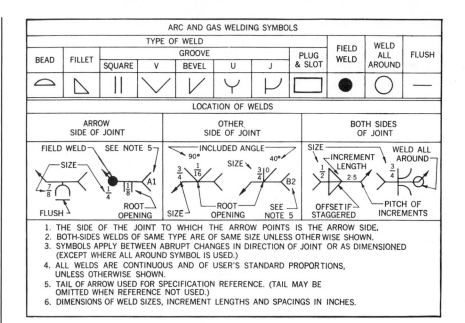

Fig. 14.44 □ ANSI Standard arc and gas welding symbols.
(Based on ANSI Y32.3–1969.)

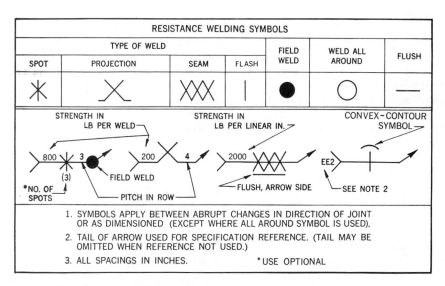

Fig. 14.45 □ Resistance welding symbols.
(Based on ANSI Y32.3–1969.)

requires two types of dimensions: *size dimensions* and *location dimensions*.

14.40 □ Size dimensions (Fig. 14.46)

Size dimensions give the size of a piece, component part, hole, or slot.

The rule for placing the three principal dimensions (width, height, and depth) on the drawing of a prism or modification of a prism is as follows: *Give two dimensions on the principal view and one dimension on one of the other views.*

The circular cylinder, which appears as a boss or shaft, requires only *the diameter and length, both of which are shown preferably on the rectangular view* (Fig. 14.52). It is better practice to dimension a hole (negative cylinder) by giving the diameter and operation as a note on the contour view with a leader to the circle (Fig. 14.46).

Cones are dimensioned by giving *the diameter of the base and the altitude on the same view*. A taper is one example of a conical shape found on machine parts (Fig. 14.83).

Pyramids, which frequently form a part of a structure, are dimensioned by giving *two dimensions on the view showing the shape of the base.*

A sphere requires only the diameter.

14.41 □ Location dimensions

Location dimensions fix the relationship of the component parts (projections, holes, slots, and other significant forms) of a piece or structure (Fig. 14.47). Particular care must be exercised in the selection and placing of location dimensions because on them depends the accuracy of the operations in making a piece and the proper mating of the piece with other parts. To select location dimensions intelligently, one must first determine the contact surfaces, finished surfaces, and center lines of the elementary geometric forms and, with the accuracy demanded and the method of production in mind, decide from what other surface or center line each should be located. Mating location dimensions must be given from the same center line or finished surface on both pieces.

Location dimensions may be from center to center, surface to center, or surface to surface (Fig. 14.48).

14.42 □ Procedure in dimensioning

The theory of dimensioning may be applied in six steps, as follows:

1 Mentally divide the object into its component geometrical shapes.

2 Place the size dimensions on each form.

3 Select the locating center lines and surfaces after giving careful

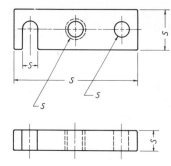

Fig. 14.46 □ Size dimensions.

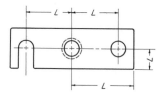

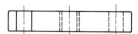

Fig. 14.47 □ Location dimensions.

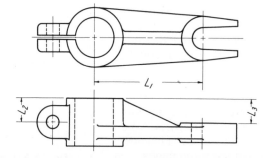

L_1 – *Center to Center*
L_2 – *Surface to Center*
L_3 – *Surface to Surface*

Fig. 14.48 □ Types of location dimensions.

consideration to mating parts and to the processes of manufacture.

4 Place the location dimensions so that each geometrical form is located from a center line or finished surface.

5 Add the overall dimensions.

6 Complete the dimensioning by adding the necessary notes.

14.43 □ Placing dimensions

Dimensions must be placed where they will be most easily understood—in the locations where the reader will expect to find them. They generally are attached to the view that shows the contour of the features to which they apply, and a majority of them usually will appear on the principal view (Fig. 14.55). Except in cases where special convenience and ease in reading are desired, or when a dimension would be so far from the form to which it referred that it might be misinterpreted, dimensions should be placed outside a view. They should appear directly on a view only when clarity demands.

All extension and dimension lines should be drawn before the arrowheads have been filled in or the dimensions, notes, and titles have been lettered. Placing dimension lines not less than $\frac{1}{2}$ in. from the view and at least $\frac{3}{8}$ in. from each other will provide spacing ample to satisfy the one rule to which there is no exception: *Never crowd dimensions.* If the location of a dimension forces a poor location on other dimensions, its shifting may allow all to be placed more advantageously without sacrificing clarity. Important location dimensions should be given where they will be conspicuous, even if a size dimension must be moved.

14.44 □ Dimensioning practices

A generally recognized system of lines, symbols, figures, and notes is used to indicate size and location. Figure 14.49 illustrates dimensioning terms and notation.

A *dimension line* is a lightweight line that is terminated at each end by an arrowhead. A numerical value, given along the dimension line, specifies the number of units for the measurement that is indicated (Fig. 14.50). When the numerals are in a single line, the dimension line is broken near the center, as shown in (*a*) and (*b*). Under no circumstances should the line pass through the numerals. When the numerals are in two lines the dimension line may be drawn without a break and one line of numerals may be placed above the dimension line and the other below, as in (*c*).

Extension lines are light, continuous lines extending from a view to indicate the extent of a measurement given by a dimension line that is located outside a view. They start $\frac{1}{16}$ in. from the view and extend $\frac{1}{8}$ in. beyond the dimension line (Fig. 14.49).

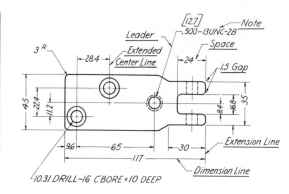

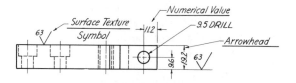

Fig. 14.49 □ Terms and dimensioning notation.
(*Dimension values are in millimeters.*)

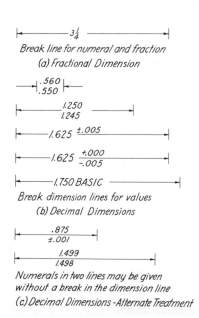

Fig. 14.50 □ A dimension line.

Arrowheads are drawn for each dimension line, before the figures are lettered. They are made with the same pen or pencil used for the lettering. The size of an arrowhead, although it may vary with the size of a drawing, should be uniform on any one drawing. To have the proper proportions, the length of an arrowhead must be approximately three times its spread (ANSI Standard). This length for average work is usually $\frac{1}{8}$ in. Figure 14.51 shows enlarged drawings of arrowheads of correct proportions. Although many draftsmen draw an arrowhead with one stroke, the beginner will get better results by using two slightly concave strokes drawn toward the point (*a*) or, as shown in (*b*), one stroke drawn to the point and one away from it.

A *leader* or *pointer* is a light, continuous line (terminated by an arrowhead) that extends from a note to the feature of a piece to which the note applies (Fig. 14.49). It should be made with a straightedge and should not be curved or made freehand.

A leader pointing to a curve should be radial, and the first $\frac{1}{8}$ in. of it should be in a line with the note (Fig. 14.49).

Dimension figures should be lettered either horizontally or vertically with the whole numbers equal in height to the capital letters in the notes and guidelines and slope lines must be used. The numerals must be legible; otherwise, they might be misinterpreted in the shop and cause errors, which would be embarrassing to the draftsman.

14.45 □ Fractional dimensioning

For ordinary work, where accuracy is relatively unimportant, shopmen work to nominal dimensions given as $\frac{1}{2}$(.50), $\frac{1}{4}$(.25), $\frac{1}{8}$(.12), $\frac{1}{16}$(.06), and so forth. When dimensions are given in this way, many large corporations specify the required accuracy through a note on the drawing that reads as follows: *Permissible variations on common fraction dimensions to machined surfaces to be ±.010 unless otherwise specified.* It should be understood that the allowable variations will differ among manufacturing concerns because of the varying degree of accuracy required for different types of work.

14.46 □ Decimal system

Since the use of the decimal system for expressing dimensional values has made rapid gains in American industry and is now accepted in the aircraft and automotive fields, many of the examples in this chapter will show decimal dimensions (Sec. A.19, Appendix A). At present, the fractional system, which dominated the whole of American industry until the Ford Motor Company adopted the decimal-inch some 40 years ago, is still used in those fields that are not under the direct influence of the automotive and aircraft companies. How the inch is treated may be immaterial in

the near future when American industry probably will be using SI(Système International) units to express dimensional and specification values. This changeover may be nearly completed within the next decade, with the automobile industry again pointing the way by adopting a dual-dimensioning system for drawings. See Sec. 14.51.

However, this need not be of great concern to the student, for he should be primarily interested in the selection and placement of dimensions. At a later time he will find it easy to use either the metric, fractional, or decimal systems as required in his field of employment. To assist one to use any of these systems, a Standard Conversion Table has been provided in Appendix E (Table 1).

In Fig. 14.52 a drawing is shown that illustrates decimal dimensioning.

The recommendation for decimal dimensioning, as given in ANSI Y14.5–1966, reads as follows:

> Optionally, decimals may be used to replace common fractions altogether. The elimination of common fractions simplifies computations; decimals can be added, subtracted, multiplied or divided more easily than fractions. . . . The following conventions are generally observed by those who have adopted the decimal system. However, the advantages of using decimals can be gained, whether or not all of these rules are followed.
>
> (*a*) Two-place decimals are used for dimensions where tolerance limits of ±.01 or more can be allowed (Fig. 14.52).
>
> (*b*) Decimals to three or more places must be used for tolerance limits less than ±.010 [see Fig. 14.50(*b*) and (*c*)].
>
> (*c*) The second decimal place in a two-place decimal should preferably be an even digit (.02, .04, .06 are preferred to .01, .03, .05, etc.), so that when divided by two, as in obtaining a radius from a diameter, the result will remain a two-place decimal. Odd two-place decimals are used where necessary for design reasons; as, to provide clearance, strength, smooth curves, etc.
>
> (*d*) Common fractions may be used to indicate standard nominal sizes of materials, punched holes, drilled holes, threads, keyways and other features produced by tools that are so designated.

Examples:

$\frac{1}{4}$–20 UNC–2A; $\frac{5}{16}$ Drill; Stock $\frac{5}{8} \times \frac{7}{8}$.

> (*e*) Where it is desired to use decimals exclusively, nominal sizes of materials, threads and other features produced by commercial tools may be expressed in decimal equivalents of commercial sizes; as, .625 HEX.

14.47 □ Use of the metric system

Since the metric system has worldwide acceptance and has been legalized for use in this country (approved by act of Congress in

1866), engineers and draftsmen in the United States will find it advantageous to be able to interpret and quite possibly prepare detail drawings using metric dimensioning. At present, there are persons who argue strongly that American industry must change over to this system if the United States is to retain leadership in world trade. In any case, a change appears to be underway and metric dimensioning is appearing more frequently on drawings made in the United States (an example has been shown in Fig. 14.53). The unit of measurement is the millimeter (mm) and, since this fact is generally known by all persons using the system, no indicating marks are needed.

Under the metric system, drawings are prepared to scales based on divisions of 10, such as 1 to 2, 1 to 5, 1 to 10, and so forth. A millimeter is one-thousandth part of a meter, which our government has established as being 39.37 in. in length. See Sec. 14.51.

General dimensioning practices

14.48 □ Selection and placement of dimensions and notes

The reasonable application of the selected dimensioning practices that follow should enable a student to dimension acceptably. The practices in boldface type should never be violated. In fact, these have been so definitely established by practice that they might be called rules.

1 Place dimensions using either of two recognized methods—aligned or unidirectional.
   a *Aligned Method.* Place the numerals for the dimension values so that they are readable from the bottom and right side of the drawing. An aligned expression is placed along and in the direction of the dimension line (Fig. 14.52).
   b *Unidirectional Method.* Place the numerals for the dimension values so that they can be read from the bottom of the drawing (see Fig. 14.53). The fraction bar for a common fraction should be parallel to the bottom of the drawing (Fig. 14.55).

2 Place dimensions outside a view, unless they will be more easily and quickly understood if shown on the view (Figs. 14.54 and 14.57).

3 Place dimensions between views unless the rules, such as the contour rule, the rule against crowding, and so forth, prevent their being so placed.

4 **Do not use an object line or a center line as a dimension line.**

5 Locate dimension lines so that they will not cross extension lines.

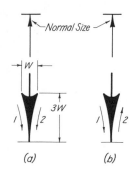

Fig. 14.51 □ Formation of arrowheads.

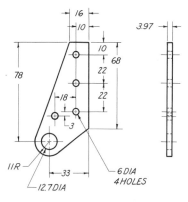

Fig. 14.53 □ Metric dimensioning. (*Dimensions are in millimeters.*)

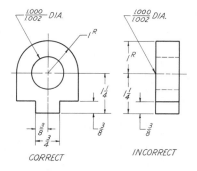

CORRECT   INCORRECT

Fig. 14.55 □ Contour principle of dimensioning.

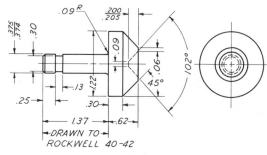

DET. 15   RIVET SET
I- TYPE "CC" STEEL
STK. 1⅜   WT. .84 LBS.
HEAT IN CYANIDE TO 1525°F
QUENCH IN OIL
DRAW AT 350°F
ROCKWELL 54-59C

Fig. 14.52 □ Decimal dimensioning. (Courtesy Ford Motor Company.)

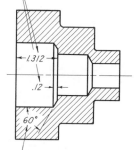

Fig. 14.54 □ Dimensions on the view.

6 If possible, avoid crossing two dimension lines.

7 A center line may be extended to serve as an extension line (Fig. 14.55).

8 Keep parallel dimensions equally spaced (usually ³⁄₈ in. apart) and the figures staggered (Fig. 14.56).

9 **Always give locating dimensions to the centers of circles that represent holes, cylindrical projections, or bosses** (Figs. 14.49 and 14.55).

10 If possible, attach the location dimensions for holes to the view on which they appear as circles (Fig. 14.57).

11 Group related dimensions on the view showing the contour of a feature (Fig. 14.55).

12 Arrange a series of dimensions in a continuous line (Fig. 14.58).

13 Dimension from a finished surface, center line, or base line that can be readily established (Figs. 14.73 and 14.100).

14 Stagger the figures in a series of parallel dimension lines to allow sufficient space for the figures and to prevent confusion (Fig. 14.56).

15 Place longer dimensions outside shorter ones so that extension lines will not cross dimension lines.

16 Give three overall dimensions located outside any other dimensions (unless the piece has cylindrical ends—see 42 and Fig. 14.77).

17 When an overall is given, one intermediate distance should be omitted unless noted (REF) as being given for reference (Fig. 14.58).

18 Do not repeat a dimension. One of the duplicated dimensions may be missed if a change is made. Give only those dimensions that are necessary to produce or inspect the part.

19 Make decimal points of a sufficient size so that dimensions cannot be misread.

20 **When dimension figures appear on a sectional view, show them in a small uncrosshatched portion so that they may be easily read.** This may be accomplished by doing the section lining after the dimensioning has been completed (Fig. 14.59).

21 **When an arc is used as a dimension line for an angular measurement, use the vertex of the angle as the center** (Fig. 14.60). It is usually undesirable to terminate the dimension line for an angle at lines that represent surfaces. It is better practice to use an extension line.

22 Place the figures of angular dimensions so they will read from the bottom of a drawing, except in the case of large angles (Fig. 14.60).

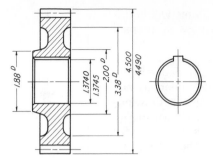

Fig. 14.56 □ **Parallel dimensions.**

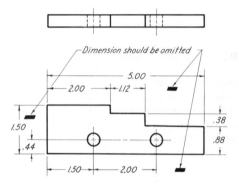

Fig. 14.58 □ **Omit unnecessary dimensions.**

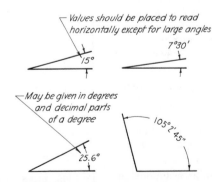

Fig. 14.60 □ **Angular dimensions.**

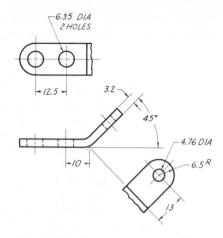

Fig. 14.57 □ **Dimensioning an angle bracket.**

(Dimension values are in millimeters.)

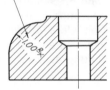

Fig. 14.59 □ **Dimension figures on a section view.**

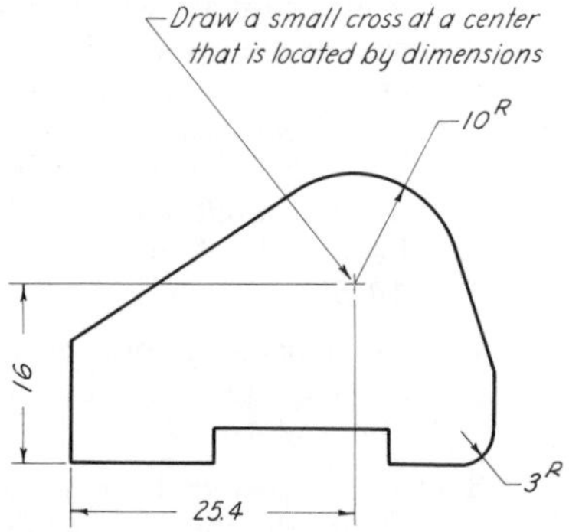

Fig. 14.61 □ **Dimensioning a circular arc.**

(Dimension values are in millimeters.)

23 Dimension an arc by giving its radius followed by the abbreviation R, and indicate the center with a small cross. [Locate the center by dimensions. (Fig. 14.61).]

24 TRUE R is added after the radius value, where the radius is dimensioned in a view that does not show the true shape of the arc (Fig. 14.62).

25 **Show the diameter of a circle, never the radius.** If it is not clear that the dimension is a diameter, the figures should be followed by the abbreviation D or DIA (Figs. 14.63 and 14.65). Often this will allow the elimination of one view.

26 When dimensioning a portion of a sphere with a radius the term SPHER. R is added (Fig. 14.64).

27 Letter all notes horizontally (Fig. 14.100).

28 **Make dimensioning complete, so that it will not be necessary for a workman to add or subtract to obtain a desired dimension or to scale the drawing.**

29 Give the diameter of a circular hole, never the radius, because all hole-forming tools are specified by diameter. If the hole does not go through the piece, the depth may be given as a note (Fig. 14.65).

30 Never crowd dimensions into small spaces. Use the practical methods suggested in Fig. 14.66.

31 When using the aligned method, place inclined dimensions so that dimensional values may be conveniently read from the bottom and right; if this is not possible, they should be read in a downward direction with the line (Figs. 14.85 and 14.86).

32 Omit superfluous dimensions. Do not supply dimensional information for the same feature in two different ways.

33 Give dimensions up to 72 in. in inches, except on structural and architectural drawings (Fig. 14.67). Omit the inch marks when all dimensions are in inches.

34 Show dimensions in feet and inches as illustrated in Fig. 14.68. Note that the use of the hyphen in (*a*) and (*b*) and the cipher in (*b*) eliminates any chance of uncertainty and misinterpretation.

35 If feasible, design a piece and its elemental parts to such dimensions as $\frac{3}{8}$, $\frac{1}{2}$, $\frac{5}{8}$, or .40 and .50 in. Avoid such fractions as $\frac{17}{32}$ and $\frac{19}{64}$ and such decimals as .19 and .53.

36 Dimension a chamfer by giving the angle and length as shown in Fig. 14.69. For a 45° angle only, it is permissible to give the needed information as a note. The word chamfer may be omitted in the note form.

37 Equally spaced holes in a circular flange may be dimensioned by giving the diameter of the bolt circle, across the circular center line, and the size and number of holes, in a note (Fig. 14.70).

A foreshortened radius may be dimensioned as:

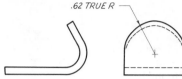

Fig. 14.62 □ **To dimension a circular arc—true R.**
(*ANSI Y14.5–1966.*)

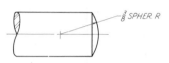

Fig. 14.64 □ **Dimensioning a piece with a spherical end.**

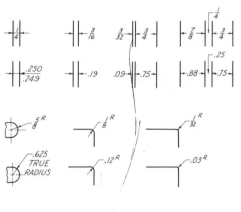

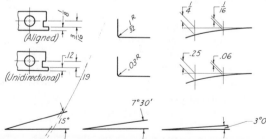

Fig. 14.66 □ **Dimensioning in limited spaces.**

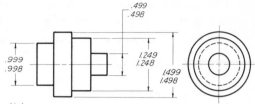

Note:
Although it is better practice to use a minimum of two views, a cylindrical part may be completely described in one view (no end view) by using the abbreviation DIA with the dimension.

Fig. 14.63 □ **Dimensioning machined cylinders.**
(*ANSI Y14.5–1966.*)

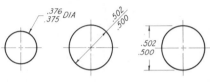

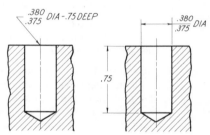

Fig. 14.65 □ **Dimensioning holes.**
(*ANSI Y14.5–1966.*)

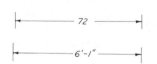

Fig. 14.67 □ **Dimension values.**

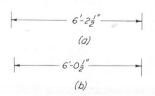

(a)

(b)

Fig. 14.68 □ **Feet and inches.**

38 When holes are unequally spaced on a circular center line, give the angles as illustrated in Fig. 14.71.

39 Holes that must be accurately located should have their location established by the coordinate method. Holes arranged in a circle may be located as shown in Fig. 14.72 rather than through the use of angular measurements. Figure 14.73 shows the application of the coordinate method to the location of holes arranged in a general rectangular form. The method with all dimensions referred to datum lines is sometimes called *base-line dimensioning*.

40 Dimension a curved line by giving offsets or radii.
   a A noncircular curve may be dimensioned by the coordinate method illustrated in Fig. 14.74. Offset measurements are given from datum lines.
   b A curved line, which is composed of circular arcs, should be dimensioned by giving the radii and locations of either the centers or points of tangency (Figs. 14.75 and 14.76).

41 Show an offset dimension line for an arc having an inaccessible center (Fig. 14.76). Locate with true dimensions the point placed in a convenient location that represents the true center.

42 Dimension a piece with cylindrical ends as required by the method of production. Give the diameters and center-to-center distance (Fig. 14.77). No overall is required.

43 The method to be used for dimensioning a piece with rounded ends is determined by the degree of accuracy required and the method of production (Figs. 14.78–14.80).
   a It has been customary to give the radii and center-to-center distance for parts and contours that would be laid out and/or machined using centers and radii. A link (Fig. 14.78) or a pad with a slot is dimensioned in this manner to satisfy the requirements of the patternmaker and machinist. An overall dimension is not needed.
   b Overall dimensions are recommended for parts having rounded ends when considerable accuracy is required. The radius is indicated but is not dimensioned when the ends are fully rounded. In Fig. 14.79, the center-to-center hole distance has been given because the hole location is critical.
   c Slots are dimensioned by giving length and width dimensions. They are located by dimensions given to their longitudinal center line and to either one end or a center line (Fig. 14.80).

44 A keyway on a shaft or hub should be dimensioned as shown in Fig. 14.81. Woodruff keyslots are dimensioned as shown in (*b*).

45 When knurls are to provide a rough surface for better grip, it is necessary to specify the pitch and kind of knurl, as shown in Fig. 14.82(*a*) and (*b*). When specifying knurling for a press fit, it is best practice to give the diameter before knurling with

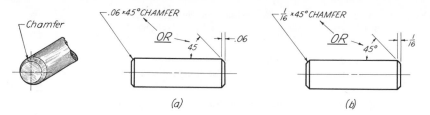

Fig. 14.69 □ **Dimensioning a chamfer.**

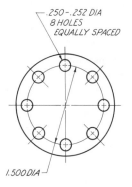

Fig. 14.70 □ **Equally spaced holes.**
(*ANSI Y14.5–1966.*)

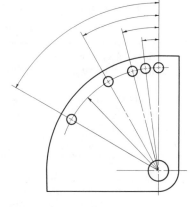

Fig. 14.71 □ **Locating holes on a circle by polar coordinates.**
(*ANSI Y14.5–1966.*)

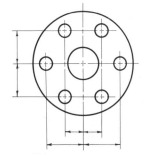

Fig. 14.72 □ **Accurate location dimensioning of holes.**
(*ANSI Y14.5–1966.*)

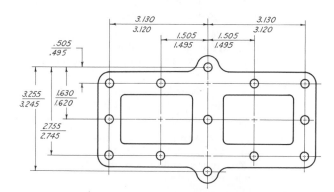

Fig. 14.73 □ **Location dimensioning of holes.**

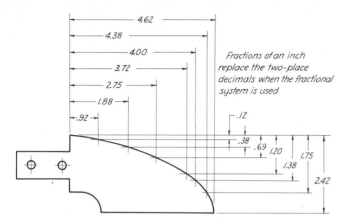

Fractions of an inch replace the two-place decimals when the fractional system is used

Fig. 14.74 □ **Dimensioning curves by offsets.**

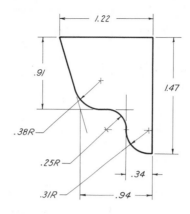

Fig. 14.75 □ **Dimensioning curves consisting of circular arcs.**
(ANSI Y14.5–1966.)

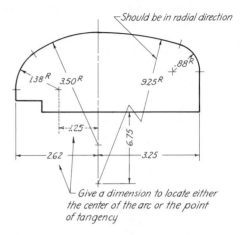

Should be in radial direction

Give a dimension to locate either the center of the arc or the point of tangency

Fig. 14.76 □ **Dimensioning curves by radii.**

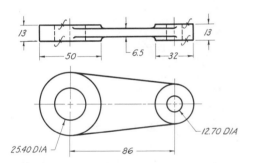

Fig. 14.77 □ **Dimensioning a part with cylindrical ends—link.**
(Dimension values are in millimeters.)

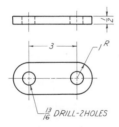

Fig. 14.78 □ **Link with rounded ends.**

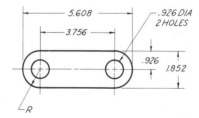

Fig. 14.79 □ **Dimensioning a part with rounded ends.**
(ANSI Y14.5–1957.)

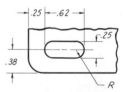

Fig. 14.80 □ **Dimensioning a slot.**

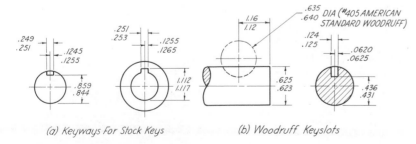

(a) Keyways For Stock Keys

(b) Woodruff Keyslots

Fig. 14.81 □ **Dimensioning keyways and keyslots.**
(ANSI Y14.5–1966.)

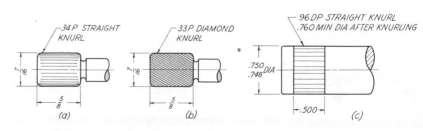

(a)   (b)   (c)

Fig. 14.82 □ **Dimensioning knurls.**

a tolerance and include the minimum diameter after knurling in the note that gives the pitch and type of knurl, as shown in (c).

46 Dimension standard and special tapers as illustrated in Fig. 14.83. Standard tapers require one diameter, the length, and a note specifying the taper by number. The usual practice is to give the diameter at the large end.

Special conical tapers may be dimensioned in different ways, as illustrated in (b), (c), and (d). The recommendation given in ANSI Y14.5–1957 reads as follows:

Conical tapers. The following dimensions may be given in different combinations, to specify the size and form of tapered conical surfaces:

(a) the diameter at each end of the taper;
(b) the length of the taper;
(c) the diameter at a selected cross-sectional plane; this plane may or may not be within the length of the tapered piece;
(d) the distance locating a cross-sectional plane at which a diameter is specified;
(e) the rate of taper;
(f) the included angle.

The dimensioning of noncritical tapers is shown in Fig. 14.83(b) and (c). The dimensions and notes given in (d) are considered adequate for tapers that engage one another permanently or intermittently.

Flat tapers may be dimensioned as shown in (e).

When given in a note, taper may be specified either in inches per foot or inches per inch [(d) and (e)].

47 A half section may be dimensioned through the use of hidden lines on the external portion of the view.

48 The fact that a dimension is out of scale may be indicated by a wavy line placed underneath the dimension value (Fig. 14.84).

49 In sheet-metal work, mold lines are used in dimensioning instead of the centers of the arcs (see Fig. 14.85). A mold line (construction line) is the line at the intersection of the plane surfaces adjoining a bend.

50 When dimensioning a gear, it is recommended that the cutting data be incorporated in an accompanying table (Fig. 14.86).

14.49 □ Dimensions from datum

When it is necessary to locate the holes and surfaces of a part with a considerable degree of accuracy, it is the usual practice to specify their position by dimensions given from a datum (Fig. 14.87) in order to avoid cumulative tolerances. By this method the different features of a part are located with respect to carefully selected datums and not with respect to each other (Fig. 14.73).

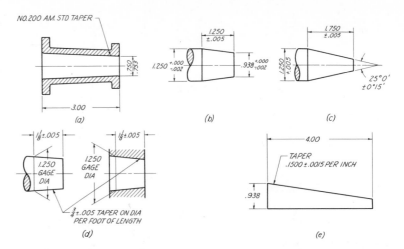

Fig. 14.83 □ Dimensioning tapers.

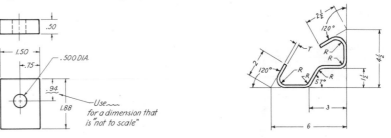

Fig. 14.84 □ Dimension out of scale.   **Fig. 14.85 □ Profile dimensioning.**

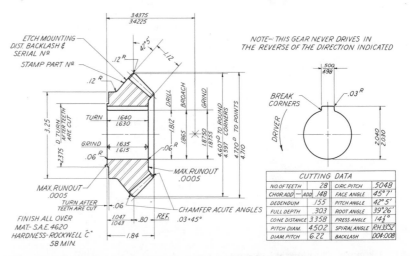

Fig. 14.86 □ Notes and specifications for a spiral bevel gear.
(Courtesy Fairfield Manufacturing Company.)

CUTTING DATA			
NO. OF TEETH	28	CIRC. PITCH	.5048
CHOR. ADD.	ADD. .148	FACE ANGLE	45°7'
DEDENDUM	.155	PITCH ANGLE	42°5'
FULL DEPTH	.303	ROOT ANGLE	39°26'
CONE DISTANCE	3.358	PRESS ANGLE	14½°
PITCH DIAM.	4.502	SPIRAL ANGLE	RH 33°52'
DIAM. PITCH	6.22	BACKLASH	.004–.008

Lines and surfaces that are selected to serve as datums must be easily recognizable and accessible during production. Corresponding datum points, lines, or surfaces must be used as datums on mating parts.

14.50 □ Notes (Fig. 14.88)

The use of properly composed notes often adds clarity to the presentation of dimensional information involving specific operations. Notes also are used to convey supplementary instructions about the kind of material, kind of fit, degree of finish, and so forth. Brevity in form is desirable for notes of general information or specific instruction. In the case of threaded parts one should use the terminology recommended by ANSI for the American–British Unified thread.

Dual dimensioning

14.51 □ Dual dimensioning

A dual-dimensioning procedure has been adopted recently by several of our large industrial organizations; the procedure calls for showing both U.S. inch units and the SI (Système International) metric units of measurements on the same drawing. See Fig. 14.89.

Dimension values, given in both inch and SI units, must insure the interchangeability of parts. The interchangeability of a part is determined by the number of decimal places retained in rounding off a converted dimension and by the degree that the tolerance limits of the conversion have been permitted to violate the limits of the original dimension. There are several reliable sources for information on conversion principles. These are: (1) Rules for Conversion and Rounding—ASTM E380 Standard Metric Practice Guide and (2) Conversion of Toleranced Linear Dimensions—SAE 1390 Dual Dimensioning Standard.

Each drawing should have noted on it how the dimension values given in inch units and those given in SI units may be identified. For linear dimensions, the SI unit is the millimeter. Two methods have been commonly employed on drawings to distinguish between values given in these different units. These methods are: (1) the position method and (2) the bracket method.

When the position method is used, the millimeter dimension may be placed either above the inch dimension as shown in Fig. 14.90(a) or to the left of the inch dimension with a slash line separating the values (d). Another method of display is similar except that the inch dimension value is placed above or to the left of the metric value.

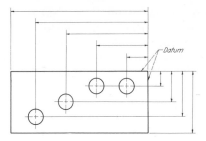

Fig. 14.87 □ Dimensions from datum lines.
(ANSI Y14.5–1957.)

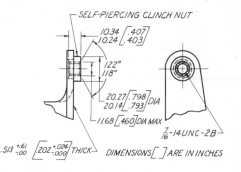

Fig. 14.89 □ A drawing showing dual-dimensioning.
(Courtesy Ford Motor Company.)

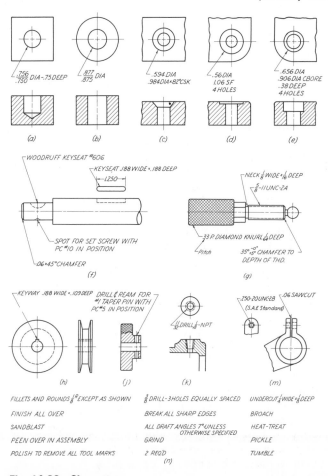

Fig. 14.88 □ Shop notes.

In using the bracket method, the inch dimension and millimeter dimension values may also be displayed in two ways but only one method should be used on a single drawing. In the first method, the millimeter dimension value is enclosed in square brackets as shown in Fig. 14.91. The bracketed millimeter dimension value may be placed either above or below or to the right or left of the inch value. However, the practice that is adopted must be followed for the entire drawing. The second and alternate method is similar except that the inch values (instead of the millimeter dimensions) are enclosed in brackets.

Many of the drawings that are made during this decade will be dual-dimensioned since it is expected that U.S. industry will be changing over to the metric system in order to remain competitive in world trade.

Limit dimensioning and geometric tolerancing

14.52 □ Limit dimensions

Present-day competitive manufacturing requires quantity production and interchangeability for many closely mating parts. The production of each of these mating parts to an exact decimal dimension, although theoretically possible, is economically unfeasible, since the cost of a part rapidly increases as an absolute correct size is approached. For this reason, the designer specifies an allowable error (tolerance) between decimal limits (Fig. 14.92). The determination of these limits depends on the accuracy and clearance required for the moving parts to function satisfactorily in the machine. Although manufacturing experience is often used to determine the proper limits for the parts of a mechanism, it is better and safer practice to adhere to the fits recommended by the American National Standards Institute in (ANSI) B4.1–1967. This standard applies to fits between plain cylindrical parts. Recommendations are made for preferred sizes, allowances, tolerances, and fits for use where applicable. Up to a diameter of 20 in. the standard is in accordance with ABC (American–British–Canadian) conference agreements.

There are many factors that a designer must take into consideration when selecting fits for a particular application. These factors might be the bearing load, speed, lubrication, materials, and length of engagement. Frequently temperature and humidity must be taken into account. Considerable practical experience is necessary to make a selection of fits or to make the subsequent adjustments that might be needed to satisfy critical functional requirements. In addition, manufacturing economy must never be overlooked.

Those interested in the selection of fits should consult texts on machine design and technical publications, for coverage of this phase of the dimensioning of cylindrical parts is not within the scope of this book. However, since it is desirable to be able to determine limits of size following the selection of a fit, attention in this section will be directed to the use of Table 29 (Appendix E). Whenever the fit to be used for a particular application has not been specified in the instructions for a problem or has not been given on the drawing, the student should consult his instructor after a tentative choice has been made based on the brief descriptions of fits as given in this section.

To compute limit dimensions it is necessary to understand the following associated terms.*

Nominal Size. The nominal size is the designation which is used for the purpose of general identification.

Basic Size. The basic size is that size from which the limits of size are derived by the application of allowances and tolerances.

Allowance. An allowance is a prescribed difference between the maximum material limits of mating parts. It is a minimum clearance (positive allowance) or maximum interference (negative allowance) between such parts.

Tolerance. A tolerance is the total permissible variation of a size. The tolerance is the difference between the limits of size.

Limits of Size. The limits of size are the applicable maximum and minimum sizes.

Fit. Fit is the general term used to signify the range of tightness or looseness which may result from the application of a specific combination of allowances and tolerances in the design of mating parts.

Clearance Fit. A clearance fit is one having limits of size so prescribed that a clearance always results when mating parts are assembled.

Interference Fit. An interference fit is one having limits of size so prescribed that an interference always results when mating parts are assembled.

Transition Fit. A transition fit is one having limits of size so prescribed that either a clearance or an interference may result when mating parts are assembled.

*Extracted from *American National Standard Preferred Limits and Fits for Cylindrical Parts* (ANSI) B4.1–1967) with permission of the publisher, The American Society of Mechanical Engineers, 345 East 47th St., New York, N.Y. 10017.

Basic Hole System. A basic hole system is a system of fits in which the design size of the hole is the basic size and the allowance, if any, is applied to the shaft.

Basic Shaft System. A basic shaft system is a system of fits in which the design size of the shaft is the basic size and the allowance, if any, is applied to the hole.

Tables 29A, B, C, D, and E in Appendix E cover three general types of fits: running fits, locational fits, and force fits. For educational purposes standard fits may be designated by means of letter symbols, as follows:

RC—Running or Sliding Clearance Fit

LC—Locational Clearance Fit

LT—Transition Clearance or Interference Fit

LN—Locational Interference Fit

FN—Force or Shrink Fit

It should be understood that these letters are not to appear on working drawings. Only the limits for sizes are shown.

When a number is added to these letter symbols a complete fit is represented. For example, FN4 specifies, symbolically, a class 4 force fit for which the limits of size for mating parts may be determined from use of Table 29E. The minimum and maximum limits of clearance or interference for a particular application may be read directly from this table.

Classes of fits as given in these tables are as follows:

Running and Sliding Fits—Classes RC1–RC9

Locational Clearance Fits—Classes LC1–LC11

Locational Transition Fits—Classes LT1–LT6

Locational Interference Fits—Classes LN1–LN3

Force and Shrink Fits—Classes FN1–FN5

*Running and Sliding Fits.** Running and sliding fits are intended to provide a similar running performance, with suitable lubrication allowance, throughout the range of sizes. The clearances for the first two classes, used chiefly as slide fits, increase more slowly with diameter than the other classes, so that accurate location is maintained even at the expense of free relative motion.

A brief description of the fits is given here. For a more complete understanding one should read and study the standard.

RC1 Close sliding fits are intended for accurate location of parts which must assemble without perceptible play.

*Extracted from *American Standard Preferred Limits and Fits for Cylindrical Parts* (ANSI B4.1–1967).

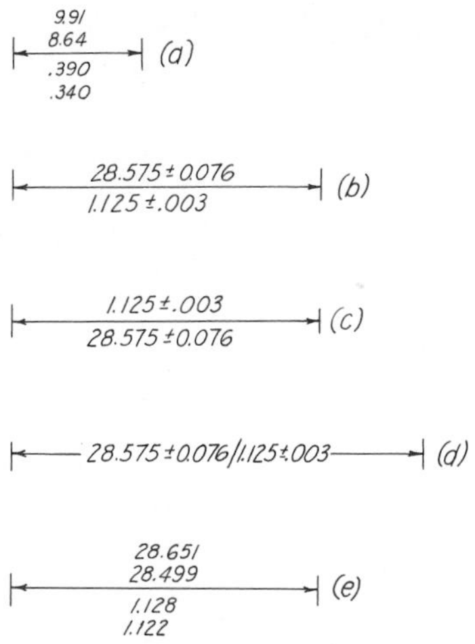

Fig. 14.90 □ **The position method applied to dual-dimensioning.**

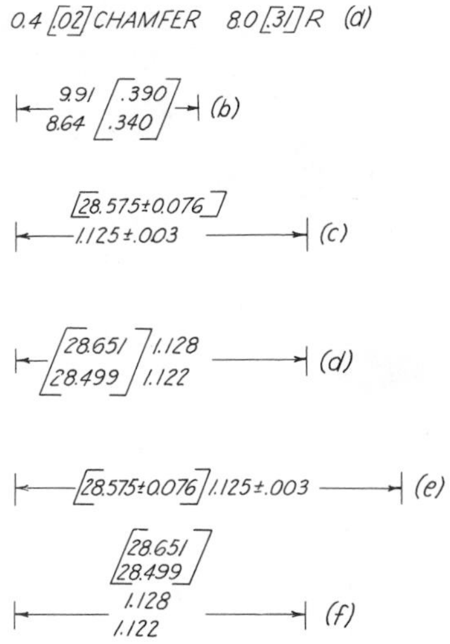

Fig. 14.91 □ **The bracket method applied to dual-dimensioning.**

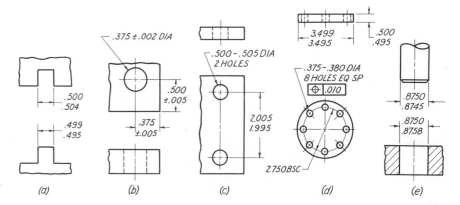

Fig. 14.92 □ **Limit dimensioning for the production of interchangeable parts.**

RC2 Sliding fits are intended for accurate location but with greater maximum clearance than Class RC1.

RC3 Precision running fits are about the closest fits which can be expected to run freely and are intended for precision work at slow speeds and light journal pressures, ...

RC4 Close running fits are intended chiefly for running fits on accurate machinery with moderate surface speeds and journal pressures, ...

RC5 ⎫ Medium running fits are intended for higher running speeds,
RC6 ⎬ or heavy journal pressures, or both.

RC7 Free running fits are intended for use where accuracy is not essential or where large temperature variations are likely to be encountered, ...

RC8 ⎫ Loose running fits are intended for use where wide commer-
RC9 ⎬ cial tolerances may be necessary, together with an allowance, on the external member.

Locational Fits. Locational fits are divided into three groups: clearance fits (LC), transition fits (LT), and interference fits (LN).

LC.* Locational clearance fits are intended for parts which are normally stationary, but which can be freely assembled or disassembled. They run from snug fits for parts requiring accuracy of location, through the medium clearance fits for parts such as ball, race and housing, to the looser fastener fits where freedom of assembly is of prime importance.

LT. Locational transition fits are a compromise between clearance and interference fits, for application where accuracy of location is important, but either a small amount of clearance or interference is permissible.

LN. Locational interference fits are used where accuracy of location is of prime importance, and for parts requiring rigidity and alignment with no special requirements for bore pressure.

*Force Fits.** Force or shrink fits constitute a special type of interference fit, normally characterized by maintenance of constant bore pressures throughout the range of sizes.

FN1 Light drive fits are those requiring light assembly pressures, and produce more or less permanent assemblies.

FN2 Medium drive fits are suitable for ordinary steel parts, or for shrink fits on light sections.

FN3 Heavy drive fits are suitable for heavier steel parts, or for shrink fits in medium sections.

FN4 ⎫ Force fits are suitable for parts which can be highly stressed,
FN5 ⎬ or for shrink fits where the heavy pressing forces required are impractical.

*Extracted from *American Standard Preferred Limits and Fits for Cylindrical Parts* (ANSI B4.1–1967).

14.53 □ Computation of limits of size for cylindrical parts

To obtain the correct fit between two engaging parts, compute limit dimensions that modify the nominal size of both. Numerical values of the modifications necessary to obtain the proper allowance and tolerances for various diameters for all fits mentioned previously are given in Tables 29A, B, C, D, and E in Appendix E.

The two systems in common use for computing limit dimensions are (1) the basic hole system, and (2) the basic shaft system. The same ANSI tables may be used conveniently for both systems.

14.54 □ Basic hole system

Because most limit dimensions are computed on the basic hole system, the illustrated example shown in Fig. 14.93 involves the use of this system. If, as is the usual case, the nominal size is known, all that is necessary to determine the limits is to convert the nominal size to the basic hole size and apply the figures given under "standard limits," adding or subtracting (according to their signs) to or from the basic size to obtain the limits for both the hole and the shaft.

Example:

Suppose that a $\frac{1}{2}$-in. shaft is to have a class RC6 fit in a $\frac{1}{2}$-in. hole [Fig. 14.93(a)]. The nominal size of the hole is $\frac{1}{2}$ in. The basic hole size is the exact theoretical size .5000.

From Table 29A it is found that the hole may vary between $+.0000$ and $+.0016$, and the shaft between $-.0012$ and $-.0022$. As can be readily observed, these values result in a variation (tolerance) of .0016 between the upper and lower limits of the hole, while the variation (tolerance) for the shaft will be .0010. The allowance (minimum clearance) is .0012, as given in the table.

The limits of the hole are

$$\frac{(.5000 + .0000)}{(.5000 + .0016)} = \frac{.5000}{.5016}$$

The limits on the shaft are

$$\frac{(.5000 - .0012)}{(.5000 - .0022)} = \frac{.4988}{.4978}$$

In the past limits have been placed in the order in which they will be approached when the part is machined [Fig. 14.93(c)]. This practice leads to the minimum limit being placed above for an internal dimension and the maximum limit above for an external dimension. However, a recent proposal requires that the maximum limit always be placed directly above the minimum limit where dimensions are associated with dimension lines (ANSI Y14.5–1966). However, when both limits are given in a horizontal line in association with a leader or note, the minimum limit is to be given first (Fig. 14.100).

14.55 □ Basic shaft system

When a number of parts requiring different fits but having the same nominal size must be mounted on a shaft, the basic shaft system is used because it is much easier to adjust the limits for the holes than to machine a shaft of one nominal diameter to a number of different sets of limits required by different fits.

For basic shaft fits the maximum size of the shaft is basic. The limits of clearance or interference are the same as those shown in Tables 29A, B, C, D, and E for the corresponding fits. The symbols for basic shaft fits are identical with those used for the standard fits with a letter S added. For example, LC4S specifies a locational clearance fit, class 4, as determined on a basic shaft basis.

Basic Shaft System—Clearance Fits. To determine the needed limits, increase each of the limits obtained, using the basic hole system, by the value given for the upper shaft limit. For example, if the same supposition (nominal size and fit requirement) is made as for the preceding illustration, the limits shown in Fig. 14.94 can be most easily obtained by adding .0012 to each of the limits shown for the hole and shaft in Fig. 14.93.

Basic Shaft System—Interference and Transition Fits. To determine the needed limits, subtract the value shown for the upper shaft limit from the basic hole limits.

Example:

The basic shaft limits are to be determined using an FN2 fit and the same nominal diameter as for the two previous illustrations.

$$\text{Hole } \frac{.5000 - .0016}{.5007 - .0016} = \frac{.4984}{.4991}$$

$$\text{Shaft } \frac{.5016 - .0016}{.0512 - .0016} = \frac{.5000}{.4996} \text{ (basic size)}$$

In brief, it can be stated that the limits for hole and shaft, as given in Tables 29A–E, are increased for clearance fits or decreased for transition or interference fits by the value shown for the upper shaft limit, which is the amount required to change the maximum shaft to basic size.

14.56 □ Tolerances

Necessary tolerances may be expressed by general notes printed on a drawing form or they may be given with definite values for specific dimensions (Fig. 14.100). When expressed in the form

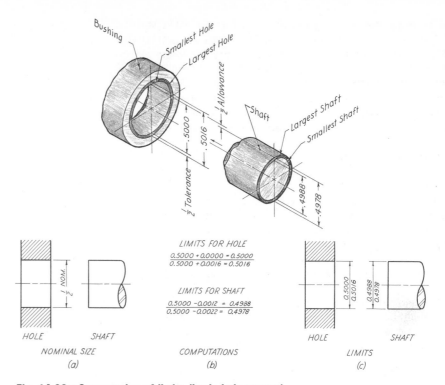

Fig. 14.93 □ Computation of limits (basic hole system).

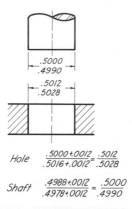

Hole $\dfrac{.5000 + .0012}{.5016 + .0012} = \dfrac{.5012}{.5028}$

Shaft $\dfrac{.4988 + .0012}{.4978 + .0012} = \dfrac{.5000}{.4990}$

Fig. 14.94 □ Computation of limits (basic shaft system).

of a printed note, the wording might be as follows: ALLOWABLE VARIATION ON ALL FRACTIONAL DIMENSIONS IS ±.010 UNLESS OTHERWISE SPECIFIED. A general note for tolerance on decimal dimensions might read: ALLOWABLE VARIATION ON DECIMAL DIMENSIONS IS ±.001. This general note would apply to all decimal dimensions where limits were not given.

The general notes on tolerances should be allowed to apply to all dimensions where it is not necessary to use specific tolerances.

14.57 □ Unilateral tolerances
Unilateral tolerances may be expressed in any one of several ways, as shown in Fig. 14.95.

Two limits may be given, as in (a), or the basic size can be shown to the required number of decimal places, followed by a plus tolerance above a minus tolerance, as in (b). Another method illustrated in (c) gives the preferred dimension with a tolerance that may be plus or minus but not both. When the dimension is given as a fraction the zero tolerance is expressed by a 0 (cipher).

14.58 □ Bilateral tolerances
Bilateral tolerances are expressed with a divided tolerance (Fig. 14.96). Whenever the plus and minus values are unequal, as in (c), the plus value is placed above the dimension line.

14.59 □ Cumulative tolerances
An undesirable condition may result when either the location of a surface or an overall dimension is affected by more than one tolerance dimension. When this condition exists, as illustrated in Fig. 14.97(a), the tolerances are said to be cumulative. In (a) surface B is located from surface A and surface C is related in turn to surface B. With the tolerances being additive, the tolerance on C is the sum of the separate tolerances (±.002). In respect to A, the position of C may vary from 1.998 to 2.002. This tolerance, as illustrated by the shaded rectangle, is .004 in. When consecutive dimensioning is used, one dimension should always be omitted to avoid serious inconsistency. The distance omitted should be the one requiring the least accuracy. To avoid the inconsistency of cumulative tolerances it is the preferred practice to locate the surfaces from a datum plane, as shown in (b), so that each surface is affected by only one dimension. The use of a datum plane makes it possible to take full advantage of permissible variations in size and still satisfy all requirements for the proper functioning of the part.

14.60 □ Specification of angular tolerances
Angular tolerances may be expressed in degrees, minutes, or seconds (see Fig. 14.98). If desired, an angle may be given in degrees

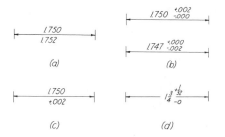

Fig. 14.95 □ Unilateral tolerances.

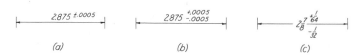

Fig. 14.96 □ Bilateral tolerances.

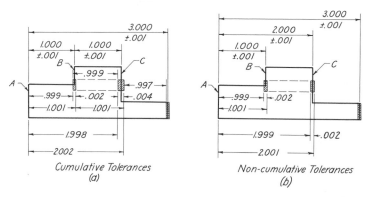

Cumulative Tolerances
(a)

Non-cumulative Tolerances
(b)

Fig. 14.97 □ Cumulative tolerances.

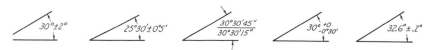

Fig. 14.98 □ Angular tolerances.

and decimal parts of a degree with the tolerance in decimal parts of a degree.

14.61 □ Geometric tolerance

Geometric tolerances specify the maximum variation that can be allowed in form or position from true geometry. Actually, a geometric tolerance is either the width or diameter of a tolerance zone within which a surface or axis of a hole or cylinder can lie with the resulting part satisfying the necessary standards of accuracy for proper functioning and interchangeability. Whenever tolerances of form are not specified on a drawing for a part, it is understood that the part as produced will be acceptable regardless of form variations. Expressions of tolerances of form control straightness, flatness, parallelism, squareness, concentricity, roundness, angular displacement, and so forth.

14.62 □ Symbols for tolerances of position and form*

The characteristic symbols shown in Fig. 14.99 have been suggested for use in lieu of, or in conjunction with, notes to express positional and form tolerances. In general, these symbols are the same as those given in Mil. Std. 8C–1962 for use by the armed services. Figure 14.100 shows typical feature control symbols applied to a drawing.

When a positional or form tolerance must be related to a datum, this relationship shall be indicated in the feature control symbol by placing the datum reference letter between the geometric characteristic symbol and the tolerance, as shown by the example at the bottom right in Fig. 14.99. It should be noted that vertical lines are used to separate the entries. Suggested dimensions for a short frame are given with the example at the left. For additional entries the frame is increased in length as may be necessary to avoid crowding. Feature control symbols may be associated with the features being toleranced by any one of the several methods shown in Fig. 14.100.

14.63 □ True-position dimensioning**

In the past, it has been the usual practice to locate points by means of rectangular dimensions given with tolerances. A point located in this manner will lie within a square tolerance zone when the positioning dimensions are at right angles to each other, as in Fig. 14.101(a). Where features are located by radial and angular di-

---

*Additional information on tolerance specifications may be found in the author's text *Fundamentals of Engineering Drawing for Design, Communication, and Numerical Control.*

**Information on true position dimensioning and the application of the MMC principle may be found in the author's text *Fundamentals of Engineering Drawing for Design, Communication, and Numerical Control.*

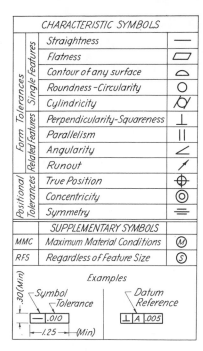

CHARACTERISTIC SYMBOLS			
Form Tolerances	Single Features	Straightness	—
		Flatness	▱
		Contour of any surface	⌒
		Roundness–Circularity	○
		Cylindricity	⌭
	Related Features	Perpendicularity-Squareness	⊥
		Parallelism	∥
		Angularity	∠
		Runout	⟋
Positional Tolerances		True Position	⊕
		Concentricity	◎
		Symmetry	≡

SUPPLEMENTARY SYMBOLS		
MMC	Maximum Material Conditions	Ⓜ
RFS	Regardless of Feature Size	Ⓢ

Examples

Symbol — Tolerance
— .010
1.25 (Min)
.30 (Min)

Datum Reference
⊥ A .005

Fig. 14.99 □ Geometric characteristic symbols.
(*Compiled from ANSI Y14.5–1966.*)

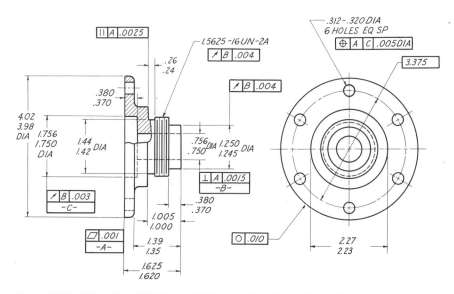

Fig. 14.100 □ Use of symbols in specifying positional and form tolerances.
(*ANSI Y14.5–1966.*)

Graphic standards □ **301**

mensions with tolerances, wedge-shaped tolerance zones result, as illustrated in Fig. 14.102.

In making a comparison of coordinate tolerancing and true-position tolerancing of circular features, it can be noted in the case of coordinate tolerancing, as illustrated in Fig. 14.101(a), that the actual position of the feature can be anywhere within the .010 square and that the maximum allowable variation from the desired position occurs along the diagonal of the square. With this allowable variation along the diagonal being 1.4 times the specified tolerance, the diameter of the cylinder for true-position tolerancing of the same feature [see (b)] could be 1.4 times the tolerance that would be used in coordinate tolerancing without any increase in the maximum allowable variation. True-position tolerancing increases the permissible tolerance in all directions, without detrimental effect on the location of the feature. True-position dimensioning takes into full account the relations that must be maintained for the interchangeable assembly of mating parts and it permits the design intent to be expressed more simply and precisely. Furthermore, the true-position approach to dimensioning corresponds to the control furnished by position and receiver gages with round pins. Such gages are commonly used for the inspection of patterns of holes in parts that being mass produced. A full discussion of the advantages of true-position tolerancing may be found in Appendix B of ANSI Y14.5–1966.

Positional tolerancing can be used for specific features of a machine part. When the part contains a number of features arranged in groups, positional tolerances can be used to relate each of the groups to one another as necessary and to tolerance the position of the features within a group independently of the features of the other groups.

The term "true position" denotes the theoretically exact position for a feature. In practice, the basic (exact) location is given with untoleranced dimensions and a positional tolerance is added to the note specifying the size and number of features (Fig. 14.103).

Basic (untoleranced) location dimensions must be excluded from the general tolerances, usually specified near the title block. This may be done in any one of several ways, as follows: (a) by a general note—UNTOLERANCED DIMENSIONS LOCATING TRUE POSITION ARE BASIC; (b) by adding the word Basic (or BSC) to each of the locating dimensions; (c) by enclosing each of the true-position-locating dimensions in a rectangular box; or (d) by reference to a separate specification document. See Fig. 14.101(b).

For true-position tolerancing, characteristic phrases or symbols are included in the hole (or feature) callout (Fig. 14.101). Either the diameter method or the radius method may be used for specifying the true-position tolerance. However, it must be realized that

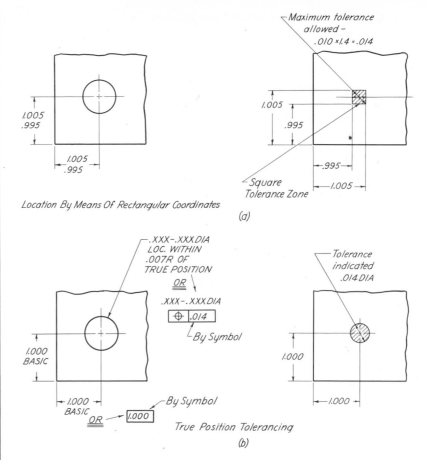

Fig. 14.101 □ Comparison between coordinate tolerancing and true position tolerancing.

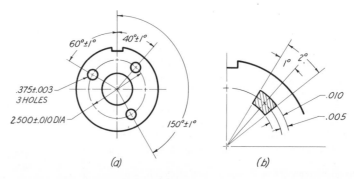

Fig. 14.102 □ Angular dimensioning.
(ANSI Y14.5–1957.)

the radius represents deviation from true position and that the tolerance is twice the radius. Since both methods achieve the same dimensional control, the expression to be included with the hole callout may be given in either of two ways:

1 LOCATED AT TRUE POSITION WITHIN .010 DIA

2 LOCATED WITHIN .005R OF TRUE POSITION

When the alignment between mating parts depends on some functional surface, this surface is selected as a datum for dimensioning, and the datum is identified in the note. Such a note for a cylindrical part might read: .XXX—.XXX DIA 6 HOLES LOC AT TRUE POS WITHIN .006 DIA IN RELATION TO SURF *A*.

The requirement of true-position dimensioning for a cylindrical feature is that the axis at all points must lie within the specified cylindrical tolerance zone having its center located at true position. This cylindrical tolerance zone also defines the limits within which variations in the squareness of the axis of the hole in relation to the flat surface must be confined.

For noncircular features, such as slots and tabs, the positional tolerance is usually applied only to surfaces related to the center plane of the feature. In applying true-position dimensioning to such features, it will be found that the principal difference is in the geometric form of the tolerance zone within which the center plane of the feature must be contained. The center plane of the tolerance zone must be located at true position. It should be noted that this tolerance zone also defines the limits within which variations in the squareness of the center plane of the slot must be contained. Notes specifying slots might read:

1 6 SLOTS EQ SP AND LOC WITHIN .005 EACH SIDE OF TRUE POS

2 .XXX–.XXX 2 SLOTS LOC AT TRUE POS WITHIN .008 TOTAL IN RELATION TO SURFACES *A* AND *B*.

Fig. 14.104 shows the application of true-position dimensioning to both holes and tabs. It should be noted that the tolerance zone for the holes has been specified by radius.

See Fig. 14.100 for the use of a feature control symbol for the true position callout for the six holes.

The illustrations used with this simplified discussion have been presented with the sole idea of showing as briefly as possible what is meant by true-position dimensioning. They are not to be thought of as being practical examples. Dotted circles have been added to the illustrations in Fig. 14.103 only to call attention to the imaginary tolerance zones. Those who may have the need for the use of this method of tolerancing should consult ANSI Y14.5 to acquire a full understanding of true-position dimensioning.

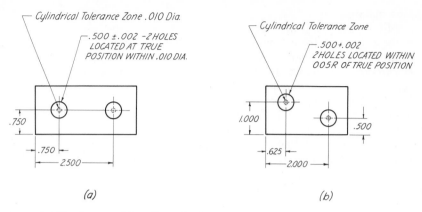

(a)   *(b)*

Fig. 14.103 □ **True position dimensioning.**

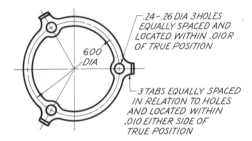

Fig. 14.104 □ **True position dimensioning—holes and tabs.**
(*ANSI Y14.5–1957.*)

14.64 □ Surface quality

The improvement in machining methods within recent years coupled with a strong demand for increased life for machined parts has caused engineers to give more attention to the quality of the surface finish. Not only the service life but also the proper functioning of the part as well may depend on obtaining the needed smoothness quality for contact surfaces.

On an engineering drawing a surface may be represented by line if shown in profile or it may appear as a bounded area in a related view. Machined and ground surfaces, however, do not have the perfect smoothness represented on a drawing. Actually a surface has three dimensions, namely, length, breadth, and curvature (waviness) as illustrated in Fig. 14.105. In addition there will be innumerable peaks and valleys of differing lengths, widths, and heights.

14.65 □ Designation of surface characteristics (Fig. 14.106)

A surface whose finish is to be specified should be marked with the finish mark having the general form of a check mark ($\sqrt{}$) so that the point of the symbol shall be on the line representing the surface, on the extension line, or on a leader pointing to the surface. Good practice dictates that the long leg and the extension shall be to the right as the drawing is read. See Fig. 14.107. Figure 14.106 illustrates the specification of roughness, waviness, and lay by listing rating values on the symbol.

Where it is desired to specify only the surface-roughness height, and the width of roughness or direction of tool marks is not important, the simplest form of the symbol may be used. The numerical value is placed in the $\sqrt{}$, as shown in Fig. 14.106.

Where it is desired to specify waviness height in addition to roughness height, a straight horizontal line must be added to the top of the simple symbol (Fig. 14.106). The numerical value of waviness height would be shown above this line.

If the nature of the preferred lay is to be shown in addition to these two characteristics, it will be indicated by the addition of a combination of lines, as shown in Fig. 14.106. Parallel or perpendicular lines indicate that the dominant lines on the surface are parallel or perpendicular to the boundary line of the surface in contact with the symbol. **X** indicates that the lay is angular in both directions to the line representing the surface to which the symbol is applied. **M** refers to multidirectional. **C** indicates lay is approximately circular relative to the center of the surface to which the symbol is applied. **R** refers to a lay approximately radial to the center of the surface.

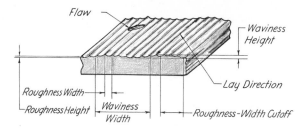

Fig. 14.105 □ **Surface texture definitions.**
(*ANSI B46.1–1962.*)

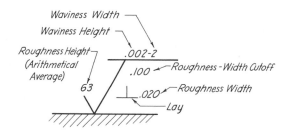

Fig. 14.106 □ **Surface texture symbol.**

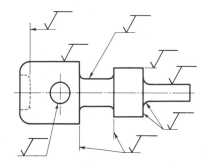

Fig. 14.107 □ **Application of surface texture symbols to a drawing of a machine part.**
(*ANSI B46.1–1962.*)

Roughness width is placed to the right of the lay symbol, as shown in Fig. 14.106.

Roughness height and roughness width are expressed in microinches (one-millionth of an inch).

The use of only one number to specify the height or width of roughness or waviness will indicate the arithmetical average.

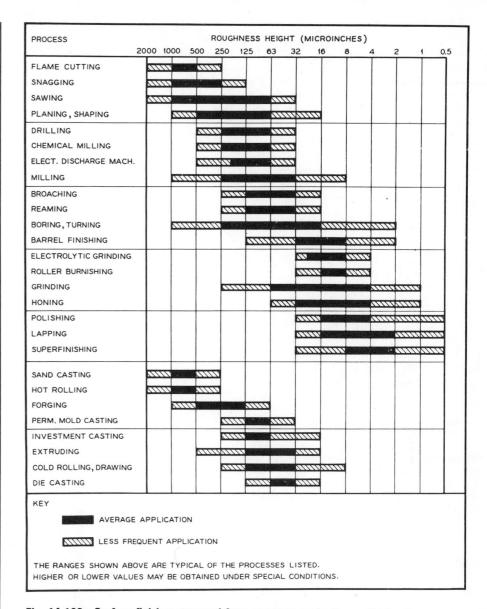

Fig. 14.108 □ Surface finishes expected from common production methods. (*ANSI B46.1–1962.*)

Problems

The following problems offer the student the opportunity to apply the rules of dimensioning given in this chapter. If it is desirable, millimeters may be used in place of fractions or decimals of an inch. Use Table 1 in Appendix E.

1 (Fig. 14.109). Reproduce the given views of an assigned part. Determine the dimensions by transferring them from the drawing to the open-divided scale by means of the dividers.

2–10 (Figs. 14.110–14.118). Make a fully dimensioned multiview sketch or drawing of an assigned part. Draw all necessary views. Give a detail title with suitable notes concerning material, number required, etc. These parts have been selected from different fields of industry—automotive, aeronautical, chemical, electrical, etc.

11 (Fig. 14.119). Make a detail drawing of an assigned part of the gear pump.

12 (Fig. 14.120). Make a fully dimensioned drawing of an assigned part of the shaft support.

13 (Fig. 14.121). Make a detail working drawing of an assigned part of the flexible joint. Compose a suitable title, giving the name of the part, the material, etc.

14 (Fig. 14.122). Make a two-view freehand detail sketch of the fan spindle. Determine dimensions by transferring them from the drawing to the open-divided scale, by means of the dividers. The material is SAE 1045, CRS. Shaft limits for the bearings are

$$\frac{.7874}{.7867} \quad \text{and} \quad \frac{.9840}{.9836}.$$

Use an RC7 fit between spindle and felt retainer, spindle and felt-retaining washer, and spindle and cone-clamp washer.

15 (Fig. 14.122). Make a complete two-view detail drawing of the fan pulley. It is suggested that a half-circular view and a full-sectional view be shown. Determine the dimensions as suggested in Problem 14. Housing limits for the given bearings are

$$\frac{1.8497}{1.8503} \quad \text{and} \quad \frac{2.4401}{2.4409}$$

16 (Fig. 14.123). Make a two-view assembly drawing of the cup center, using the given details. Use the schematic symbol for screw threads. Study the pictorial drawing carefully before starting the views.

17 (Figs. 14.124–14.126). Make an assembly drawing of the hand grinder.

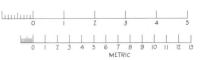

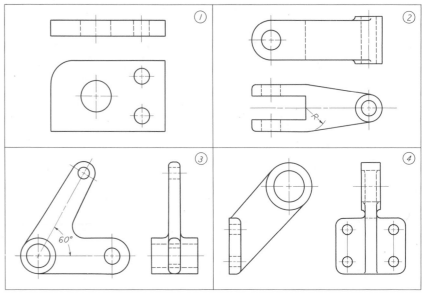

Fig. 14.109 □ **Dimensioning problems.**

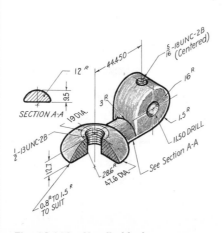

Fig. 14.110 □ **Handle block.**
(Dimension values are in millimeters.)

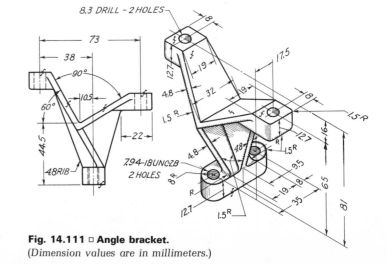

Fig. 14.111 □ **Angle bracket.**
(Dimension values are in millimeters.)

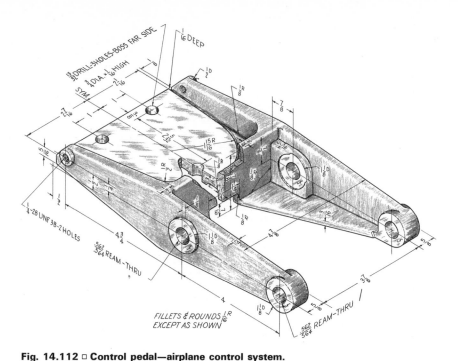

Fig. 14.112 □ **Control pedal—airplane control system.**

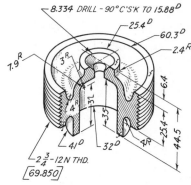

Fig. 14.113 □ **Elevator bracket.**

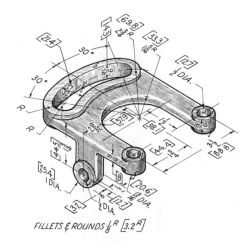

Fig. 14.114 □ **Valve seat.**
(Dimension values are in millimeters.)

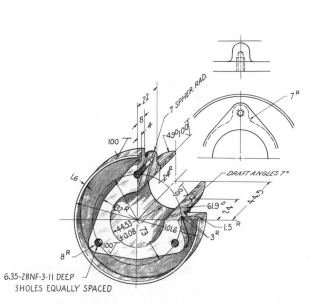

Fig. 14.115 □ **Inlet flange—airplane cooling system.**
(Dimension values are in millimeters.)

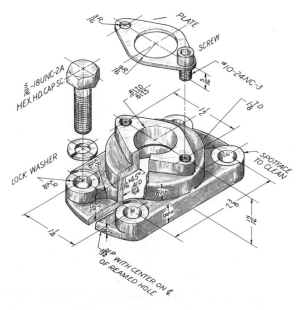

Fig. 14.116 □ **Cover—mixing machine.**

Fig. 14.117 □ **Shifter arm.**
(Dimension values shown in [ ] are in millimeters.)

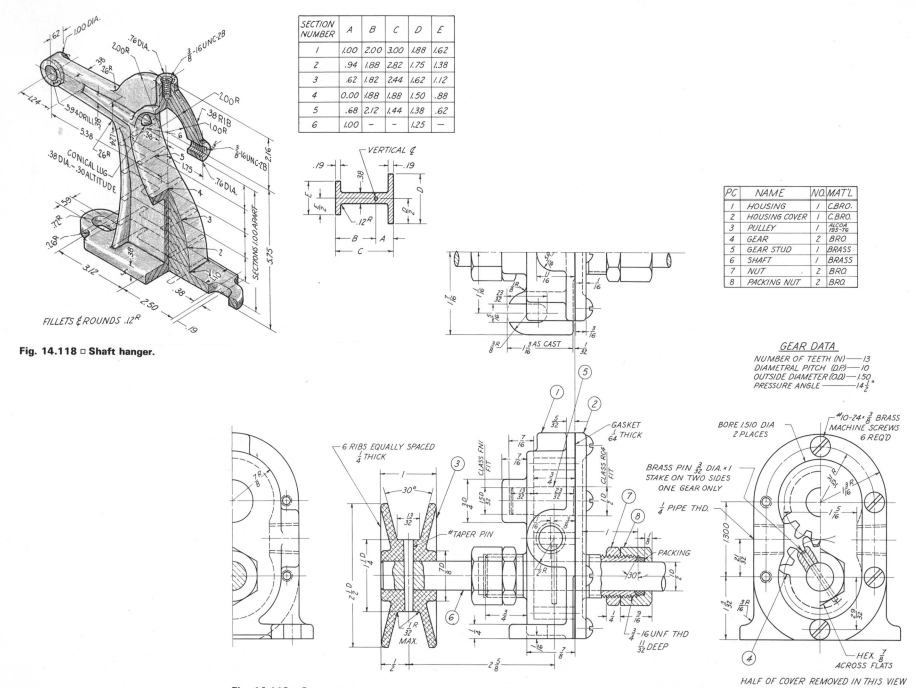

SECTION NUMBER	A	B	C	D	E
1	1.00	2.00	3.00	1.88	1.62
2	.94	1.88	2.82	1.75	1.38
3	.62	1.82	2.44	1.62	1.12
4	0.00	1.88	1.88	1.50	.88
5	.68	2.12	1.44	1.38	.62
6	1.00	—	—	1.25	—

Fig. 14.118 □ Shaft hanger.

PC	NAME	NO.	MAT'L
1	HOUSING	1	C.BRO.
2	HOUSING COVER	1	C.BRO.
3	PULLEY	1	ALCOA 195-T6
4	GEAR	2	BRO.
5	GEAR STUD	1	BRASS
6	SHAFT	1	BRASS
7	NUT	2	BRO.
8	PACKING NUT	2	BRO.

GEAR DATA

NUMBER OF TEETH (N) —— 13
DIAMETRAL PITCH (D.P.) —— 10
OUTSIDE DIAMETER (O.D.) —— 1.50
PRESSURE ANGLE —— $14\frac{1}{2}°$

Fig. 14.119 □ Gear pump.

HALF OF COVER REMOVED IN THIS VIEW

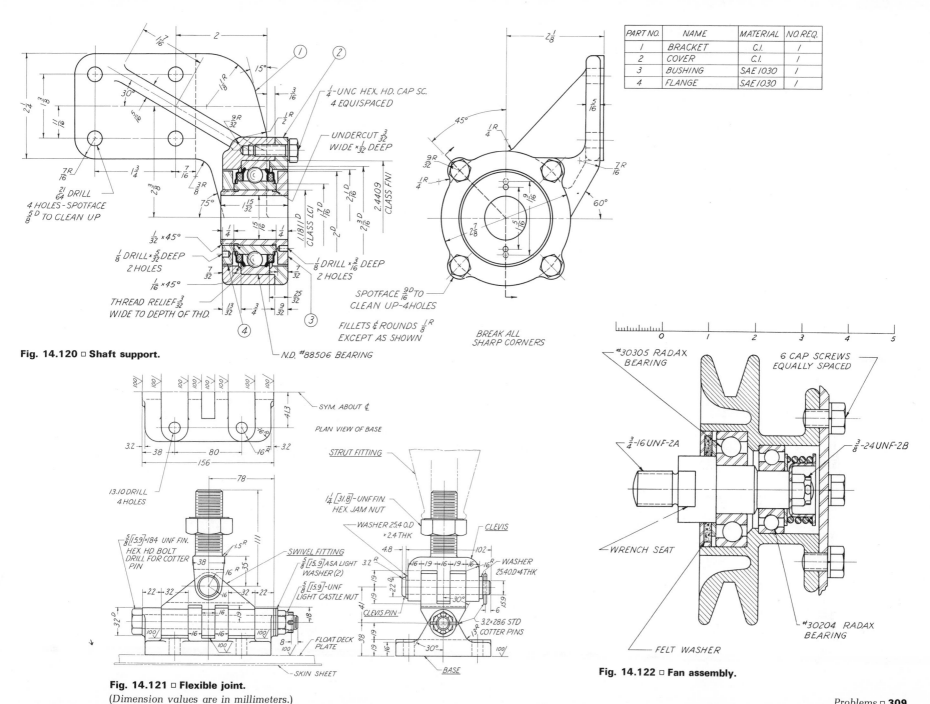

PART NO.	NAME	MATERIAL	NO. REQ.
1	BRACKET	C.I.	1
2	COVER	C.I.	1
3	BUSHING	SAE 1030	1
4	FLANGE	SAE 1030	1

Fig. 14.120 □ Shaft support.

Fig. 14.121 □ Flexible joint.
(Dimension values are in millimeters.)

Fig. 14.122 □ Fan assembly.

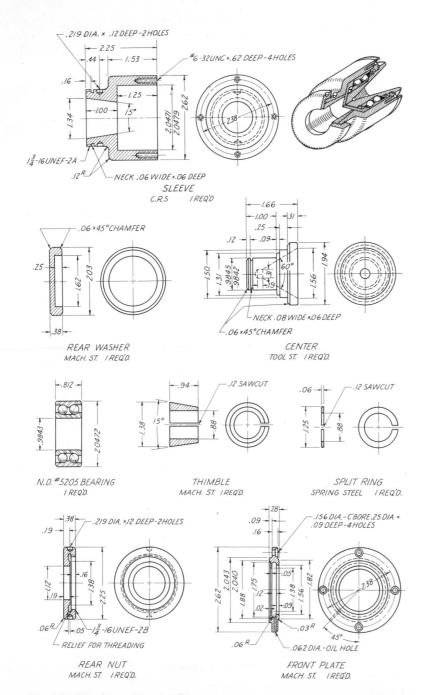

.219 DIA. × .12 DEEP-2 HOLES

2.25

.44 ── 1.53

#6-32UNC×.62 DEEP-4HOLES

.16

.16

1.25

1.34

1.00

15°

2.0471
2.0479

2.62

2.38

1¾-16UNEF-2A

.12 R

NECK .06 WIDE ×.06 DEEP

SLEEVE

C.R.S   1 REQ'D

.06 × 45° CHAMFER

.25

2.03

1.62

.38

REAR WASHER

MACH. ST. 1 REQ'D.

1.66

1.00   .31

.25

.12   .09

1.50

1.31

.9845
.9842

60°

.31

.19

1.56

1.94

NECK .08 WIDE ×.06 DEEP

.06 × 45° CHAMFER

CENTER

TOOL ST.   1 REQ'D.

.812

.9843

2.0472

N.D. #5205 BEARING

1 REQ'D.

.94

.12 SAWCUT

1.38

15°

.88

THIMBLE

MACH. ST. 1 REQ'D.

.06

.12 SAWCUT

1.25

.88

SPLIT RING

SPRING STEEL   1 REQ'D.

.38

.219 DIA.×.12 DEEP-2 HOLES

.19

.16

1.12

.16

1.38

.19

2.25

.06 R

.05   1¾-16UNEF-2B

RELIEF FOR THREADING

REAR NUT

MACH. ST.   1 REQ'D.

.28

.09

.16

.156 DIA.-C'BORE .25 DIA. ×
.09 DEEP-4 HOLES

.05

2.62

2.043

2.040

.175

1.88

.12

1.38

1.56

1.82

.02

.09

2.38

.03 R

45°

.06 R

.062 DIA.-OIL HOLE

FRONT PLATE

MACH. ST.   1 REQ'D.

Fig. 14.123 □ Cup center details.

Fig. 14.124 □ Hand grinder.

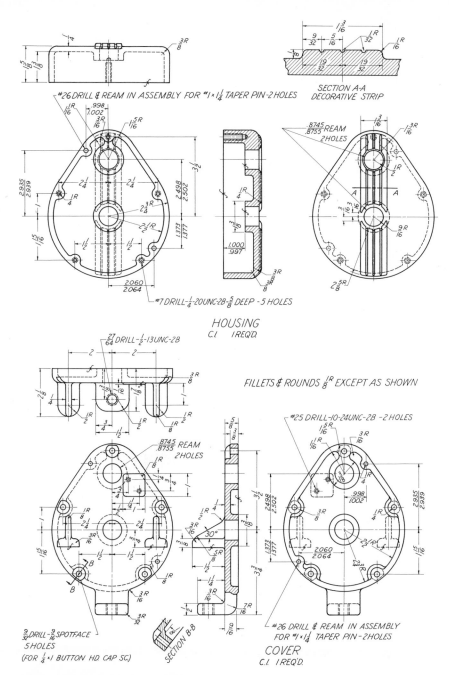

Fig. 14.125 □ Hand grinder details.
(See Fig. 14.124.)

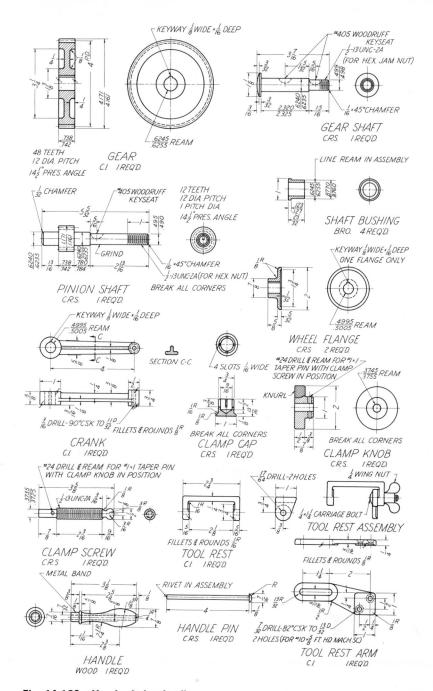

Fig. 14.126 □ Hand grinder details.

Appendix

Use of instruments

A.1 □ Introduction

The instruments and materials needed for making ordinary engineering drawings are shown in Fig. A.2. The instruments in the plush-lined case should be particularly well made, for with inferior ones it is often difficult to produce accurate drawings of professional quality.

A.2 □ List of equipment and materials

The following list is a practical selection of equipment and materials necessary for making pencil drawings.

1 Case of drawing instruments

2 Drawing board

3 T-square

4 45° triangle*

5 10-in. 30°–60° triangle

6 French curve

7 Scales (Sec. A.19)

8 Drawing pencils

9 Pencil pointer (file or sandpaper pad)

10 Thumbtacks, brad machine, or Scotch tape

11 Pencil eraser

*A 6-in. 45° Braddock lettering triangle, which may be used as either a triangle or a lettering instrument, may be substituted for this item.

12 Cleaning eraser

13 Erasing shield

14 Dusting brush

15 Protractor

Large-bow sets are preferred by many persons, especially in the aircraft and automotive fields. A complete set might include the following: large bow (compass), beam compass with an extension bar for drawing very large circles, dividers, small bow, and a slip-handle ruling pen. Since the large bow (Fig. A.30) is particularly suited for drawing circles of very small size (up to 10 in. in diameter) its range covers that of both the bow instruments and the compass (without the extension bar) of the standard set (not shown).

The large bow is preferred by its advocates because its sturdy construction permits the draftsman to exert the pressure necessary to secure black, opaque lines on pencil drawings that are to be used for making prints.

A.3 □ The Protractor

The protractor (Fig. A.3) is used for measuring and laying off angles.

A.4 □ Artist's pencils

Because of the time consumed in cutting back the wood to repoint an ordinary drawing pencil, some designers and draftsmen favor the use of artist's automatic pencils (Fig. A.4). Separate leads for these pencils may be purchased in any of the 17 degrees of hardness obtainable for regular drawing pencils.

A.5 □ Special templates

The use of templates (Fig. A.5) can save valuable time in the drawing of standard figures and symbols on plans and drawings.

A.6 □ The drafting machine

The drafting machine (Fig. A.6) is designed to combine the functions of the T-square, triangles, scale, and protractor. Drafting machines are used extensively in commercial drafting rooms because it has been estimated that their use leads to a 25–50% saving in time.

A.7 □ Pencils

The student and professional man should be equipped with a selection of good, well-sharpened pencils with leads of various degrees of hardness, such as 9H, 8H, 7H, and 6H (hard); 5H and 4H (medium hard); 3H and 2H (medium); and H and F (medium soft).

Fig. A.1 □ A draftsman discusses a problem with a designer. It should be noted that they are reviewing previous designs that have been filed on microfilm. The microfilm reader-printer can be seen in the background. (*Courtesy 3M Company.*)

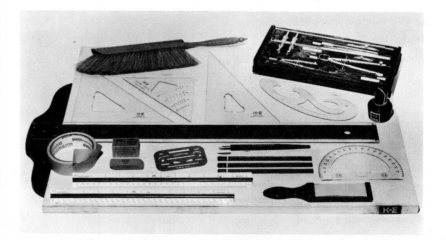

Fig. A.2 □ Essential drafting equipment.

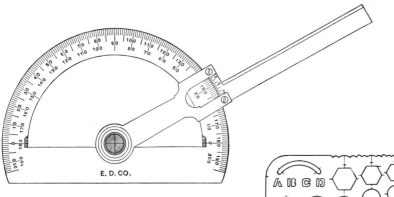

Fig. A.3 □ **Protractor.**

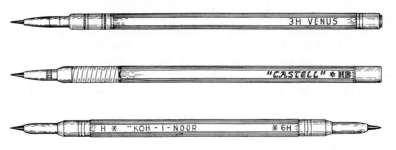

Fig. A.4 □ **Artist's pencils.**

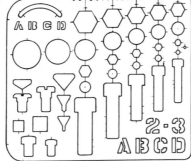

TOOLING TEMPLATE

Fig. A.5 □ **Special templates.**
(*Courtesy Frederick Post Company.*)

Fig. A.6 □ **Drafting machine.**
(*Courtesy Keuffel & Esser Company.*)

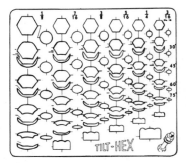

TILT-HEX DRAFTING TEMPLATE

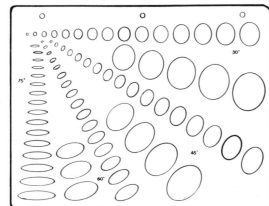

ELLIPSES

ELECTRO SYMBOL TEMPLATE

The grade of pencil to be used for various purposes depends on the type of line desired, the kind of paper employed, and the humidity, which affects the surface of the paper. Standards for line quality usually will govern the selection. As a minimum, however, the student should have available a 6H pencil for the light construction lines in layout work where accuracy is required, a 4H for repenciling light finished lines (dimension lines, center lines, and invisible object lines), a 2H for visible object lines, and an F or H for all lettering and freehand work.

A.8 □ Tracing paper and tracing cloth

White lightweight tracing paper, on which pencil drawings can be made and from which blueprints can be produced, is used in most commercial drafting rooms in order to keep labor costs at a minimum. Pencil cloth is a white cloth with a surface specially prepared to take pencil marks readily.

A.9 □ Pointing the pencil

Many persons prefer the conical point for general use (Fig. A.7), while others find the wedge point more suitable for straight-line work, as it requires less sharpening and makes a denser line (Fig. A.8).

When sharpening a pencil, the wood should be cut away (on the unlettered end) with a knife or a pencil sharpener equipped with draftsman's cutters. About $\frac{3}{8}$ in. of the lead should be exposed and should form a cut, including the wood, about $1\frac{1}{2}$ in. long. The lead then should be shaped to a conical point on the pointer (file or sandpaper pad). This is done by holding the file stationary in the left hand and drawing the lead toward the handle while rotating the pencil against the movement (Fig. A.7). All strokes should be made in the same manner, a new grip being taken each time so that each stroke starts with the pencil in the same rotated position as at the end of the preceding stroke.

A.10 □ Drawing pencil lines

Pencil lines should be sharp and uniform along their entire length and should be sufficiently distinct to fulfill their ultimate purposes. Construction lines (preliminary lines) should be drawn *very* lightly so that they may be easily erased. Finished lines should be made boldly and distinctly, so that there will be definite contrast between visible and invisible object lines and auxiliary lines, such as dimension lines, center lines, and section lines. To give this contrast, which is necessary for clearness and ease in reading, object lines should be very black, and of medium width, invisible lines black and not so wide, and auxiliary lines dark and thin.

When drawing a line, the pencil should be inclined slightly (about 60°) in the direction in which the line is being drawn (Fig. A.10).

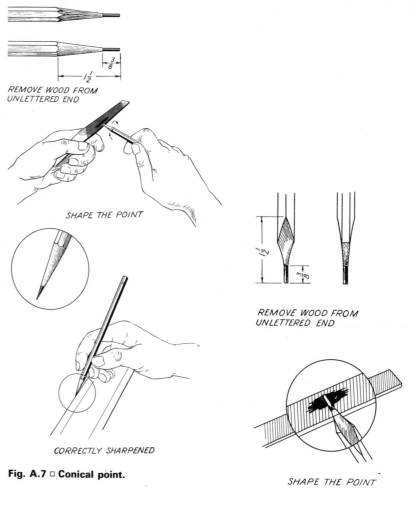

REMOVE WOOD FROM UNLETTERED END

SHAPE THE POINT

CORRECTLY SHARPENED

Fig. A.7 □ Conical point.

REMOVE WOOD FROM UNLETTERED END

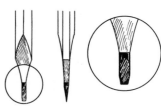

SHAPE THE POINT

CORRECTLY SHARPENED

Fig. A.8 □ Wedge point.

The pencil should be "pulled" (never pushed) at the same inclination for the full length of the line. If it is rotated (twirled) slowly between the fingers as the line is drawn, a symmetrical point will be maintained and a straight uniform line will be insured.

A.11 □ Placing and fastening the paper
For accuracy and ease in manipulating the T-square, the drawing paper should be located well up on the board and near the left-hand edge. The lower edge of the sheet (if plain) or the lower border line (if printed) should be aligned along the working edge of the T-square before the sheet is fastened down at all four corners with thumbtacks, Scotch tape, or staples.

A.12 □ The T-square
The T-square is used primarily for drawing horizontal lines and for guiding the triangles when drawing vertical and inclined lines. It is manipulated by sliding the guiding edge (inner face) of the head along the left edge of the board [Fig. A.9(a)] until the blade is in the required position. The left hand then should be shifted to a position near the center of the blade to hold it in place and to prevent its deflection while drawing the line. Experienced draftsmen hold the T-square, as shown in Fig. A.9(b), with the fingers pressing on the blade and the thumb on the paper. Small adjustments may be made with the hand in this position by sliding the blade with the fingers.

Horizontal lines are drawn from left to right along the upper edge of the T-square (Fig. A.10). (*Exception:* Left-handed persons should use the T-square head at the right side of the board and draw from right to left.) While drawing the line, the ruling hand should slide along the blade on the little finger.

A.13 □ The triangles
The 45° and the 30° × 60° triangles (Fig. A.12) are the ones commonly used for ordinary work. A triangle may be checked for nicks by sliding the thumbnail along the ruling edges.

A.14 □ Vertical lines
Vertical lines are drawn upward along the vertical leg of a triangle whose other (horizontal) leg is supported and guided by the T-square blade. The blade is held in position with the palm and thumb of the left hand, and the triangle is adjusted and held by the fingers, as shown in Fig. A.11. In the case of a right-handed person, the triangle should be to the right of the line to be drawn.

Either the 30° × 60° or the 45° triangle may be used since both triangles have a right angle. However, the 30° × 60° is generally preferred because it usually has a longer perpendicular leg.

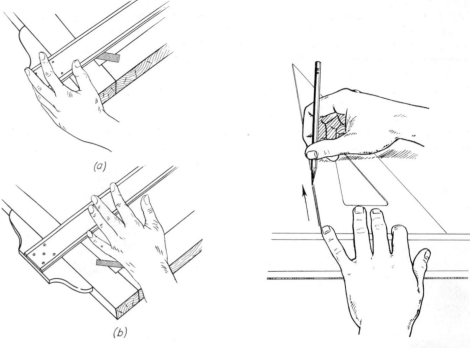

(a)

(b)

Fig. A.9 □ Manipulating the T-square.

Fig. A.11 □ Drawing vertical lines.

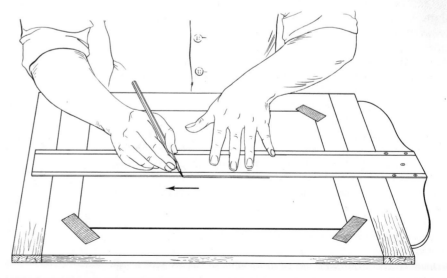

Fig. A.10 □ Drawing horizontal lines.

A.15 □ Inclined lines

Triangles also are used for drawing inclined lines. Lines that make angles of 30°, 45°, or 60° with the horizontal may be drawn with the 30° × 60° or the 45° triangle in combination with the T-square, as shown in Fig. A.12. If the two triangles are combined, lines that make 15° or a multiple of 15° may be drawn with the horizontal. Several possible arrangements and the angles that result are shown in Fig. A.13.

The triangles used singly or in combination offer a useful method for dividing a circle into 4, 6, 8, 12, or 24 equal parts (Fig. A.14). For angles other than those divisible by 15, a protractor must be used.

A.16 □ Parallel lines

The triangles are used in combination to draw a line parallel to a given line. To draw such a line, place a ruling edge of a triangle, supported by a T-square or another triangle, along the given line; then slip the triangle, as shown in Fig. A.15, to the required position and draw the parallel line along the same ruling edge that previously coincided with the given line.

A.17 □ Perpendicular lines

Either the sliding triangle method [Fig. A.16(a)] or the revolved triangle method [Fig. A.16(b)] may be used to draw a line perpendicular to a given line. When using the sliding triangle method, adjust to the given line a side of a triangle that is adjacent to the right angle. Guide the side opposite the right angle with a second triangle, as shown in Fig. A.16(a); then slide the first triangle along the guiding triangle until it is in the required position for drawing the perpendicular along the other edge adjacent to the right angle.

Although the revolved triangle method [Fig. A.16(b)] is not so quickly done, it is widely used. To draw a perpendicular using this method, align along the given line the hypotenuse of a triangle, one leg of which is guided by the T-square or another triangle; then hold the guiding member in position and revolve the triangle about the right angle until the other leg is against the guiding edge. The new position of the hypotenuse will be perpendicular to its previous location along the given line and, when moved to the required position, may be used as a ruling edge for the desired perpendicular.

A.18 □ Inclined lines making 15°, 30°, 45°, 60°, or 75° with an oblique line

A line making an angle with an oblique line equal to any angle of a triangle may be drawn with the triangles. The two methods previously discussed for drawing perpendicular lines are applicable with slight modifications. To draw an oblique line using the re-

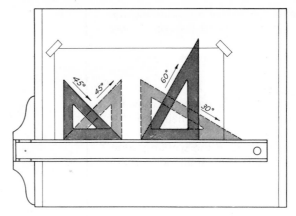

Fig. A.12 □ Inclined lines.

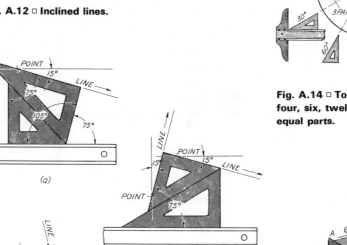

(a)

(b)

(c)

(d)

Fig. A.13 □ Drawing inclined lines with triangles.

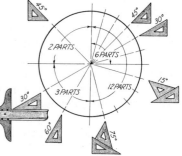

Fig. A.14 □ To divide a circle into four, six, twelve, or twenty-four equal parts.

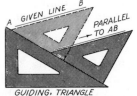

Fig. A.15 □ To draw a line parallel to a given line.

volved triangle method [Fig. A.17(a)], adjust along the given line the edge that is opposite the required angle; then revolve the triangle about the required angle, slide it into position, and draw the required line along the side opposite the required angle.

To use the sliding triangle method [Fig. A.17(b)], adjust to the given line one of the edges adjacent to the required angle, and guide the side opposite the required angle with a straight edge; then slide the triangle into position and draw the required line along the other adjacent side.

To draw a line making 75° with a given line, place the triangles together so that the sum of a pair of adjacent angles equals 75°, and adjust one side of the angle thus formed to the given line; then slide the triangle, whose leg forms the other side of the angle, across the given line into position and draw the required line, as shown in Fig. A.18(a).

To draw a line making 15° with a given line, select any two angles whose difference is 15°. Adjust to the given line a side adjacent to one of these angles, and guide the side adjacent with a straight edge. Remove the first triangle and substitute the other so that one adjacent side of the angle to be subtracted is along the guiding edge, as shown in Fig. A.18(b); then slide it into position and draw along the other adjacent side.

A.19 □ Scales

A number of kinds of scales are available for varied types of engineering design. For convenience, however, all scales may be classified according to their use as mechanical engineers' scales (both fractional and decimal), civil engineers' scales, architects' scales, or metric scales.

The mechanical engineers' scales are generally of the full-divided type, graduated proportionally to give reductions based on inches. On one form (Fig. A.19) the principal units are divided into the common fractions of an inch (4, 8, 16, and 32 parts). The scales are indicated on the stick as eighth size (1½ in. = 1 ft), quarter size (3 in. = 1 ft), half size (6 in. = 1 ft), and full size.

The use of decimal scales has become widespread. The full-size scale (Fig. A.20), which has the principal units (inches) divided into fiftieths, is particularly suited for use with the two-place decimal system. The half-size, three-eighths size, and quarter-size scales (Fig. A.23) have the principal units divided into tenths.

The civil engineers' (chain) scales (Fig. A.21) are full-divided, and are graduated in decimal parts, usually 10, 20, 30, 40, 50, 60, 80, and 100 divisions to the inch.

Architects' scales (Fig. A.22) differ from mechanical engineers' scales in that the divisions represent a foot, and the end units are divided into inches, half-inches, quarter-inches, and so forth (6, 12, 24, 48, or 96 parts). The usual scales are ⅛ in. = 1 ft,

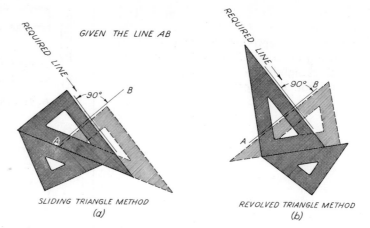

Fig. A.16 □ To draw a line perpendicular to another line.

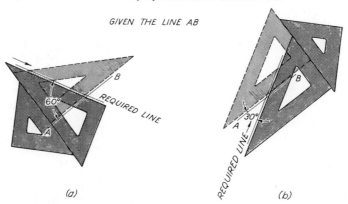

Fig. A.17 □ To draw lines making 30°, 45°, or 60° with a given line.

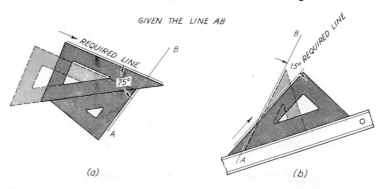

Fig. A. 18 □ To draw lines making 15° or 75° with a given line.

¼ in. = 1 ft,   ⅜ in. = 1 ft,   ½ in. = 1 ft,   1 in. = 1 ft, 1½ in. = 1 ft, and 3 in. = 1 ft.

The sole purpose of the scale is to reproduce the dimensions of an object full-size on a drawing or to reduce or enlarge them to some regular proportion, such as eighth-size, quarter-size, half-size, or double-size. The scales of reduction most frequently used are as follows:

Fractional

Mechanical engineers' scales

Full-size	(1″ = 1″)
Half-size	(½″ = 1″)
Quarter-size	(¼″ = 1″)
Eighth-size	(⅛″ = 1″)

Architects' or mechanical engineers' scales

Full-size	(12″ = 1′-0)
Half-size	(6″ = 1′-0)
Quarter-size	(3″ = 1′-0)
Eighth-size	(1½″ = 1′-0)
1″ = 1′-0	¼″ = 1′-0
¾″ = 1′-0	3/16″ = 1′-0
½″ = 1′-0	⅛″ = 1′-0
⅜″ = 1′-0	3/32″ = 1′-0

Decimal

Mechanical engineers' scales

Full-size	(1.00″ = 1.00″)
Half-size	(0.50″ = 1.00″)
Three-eighths-size	(0.375″ = 1.00″)
Quarter-size	(0.25″ = 1.00″)

Civil engineers' scales

10 scale:	1″ = 1′;	1″ = 10′;
	1″ = 100′;	1″ = 1000′
20 scale:	1″ = 2′;	1″ = 20′;
	1″ = 200′;	1″ = 2000′
30 scale:	1″ = 3′;	1″ = 30′;
	1″ = 300′;	1″ = 3000′
40 scale:	1″ = 4′;	1″ = 40′;
	1″ = 400′;	1″ = 4000′
50 scale:	1″ = 5′;	1″ = 50′;
	1″ = 500′;	1″ = 5000′
60 scale:		1″ = 60′; etc.
80 scale:		1″ = 80′; etc.

Fig. A. 19 □ **Mechanical engineers' scale. Full-divided.**

Fig. A.20 □ **Engineers' decimal scale.**

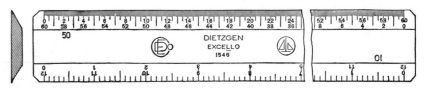

Fig. A.21 □ **Civil engineers' scale.**

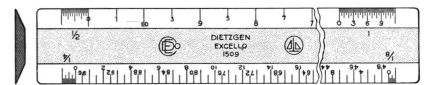

Fig. A.22 □ **Architects' scale. Open-divided.**

The first four scales, full-size, half-size, quarter-size, and eighth-size, are the ones most frequently selected for drawing machine parts, although other scales can be used. Since objects drawn by structural draftsmen and architects vary from small to very large, scales from full-size to $\frac{3}{32}$ in. = 1 ft ($\frac{1}{128}$-size) are commonly encountered. For maps, the civil engineers' decimal scales having 10, 20, 30, 40, 50, 60, and 80 divisions to the inch are used to represent 10, 20, 30 ft, and so forth, to the inch.

On a machine drawing, it is considered good practice to omit the inch ('') marks in a scale specification. For example, a scale may be specified as: FULL SIZE, 1.00 = 1.00, or 1 = 1; HALF SIZE, .50 = 1.00, or $\frac{1}{2}$ = 1; and so forth.

The decimal scales shown in Fig. A.23 have been approved by the American National Standards Institute for making machine drawings when the decimal system is used.

The metric scale (Fig. A.24) is used in those countries where the meter is the standard of linear measurement. As of this time, the International System of Units (Systéme International d'Unites)(SI) for the measurement of length, measurement of surface, measurement of volume, measurement of weight, and so forth is used as a world standard in every industrialized country except the United States. We stand alone with our inches, gallons, and pounds in spite of the fact that the metric system was legalized for use in this country by an Act of Congress more than one-hundred years ago (1866). The metric system is legal in all states. Even though the cost of conversion to the metric system will be enormous, there can be little doubt that a changeover will be made in the near future. As a practical matter, it is probable that American industry will shift gradually to the metric system without prodding by an Act of Congress since even now the automobile industry has adopted and is using a dual-dimensioning system by which dimensions are given on drawings in both inches and millimeters. Read Sec. 14.51 and see Figs. 14.90 and 14.91. It is for this reason that the dimensional values for some of the problems in each chapter of this book have been given in millimeters. It is suggested that the dual-dimensioning system be used now and then by a student when he prepares detail drawings. If desired, problems given using the inch system may be prepared showing linear measurements in millimeters by utilizing the conversion tables in Appendix E.

It is essential that a design draftsman always think and speak of each dimension as full-size when scaling measurements, because the dimension figures given on the finished drawing indicate full-size measurements of the finished piece, regardless of the scale used.

The reading of an open-divided scale is illustrated in Fig. A.25 with the eighth-size ($1\frac{1}{2}$ in. = 1 ft) scale shown. The dimension

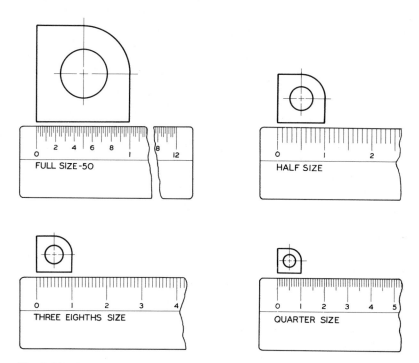

Fig. A.23 □ Decimal scales.

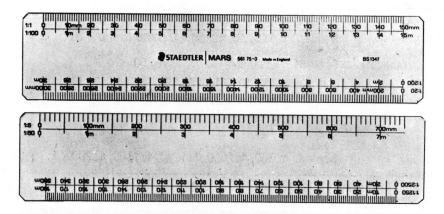

Fig. A.24 □ A flat metric scale (front and reverse sides).
(Courtesy J. S. Staedtler, Inc.)

can be read directly as 21 in., the 9 in. being read in the divided segment to the left of the cipher. Each long open division represents 12 in. (1 ft).

To lay off a measurement, using a scale starting at the left of the stick, align the scale in the direction of the measurement with the zero of the scale being used toward the left. After it has been adjusted to the correct location, make short marks opposite the divisions on the scale that establish the desired distance (Fig. A.26). For ordinary work most draftsmen use the same pencil used for the layout.

The reading of the full-size decimal scale is illustrated in Fig. A.27. The largest division indicated in the illustration represents one inch, which is subdivided into tenths and fiftieths (.02 in.). In Fig. A.23, the largest divisions on the half-size, three-eighths-size, and quarter-size decimal scales represent one inch.

A.20 □ The compass or large bow

The compass or large bow is used for drawing circles and circle arcs. For drawing pencil circles, the style of point illustrated in Fig. A.28(c) should be used because it gives more accurate results and is easier to maintain than most other styles. This style of point is formed by first sharpening the outside of the lead on a file (Fig. A.29) or sandpad to a long flat bevel approximately $\frac{1}{4}$ in. long [Fig. A.28(a)] and then finishing it [Fig. A.28(b)] with a slight rocking motion to reduce the width of the point. Although a hard lead (4H–6H) will maintain a point longer without resharpening, it gives a finished object line that is too light in color. Soft lead (F or H) gives a darker line but quickly loses its edge and, on larger circles, gives a thicker line at the end than at the beginning. Some draftsmen have found that a medium-grade (2H–3H) lead is a satisfactory compromise for ordinary working drawings. For design drawings, layout work, and graphical solutions, however, a harder lead will give better results.

The needle point should have the shouldered end out and should be adjusted approximately $\frac{3}{8}$ in. beyond the end of the split sleeve (Fig. A.28).

A.21 □ Using the compass or large bow

To draw a circle, it is first necessary to draw two intersecting center lines at right angles and mark off the radius. The pivot point should be guided accurately into position at the center. After the pencil point has been adjusted to the radius mark, the circle is drawn in a clockwise direction, as shown in Fig. A.30. While drawing the circle, the instrument should be inclined slightly forward. If the pencil line is not dark enough, it may be drawn around again.

The beam compass is manipulated by steadying the instrument

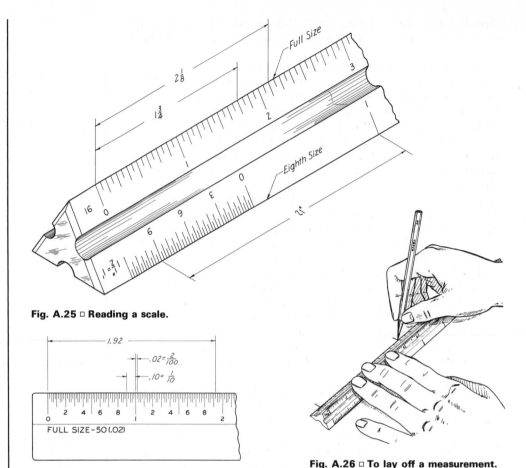

Fig. A.25 □ Reading a scale.

Fig. A.27 □ Reading the decimal scale.

Fig. A.26 □ To lay off a measurement.

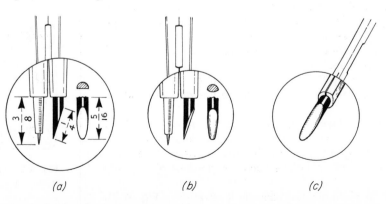

(a)   *(b)*   *(c)*

Fig. A.28 □ Shaping the compass lead.

at the pivot leg with one hand while rotating the marking leg with the other (Fig. A.31).

A.22 □ The dividers

The dividers are used principally for dividing curved and straight lines into any number of equal parts and for transferring measurements. If the instrument is held with one leg between the forefinger and second finger, and the other leg between the thumb and third finger, as illustrated in Fig. A.32, an adjustment may be made quickly and easily with one hand. The second and third fingers are used to "open out" the legs, and the thumb and forefinger to close them. This method of adjusting may seem awkward to the beginner at first, but with practice absolute control can be developed.

A.23 □ Use of the dividers

The trial method is used to divide a line into a given number of equal parts (Fig. A.33). To divide a line into a desired number of equal parts, open the dividers until the distance between the points is estimated to be equal to the length of a division, and step off the line *lightly*. If the last prick mark misses the end point, increase or decrease the setting by an amount estimated to be equal to the error divided by the number of divisions, before lifting the dividers from the paper. Step off the line again. Repeat this procedure until the dividers are correctly set, then space the line again and indent the division points. When stepping off a line, the dividers are rotated alternately in an opposite direction on either side of the line, each half-revolution, as shown in Fig. A.33.

  Although the dividers are used to transfer a distance on a drawing, they should never be used to transfer a measurement from the scale, as the method is slow and inaccurate and results in serious damage to the graduation marks. Care should be taken to avoid pricking large unsightly holes with the divider points. It is the common practice of many expert draftsmen to draw a small freehand circle around a very light indentation to establish its location.

A.24 □ Use of the bow instruments

The small bow pencil is convenient for drawing circles having a radius of 1 in. or less. The needle point should be adjusted slightly longer than the marking point, as in the case of the large bow.

A.25 □ Use of the French curve

A French curve is used for drawing irregular curves that are not circle arcs. After sufficient points have been located, the French curve is applied so that a portion of its ruling edge passes through at least three points, as shown in Fig. A.34. It should be so placed

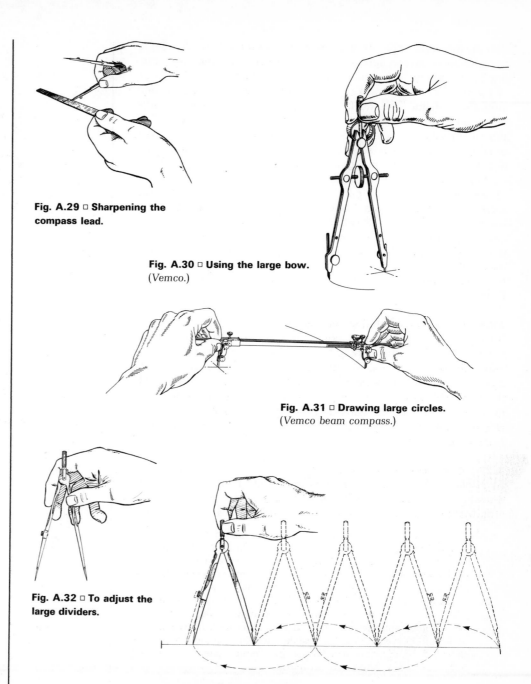

Fig. A.29 □ Sharpening the compass lead.

Fig. A.30 □ Using the large bow. (*Vemco.*)

Fig. A.31 □ Drawing large circles. (*Vemco beam compass.*)

Fig. A.32 □ To adjust the large dividers.

Fig. A.33 □ Use of the dividers.

that the increasing curvature of the section of the ruling edge being used follows the direction of that part of the curve which is changing most rapidly. To ensure that the finished curve will be free of humps and sharp breaks, the first line drawn should start and stop short of the first and last points to which the French curve has been fitted. Then the curve is adjusted in a new position with the ruling edge coinciding with a section of the line previously drawn. Each successive segment should stop short of the last point matched by the curve. In Fig. A.34, the curve fits the three points, A, 1, and 2. A line is drawn from between point A and point 1 to between point 1 and point 2. Then, the curve is shifted, as shown, to fit again points 1 and 2 with an additional point 3, and the line is extended to between point 2 and point 3.

Some people sketch a smooth continuous curve through the points in pencil before drawing the mechanical line. This procedure makes the task of drawing the curve less difficult, since it is easier to adjust the ruling edge to segments of the freehand curve than to the points.

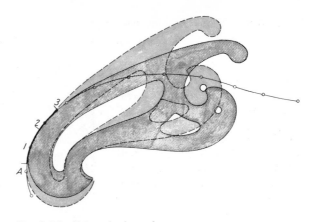

Fig. A.34 □ Using the irregular curve.

A.26 □ Use of the erasing shield and eraser
An erasure is made on a drawing by placing an opening in the erasing shield over the work to be erased and rubbing with a pencil eraser (never an ink eraser) until it is removed (Fig. A.35). Excessive pressure should not be applied to the eraser because, although the lines will disappear more quickly, the surface of the paper is likely to be permanently damaged. The fingers holding the erasing shield should rest partly on the drawing paper to prevent the shield from slipping.

Fig. A.35 □ Using the erasing shield.

A.27 □ Conventional line symbols
Symbolic lines of various weights are used in making technical drawings. The recommendations of the American National Standards Institute, as given in ANSI Y14.2–1957, are the following:

> Three widths of lines—thick, medium, and thin—are recommended for use on drawings (Fig. A.36). Pencil lines in general should be in proportion to the ink lines except that the thicker pencil lines will be necessarily thinner than the corresponding ink lines but as thick as practicable for pencil work. Exact thicknesses may vary according to the size and type of drawing. For example, where lines are close together, the lines may be slightly thinner.
>
> Pencil lines may be further simplified, if desired, to two widths of lines: *medium-thick* for visible, hidden, cutting-plane, and short-break lines; and *thin* for section, center, extension, dimension, long-break, and phantom lines.

The lines illustrated in Fig. A.36 are shown full-size. When symbolic lines are used on a pencil drawing they should not vary in color. For example, center lines, extension lines, dimension lines, and section lines should differ from object lines only in width. The resulting contrast makes a drawing easier to read. All lines, except construction lines, should be very dark and bright to give the drawing the "snap" that is needed for good appearance. If the drawing is on tracing paper the lead must be "packed on" so that a satisfactory print can be obtained. Construction lines should be drawn very fine so as to be unnoticeable on the finished drawing.

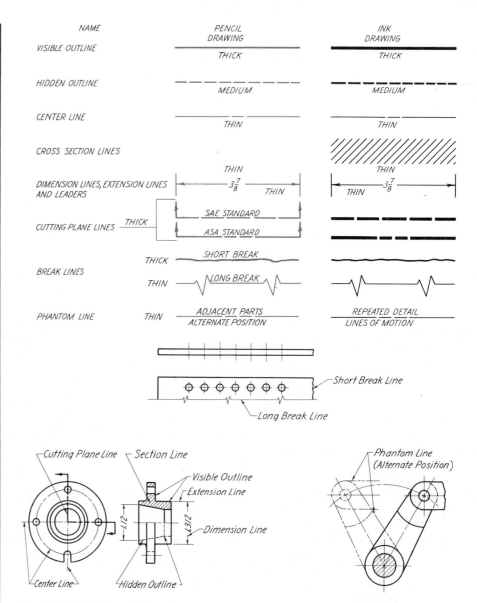

Fig. A.36 ◻ Alphabet of lines (finished weight).

B.1 □ Introduction

To impart to the men in the shops all the necessary information for the complete construction of a machine or structure, the shape description, which is conveyed graphically by the views, must be accompanied by size descriptions and instructive specifications in the form of figured dimensions and notes (Fig. B.1).

All dimensions and notes should be lettered freehand in a plain, legible style that can be rapidly executed. Poor lettering detracts from the appearance of a drawing and often impairs its usefulness, regardless of the quality of the line work.

B.2 □ Single-stroke letters (Reinhardt)

Single-stroke letters are now used universally for technical drawings. This style is suitable for most purposes because it possesses the qualifications necessary for legibility and speed. On commercial drawings it appears in slightly modified forms, however, since each person finally develops a style that reflects his own individuality.

The expression ''single-stroke'' means that the width of the straight and curved lines that form the letters are the same width as the stroke of the pen or pencil.

B.3 □ The general proportions of letters

Although there is no fixed standard for the proportions of the letters, certain definite rules must be observed in their design if one wishes to have his lettering appear neat and pleasing. The recognized characteristics of each letter should be carefully studied and then thoroughly learned through practice.

Freehand technical lettering

It is advisable for the beginner, instead of relying on his untrained eye for proportions, to follow the fixed proportions given in Figs. B.7–B.13. Otherwise, his lettering will most likely be displeasing to the trained eye of the professional man. Later, after he has thoroughly mastered the art of lettering, his individuality will be revealed naturally by slight variations in the shapes and proportions of some of the letters.

It is often desirable to increase or decrease the width of letters in order to make a word or group of words fill a certain space. Letters narrower than normal letters of the same height are called *compressed letters*; those that are wider are called *extended letters* (Fig. B.2).

B.4 □ Lettering pencils

Pencil lettering is usually done with a medium-soft pencil. Since the degree of hardness of the lead required to produce a dark opaque line will vary with the type of paper used, a pencil should be selected only after drawing a few trial lines. In order to obtain satisfactory lines, the pencil should be sharpened to a long conical point and then rounded slightly on a piece of scratch paper. To keep the point symmetrical while lettering, the pencil should be rotated a partial revolution before each new letter is started.

B.5 □ Devices for drawing guide lines and slope lines

Devices for drawing guide lines are available in a variety of forms. The two most popular are the *Braddock lettering triangle* (Fig. B.3), and the *Ames lettering instrument* (Fig. B.4).

The Braddock lettering triangle is provided with sets of grouped countersunk holes that may be used to draw guide lines by inserting a sharp-pointed pencil (4H or 6H) into the holes and sliding the triangle back and forth along the guiding edge of a T-square or a triangle supported by a T-square (Fig. B.3). The holes are grouped to give guide lines for capitals and lowercase letters. The numbers below each set indicate the height of the capitals in thirty-seconds of an inch. For example, the No. 3 set is for capitals $\frac{3}{32}$ in. high, the No. 4 set is for capitals $\frac{1}{8}$ in. high, the No. 5 is for capitals $\frac{5}{32}$ in. high, and so on.

B.6 □ Uniformity in lettering

Uniformity in height, inclination, spacing, and strength of line is essential for good lettering (Fig. B.5). Professional appearance depends as much on uniformity as on the correctness of the proportion and shape of the individual letters. Uniformity in height and inclination is assured by the use of guide lines and slope lines and uniformity in weight and color, by the skillful use of the pencil and proper control of the pressure of its point on the paper. The

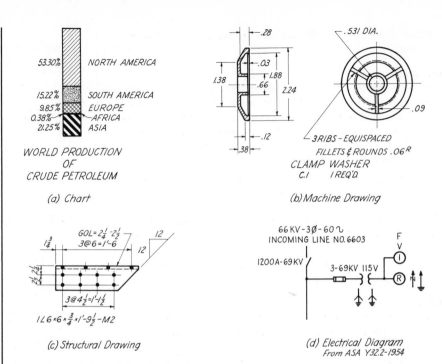

WORLD PRODUCTION
OF
CRUDE PETROLEUM

53.30% NORTH AMERICA
15.22% SOUTH AMERICA
9.85% EUROPE
0.38% AFRICA
21.25% ASIA

(a) Chart

CLAMP WASHER
C.I 1 REQ'D.

(b) Machine Drawing

(c) Structural Drawing

(d) Electrical Diagram
From ASA Y32.2-1954

Fig. B.1 □ Technical drawings.

NORMAL LETTERS
COMPRESSED LETTERS
EXTENDED LETTERS

Fig. B.2 □ Compressed and extended letters.

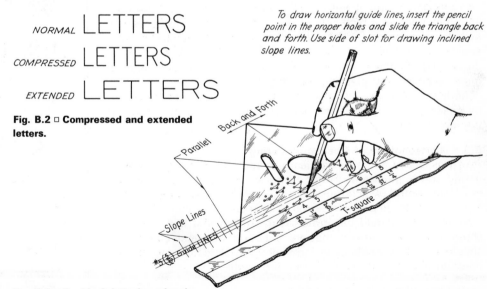

To draw horizontal guide lines, insert the pencil point in the proper holes and slide the triangle back and forth. Use side of slot for drawing inclined slope lines.

Fig. B.3 □ Braddock lettering triangle.

ability to space letters correctly becomes easy after continued thoughtful practice.

B.7 □ Composition

In combining letters into words, the spaces for the various combinations of letters are arranged so that the areas appear to be equal (Fig. B.6). For standard lettering, this area should be about equal to one-half the area of the letter M. If the adjacent sides are stems, this area is obtained by making the distance between the letters slightly greater than one-half the height of a letter; a smaller amount, depending on the contours, is used for other combinations. Examples of good and poor composition are shown in Fig. B.6.

The space between words should be equal to or greater than the height of a letter but not more than twice the height. The space between sentences should be somewhat greater. The distance between lines of lettering may vary from one-half the height of the capitals to $1\frac{1}{2}$ times their height.

B.8 □ Stability

If the areas of the upper and lower portions of certain letters are made equal, an optical illusion will cause them to appear to be unstable and top-heavy. To overcome this effect, the upper portions of the letters B, E, F, H, K, S, X, and Z and the figures 2, 3, and 8 must be reduced slightly in size.

An associated form of illusion is the phenomenon that a horizontal line drawn across a rectangle at the vertical center will appear to be below the center. Since the letters B, E, F, and H are particularly subject to this illusion, their central horizontal strokes must be drawn slightly above the vertical center in order to give them a more balanced and pleasing appearance.

The letters K, S, X, Z and the figures 2, 3, and 8 are stabilized by making the width of the upper portion less than the width of the lower portion.

B.9 □ The technique of freehand lettering

Any prospective engineer can learn to letter if he practices intelligently and is persistent in his desire to improve. The necessary muscular control, which must accompany the knowledge of lettering, can be developed only through constant repetition.

Pencil letters should be formed with strokes that are dark and sharp, never with strokes that are gray and indistinct. Beginners should avoid the tendency to form letters by sketching, as strokes made in this manner vary in color and width.

B.10 □ Inclined and vertical capital letters

The letters shown in Figs. B.7 and B.8 have been arranged in related groups. In laying out the characters, the number of widths

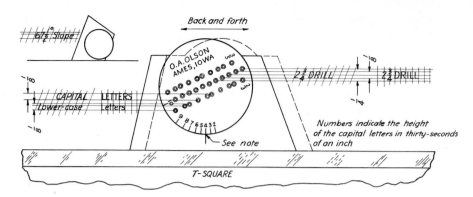

Fig. B.4 □ Ames lettering instrument.

UNIFORMITY IN HEIGHT, INCLINATION, AND STRENGTH OF LINE IS ESSENTIAL FOR GOOD LETTERING

Fig. B.5 □ Uniformity in lettering.

Fig. B.6 □ Letter areas.

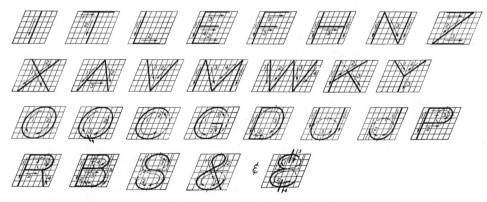

Fig. B.7 □ Inclined capital letters.

has been reduced to the smallest number consistent with good appearance; similarities of shape have been emphasized and minute differences have been eliminated. Each letter is drawn to a large size on a cross-section grid that is 2 units wider, to facilitate the study of its characteristic shape and proportions. Arrows with numbers indicate the order and direction of the strokes. The curves of the inclined capital letters are portions of ellipses, while the curves of the vertical letters are parts of circles.

B.11 □ Inclined and vertical numerals
The numerals shown in Figs. B.9 and B.10 have been arranged in related groups in accordance with the common characteristics that can be recognized in their construction.

B.12 □ Single-stroke lowercase letters
Single-stroke lowercase letters, either vertical or inclined, are commonly used on map drawings, topographic drawings, structural drawings, and in survey field books. They are particularly suitable for long notes and statements because, first, they can be executed much faster than capitals and, second, words and statements formed with them can be read more easily.

The construction of inclined lowercase letters (Fig. B.12) is based on the straight line and the ellipse. This basic principle of forming letters is followed more closely for lowercase letters than for capitals. The body portions are two-thirds the height of the related capitals. As shown in Fig. B.11, ascenders extend to the cap line, and descenders descend to the drop line. For lowercase letters based on a capital letter 6 units high, the waistline is 2 units down from the top and the drop line 2 units below the base line.

The order of stroke, direction of stroke, and formation of the letters follow the same principles as for the capitals. The letters are presented in family groups having related characteristics, to enable the beginner to understand their construction. The vertical lowercase letters illustrated in Fig. B.13, are constructed in the same manner as inclined letters.

B.13 □ Fractions
The height of the figures in the numerator and denominator is equal to three-fourths the height of the whole number, and the total height of the fraction is twice the height of the whole number. The division bar should be horizontal and centered between the fraction numerals, as shown in Fig. B.14. It should be noted that the sloping center line of the fraction bisects both the numerator and denominator and is parallel to the sloping center line of the whole number.

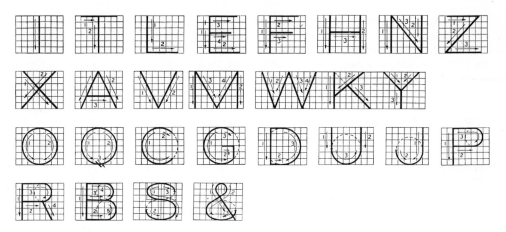

Fig. B.8 □ Vertical capital letters.

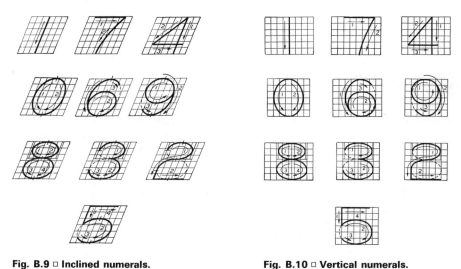

Fig. B.9 □ Inclined numerals.

Fig. B.10 □ Vertical numerals.

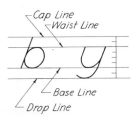

Fig. B.11

B.14 □ Large caps and small caps in combination

Many commercial draftsmen use a combination of large caps and small caps in forming words, as illustrated in Fig. B.15. When this style is used, the height of the small caps should be approximately three-fifths the height of the first capital letter of the word.

B.15 □ Titles

Every drawing, sketch, graph, chart, or diagram has some form of descriptive title to impart certain necessary information and to identify it. On machine drawings, where speed and legibility are prime requirements, titles are usually single-stroke. On display drawings, maps, and so on, which call for an artistic effect, the titles are usually composed of "built-up" ornate letters.

Figure B.16 shows a title block that might be used on a machine drawing. It should be noted that the important items are made more prominent by the use of larger letters formed with heavier lines. Less important data, such as the scale, date, drafting information, and so on, are given less prominence.

To be pleasing in appearance, a title should be symmetrical about a vertical center line and should have some simple geometric form.

B.16 □ Mechanical lettering devices and templates

Although mechanical lettering devices produce letters that may appear stiff to an expert, they are used in many drafting rooms for the simple reason that they enable even the unskilled to do satisfactory lettering with ink. The average person rightly prefers stiff uniformity to wavy lines and irregular shapes. One of the oldest instruments of this sort on the market is the *Wrico* outfit. With a satisfactory set of Wrico pens and templates, letters ranging in size from $\frac{3}{32}-\frac{1}{2}$ in. in height may be executed. The letters are formed by a stylographic pen that is guided around the sides of openings in a template made of transparent pyralin.

The *Leroy* device, shown in Fig. B.17, is possibly even more efficient than the Wrico, for it does not require the sliding of a template to complete a letter.

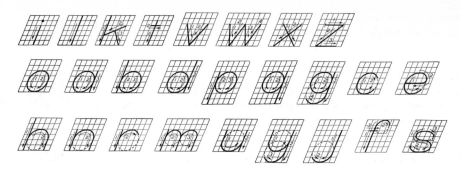

Fig. B.12 □ Inclined lower-case letters.

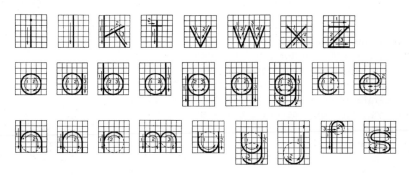

Fig. B.13 □ Vertical lower-case letters.

Fig. B.14

HANDLE PIN
C.R.S. 1 REQ'D.

Fig. B.15 □ Use of large and small caps.

APPROVED BY CUSTOMER	*ROSS-SHOER CO.*	
ADAMS MFG. CO. AURORA, ILL.		
ASSEMBLY STANDARD DUTY FLANGE UNIT		
SCALE *FULL SIZE*	DATE *JAN. 3, 1959*	
DRAWN BY *CARL KING*	CHECKED BY *J.B. JONES*	
TRACED BY *K. BODINE*	APPROVED BY *W.L. KOE*	
ORDER NO. *M-L-24631*	DRG. *5-19-9612*	

Fig. B.16 □ A machine-drawing title block.

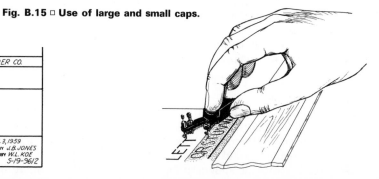

Fig. B.17 □ Leroy lettering device.

Geometry of design

C.1 □ Introduction

The simplified geometrical constructions presented in this section are those with which an engineer should be familiar, for they frequently occur in engineering design drawing. The methods are applications of the principles found in textbooks on plane geometry. The constructions have been modified to take advantage of time-saving methods made possible by the use of drawing instruments.

C.2 □ To bisect an angle (Fig. C.1)

(a) Given the angle BAC. Use any radius with the vertex A as a center, and strike an arc that intersects the sides of the angle at D and E. With D and E as centers and a radius larger than one-half of DE, draw intersecting arcs. Draw AF. Angle BAF equals angle FAC.

(b) Given an angle formed by the lines KL and MN having an inaccessible point of intersection. Draw BA parallel to KL and CA parallel to MN at the same distance from MN as BA is from KL. Bisect angle BAC using the method explained in part (a). The bisector FA of angle BAC bisects the angle between the lines KL and MN.

C.3 □ To divide a straight line into a given number of equal parts (Fig. C.2)

Given the line LM, which is to be divided into five equal parts.

(a) Step off, with the dividers, five equal divisions along a line making any convenient angle with LM. Connect the last point P

with M, and through the remaining points draw lines parallel to MP intersecting the given line. These lines divide LM into five equal parts.

(b) Some commercial draftsmen prefer a modification of this construction known as the scale method. For the first step, draw a vertical PM through point M. Place the scale so that the first mark of five equal divisions is at L and the last mark falls on PM. Locate the four intervening division points, and through these draw verticals intersecting the given line. The verticals will divide LM into five equal parts.

C.4 □ To divide a line proportionally (Fig. C.3)

Given the line AB. Draw BC perpendicular to AB. Place the scale across A and BC so that the number of divisions intercepted is equal to the sum of the numbers representing the proportions. Mark off these proportions and draw lines parallel to BC to divide AB as required. The proportions in Fig. C.3 are $1:2:3$.

C.5 □ To construct an angle equal to a given angle (Fig. C.4)

Given the angle BAC and the line $A'C'$ that forms one side of the transferred angle. Use any convenient radius with the vertex A as a center, and strike the arc that intersects the sides of the angle at D and E. With A' as a center, strike the arc intersecting $A'C'$ at E'. With E' as a center and the chord distance DE as a radius, strike a short intersecting arc to locate D'. $A'B'$ drawn through D' makes angle $B'A'C'$ equal angle BAC.

C.6 □ To draw a line through a given point and the inaccessible intersection of two given lines (Fig. C.5)

Given the lines KL and MN, and the point P. Construct any triangle, such as PQR, having its vertices falling on the given lines and the given point. At some convenient location construct triangle STU similar to PQR, by drawing SU parallel to PR, TU parallel to QR, and ST parallel to PQ. PS is the required line.

C.7 □ To construct an angle, tangent method (Fig. C.6)

Design draftsmen often find it necessary to draw long lines having an angle between them that is not equal to an angle of a triangle. Such an angle may be laid off with a protractor, but it should be remembered that as the lines are extended any error is multiplied. To avoid this situation, the tangent method may be used. The tangent method involves trigonometry but, since it is frequently used, a discussion of it here is pertinent. (See Table 7 of Appendix E.)

In this method, a distance D_1 is laid off along a line that is to form one side of the angle, and a distance D_2, equal to D_1 times

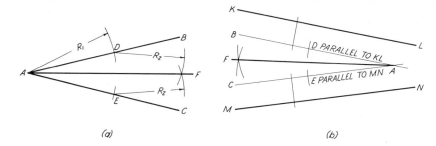

(a)   (b)

Fig. C.1 □ To bisect an angle.

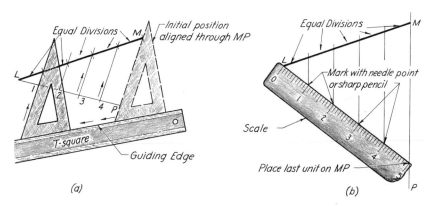

(a)   (b)

Fig. C.2 □ To divide a straight line into a number of equal parts.

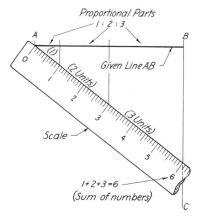

Fig. C.3 □ To divide a line proportionally.

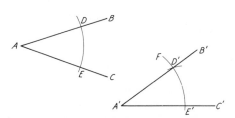

Fig. C.4 □ To construct an angle equal to a given angle.

the natural tangent of the angle, is marked off along a perpendicular through point P. A line through point X is the required line, and angle A is the required angle. In laying off the distance D_1, unnecessary multiplication will be eliminated if the distance is arbitrarily made 10 in. When the use of 10 in. for D_1 makes P fall off the drawing, a temporary auxiliary sheet will furnish space needed to carry out the construction. However, in order to keep the construction on the drawing, the draftsman may decide to lay off the 10-in. length D_1 using either the half-size or quarter-size scale.

This method is also used for angles formed by short lines whenever a protractor is not available.

In using the metric scale, the distance D_1 may be made 10 centimeters (100 mm).

C.8 □ To construct a triangle given its three sides (Fig. C.7)

Given the three sides AB, AC, and BC. Draw the side AB in its correct location. Using its end points A and B as centers and radii equal to AC and BC, respectively, strike the two intersecting arcs locating point C. ABC is the required triangle. This construction is particularly useful for developing the surface of a transition piece by triangulation.

C.9 □ To construct a regular pentagon (Fig. C.8)

Given the circumscribing circle. Draw the perpendicular diameters AB and CD. Bisect OB and, with its mid-point E as a center and EC as a radius, draw the arc CF. Using C as a center and CF as a radius, draw the arc FG. The line CG is one of the equal sides of the required pentagon. Locate the remaining vertices by striking off this distance around the circumference.

If the length of one side of a pentagon is given, the construction described in Sec. C.12 should be used.

C.10 □ To construct a regular hexagon (Fig. C.9)

(a) Given the distance AB across corners. Draw a circle having AB as a diameter. Using the same radius and with points A and B as centers, strike arcs intersecting the circumference. Join these points to complete the construction.

(b) Given the distance AB across corners. Using a 30°–60° triangle and a T-square, draw the lines in the order indicated by the numbers on the figure.

(c) Given the distance across flats. Draw a circle whose diameter equals the distance across flats. Using a 30°–60° triangle and a T-square, as shown, draw the tangents that establish the sides and vertices of the required hexagon.

This construction is used in drawing hexagonal bolt heads and nuts.

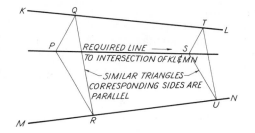

Fig. C.5 □ To draw a line through a given point and the inaccessible intersection of two given lines.

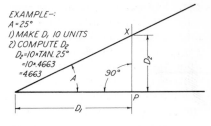

Fig. C.6 □ To construct an angle, tangent method.

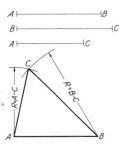

Fig. C.7 □ To construct a triangle, given its three sides.

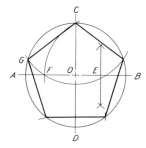

Fig. C.8 □ To construct a regular pentagon.

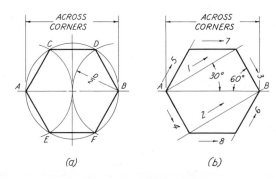

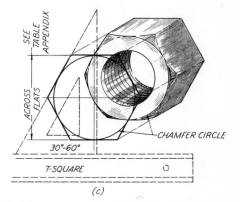

Fig. C.9 □ To construct a regular hexagon.

C.11 □ To construct a regular octagon (Fig. C.10)

(a) Given the distance across flats. Draw the circumscribed square and its diagonals. Using the corners as centers and one-half the diagonal as a radius, strike arcs across the sides of the square. Join these points to complete the required octagon.

(b) Given the distance across flats. Draw the inscribed circle; then, using a 45° triangle and T-square, draw the tangents that establish the sides and vertices of the required octagon.

C.12 □ To construct any regular polygon, given one side (Fig. C.11)

Given the side *LM*. With *LM* as a radius, draw a semicircle and divide it into the same number of equal parts as the number of sides needed for the polygon. Suppose the polygon is to be seven-sided. Draw radial lines through points 2, 3, and so forth. Point 2 (the second division point) is always one of the vertices of the polygon, and line *L2* is a side. Using point *M* as a center and *LM* as a radius, strike an arc across the radial line *L6* to locate point *N*. Using the same radius with *N* as a center, strike another arc across *L5* to establish *O* on *L5*. Although this procedure may be continued with point *O* as the next center, more accurate results will be obtained if point *R* is used as a center for the arc to locate *Q*, and *Q* as a center for *P*.

C.13 □ To divide the area of a triangle or trapezoid into a given number of equal parts (Fig. C.12)

(a) Given the triangle *ABC*. Divide the side *AC* into (say, five) equal parts, and draw a semicircle having *AC* the diameter. Through the division points (1, 2, 3, and 4) draw perpendicular lines to points of intersection with the semicircle (5, 6, 7, and 8). Using *C* as a center, strike arcs through these points (5, 6, 7, and 8) that will cut *AC*. To complete the construction, draw lines parallel to *AB* through the points (9, 10, 11, and 12) at which the arcs intersect the side *AC*.

(b) Given the trapezoid *DEBA*. Extend the sides of the trapezoid to form the triangle *ABC* and draw a semicircle on *AC* with *AC* as a diameter. Using *C* as a center and *CD* as a radius, strike an arc cutting the semicircle at point *P*. Through *P* draw a perpendicular to *AC* to locate point *Q*. Divide *QA* into the same number of equal parts as the number of equal areas required (in this case, four), and proceed using the construction explained in (a) for dividing the area of a triangle into a given number of equal parts.

C.14 □ To find the center of a circle through three given points not in a straight line (Fig. C.13)

Given the three points *A*, *B*, and *C*. Join the points with straight

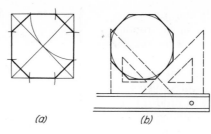

Fig. C.10 □ To construct a regular octagon.

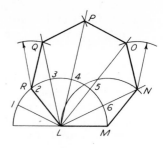

Fig. C.11 □ To construct any regular polygon, given one side.

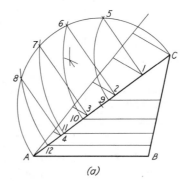

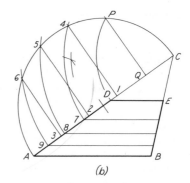

Fig. C.12 □ To divide the area of a triangle or trapezoid into a given number of equal parts.

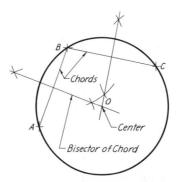

Fig. C.13 □ To find the center of a circle through three points.

lines (which will be chords of the required circle), and draw the perpendicular bisectors. The point of intersection O of the bisectors is the center of the required circle, and OA, OB, or OC is its radius.

C.15 □ To draw a circular arc of radius R
tangent to two lines (Fig. C.14)

(a) Given the two lines AB and CD at right angles to each other, and the radius of the required arc R. Using their point of intersection X as a center and R as a radius, strike an arc cutting the given lines at T_1 and T_2 (tangent points). With T_1 and T_2 as centers and the same radius, strike the intersecting arcs locating the center O of the required arc.

(b),(c) Given the two lines AB and CD, not at right angles, and the radius R. Draw lines EF and GH parallel to the given lines at a distance R. Since the point of intersection of these lines is distance R from both given lines, it will be the center O of the required arc. Mark the tangent points T_1 and T_2 that lie along perpendiculars to the given lines through O.

C.16 □ To draw a circular arc of radius R_1
tangent to a given circular arc
and a given straight line (Fig. C.15)

Given the line AB and the circular arc with center O.

(a),(b) Draw line CD parallel to AB at a distance R_1. Using the center O of the given arc and a radius equal to its radius plus or minus the radius of the required arc (R_2 plus or minus R_1), swing a parallel arc intersecting CD. Since the line CD and the intersecting arc will be the loci of centers for all circles of radius R_1, tangent respectively to the given line AB and the given arc, their point of intersection P will be the center of the required arc. Mark the points of tangency T_1 and T_2. T_1 lies along a perpendicular to AB through the center P, and T_2 along a line joining the centers of the two arcs.

This construction is also useful for drawing fillets and rounds on views of machine parts.

C.17 □ To draw a circular arc of a given radius R_1
tangent to two given circular arcs (Fig. C.16)

Given the circular arcs AB and CD with centers O and P, and radii R_2 and R_3, respectively.

(a) Using O as a center and R_2 plus R_1 as a radius, strike an arc parallel to AB. Using P as a center and R_3 plus R_1 as a radius, strike an intersecting arc parallel to CD. Since each of these intersecting arcs is the locus of centers for all circular arcs of radius R_1 tangent to the given arc to which it is parallel, their point of intersection S will be the center for the required arc that is tangent

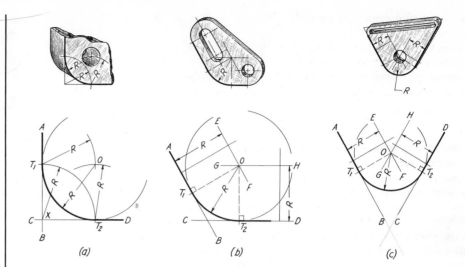

Fig. C.14 □ To draw a circular arc tangent to two lines.

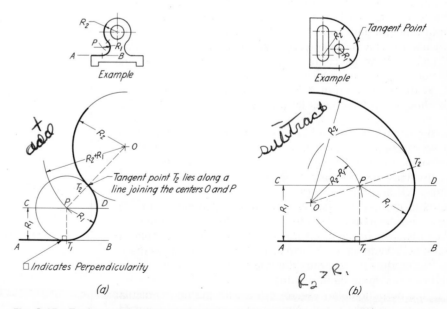

Fig. C.15 □ To draw a circular arc tangent to a given circular arc and a line.

to both. Mark the points of tangency T_1 and T_2 that lie on the lines of centers PS and OS.

(b) Using O as a center and R_2 plus R_1 as a radius, strike an arc parallel to AB. Using P as a center and R_3 minus R_1 as a radius, strike an intersecting arc parallel to CD. The point of intersection of these arcs is the center for the required arc.

C.18 □ To draw a reverse (Ogee) curve (Fig. C.17)

(a) *Reverse (Ogee) Curve Connecting Two Parallel Lines.* Given the two parallel lines AB and CD. At points B and C, the termini and tangent points of the reverse curve, erect perpendiculars. Join B and C with a straight line and assume a point E as the point at which the curves will be tangent to each other. Draw the perpendicular bisectors of BE and EC. Since an arc tangent to AB at B must have its center on the perpendicular BP, the point of intersection P of the bisector and the perpendicular is the center for the required arc that is to be tangent to the line at B and the other required arc at point E. For the same reason, point Q is the center for the other required arc.

This construction is useful to engineers in laying out center lines for railroad tracks, pipe lines, and so forth.

(b) *Reverse (Ogee) Curve Connecting Two Nonparallel Lines.* Given the two nonparallel lines AB and CD. At points B and C, the termini and tangent points, erect perpendiculars. Along the perpendicular at B lay off the given (or selected) radius R and draw the arc having P as its center. Then draw a construction line through point P perpendicular to CD to establish the location of point X. With the position of X known, join points X and C with a straight line along which will lie the chords of the arcs forming the ogee curve between points X and C. The broken line XY (not a part of the construction) has been added to show that the procedure to be followed in completing the required curve will be as previously explained for drawing a reverse curve joining two parallel lines. In this case the parallel lines are XY and CD, instead of the lines AB and CD as in (a).

An alternate method for establishing the needed center for the required arc has been added to the illustration in (b). In this method the radius distance R is laid off upward along a perpendicular to CD through C. With point S established by this measurement, the line PS, as drawn, becomes the chord of an arc (not shown) that will have the same center as the required arc EC. The intersection of the perpendicular bisector of PS with the perpendicular erected downward from C will establish the position of point Q, the center of concentric arcs having chords PS and EC.

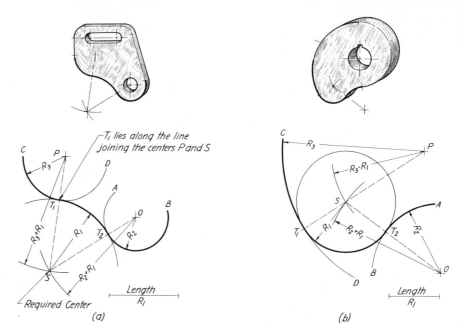

Fig. C.16 □ To draw a circular arc tangent to two given arcs.

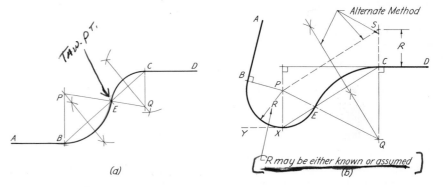

Fig. C.17 □ To draw a reverse curve.

C.19 □ To draw a reverse curve
tangent to three given lines (Fig. C.18)
Given the lines AB and CD that are intersected by a third line BC at points B and C. Assume the position of point E (point of tangency) along BC and locate the termini points T_1 and T_2 by making CT_1 equal to CE and BT_2 equal to BE. The intersections of the perpendiculars erected at points T_1, E, and T_2 establish the centers P and Q of the arcs that form the reversed curve.

C.20 □ To draw a line tangent to a circle
at a given point on the circumference (Fig. C.19)
Given a circle with center O and point P on its circumference. Place a triangle supported by a T-square or another triangle in such a position that one side passes through the center O and point P. When using the method illustrated in (a), align the hypotenuse of one triangle on the center of the circle and the point of tangency; then, with the guiding triangle held in position, revolve the triangle about the 90° angle and slide into position for drawing the required tangent line.

Another procedure is shown in (b). To draw the tangent by this method, align one leg of a triangle, which is adjacent to the 90° angle, through the center of the circle and the point of tangency; then slide it along the edge of a guiding triangle into position.

This construction satisfies the geometric requirement that a tangent must be perpendicular to a radial line drawn to the point of tangency.

C.21 □ To draw a line tangent to a circle
through a given point outside the circle (Fig. C.20)
Given a circle with center O and an external point P. Join the point P and the center O with a straight line, and bisect it to locate point S. Using S as a center and SO (one-half PO) as a radius, strike an arc intersecting the circle at point T (point of tangency). Line PT is the required tangent.

C.22 □ To draw a line tangent to a circle
through a given point outside the circle (Fig. C.21)
Place a triangle supported by a T-square or another triangle in such a position that one leg passes through point P tangent to the circle, and draw the tangent. Slide the triangle along the guiding edge until the other leg coincides with the center O, and mark the point of tangency. Although this method is not as accurate as the geometric one explained in Sec. C.21, it is frequently employed by commercial draftsmen.

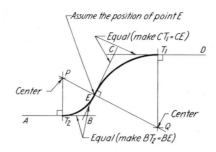

Fig. C.18 □ To draw a reverse curve tangent to three lines.

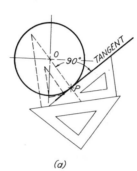

(a)

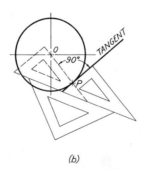

(b)

Fig. C.19 □ To draw a line tangent to a circle at a point on the circumference.

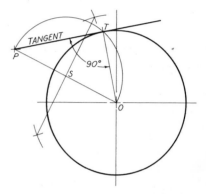

Fig. C.20 □ To draw a line tangent to a circle through a given point outside.

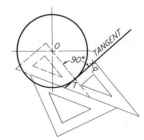

Fig. C.21 □ To draw a line tangent to a circle through a given point outside.

C.23 □ To draw a line tangent to two given circles (Fig. C.22)
Given two circles with centers O and P and radii R_1 and R.
(a) *Open Belt.* Using P as a center and a radius equal to R minus R_1, draw an arc. Through O draw a tangent to this arc using the method explained in Sec. C.21. With the location of tangent point T established, draw line PT and extend it to locate T_1. Draw OT_2 parallel to PT_1. The line from T_2 to T_1 is the required tangent to the given circles.
(b) *Crossed Belt.* Using P as a center and a radius equal to R plus R_1, draw an arc. With the location of tangent point T determined through use of the method shown in Fig. C.20, locate tangent point. T_1 on line TP and draw OT_2 parallel to PT. The line T_1T_2, drawn parallel to OT, is the required tangent.

C.24 □ To lay off the approximate length of a circular arc on its tangent (Fig. C.23)
Given the arc AB.
(a) Draw the tangent through A and extend the chord BA. Locate point C by laying off AC equal to one-half the length of the chord AB. With C as a center and a radius equal to CB, strike an arc intersecting the tangent at D. The length AD along the tangent is slightly shorter than the true length of the arc AB by an amount that may be disregarded, for when the angle is less than 60°, the length of AD differs from the true length of the arc AB by less than 6 ft in 1 mile; when 30°, the error is $4\frac{1}{2}$ in. in 1 mile.
(b) Draw the tangent through A. Using the dividers, start at B and step off equal chord distances around the arc until the point nearest A is reached. From this point (without raising the dividers) step off along the tangent an equal number of distances to locate point C. If the point nearest A is indented into the tangent instead of the arc, the almost negligible error in the length of AC will be still less.
Since the small distances stepped off are in reality the chords of small arcs, the length AC will be slightly less than the true length of the arc. For most practical purposes the difference may be disregarded.
When the central angle (θ) and the radius of an arc are known, the length of the arc may be computed by the formula $L = 2\pi R(\theta/360°) = 0.01745R\theta$.

C.25 □ To lay off a specified length along a given circle arc (Fig. C.24)
On the tangent to the arc, lay off the distance DE representing the specified length of arc. Divide DE into four equal parts. Then, using point 1 as a center and with the length $1-E$ as the radius R, strike an arc intersecting the given arc at F. The arc DF is

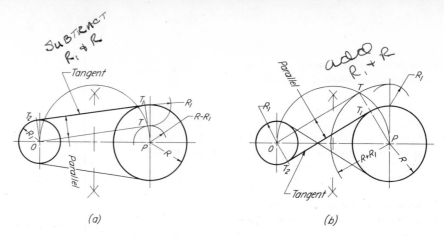

(a)   (b)

Fig. C.22 □ To draw a line tangent to two given circles.

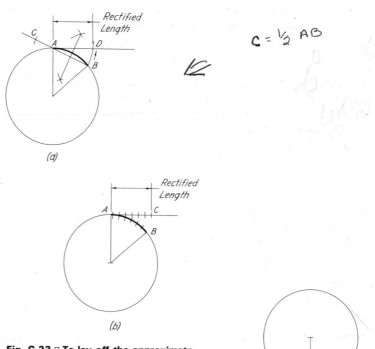

(a)

(b)

Fig. C.23 □ To lay off the approximate length of a circular arc on its tangent.

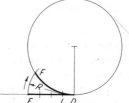

Fig. C.24 □ To lay off a specified length along an arc.

approximately equal in length to the line DE. For large angles, it is advisable to make the construction for one-half of DE.

C.26 □ Conic sections (Fig. C.25)

When a right circular cone of revolution is cut by planes at different angles, four curves of intersection are obtained that are called *conic sections*.

> When the intersecting plane is perpendicular to the axis, the resulting curve of intersection is a *circle*.
>
> If the plane makes a greater angle with the axis than do the elements, the intersection is an *ellipse*.
>
> If the plane makes the same angle with the axis as the elements, the resulting curve is a *parabola*.
>
> Finally, if the plane makes a smaller angle with the axis than do the elements or is parallel to the axis, the curve of intersection is a *hyperbola*.

The geometric methods for constructing the ellipse, parabola, and hyperbola are discussed in succeeding sections.

C.27 □ The ellipse (Fig. C.25)

Mathematically the ellipse is a curve generated by a point moving so that at any position the sum of its distances from two fixed points (foci) is a constant (equal to the major diameter). It is encountered very frequently in orthographic drawing when holes and circular forms are viewed obliquely. Ordinarily, the major and minor diameters are known.

C.28 □ To construct an ellipse, foci method (Fig. C.26)

Draw the major and minor axes (AB, CD) and locate the foci F_1 and F_2 by striking arcs, centered at C and having a radius equal to OA (one-half of the major diameter). The construction is as follows: Determine the number of points needed along the circumference of each quadrant of the ellipse for a relatively accurate layout (say, four) and mark off this number of division points (P, Q, R, and S) between O and F_1 on the major axis. In many cases it may be desirable to use additional points spaced closer together nearer F_1 in order to form accurately the sharp curvature at the end of the ellipse. Next, with F_1 and F_2 as centers and the distances AP and BP as radii, respectively, strike intersecting arcs to locate P' on the circumference of the ellipse. Distances AQ and BQ are radii for locating points Q'. Locate points R' and S' in a similar manner and complete the ellipse using a French curve.

This method is sometimes known as the definition method, since it is based on the mathematical definition of the ellipse as given in Sec. C.27.

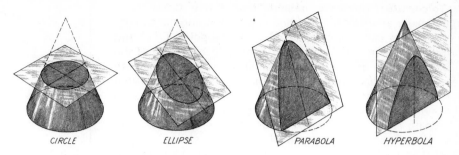

Fig. C.25 □ Conic sections.

CIRCLE   ELLIPSE   PARABOLA   HYPERBOLA

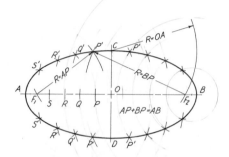

Fig. C.26 □ To construct an ellipse, foci (definition) method.

C.29 □ To construct an ellipse, trammel method (Fig. C.27)

Given the major axis AB and the minor axis CD. Along the straight edge of a strip of paper or cardboard, locate the points O, C, and A, so that the distance OA is equal to one-half the length of the major axis, and the distance OC is equal to one-half the length of the minor axis. Place the marked edge across the axes so that point A is on the minor axis and point C is on the major axis. *Point O will fall on the circumference of the ellipse.* Move the strip, keeping A on the minor axis and C on the major axis, and mark at least five other positions of O on the ellipse in each quadrant. Using a French curve, complete the ellipse by drawing a smooth curve through the points. The ellipsograph, which draws ellipses mechanically, is based on this same principle. The trammel method is an accurate method.

An alternate method for marking off the location of points A, O, and C is given in Fig. C.27.

C.30 □ To construct an ellipse, concentric circle method (Fig. C.28)

Given the major axis AB and the minor axis CD. Using the center of the ellipse (point O) as a center, describe circles having the major and minor axes as diameters. Divide the circles into equal central angles and draw diametrical lines such as P_1P_2. From point P_1 on the circumference of the larger circle, draw a line parallel to CD, the minor axis, and from point P_1' at which the diameter P_1P_2 intersects the inner circle, draw a line parallel to AB, the major axis. The point of intersection of these lines, point E, is on the required ellipse. At points P_2 and P_2' repeat the same procedure and locate point F. Thus, two points are established by the line P_1P_2. Locate at least five points in each of the four quadrants. The ellipse is completed by drawing a smooth curve through the points.

This is one of the most accurate methods used to form ellipses.

C.31 □ To construct an ellipse, parallelogram method (Fig. C.29)

Given the major axis AB and the minor axis CD. Construct the circumscribing parallelogram. Divide AO and AE into the same number of equal parts (say, four) and number the division points from A. From C draw a line through point 3 on line AE, and from D draw a line through point 3 on line AO. The point of intersection of these lines is on the required ellipse. Similarly, the intersections of lines from C and D through points numbered 1 and 2 are on the ellipse. A similar construction will locate points in the other three quadrants of the ellipse. Use of a French curve will permit a smooth curve to be drawn through the points.

Had the circumscribing parallelogram not been a rectangle as in Fig. C.29, the completed construction would appear as in Fig.

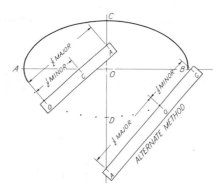

Fig. C.27 □ To construct an ellipse, trammel method.

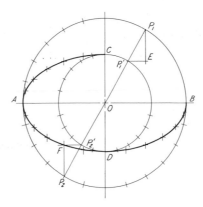

Fig. C.28 □ To construct an ellipse, concentric circle method.

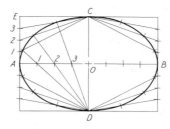

Fig. C.29 □ To construct an ellipse, parallelogram method.

C.30, and AB and CD would be conjugate axes. To establish the major and minor axes, draw a semicircle on CD as a diameter, intersecting the ellipse at E. FG, running parallel to CE through the center of the ellipse, will be the required minor axis. HK, running through the center of the ellipse parallel to DE and perpendicular to FG, will be the major axis.

C.32 □ To draw a tangent to an ellipse at any given point (Fig. C.31)

Given any point, such as P, on the perimeter of the ellipse $ABCD$. Using C as a center and a radius equal to OA (one-half the major diameter), strike arcs across the major axis at F_1 and F_2. From these points, which are foci of the ellipse, draw F_1P and F_2G. The bisector of the angle GPF_1 is the required tangent to the ellipse.

C.33 □ To draw a tangent to an ellipse from a given point P outside the ellipse (Fig. C.32)

With the end of the minor axis as a center and a radius R equal to one-half the length of the major axis, strike an arc to find the foci F_1 and F_2. With point P as a center and the distance PF_2 as a radius, draw an arc. Using F_1 as a center and the length AB as a radius, strike arcs cutting the arc with center P at points G and H. Draw lines GF_1 and HF_1 to establish the location of the tangent points T_1 and T_2. Draw the required tangent.

C.34 □ The parabola (Fig. C.25)

Mathematically the parabola is a curve generated by a point moving so that at any position its distance from a fixed point (the focus) is always exactly equal to its distance to a fixed line (the directrix). The construction shown in Fig. C.33 is based on this definition.

In engineering design, the parabola is used for parabolic sound and light reflectors, for vertical curves on highways, and for bridge arches.

C.35 □ To construct a parabola (Fig. C.33)

Given the focus F and the directrix AB. Draw the axis of the parabola perpendicular to the directrix. Through any point on the axis, for example, point C, draw a line parallel to the directrix AB. Using F as a center and the distance OC as a radius, strike arcs intersecting the line at points P_4 and P_4'. Repeat this procedure until a sufficient number of additional points have been located to determine a smooth curve. The vertex V is located at a point midway between O and F.

To construct a tangent to a parabola, say, at point P_6, draw the line P_6D parallel to the axis; then bisect the angle DP_6F. The bisector of the angle is the required tangent.

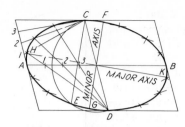

Fig. C.30 □ To draw the major and minor axes of an ellipse, given the conjugate diameters.

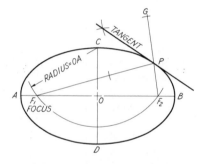

Fig. C.31 □ To draw a tangent to an ellipse.

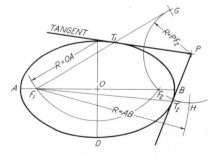

Fig. C.32 □ To draw a tangent to an ellipse through a point outside the ellipse.

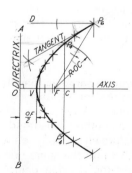

Fig. C.33 □ To construct a parabola.

C.36 □ To construct a parabola, tangent method (Fig. C.34)
Given the points A and B and the distance CD from AB to the vertex. Extend the axis CD, and set off DE equal to CD. EA and EB are tangents to the parabola at A and B, respectively.

Divide EA and EB into the same number of equal parts (say, six), and number the division points as shown. Connect the corresponding points 1 and 1, 2 and 2, 3 and 3, and so forth. These lines, as tangents of the required parabola, form its envelope. Draw the tangent curve.

C.37 □ To construct a parabola, offset method (Fig. C.35)
Given the enclosing rectangle $A'ABB'$. Divide VA' into any number of equal parts (say, ten) and draw from the division points the perpendiculars parallel to VC, along which offset distances are to be laid off. The offsets vary as the square of their distances from V. For example, since V to 2 is two-tenths of the distance from V to A', $2-2'$ will be $(.2)^2$ or .04 of $A'A$. Similarly, $6-6'$ will be $(.6)^2$ or .36 of $A'A$; and $8-8'$ will be .64 of $A'A$. To complete the parabola, lay off the computed offset values along the perpendiculars and form the figure with a French curve.

The entire construction can be done graphically (as illustrated) by first calculating the values for the squared distances and then dividing the depth distance (along the axis) proportionally using these values. The graphical method shown in Fig. C.3 was used.

The offset method is preferred by civil engineers for laying out parabolic arches and computing vertical curves for highways. The parabola shown in Fig. C.35 could represent a parabolic reflector.

C.38 □ To construct a curve of parabolic form through two given points (Fig. C.36)
Given the points A and B. Assume a point C. Draw the tangents CA and CB, and construct the parabolic curve using the tangent method shown in Fig. C.34. This method is frequently used in machine design to draw curves that are more pleasing than circular arcs.

C.39 □ To construct a parabola, parallelogram method
Since the dimensions for a parallelogram that will enclose a given parabola are generally known, a parabola may be constructed by the parallelogram method illustrated in Fig. C.37. With the enclosing rectangle drawn to the given width and depth dimensions, divide VA and AB into the same number of equal divisions (say, five) and number the division points as shown in Fig. C.37(a). Draw light construction lines from point V to each of the division points along AB. Then draw lines parallel to the axis from points 1, 2, 3, and 4 on VA. The intersection of the construction lines from points numbered 1 is on the parabola.

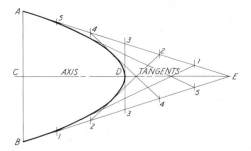

Fig. C.34 □ To construct a parabola, tangent method.

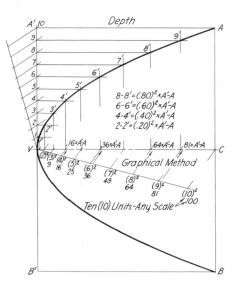

Fig. C.35 □ To construct a parabola, offset method.

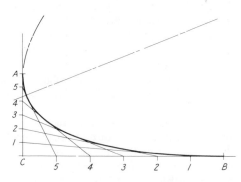

Fig. C.36 □ To construct a curve of parabolic form.

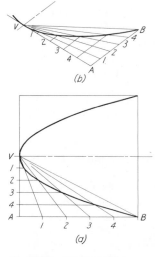

Fig. C.37 □ To construct a parabola, parallelogram method.

Likewise, the intersection of the lines from points numbered 2 is on the parabola. The complete parabolic outline passes through the additional points at the intersection of the lines from points numbered 3 and the lines from points numbered 4. The method as explained for a parabola enclosed in a rectangle may be applied to a nonrectangular shape, as shown in Fig. C.37(*b*).

C.40 □ To locate the directrix and focus of a given parabolic curve (Fig. C.38)

With the location of the axis known [see (*a*)] draw the tangent PY by locating point Y on the axis extended at a distance D_1 from the vertex V. Draw PX parallel to the axis and construct angle FPY equal to YPX. The point at which the line forming the newly constructed angle cuts the given axis is the focus F. The directrix through O is perpendicular to the axis at the same distance from the vertex V as the focus F.

When the position of the axis is not known the procedure illustrated in (*b*) will establish the location of the axis, focus, and directrix. As the initial step, draw two parallel chords at random and locate the midpoint of each, points R and S. Draw line RS to establish the direction of the axis that will be parallel to RS. Next, draw a chord perpendicular to RS at any location and through the midpoint M draw the axis of the parabola as required. The position of the vertex is now known and one might follow the procedure given in (*a*) to locate the focus and directrix. However, the position of the directrix may be found easily and quickly merely by drawing a tangent to the parabola at 45° with the axis. The directrix passes through O at the intersection of the tangent and the axis extended.

C.41 □ The hyperbola (Fig. C.25)

Mathematically the hyperbola can be described as a curve generated by a point moving so that at any position the difference of its distances from two fixed points (foci) is a constant (equal to the transverse axis of the hyperbola). This definition is the basis for the construction shown in Fig. C.39.

C.42 □ To construct a hyperbola (Fig. C.39)

Given the foci F_1 and F_2, and the transverse axis AB. Using F_1 and F_2 as centers, and any radius R_1 greater than F_1B, strike arcs. With these same centers and a radius equal to $R_1 - AB$, strike arcs intersecting the first arcs. These intersecting arcs establish the positions of four symmetrically located points (P_1, P_2, P_3, and P_4) using a single pair of radii. Additional sets of four points are obtained by assuming a different initial radius each time. Repeat this procedure, as outlined, until a sufficient number of points have been located to determine a smooth curve.

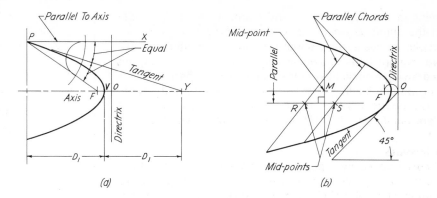

(a)   (b)

Fig. C.38 □ To locate the directrix and focus of a given parabolic curve.

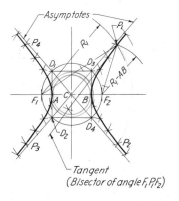

Fig. C.39 □ To construct a hyperbola.

The tangent to the hyperbola at any point, such as P_1, is the bisector of the angle between the focal radii F_1P_1 and F_2P_1.

As hyperbolic curves extend toward infinity they gradually approach two straight lines known as *asymptotes*. These may be located by drawing a circle having the distance F_1–F_2 as a diameter and erecting perpendiculars to the transverse axis through points A and B. The points at which these perpendicular lines intersect the circle are points (D_1, D_2, D_3, and D_4) on the asymptotes.

C.43 □ An involute

The spiral curve traced by a point on a chord as it unwinds from around a circle or a polygon is an *involute curve*. Figure C.40 (a) shows an involute of a circle, while (b) shows that of a square. The involute of a polygon is obtained by extending the sides and drawing arcs using the corners, in order, as centers. The circle in (a) may be considered to be a polygon having an infinite number of sides.

C.44 □ To draw an involute of a circle [Fig. C.40(a)]

Divide the circumference into a number of equal parts. Draw tangents through the division points. Then, along each tangent, lay off the rectified length of the corresponding circular arc, from the starting point to the point of tangency. The involute curve is a smooth curve through these points. The involute of a circle is used in the development of tooth profiles in gearing.

C.45 □ To draw the involute of a polygon [Fig. C.40(b)]

Extend the sides of the polygon as shown in (b). With the corners as centers, in order around the polygon, draw arcs terminating on the extended sides. The first radius is equal to the length of one side of the polygon. The radius of each successive arc is the distance from the center to the terminating point of the previous arc.

C.46 □ A cycloid

A cycloid is the curve generated by a point on the circumference of a moving circle when the circle rolls in a plane along a straight line, as shown in Fig. C.41.

C.47 □ To draw a cycloid (Fig. C.41)

Draw the generating circle and the line AB tangent to it. The length AB should be made equal to the circumference of the circle. Divide the circle and the line AB into the same number of equal parts. With this much of the construction completed, the next step is to draw the line of centers CD through point O and project the division points along AB to CD by drawing perpendiculars. Using

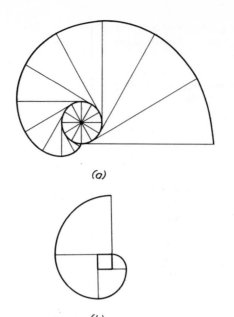

(a)

(b)

Fig. C.40 □ The involute.

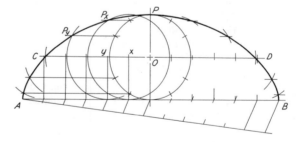

Fig. C.41 □ A cycloid.

these points as centers for the various positions of the moving circle, draw circle arcs. For the purpose of illustration, assume the circle is moving to the left. When the circle has moved along CD to x, point P will have moved to point P_x. Similarly, when the center is at y, P will be at P_y. To locate positions of P along the cycloidal curve, project the division points of the divided circle in their proper order, across to the position circles. A smooth curve through these points will be the required cycloid.

C.48 □ An epicycloid (Fig. C.42)

An epicycloid is the curve generated by a point on the circumference of a circle that rolls in a plane on the outside of another circle. The method used in drawing an epicycloid is similar to the one used in drawing the cycloid.

C.49 □ A hypocycloid (Fig. C.43)

A hypocycloid is the curve generated by a point on the circumference of a circle that rolls in a plane on the inside of another circle. The method used to draw a hypocycloid is similar to the method used to draw the cycloid.

Additional information on the use of cycloidal curves to form the outlines of cycloidal gear teeth may be found in Chapter 10.

C.50 □ Spiral of Archimedes

Archimedes' spiral is a plane curve generated by a point moving uniformly around and away from a fixed point. In order to define this curve more specifically, it can be said that it is generated by a point moving uniformly along a straight line while the line revolves with uniform angular velocity about a fixed point.

The definition of the spiral of Archimedes is applied in drawing this curve as illustrated in Fig. C.44. To find a sufficient number of points to allow the use of an irregular curve for drawing the spiral it is the practice to divide the given circle into a number of equal parts (say, 12) and draw radial lines to the division points. Next, divide a radial line into the same number of equal parts as the circle and number the division points on the circumference of the circle beginning with the radial line adjacent to the divided one. With the center of the circle as a center draw concentric arcs that in each case will start at a numbered division point on the divided radial line and will end at an intersection with the radial line that is numbered correspondingly. The arc starting at point 1 gives a point on the curve at its intersection with radial line 1, the arc starting at 2 gives an intersection point on radial line 2, etc. The spiral is a smooth curve drawn through these intersection points.

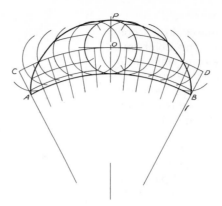

Fig. C.42 □ An epicycloid.

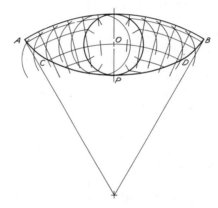

Fig. C.43 □ A hypocycloid.

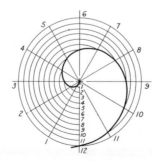

Fig. C.44 □ Spiral of Archimedes.

C.51 □ The helix (Fig. C.45)

The cylindrical helix is a space curve that is generated by a point moving uniformly on the surface of a cylinder. The point must travel parallel to the axis with uniform linear velocity while at the same time it is moving with uniform angular velocity around the axis. The curve can be thought of as being generated by a point moving uniformly along a straight line while the line is revolving with uniform angular velocity around the axis of the given cylinder. Study the pictorial drawing, Fig. C.45.

The first step in drawing a cylindrical helix is to lay out the two views of the cylinder. Next, the lead should be measured along a contour element and divided into a number of equal parts (say, 12). Divide the circular view of the cylinder into the same number of parts and number the division points.

The division lines of the lead represent the various positions of the moving point as it travels in a direction parallel to the axis of the cylinder along the moving line. The division points on the circular view are the related position of the moving line. For example, when the line has moved from the 0 to the 1 position, the point has traveled along the line a distance equal to one twelfth of the lead; when the line is in the 2 position, the point has traveled one-sixth of the lead. (See pictorial drawing, Fig. C.45.) In constructing the curve the necessary points are found by projecting from a numbered point on the circular view to the division line of the lead that is numbered similarly.

A helix may be either right-hand or left-hand. The one shown in Fig. C.45 is a left-hand helix.

When the cylinder is developed, the helix becomes a straight line on the development, as shown. It is inclined to the base line at an angle known as the "helix angle." A screw thread is an example of a practical application of the cylindrical helix.

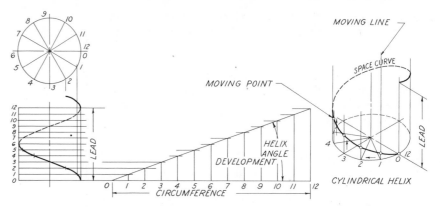

Fig. C.45 □ The helix.

Geometry of developments

D.1 □ Introduction

A layout of the complete surface of an object is called a *development* or *pattern*. The development of an object bounded by plane surfaces may be thought of as being obtained by turning the object, as illustrated in Figs. D.1 and D.2, so as to unroll the imaginary enclosing surface upon a plane. Practically, the drawing operation consists of drawing the successive surfaces in their true size with their common edges joined.

The surfaces of cones and cylinders also may be unrolled upon a plane. The development of a right cylinder (Fig. D.3) is a rectangle having a width equal to the altitude of the cylinder and a length equal to the cylinder's computed circumference (πd). The development of a right circular cone (Fig. D.4) is a sector of a circle having a radius equal to the slant height of the cone and an arc length equal to the circumference of its base.

Warped and double-curved surfaces cannot be developed accurately, but they may be developed by some approximate method. Ordinarily, an approximate pattern will prove to be sufficiently accurate for practical purposes if the material of which the piece is to be made is somewhat flexible.

Plane and single-curved surfaces (prisms, pyramids, cylinders, and cones), which can be accurately developed, are said to be developable. Warped and double-curved surfaces, which can be only approximately developed, are said to be nondevelopable.

D.2 □ Practical developments

On many industrial drawings, a development must be shown to furnish the necessary information for making a pattern to facilitate the cutting of a desired shape from sheet metal. Because of the rapid advance of the art of manufacturing an ever-increasing number of pieces by folding, rolling, or pressing cut sheet-metal shapes, one must have a broad knowledge of the methods of constructing varied types of developments. Patterns also are used in stone cutting as guides for shaping irregular faces.

A development of a surface should be drawn with the inside face up, as it theoretically would be if the surface were unrolled or unfolded as illustrated in Figs. D.1–D.4. This practice is further justified because sheet-metal workers must make the necessary punch marks for folding on the inside surface.

Although in actual sheet-metal work extra metal must be allowed for lap at seams, no allowance will be shown on the developments that follow. Many other practical considerations have been purposely ignored, as well.

D.3 □ To develop a right truncated prism

Before the development of the lateral surface of a prism can be drawn, the true lengths of the edges and the true size of a right section must be determined. In the right truncated prism, shown in Fig. D.5, the true lengths of the prism edges are shown in the front view and the true size of the right section is shown in the top view.

The lateral surface is "unfolded" by first drawing a "stretch-out line" and marking off the widths of the faces (distances 1–2, 2–3, 3–4, and so on, from the top view) along it in succession. Through these points light construction lines are then drawn perpendicular to the line 1_D1_D, and the length of the respective edge is set off on each by projecting from the front view. When projecting edge lengths to the development, the points should be taken in a clockwise order around the perimeter as indicated by the order of the figures in the top view. The outline of the development is completed by joining these points. Thus far, nothing has been said about the lower base or the inclined upper face. These may be joined to the development of the lateral surface, if so desired.

In sheet-metal work, it is usual practice to make the seam on the shortest element in order to save time and conserve solder or rivets.

D.4 □ To develop an oblique prism

The lateral surface of an oblique prism, such as the one shown in Fig. D.6, is developed by the same general method used for a right prism. Similarly, the true lengths of the edges are shown in the front view, but it is necessary to find the true size of the right section by auxiliary plane construction. The widths of the faces, as taken from the auxiliary right section, are set off along the stretch-out line, and perpendicular construction lines repre-

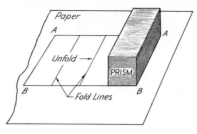

Fig. D.1 □ **The development of a prism.**

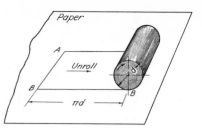

Fig. D.3 □ **The development of a cylinder.**

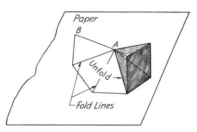

Fig. D.2 □ **The development of a pyramid.**

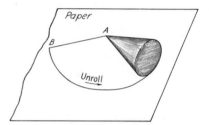

Fig. D.4 □ **The development of a cone.**

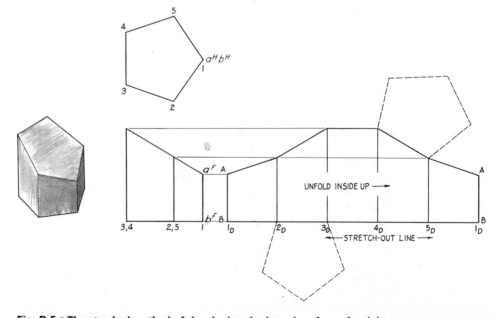

Fig. D.5 □ **The standard method of developing the lateral surface of a right prism.**

350

senting the edges are drawn through the division points. The lengths of the portions of each respective edge, above and below plane XX, are transferred to the corresponding line in development. Distances above plane XX are laid off above the stretch-out line, and distances below XX are laid off below it. The development of the lateral surface is then completed by joining the end-points of the edges by straight lines. Since an actual fold will be made at each edge line when the prism is formed, it is the usual practice to heavy these edge (fold) lines on the development.

The stretch-out line might well have been drawn in a position perpendicular to the edges of the front view, so that the length of each edge might be projected to the development (as in the case of the right prism).

D.5 □ To develop a right cylinder

When the lateral surface of a right cylinder is rolled out upon a plane, the base develops into a straight line (Fig. D.7). The length of this line, which is equal to the circumference of a right section ($\pi \times$ diam.), may be calculated and laid off as the stretch-out line $1_D 1_D$.

Since the cylinder can be thought of as being a many-sided prism, the development may be constructed in a manner similar to the method illustrated in Fig. D.5. The elements drawn on the surface of the cylinder serve as edges of the many-sided prism. Twelve or twenty-four of these elements ordinarily are used, the number depending upon the size of the cylinder. Usually they are spaced by dividing the circumference of the base, as shown by the circle in the top view, into an equal number of parts. The stretch-out line is divided into the same number of equal parts, and perpendicular elements are drawn through each division point. Then the true length of each element is projected to its respective representation on the development, and the development is completed by joining the points with a smooth curve. In joining the points, it is advisable to sketch the curve in lightly, freehand, before using the French curve. Since the surface of the finished cylindrical piece forms a continuous curve, the elements on the development are not heavied. When the development is symmetrical, as in this case, only one-half need be drawn.

D.6 □ To develop an oblique cylinder

Since an oblique cylinder theoretically may be thought of as enclosing a regular oblique prism having an infinite number of sides, the development of the lateral surface of the cylinder shown in Fig. D.8 may be constructed by using a method similar to the method illustrated in Fig. D.6. The circumference of the right section becomes stretch-outline $1_D 1_D$ for the development.

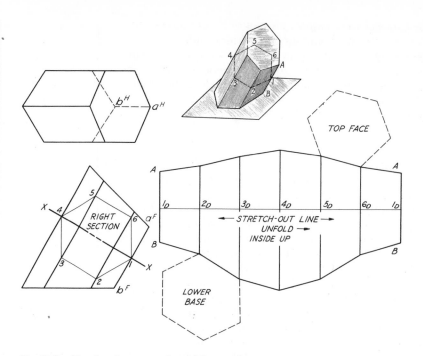

Fig. D.6 □ The development of an oblique prism.

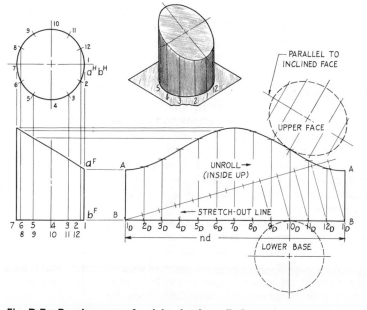

Fig. D.7 □ Development of a right circular cylinder.

D.7 □ To determine the true length of a line

In order to construct the development of the lateral surface of some objects, it frequently is necessary to determine the true lengths of oblique lines that represent the edges. The general method for determining the true lengths of lines inclined to all of the coordinate planes of projection has been explained in detail in Sec. 9.21. This article should be reviewed before reading the discussion that follows.

D.8 □ True-length diagrams

When it is necessary in developing a surface to find the true lengths of a number of edges or elements, some confusion may be avoided by constructing a true-length diagram adjacent to the orthographic view as shown in Fig. D.9. The elements were revolved into a position parallel to the F (frontal) plane so that their true lengths show in the diagram. This practice prevents the front view in the illustration from being cluttered with lines, some of which would represent elements and others their true lengths.

  Figure D.11 shows a diagram that gives the true lengths of the edges of the pyramid. Each line representing the true length of an edge is the hypotenuse of a right triangle whose altitude is the altitude of the edge in the front view and whose base is equal to the length of the projection of the edge in the top view. The lengths of the top projections of the edges of the pyramid are laid off horizontally from the vertical line, which could have been drawn at any distance from the front view. Since all the edges have the same altitude, this line is a common vertical leg for all the right triangles in the diagram. The true-length diagram shown in Fig. D.9 could very well have been constructed by this method.

D.9 □ To develop a right pyramid

To develop (unfold) the lateral surface of a right pyramid, it is first necessary to determine the true lengths of the edges and the true size of the base. With this information, the development can be constructed by laying out the faces in successive order with their common edges joined. If the surface is imagined to be unfolded by turning the pyramid, as shown in Fig. D.2, each triangular face is revolved into the plane of the paper about the edge that is common to it and the preceding face.

  Since the edges of the pyramid shown in Fig. D.10 are all equal in length, it is necessary only to find the length of the one edge $A1$ by revolving it into the position a^F1r. The edges of the base, 1–2, 2–3, and so on, are parallel to the horizontal plane of projection and consequently show in their true length in the top view. With this information, the development is easily completed by constructing the four triangular surfaces.

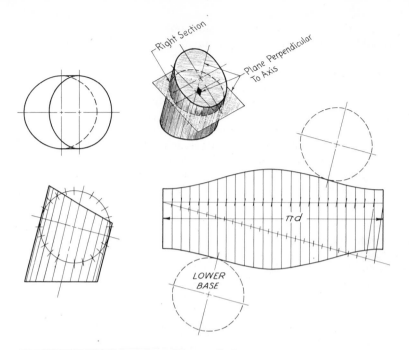

Fig. D.8 □ Development of an oblique cylinder.

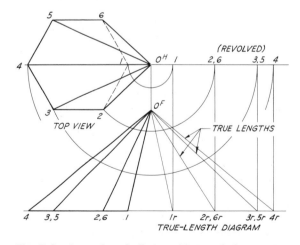

Fig. D.9 □ A true-length diagram (the revolution method).

D.10 □ To develop the surface of a frustum of a pyramid

To develop the lateral surface of the frustum of a pyramid (Fig. D.11), it is necessary to determine the true lengths of edges of the complete pyramid as well as the true lengths of edges of the frustum. The desired development is obtained by first constructing the development of the complete pyramid and then laying off the true lengths of the edges of the frustum on the corresponding lines of the development.

It may be noted with interest that the true length of the edge $B3$ is equal to the length $b'3'$ on the true-length line a^F3', and that the location of point b' can be established by the short-cut method of projecting horizontally from point b^F. Point b' on a^F3' is the true revolved position of point B, because the path of point B is in a horizontal plane that projects as a line in the front view.

D.11 □ To develop a right cone

As previously explained in Sec. D.1, the development of a regular right circular cone is a sector of a circle. The development will have a radius equal to the slant height of the cone and an included angle at the center equal to $(r/s) \times 360°$ (Fig. D.12). In this equation, r is the radius of the base and s is the slant height.

D.12 □ To develop a right truncated cone

The development of a right truncated cone must be constructed by a modified method of triangulation, in order to develop the outline of the elliptical inclined surface. This commonly used method is based upon the theoretical assumption that a cone is a pyramid having an infinite number of sides. The development of the incomplete right cone shown in Fig. D.13 is constructed upon a layout of the whole cone by a method similar to the standard method illustrated for the frustum of a pyramid in Fig. D.11.

Elements are drawn on the surface of the cone to serve as edges of the many-sided pyramid. Either twelve or twenty-four are used, depending upon the size of the cone. Their location is established upon the developed sector by dividing the arc representing the unrolled base into the same number of equal divisions, into which the top view of the base has been divided. At this point in the procedure, it is necessary to determine the true lengths of the elements of the frustum in the same manner that the true lengths of the edges of the frustum of a pyramid were obtained in Fig. D.11. With this information, the desired development can be completed by setting off the true lengths on the corresponding lines of the development and joining the points thus obtained with a smooth curve.

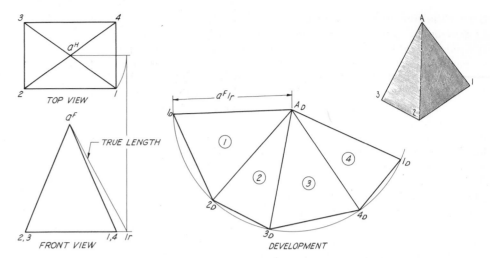

Fig. D.10 □ The development of a rectangular right pyramid.

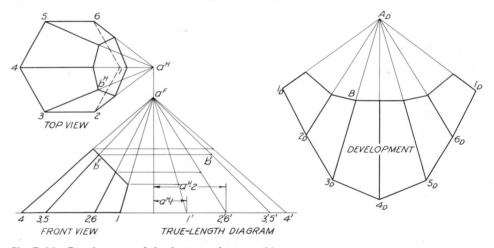

Fig. D.11 □ Development of the frustum of a pyramid.

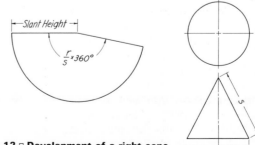

Fig. D.12 □ Development of a right cone.

D.13 □ The triangulation method
of developing approximately developable surfaces

A nondevelopable surface may be developed approximately if the surface is assumed to be composed of a number of small developable surfaces (Fig. D.14). The particular method ordinarily used for warped surfaces and the surfaces of oblique cones is known as the triangulation method. The procedure consists of completely covering the lateral surface with numerous small triangles that will lie approximately on the surface (Fig. D.15). These triangles, when laid out in their true size with their common edges joined, produce an approximate development that is accurate enough for most practical purposes.

Although this method of triangulation is sometimes used to develop the lateral surface of a right circular cone, it is not recommended for such a purpose. The resulting development is not as accurate as it would be if constructed by one of the standard methods (see Secs. D.11 and D.12).

D.14 □ To develop an oblique cone
using the triangulation method

A development of the lateral surface of an oblique cone is constructed by a method similar to that used for an oblique pyramid. The surface is divided into a number of unequal triangles having sides that are elements on the cone and bases that are the chords of short arcs of the base.

The first step in developing an oblique cone (Fig. D.16) is to divide the circle representing the base into a convenient number of equal parts and draw elements on the surface of the cone through the division points (1, 2, 3, 4, 5, and so on). To construct the triangles forming the development, it is necessary to know the true lengths of the elements (sides of the triangles) and chords. In the illustration, all the chords are equal. Their true lengths are shown in the top view. The true lengths of the oblique elements may be determined by one of the standard methods explained in Sec. D.8.

Since the seam should be made along the shortest element, $A1$ will lie on the selected starting line for the development and $A7$ will be on the center line. To obtain the development, the triangles are constructed in order, starting with the triangle A–1–2 and proceeding around the cone in a clockwise direction (as shown by the arrow in the top view). The first step in constructing triangle A–1–2 is to set off the true length $a^F 1'$ along the starting line. With point A_D of the development as a center, and with a radius equal to a $a^F 2'$, strike an arc; then, with point 1_D as a center, and with a radius equal to the chord 1–2, strike an arc across the first arc to locate point 2_D. The triangle $A_D 2_D 3_D$ and the

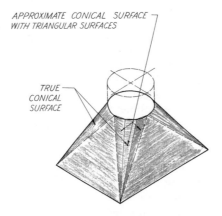

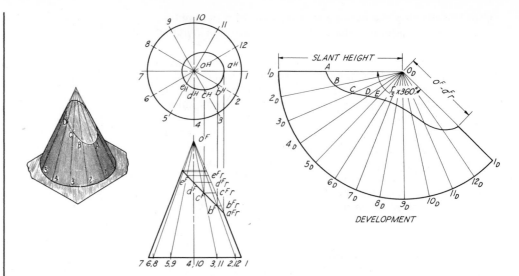

Fig. D.13 □ Development of a truncated cone.

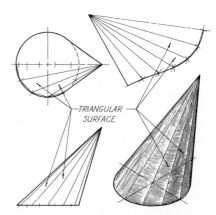

Fig. D.14 □ Triangulation of a surface.

Fig. D.15 □ Triangulation of an oblique cone.

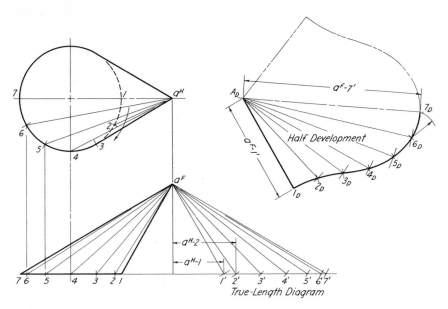

Fig. D.16 □ Development of an oblique cone.

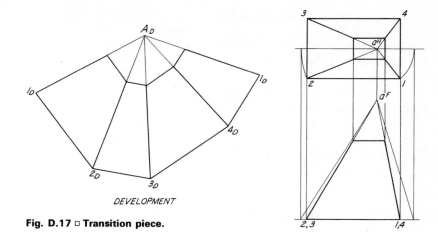

DEVELOPMENT

Fig. D.17 □ Transition piece.

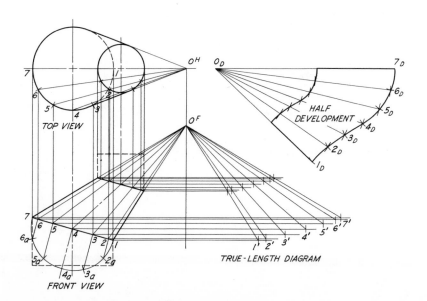

Fig. D.18 □ Transition piece connecting two pipes.

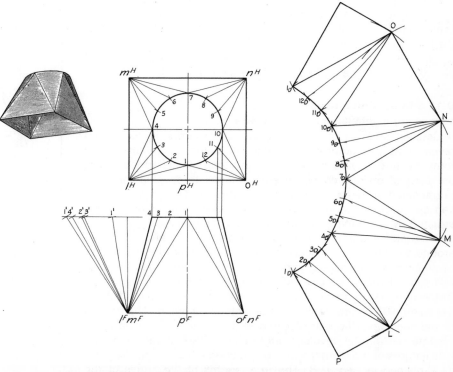

Fig. D.19 □ Transition piece connecting a circular and square pipe.

remaining triangles are formed in exactly the same manner. When all the triangles have been laid out, the development of the whole conical surface is completed by drawing a smooth curve through the end-points of the elements.

D.15 □ To develop a transition piece
connecting rectangular pipes

The transition piece shown in Fig. D.17 is designed to connect two rectangular pipes of different sizes on different axes. Since the piece is a frustum of a pyramid, it can be accurately developed by the method explained in Sec. D.10.

D.16 □ To develop a transition piece
connecting two circular pipes

The transition piece shown in Fig. D.18 connects two circular pipes on different axes. Since the piece is a frustum of an oblique cone, the surface must be triangulated, as explained in Sec. D.14, and the development must be constructed by laying out the triangles in their true size in regular order. The general procedure is the same as that illustrated in Fig. D.16. In this case, however, since the true size of the base is not shown in the top view, it is necessary to construct a partial auxiliary view to find the true lengths of chords between the end-points of the elements.

D.17 □ To develop a transition piece
connecting a circular and a square pipe

A detailed analysis of the transition piece shown in Fig. D.19 reveals that it is composed of four isosceles triangles whose bases form the square base of the piece and four conical surfaces that are parts of oblique cones. It is not difficult to develop this type of transition piece because, since the whole surface may be "broken up" into component surfaces, the development may be constructed by developing the first and then each succeeding component surface separately (Fig. D.14). The surfaces are developed around the piece in a clockwise direction, in such a manner that each successive surface is joined to the preceding surface at their common element. In the illustration, the triangles 1LO, 4LM, 7MN, and 10NO are clearly shown in the top view. Two of these 1LO and 10NO, are visible on the pictorial drawing. The apexes of the conical surfaces are located at the corners of the base.

Before starting the development, it is necessary to determine the true lengths of the elements by constructing a true-length diagram as explained in Sec. D.8. The true lengths of the edges of the lower base (LM, MN, NO, and OL) and the true lengths of the chords (1–2, 2–3, 3–4, and so on) of the short arcs of the upper base are shown in the top view. The development is constructed

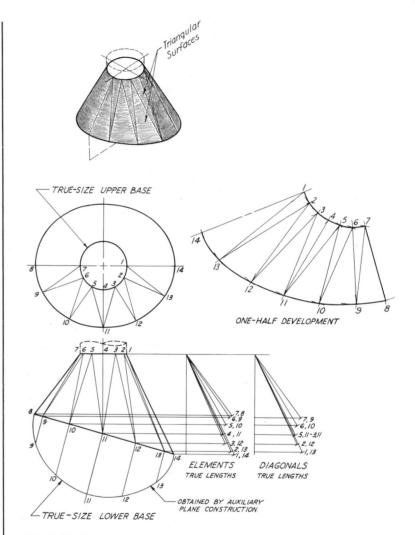

Fig. D.20 □ Development of transition piece by triangulation.

in the following manner: First, the triangle $1_D PL$ is constructed, using the length $p^H I^H$ taken from the top view, and true lengths from the diagram. Next, using the method explained in Sec. D.14, the conical surface whose apex is at L is developed in an attached position. Triangle $4_D LM$ is then added, and so on, until all component surfaces have been drawn.

D.18 □ To develop a transition piece having an approximately developable surface by the triangulation method

Figure D.20 shows a half development of a transition piece that has a warped surface instead of a partially conical one like that discussed in Sec. D.17. The method of constructing the development is somewhat similar, however, in that it is formed by laying out, in true size, a number of small triangles that approximate the surface. The true size of the circular intersection is shown in the top view, and the true size of the elliptical intersection is shown in the auxiliary view, which was constructed for that purpose.

The front half of the circle in the top view should be divided into the same number of equal parts as the half-auxiliary view. By joining the division points, the lateral surface may be initially divided into narrow quadrilaterals. These in turn may be subdivided into triangles by drawing diagonals which, though theoretically they are curved lines, are assumed to be straight. The true lengths of the elements and the diagonals are found by constructing two separate true-length diagrams by the method illustrated in Fig. D.11.

D.19 □ To develop a sphere

The surface of a sphere is a double-curved surface that can be developed only by some approximate method. The standard methods commonly used are illustrated in Fig. D.21.

In (a) the surface is divided into a number of equal meridian sections of cylinders. The developed surfaces of these form an approximate development of the sphere. In drawing the development it is necessary to develop the surface of only one section, for this can be used as a pattern for the developed surface of each of the others.

In (b) the sphere is cut by parallel planes, which divide it into a number of horizontal sections, the surfaces of which approximate the surface of the sphere. Each of these sections may be considered the frustum of a right cone whose apex is located at the intersection of the chords extended.

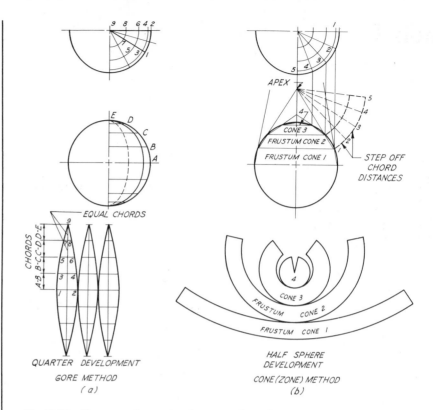

Fig. D.21 □ The approximate development of a sphere.

Appendix E

Tables

Table 1 □ Inch-millimeter equivalents

Common fractions					Decimal			Millimeters
4ths	8ths	16ths	32nds	64ths	To 4 places	To 3 places	To 2 places	To 4 places
				$\frac{1}{64}$	.0156	.016	.02	0.3969
			$\frac{1}{32}$		.0312	.031	.03	0.7938
				$\frac{3}{64}$	.0469	.047	.05	1.1906
		$\frac{1}{16}$			.0625	.062	.06	1.5875
				$\frac{5}{64}$	.0781	.078	.08	1.9844
			$\frac{3}{32}$		.0938	.094	.09	2.3813
				$\frac{7}{64}$	.1094	.109	.11	2.7781
	$\frac{1}{8}$				.1250	.125	.12	3.1750
				$\frac{9}{64}$	.1406	.141	.14	3.5719
			$\frac{5}{32}$		.1562	.156	.16	3.9688
				$\frac{11}{64}$	.1719	.172	.17	4.3656
		$\frac{3}{16}$			.1875	.188	.19	4.7625
				$\frac{13}{64}$	.2031	.203	.20	5.1594
			$\frac{7}{32}$		.2188	.219	.22	5.5563
				$\frac{15}{64}$	.2344	.234	.23	5.9531
$\frac{1}{4}$					.2500	.250	.25	6.3500
				$\frac{17}{64}$	.2656	.266	.27	6.7469
			$\frac{9}{32}$		.2812	.281	.28	7.1438
				$\frac{19}{64}$	.2969	.297	.30	7.5406
		$\frac{5}{16}$			.3125	.312	.31	7.9375
				$\frac{21}{64}$	.3281	.328	.33	8.3344
			$\frac{11}{32}$		.3438	.344	.34	8.7313
				$\frac{23}{64}$	.3594	.359	.36	9.1281
	$\frac{3}{8}$				.3750	.375	.38	9.5250
				$\frac{25}{64}$	.3906	.391	.39	9.9219
			$\frac{13}{32}$		.4062	.406	.41	10.3188
				$\frac{27}{64}$	.4219	.422	.42	10.7156
		$\frac{7}{16}$			.4375	.438	.44	11.1125
				$\frac{29}{64}$	.4531	.453	.45	11.5094
			$\frac{15}{32}$		.4688	.469	.47	11.9063
				$\frac{31}{64}$	.4844	.484	.48	12.3031
					.5000	.500	.50	12.7000

Table 1 □ (cont.) Inch-millimeter equivalents

Common fractions					Decimal			Millimeters
4ths	8ths	16ths	32nds	64ths	To 4 places	To 3 places	To 2 places	To 4 places
				$^{33}/_{64}$	.5156	.516	.52	13.0969
			$^{17}/_{32}$		.5312	.531	.53	13.4938
				$^{35}/_{64}$	.5469	.547	.55	13.8906
		$^{9}/_{16}$			.5625	.562	.56	14.2875
				$^{37}/_{64}$	.5781	.578	.58	14.6844
			$^{19}/_{32}$		.5938	.594	.59	15.0813
				$^{39}/_{64}$	.6094	.609	.61	15.4781
	$^{5}/_{8}$				.6250	.625	.62	15.8750
				$^{41}/_{64}$	.6406	.641	.64	16.2719
			$^{21}/_{32}$		.6562	.656	.66	16.6688
				$^{43}/_{64}$	.6719	.672	.67	17.0656
		$^{11}/_{16}$			.6875	.688	.69	17.4625
				$^{45}/_{64}$	.7031	.703	.70	17.8594
			$^{23}/_{32}$		.7188	.719	.72	18.2563
				$^{47}/_{64}$	.7344	.734	.73	18.6531
$^{3}/_{4}$					.7500	.750	.75	19.0500
				$^{49}/_{64}$	.7656	.766	.77	19.4469
			$^{25}/_{32}$		.7812	.781	.78	19.8438
				$^{51}/_{64}$	.7969	.797	.80	20.2406
		$^{13}/_{16}$			.8125	.812	.81	20.6375
				$^{53}/_{64}$	.8281	.828	.83	21.0344
			$^{27}/_{32}$		.8438	.844	.84	21.4313
				$^{55}/_{64}$	.8594	.859	.86	21.8281
	$^{7}/_{8}$				.8750	.875	.88	22.2250
				$^{57}/_{64}$	.8906	.891	.89	22.6219
			$^{29}/_{32}$		.9062	.906	.91	23.0188
				$^{59}/_{64}$	.9219	.922	.92	23.4156
		$^{15}/_{16}$			.9375	.938	.94	23.8125
				$^{61}/_{64}$	.9531	.953	.95	24.2094
			$^{31}/_{32}$		.9688	.969	.97	24.6063
				$^{63}/_{64}$	.9844	.984	.98	25.0031
					1.0000	1.000	1.00	25.4000

Table 2 □ Inches to millimeters*

Inches

Inches	0	1	2	3	4	5	6	7	8	9
0–9	0	25.4	50.8	76.2	101.6	127.0	152.4	177.8	203.2	228.6
10–19	254.0	279.4	304.8	330.2	355.6	381.0	406.4	431.8	457.2	482.6
20–29	508.0	533.4	558.8	584.2	609.6	635.0	660.4	685.8	711.2	736.6
30–39	762.0	787.4	812.8	838.2	863.6	889.0	914.4	939.8	965.2	990.6
40–49	1016.0	1041.4	1066.8	1092.2	1117.6	1143.0	1168.4	1193.8	1219.2	1244.6
50–59	1270.0	1295.4	1320.8	1346.2	1371.6	1397.0	1422.4	1447.8	1473.2	1498.6
60–69	1524.0	1549.4	1574.8	1600.2	1625.6	1651.0	1676.4	1701.8	1727.2	1752.6
70–79	1778.0	1803.4	1828.8	1854.2	1879.6	1905.0	1930.4	1955.8	1981.2	2006.6
80–89	2032.0	2057.4	2082.8	2108.2	2133.6	2159.0	2184.4	2209.8	2235.2	2260.6
90–99	2286.0	2311.4	2336.8	2362.2	2387.6	2413.0	2438.4	2463.8	2489.2	2514.6

*Based on 1 in. = 25.4 millimeters

Examples: To obtain the millimeter equivalent of 14 in., read the value below 4 in. in the horizontal line of values for 10–19 in. (355.6). The equivalent value of 52.4 in. is: 1320.8 plus 10.16 = 1330.96, the 10.16 value being obtained by moving the decimal point left in the equivalent value for 4 in.

Table 3 □ Metric equivalents
millimeters to decimal inches*

Mm = in.	Mm = in.	Mm = in.	Mm = in.	Mm = in.
1 = 0.0394	21 = 0.8268	41 = 1.6142	61 = 2.4016	81 = 3.1890
2 = 0.0787	22 = 0.8661	42 = 1.6535	62 = 2.4409	82 = 3.2283
3 = 0.1181	23 = 0.9055	43 = 1.6929	63 = 2.4803	83 = 3.2677
4 = 0.1575	24 = 0.9449	44 = 1.7323	64 = 2.5197	84 = 3.3071
5 = 0.1969	25 = 0.9843	45 = 1.7717	65 = 2.5591	85 = 3.3465
6 = 0.2362	26 = 1.0236	46 = 1.8110	66 = 2.5984	86 = 3.3858
7 = 0.2756	27 = 1.0630	47 = 1.8504	67 = 2.6378	87 = 3.4252
8 = 0.3150	28 = 1.1024	48 = 1.8898	68 = 2.6772	88 = 3.4646
9 = 0.3543	29 = 1.1417	49 = 1.9291	69 = 2.7165	89 = 3.5039
10 = 0.3937	30 = 1.1811	50 = 1.9685	70 = 2.7559	90 = 3.5433
11 = 0.4331	31 = 1.2205	51 = 2.0079	71 = 2.7953	91 = 3.5827
12 = 0.4724	32 = 1.2598	52 = 2.0472	72 = 2.8346	92 = 3.6220
13 = 0.5118	33 = 1.2992	53 = 2.0866	73 = 2.8740	93 = 3.6614
14 = 0.5512	34 = 1.3386	54 = 2.1260	74 = 2.9134	94 = 3.7008
15 = 0.5906	35 = 1.3780	55 = 2.1654	75 = 2.9528	95 = 3.7402
16 = 0.6299	36 = 1.4173	56 = 2.2047	76 = 2.9921	96 = 3.7795
17 = 0.6693	37 = 1.4567	57 = 2.2441	77 = 3.0315	97 = 3.8189
18 = 0.7087	38 = 1.4961	58 = 2.2835	78 = 3.0709	98 = 3.8583
19 = 0.7480	39 = 1.5354	59 = 2.3228	79 = 3.1102	99 = 3.8976
20 = 0.7874	40 = 1.5748	60 = 2.3622	80 = 3.1496	100 = 3.9370

*To nearest fourth decimal place

Table 4 □ Metric tolerance equivalents conversion of tolerances—decimal inch to millimeters

Design practice		Calculated equivalents of decimal inch tolerances*	Design practice		Calculated equivalents of decimal inch tolerances*
Decimal in.	Metric mm		Decimal in.	Metric mm	
0.0001	0.0025	0.00254	0.004	0.10	0.1016
0.0002	0.005	0.00508	0.005	0.13	0.1270
0.0003	0.008	0.00762	0.006	0.15	0.1524
0.0004	0.010	0.01016	0.007	0.18	0.1778
0.0005	0.013	0.01270	0.008	0.20	0.2032
0.0006	0.015	0.01524	0.009	0.23	0.2286
0.0007	0.018	0.01778	0.010	0.25	0.254
0.0008	0.020	0.02032	0.015	0.38	0.381
0.0009	0.023	0.02286	0.02	0.5	0.508
0.001	0.025	0.0254	0.03	0.8	0.762
0.0015	0.038	0.0381	0.04	1.0	1.016
0.002	0.05	0.0508	0.06	1.5	1.524
0.0025	0.06	0.0635	0.08	2.0	2.032
0.003	0.08	0.0762	0.10	2.5	2.540

*Calculated on the basis of 1 in. = 25.4 millimeters
Additional information relating to the conversion of tolerances may be obtained from the SAE Standard-Dual Dimensioning-SAE J390

Table 5 □ English-metric equivalents

Measures of length

1 Meter = 39.37 Inches = 3.281 Feet = 1.094 Yards
1 Centimeter = 0.3937 Inch
1 Millimeter = $\frac{1}{25}$ Inch (approximately)
1 Kilometer = 0.621 Mile
1 Inch = 2.540 Centimeters = 25.400 Millimeters
1 Foot = 0.305 Meter
1 Mile = 1.609 Kilometers

Measures of surface

1 Square Meter = 10.764 Square Feet = 1.196 Square Yards
1 Square Centimeter = 0.155 Square Inch
1 Square Millimeter = 0.00155 Square Inch
1 Square Yard = 0.836 Square Meter
1 Square Foot = 0.0929 Square Meter
1 Square Inch = 6.452 Square Centimeters = 645.2 Square Millimeters

Measures of volume

1 Cubic Meter = 35.314 Cubic Feet = 1.308 Cubic Yards
1 Cubic Decimeter = 61.023 Cubic Inches = 0.0353 Cubic Foot
1 Cubic Centimeter = 0.061 Cubic Inch
1 Liter = 61.223 Cubic Inches = 0.0353 Cubic Foot = 0.2642 (U.S.) Gallons
1 Cubic Foot = 28.317 Cubic Decimeters = 28.317 Liters
1 Cubic Inch = 16.387 Cubic Centimeters

Measures of weight

1 Kilogram = 2.2046 Pounds
1 Metric Ton = 2,204.6 Pounds = .9842 Ton (2240 pounds)
1 Ounce (avoirdupois) = 28.35 Grams
1 Pound = 0.4536 Kilograms

Speed measurements

1 Kilometer Per Hour = 0.621 Miles Per Hour

Table 6 □ English-metric conversion multipliers

To convert from	To	Multiplier
Centigrade (°C)	Fahrenheit (°F)	$\frac{9}{5}(°C) + 32$
Centimeters (CM)	feet	0.0328
Centimeters	inches	0.3937
Centimeters	meters	0.01
Cubic centimeters (CU CM)	cubic feet	0.00003531
Cubic centimeters	cubic inches	0.06102
Cubic centimeters	liters	0.0010
Cubic centimeters	cubic meter	0.0000010
Cubic inches (CU IN)	cubic centimeters	16.3872
Cubic inches	cubic meters	0.000016
Cubic feet	cubic centimeters	28,317.08
Cubic feet	cubic meters	0.0283
Cubic meters	cubic feet	35.3133
Cubic millimeters (CU MM)	cubic centimeters	0.001
Cubic yards	cubic meters	0.7646
Degrees (arc)	radians	0.01745
Fahrenheit (°F)	Centigrade (°C)	$\frac{5}{9}(°F-32)$
Feet (FT)	centimeters	30.4801
Feet	meters	0.3048
Foot-pound-force	joule (J)	1.3558
Foot^2	meter^2	0.0929
Foot^3	meter^3	0.0283
Foot/second	meter/second	0.3048
Gallons (U.S.)	cubic meters	0.0038
Gallons (U.S.)	liters	3.7878
Grams (G)	kilograms	0.0010
Grams	milligrams	1,000
Grams/cubic centimeter	kilograms/cubic meter	1,000
Hectometers (HM)	meters	100
Horsepower (HP)	kilogram-meters/second	76.042
Horsepower	metric horsepower	1.0139
Horsepower, metric	kilogram-meters/second	75.0
Inches (IN.)	centimeters	2.54
Inches	meters	0.0254
Inches^2	meters^2	0.0006452
Inches^3	meters^3	0.00001639
Inches of mercury	grams/square centimeter	34.542

Table 6 □ (cont.) English-metric conversion multipliers

To convert from	To	Multiplier
Kilogram-meters/second	horsepower	0.01305
Kilogram-meters/second	horsepower, metric	0.01333
Kilograms (KG)	grams	1,000
Kilograms	tons	0.0011
Kilograms	tons, metric	0.001
Kilogram-force (KGF)	newton (N)	9.8066
Kiloliters (KL)	liters	1,000
Kilometers (KM)	meters	1,000
Kilometers	statute miles	0.6214
Liters (L)	cubic centimeters	1,000
Liters	cubic feet	0.035313
Liters	cubic inches	61.02398
Liters	U.S. gallons	0.2641
Megameters	meters	100,000
Meters (M)	U.S. miles	0.000622
Microns	meters	0.000001
Miles, statute (MI)	kilometers	1.6093
Miles, statute	meters	1,609.34
Miles/hour (MPH)	kilometers/hour	1.6093
Miles/hour	meters/second	0.4470
Milligrams	grams	0.001
Millimeters (MM)	inches	0.03937
Millimeters	meters	0.001
Millimeters	microns	1,000
Ounces, avoirdupois	grams	28.3495
Ounces, fluid	milliliters	29.57
Ounces, U.S. fluid	liters	0.0296
Pints, liquid	liters	0.4732
Pounds, avoirdupois	grams	453.5924
Poundal	newton (N)	0.1383
Poundal/foot^2	pascal (Pa)	1.4882
Pound-force (avoirdupois)	newton (N)	4.4482
Pound-force-foot	newton-meter (Nm)	1.3558
Pound-force/foot^2	pascal (Pa)	47.88
Pound-force/inch^2 (psi)	pascal (Pa)	6,894.8
Pound-mass (avoirdupois)	kilogram	0.4536
Pound-mass/foot^2	kilogram/meter^2	4.8824
Pound-mass/foot^3	kilogram/meter^3	16.0185
Pound-mass/inch^3	kilogram/meter^3	2,767.99

Table 6 □ (cont.) English-metric conversion multipliers

To convert from	To	Multiplier
Quarts, liquid	liters	0.9464
Quintals	grams	100,000
Radians	degrees, arc	57.2958
Square centimeters	square feet	0.001076
Square centimeters	square inches	0.1550
Square centimeters	square millimeters	100
Square decameters	square meters	100
Square decimeters	square meters	0.01
Square feet	square centimeters	929.0341
Square hectometers	square meters	10,000
Square inches	square centimeters	6.4516
Square inches	square millimeters	645.1625
Square kilometers	hectares	100
Square kilometers	square meters	1,000,000
Square kilometers	square miles	0.3861
Square meters	square feet	10.7639
Square meters	square yards	1.1960
Square miles	square kilometers	2.590
Square millimeters (SQ MM)	square inches	0.00155
Square millimeters	square meters	0.000001
Square yards	square meters	0.8361
Tons, long	kilograms	1,016.0470
Tons, metric	kilograms	1,000
Tons, short	kilograms	907.18
Yards	meters	0.9144

*A complete listing of conversion factors prepared for use with the computer may be found in the ASTM–Metric Practice Guide (E380-72). Definitions of derived units of the International System, such as the joule, the newton, and the pascal listed in this table, will be found in this publication.

Energy—the joule is the meter-kilogram-second unit of work that is equal to the work done by a force of one newton when the point of application is displaced one meter in the direction of the force (Nm).

Force—the newton is the standard meter-kilogram-second unit of force. It is equal to the force that produces an acceleration of one meter per second per second when applied to a body having a mass of one kilogram ($kg \cdot m/s^2$).

Stress or pressure—the pascal is the International System (SI) unit of pressure or stress. The pascal (Pa) is the pressure or stress of one newton per square meter (N/m^2).

Table 7 □ Trigonometric functions

Angle	Sine	Cosine	Tan	Co-Tan	Angle
0°	0.0000	1.0000	0.0000	∞	90°
1°	0.0175	0.9998	0.0175	57.290	89°
2°	.0349	.9994	.0349	28.636	88°
3°	.0523	.9986	.0524	19.081	87°
4°	.0698	.9976	.0699	14.301	86°
5°	.0872	.9962	.0875	11.430	85°
6°	.1045	.9945	.1051	9.5144	84°
7°	.1219	.9925	.1228	8.1443	83°
8°	.1392	.9903	.1405	7.1154	82°
9°	.1564	.9877	.1584	6.3138	81°
10°	.1736	.9848	.1763	5.6713	80°
11°	.1908	.9816	.1944	5.1446	79°
12°	.2079	.9781	.2126	4.7046	78°
13°	.2250	.9744	.2309	4.3315	77°
14°	.2419	.9703	.2493	4.0108	76°
15°	.2588	.9659	.2679	3.7321	75°
16°	.2756	.9613	.2867	3.4874	74°
17°	.2924	.9563	.3057	3.2709	73°
18°	.3090	.9511	.3249	3.0777	72°
19°	.3256	.9455	.3443	2.9042	71°
20°	.3420	.9397	.3640	2.7475	70°
21°	.3584	.9336	.3839	2.6051	69°
22°	.3746	.9272	.4040	2.4751	68°
23°	.3907	.9205	.4245	2.3559	67°
24°	.4067	.9135	.4452	2.2460	66°
25°	.4226	.9063	.4663	2.1445	65°
26°	.4384	.8988	.4877	2.0503	64°
27°	.4540	.8910	.5095	1.9626	63°
28°	.4695	.8829	.5317	1.8807	62°
29°	.4848	.8746	.5543	1.8040	61°
30°	.5000	.8660	.5774	1.7321	60°
31°	.5150	.8572	.6009	1.6643	59°
32°	.5299	.8480	.6249	1.6003	58°
33°	.5446	.8387	.6494	1.5399	57°
34°	.5592	.8290	.6745	1.4826	56°
35°	.5736	.8192	.7002	1.4281	55°
36°	.5878	.8090	.7265	1.3764	54°
37°	.6018	.7986	.7536	1.3270	53°
38°	.6157	.7880	.7813	1.2799	52°
39°	.6293	.7771	.8098	1.2349	51°
40°	.6428	.7660	.8391	1.1918	50°
41°	.6561	.7547	.8693	1.1504	49°
42°	.6691	.7431	.9004	1.1106	48°
43°	.6820	.7314	.9325	1.0724	47°
44°	.6947	.7193	.9657	1.0355	46°
45°	.7071	.7071	1.0000	1.0000	45°
Angle	Cosine	Sine	Co-Tan	Tan	Angle

Table 8 □ Logarithms of numbers

	0	1	2	3	4	5	6	7	8	9	1	2	3	4	5	6	7	8	9
10	0000	0043	0086	0128	0170	0212	0253	0294	0334	0374	4	8	12	17	21	25	29	33	37
11	0414	0453	0492	0531	0569	0607	0645	0682	0719	0755	4	8	11	15	19	23	26	30	34
12	0792	0828	0864	0899	0934	0969	1004	1038	1072	1106	3	7	10	14	17	21	24	28	31
13	1139	1173	1206	1239	1271	1303	1335	1367	1399	1430	3	6	10	13	16	19	23	26	29
14	1461	1492	1523	1553	1584	1614	1644	1673	1703	1732	3	6	9	12	15	18	21	24	27
15	1761	1790	1818	1847	1875	1903	1931	1959	1987	2014	3	6	8	11	14	17	20	22	25
16	2041	2068	2095	2122	2148	2175	2201	2227	2253	2279	3	5	8	11	13	16	18	21	24
17	2304	2330	2355	2380	2405	2430	2455	2480	2504	2529	2	5	7	10	12	15	17	20	22
18	2553	2577	2601	2625	2648	2672	2695	2718	2742	2765	2	5	7	9	12	14	16	19	21
19	2788	2810	2833	2856	2878	2900	2923	2945	2967	2989	2	4	7	9	11	13	16	18	20
20	3010	3032	3054	3075	3096	3118	3139	3160	3181	3201	2	4	6	8	11	13	15	17	19
21	3222	3243	3263	3284	3304	3324	3345	3365	3385	3404	2	4	6	8	10	12	14	16	18
22	3424	3444	3464	3483	3502	3522	3541	3560	3579	3598	2	4	6	8	10	12	14	15	17
23	3617	3636	3655	3674	3692	3711	3729	3747	3766	3784	2	4	6	7	9	11	13	15	17
24	3802	3820	3838	3856	3874	3892	3909	3927	3945	3962	2	4	5	7	9	11	12	14	16
25	3979	3997	4014	4031	4048	4065	4082	4099	4116	4133	2	3	5	7	9	10	12	14	15
26	4150	4166	4183	4200	4216	4232	4249	4265	4281	4298	2	3	5	7	8	10	11	13	15
27	4314	4330	4346	4362	4378	4393	4409	4425	4440	4456	2	3	5	6	8	9	11	13	14
28	4472	4487	4502	4518	4533	4548	4564	4579	4594	4609	2	3	5	6	8	9	11	12	14
29	4624	4639	4654	4669	4683	4698	4713	4728	4742	4757	1	3	4	6	7	9	10	12	13
30	4771	4786	4800	4814	4829	4843	4857	4871	4886	4900	1	3	4	6	7	9	10	11	13
31	4914	4928	4942	4955	4969	4983	4997	5011	5024	5038	1	3	4	6	7	8	10	11	12
32	5051	5065	5079	5092	5105	5119	5132	5145	5159	5172	1	3	4	5	7	8	9	11	12
33	5185	5198	5211	5224	5237	5250	5263	5276	5289	5302	1	3	4	5	6	8	9	10	12
34	5315	5328	5340	5353	5366	5378	5391	5403	5416	5428	1	3	4	5	6	8	9	10	11
35	5441	5453	5465	5478	5490	5502	5514	5527	5539	5551	1	2	4	5	6	7	9	10	11
36	5563	5575	5587	5599	5611	5623	5635	5647	5658	5670	1	2	4	5	6	7	8	10	11
37	5682	5694	5705	5717	5729	5740	5752	5763	5775	5786	1	2	3	5	6	7	8	9	10
38	5798	5809	5821	5832	5843	5855	5866	5877	5888	5899	1	2	3	5	6	7	8	9	10
39	5911	5922	5933	5944	5955	5966	5977	5988	5999	6010	1	2	3	4	5	7	8	9	10
40	6021	6031	6042	6053	6064	6075	6085	6096	6107	6117	1	2	3	4	5	6	8	9	10
41	6128	6138	6149	6160	6170	6180	6191	6201	6212	6222	1	2	3	4	5	6	7	8	9
42	6232	6243	6253	6263	6274	6284	6294	6304	6314	6325	1	2	3	4	5	6	7	8	9
43	6335	6345	6355	6365	6375	6385	6395	6405	6415	6425	1	2	3	4	5	6	7	8	9
44	6435	6444	6454	6464	6474	6484	6493	6503	6513	6522	1	2	3	4	5	6	7	8	9
45	6532	6542	6551	6561	6571	6580	6590	6599	6609	6618	1	2	3	4	5	6	7	8	9
46	6628	6637	6646	6656	6665	6675	6684	6693	6702	6712	1	2	3	4	5	6	7	7	8
47	6721	6730	6739	6749	6758	6767	6776	6785	6794	6803	1	2	3	4	5	5	6	7	8
48	6812	6821	6830	6839	6848	6857	6866	6875	6884	6893	1	2	3	4	4	5	6	7	8
49	6902	6911	6920	6928	6937	6946	6955	6964	6972	6981	1	2	3	4	4	5	6	7	8
50	6990	6998	7007	7016	7024	7033	7042	7050	7059	7067	1	2	3	3	4	5	6	7	8
51	7076	7084	7093	7101	7110	7118	7126	7135	7143	7152	1	2	3	3	4	5	6	7	8
52	7160	7168	7177	7185	7193	7202	7210	7218	7226	7235	1	2	2	3	4	5	6	7	7
53	7243	7251	7259	7267	7275	7284	7292	7300	7308	7316	1	2	2	3	4	5	6	6	7
54	7324	7332	7340	7348	7356	7364	7372	7380	7388	7396	1	2	2	3	4	5	6	6	7

Table 8 (cont.) □ Logarithms of numbers

	0	1	2	3	4	5	6	7	8	9	1	2	3	4	5	6	7	8	9
55	7404	7412	7419	7427	7435	7443	7451	7459	7466	7474	1	2	2	3	4	5	5	6	7
56	7482	7490	7497	7505	7513	7520	7528	7536	7543	7551	1	2	2	3	4	5	5	6	7
57	7559	7566	7574	7582	7589	7597	7604	7612	7619	7627	1	2	2	3	4	5	5	6	7
58	7634	7642	7649	7657	7664	7672	7679	7686	7694	7701	1	1	2	3	4	4	5	6	7
59	7709	7716	7723	7731	7738	7745	7752	7760	7767	7774	1	1	2	3	4	4	5	6	7
60	7782	7789	7796	7803	7810	7818	7825	7832	7839	7846	1	1	2	3	4	4	5	6	6
61	7853	7860	7868	7875	7882	7889	7896	7903	7910	7917	1	1	2	3	4	4	5	6	6
62	7924	7931	7938	7945	7952	7959	7966	7973	7980	7987	1	1	2	3	3	4	5	6	6
63	7993	8000	8007	8014	8021	8028	8035	8041	8048	8055	1	1	2	3	3	4	5	5	6
64	8062	8069	8075	8082	8089	8096	8102	8109	8116	8122	1	1	2	3	3	4	5	5	6
65	8129	8136	8142	8149	8156	8162	8169	8176	8182	8189	1	1	2	3	3	4	5	5	6
66	8195	8202	8209	8215	8222	8228	8235	8241	8248	8254	1	1	2	3	3	4	5	5	6
67	8261	8267	8274	8280	8287	8293	8299	8306	8312	8319	1	1	2	3	3	4	5	5	6
68	8325	8331	8338	8344	8351	8357	8363	8370	8376	8382	1	1	2	3	3	4	4	5	6
69	8388	8395	8401	8407	8414	8420	8426	8432	8439	8445	1	1	2	2	3	4	4	5	6
70	8451	8457	8463	8470	8476	8482	8488	8494	8500	8506	1	1	2	2	3	4	4	5	6
71	8513	8519	8525	8531	8537	8543	8549	8555	8561	8567	1	1	2	2	3	4	4	5	5
72	8573	8579	8585	8591	8597	8603	8609	8615	8621	8627	1	1	2	2	3	4	4	5	5
73	8633	8639	8645	8651	8657	8663	8669	8675	8681	8686	1	1	2	2	3	4	4	5	5
74	8692	8698	8704	8710	8716	8722	8727	8733	8739	8745	1	1	2	2	3	4	4	5	5
75	8751	8756	8762	8768	8774	8779	8785	8791	8797	8802	1	1	2	2	3	3	4	5	5
76	8808	8814	8820	8825	8831	8837	8842	8848	8854	8859	1	1	2	2	3	3	4	5	5
77	8865	8871	8876	8882	8887	8893	8899	8904	8910	8915	1	1	2	2	3	3	4	4	5
78	8921	8927	8932	8938	8943	8949	8954	8960	8965	8971	1	1	2	2	3	3	4	4	5
79	8976	8982	8987	8993	8998	9004	9009	9015	9020	9025	1	1	2	2	3	3	4	4	5
80	9031	9036	9042	9047	9053	9058	9063	9069	9074	9079	1	1	2	2	3	3	4	4	5
81	9085	9090	9096	9101	9106	9112	9117	9122	9128	9133	1	1	2	2	3	3	4	4	5
82	9138	9143	9149	9154	9159	9165	9170	9175	9180	9186	1	1	2	2	3	3	4	4	5
83	9191	9196	9201	9206	9212	9217	9222	9227	9232	9238	1	1	2	2	3	3	4	4	5
84	9243	9248	9253	9258	9263	9269	9274	9279	9284	9289	1	1	2	2	3	3	4	4	5
85	9294	9299	9304	9309	9315	9320	9325	9330	9335	9340	1	1	2	2	3	3	4	4	5
86	9345	9350	9355	9360	9365	9370	9375	9380	9385	9390	1	1	2	2	3	3	4	4	5
87	9395	9400	9405	9410	9415	9420	9425	9430	9435	9440	0	1	1	2	2	3	3	4	4
88	9445	9450	9455	9460	9465	9469	9474	9479	9484	9489	0	1	1	2	2	3	3	4	4
89	9494	9499	9504	9509	9513	9518	9523	9528	9533	9538	0	1	1	2	2	3	3	4	4
90	9542	9547	9552	9557	9562	9566	9571	9576	9581	9586	0	1	1	2	2	3	3	4	4
91	9590	9595	9600	9605	9609	9614	9619	9624	9628	9633	0	1	1	2	2	3	3	4	4
92	9638	9643	9647	9652	9657	9661	9666	9671	9675	9680	0	1	1	2	2	3	3	4	4
93	9685	9689	9694	9699	9703	9708	9713	9717	9722	9727	0	1	1	2	2	3	3	4	4
94	9731	9736	9741	9745	9750	9754	9759	9763	9768	9773	0	1	1	2	2	3	3	4	4
95	9777	9782	9786	9791	9795	9800	9805	9809	9814	9818	0	1	1	2	2	3	3	4	4
96	9823	9827	9832	9836	9841	9845	9850	9854	9859	9863	0	1	1	2	2	3	3	4	4
97	9868	9872	9877	9881	9886	9890	9894	9899	9903	9908	0	1	1	2	2	3	3	4	4
98	9912	9917	9921	9926	9930	9934	9939	9943	9948	9952	0	1	1	2	2	3	3	4	4
99	9956	9961	9965	9969	9974	9978	9983	9987	9991	9996	0	1	1	2	2	3	3	3	4

Table 9 □ Unified—American Thread Series*

Nominal size diameter		Coarse (NC) (UNC)		Fine (NF) (UNF)		Extra-fine (NEF) (UNEF)	
		Threads per inch	Tap drill†	Threads per inch	Tap drill†	Threads per inch	Tap drill†
.060	0	—	—	80	$\frac{3}{64}$	—	—
.073	1	64	No. 53	72	No. 53	—	—
.086	2	56	No. 50	64	No. 50	—	—
.099	3	48	No. 47	56	No. 45	—	—
.112	4	**40**	No. 43	48	No. 42	—	—
.125	5	40	No. 38	44	No. 37	—	—
.138	6	**32**	No. 36	40	No. 33	—	—
.164	8	**32**	No. 29	36	No. 29	—	—
.190	10	**24**	No. 25	**32**	No. 21	—	—
.216	12	24	No. 16	28	No. 14	32	No. 13
.250	$\frac{1}{4}$	**20**	No. 7	**28**	No. 3	32	No. 2
.3125	$\frac{5}{16}$	**18**	F	**24**	I	32	K
.375	$\frac{3}{8}$	**16**	$\frac{5}{16}$	**24**	Q	32	S
.4375	$\frac{7}{16}$	**14**	U	**20**	$\frac{25}{64}$	**28**	Y
.500	$\frac{1}{2}$	**13**	$\frac{27}{64}$	**20**	$\frac{29}{64}$	**28**	$\frac{15}{32}$
.5625	$\frac{9}{16}$	**12**	$\frac{31}{64}$	**18**	$\frac{33}{64}$	24	$\frac{17}{32}$
.625	$\frac{5}{8}$	**11**	$\frac{17}{32}$	**18**	$\frac{37}{64}$	24	$\frac{19}{32}$
.6875	$\frac{11}{16}$	—	—	—	—	24	$\frac{41}{64}$
.750	$\frac{3}{4}$	**10**	$\frac{21}{32}$	**16**	$\frac{11}{16}$	**20**	$\frac{45}{64}$
.8125	$\frac{13}{16}$	—	—	—	—	20	$\frac{49}{64}$
.875	$\frac{7}{8}$	**9**	$\frac{49}{64}$	**14**	$\frac{13}{16}$	20	$\frac{53}{64}$
.9375	$\frac{15}{16}$	—	—	—	—	20	$\frac{57}{64}$
1.000	1	**8**	$\frac{7}{8}$	**12**	$\frac{59}{64}$	20	$\frac{61}{64}$
1.0625	$1\frac{1}{16}$	—	—	—	—	18	1
1.125	$1\frac{1}{8}$	**7**	$\frac{63}{64}$	**12**	$1\frac{3}{64}$	18	$1\frac{5}{64}$
1.1875	$1\frac{3}{16}$	—	—	—	—	18	$1\frac{9}{64}$
1.250	$1\frac{1}{4}$	**7**	$1\frac{7}{64}$	**12**	$1\frac{11}{64}$	18	$1\frac{13}{64}$
1.3125	$1\frac{5}{16}$	—	—	—	—	18	$1\frac{17}{64}$
1.375	$1\frac{3}{8}$	**6**	$1\frac{13}{64}$	**12**	$1\frac{19}{64}$	18	$1\frac{5}{16}$
1.4375	$1\frac{7}{16}$	—	—	—	—	18	$1\frac{3}{8}$
1.500	$1\frac{1}{2}$	**6**	$1\frac{21}{64}$	**12**	$1\frac{27}{64}$	18	$1\frac{29}{64}$
1.5625	$1\frac{9}{16}$	—	—	—	—	18	$1\frac{1}{2}$
1.625	$1\frac{5}{8}$	—	—	—	—	18	$1\frac{9}{16}$
1.6875	$1\frac{11}{16}$	—	—	—	—	18	$1\frac{5}{8}$
1.750	$1\frac{3}{4}$	**5**	$1\frac{35}{64}$	—	—	**16**	$1\frac{11}{16}$
2.000	2	**4½**	$1\frac{25}{32}$	—	—	**16**	$1\frac{15}{16}$
2.250	$2\frac{1}{4}$	**4½**	$2\frac{1}{32}$	—	—	—	—
2.500	$2\frac{1}{2}$	**4**	$2\frac{1}{4}$	—	—	—	—
2.750	$2\frac{3}{4}$	**4**	$2\frac{1}{2}$	—	—	—	—
3.000	3	**4**	$2\frac{3}{4}$	—	—	—	—
3.250	$3\frac{1}{4}$	**4**	3	—	—	—	—
3.500	$3\frac{1}{2}$	**4**	$3\frac{1}{4}$	—	—	—	—
3.750	$3\frac{3}{4}$	**4**	$3\frac{1}{2}$	—	—	—	—
4.000	4	**4**	$3\frac{3}{4}$	—	—	—	—

*ANSI B1.1–1960.
Bold type indicates Unified threads. To be designated UNC or UNF.
Unified Standard—Classes 1A, 2A, 3A, 1B, 2B, and 3B.
For recommended hole-size limits before threading see Tables 38 and 39, ANSI B1.1–1960.
†Tap drill for a 75% thread (not Unified—American Standard).
Bold type sizes smaller than $\frac{1}{4}$ in. are accepted for limited applications by the British, but the symbols NC or NF, as applicable, are retained.

Table 10 □ Unified–American special threads* (8-pitch, 12-pitch, and 16-pitch series)

Nominal size diameter		Threads per inch			Nominal size diameter		Threads per inch		
.500	½	—	**12**	—	1.750	1¾	**8**	**12**	**16**
.5625	9/16	—	**12**	—	1.8125	1 13/16	—	—	16
.625	5/8	—	12	—	1.875	1⅞	8	12	16
.6875	11/16	—	12	—	1.9375	1 15/16	—	—	16
.750	¾	—	12	**16**	2.000	2	**8**	**12**	**16**
.8125	13/16	—	12	16	2.0625	2 1/16	—	—	16
.875	⅞	—	12	16	2.125	2⅛	8	12	16
.9375	15/16	—	12	16	2.1875	2 3/16	—	—	16
1.000	1	8	**12**	**16**	2.250	2¼	**8**	**12**	**16**
1.0625	1 1/16	—	**12**	**16**	2.3125	2 5/16	—	—	16
1.125	1⅛	8	**12**	**16**	2.375	2⅜	—	12	16
1.1875	1 3/16	—	**12**	**16**	2.4375	2 7/16	—	—	16
1.250	1¼	8	**12**	**16**	2.500	2½	**8**	**12**	**16**
1.3125	1 5/16	—	**12**	**16**	2.625	2⅝	—	12	16
1.375	1⅜	8	**12**	**16**	2.750	2¾	**8**	**12**	**16**
1.4375	1 7/16	—	**12**	**16**	2.875	2⅞	—	12	16
1.500	1½	8	**12**	**16**	3.000	3	**8**	**12**	**16**
1.5625	1 9/16	—	—	16	3.125	3⅛	—	12	16
1.625	1⅝	8	12	16	3.250	3¼	**8**	**12**	**16**
1.6875	1 11/16	—	—	16	3.375	3⅜	—	12	16

*ANSI B1.I–1960. Bold type indicates Unified threads (UN).

Table 11 □ American national standard acme and stub acme threads*

Nominal size diameter		Threads per inch	Nominal size diameter		Threads per inch
.250	¼	16	1.250	1¼	5
.3125	5/16	14	1.375	1⅜	4
.375	⅜	12	1.500	1½	4
.4375	7/16	12	1.750	1¾	4
.500	½	10	2.000	2	4
.625	⅝	8	2.250	2¼	3
.750	¾	6	2.500	2½	3
.875	⅞	6	2.750	2¾	3
1.000	1	5	3.000	3	2
1.125	1⅛	5			

*ANSI B1.5 and B1.8–1952.

Table 12 □ Buttress threads* (suggested combinations of diameters and pitches)

Nominal size diameter				Associated pitches (tpi)						
.500	.5625	.625	.6875	20	16	12				
.750	.875	1.000	—	—	16	12	10			
1.125	1.250	1.375	1.500	—	16	12	10	8	6	
1.750	3.000	2.250	2.500	—	16	12	10	8	6	5	4
2.750	3.000	3.500	4.000	—	16	12	10	8	6	5	4

*ANSI B1.9–1953.

Table 13 □ American national standard wrench-head bolts and nuts—regular series*

Bolt Diameter Nominal size		Bolt heads: Width across flats	Bolt heads: Height of head			Nuts: Width across flats	Nuts: Thickness			
							Regular		Jam	
		Unfinished and semifinished square and hexagon†	*Unfinished square*	*Unfinished hexagon*	*Semifinished hexagon*	*Unfinished hexagon, semifinished hexagon and hexagon jam*	*Unfinished square and hexagon*	*Semifinished hexagon and hexagon slotted*	*Unfinished hexagon*	*Semifinished hexagon*
.250	1/4	[3/8 sq. / 7/16 hex.]	11/64	11/64	5/32	7/16	7/32	7/32	5/32	5/32
.3125	5/16	1/2	13/64	7/32	13/64	1/2	17/64	17/64	3/16	3/16
.375	3/8	9/16	1/4	1/4	15/64	9/16	21/64	21/64	7/32	7/32
.4375	7/16	5/8	19/64	19/64	9/32	11/16	3/8	3/8	1/4	1/4
.500	1/2	3/4	21/64	11/32	5/16	3/4	7/16	7/16	5/16	5/16
.5625	9/16	13/16	—	—	23/64	7/8	31/64	31/64	5/16	5/16
.625	5/8	15/16	27/64	27/64	25/64	15/16	35/64	35/64	3/8	3/8
.750	3/4	1 1/8	1/2	1/2	15/32	1 1/8	21/32	41/64	7/16	27/64
.875	7/8	1 5/16	19/32	37/64	35/64	1 5/16	49/64	3/4	1/2	31/64
1.000	1	1 1/2	21/32	43/64	39/64	1 1/2	7/8	55/64	9/16	35/64
1.125	1 1/8	1 11/16	3/4	3/4	11/16	1 11/16	1	31/32	5/8	39/64
1.250	1 1/4	1 7/8	27/32	27/32	25/32	1 7/8	1 3/32	1 1/16	3/4	23/32
1.375	1 3/8	2 1/16	29/32	29/32	27/32	2 1/16	1 13/64	1 11/64	13/16	25/32
1.500	1 1/2	2 1/4	1	1	15/16	2 1/4	1 5/16	1 9/32	7/8	27/32
1.625	1 5/8	2 7/16	1 3/32	—	—	2 7/16	—	1 25/64	—	29/32
1.750	1 3/4	2 5/8	—	1 5/32	1 3/32	2 5/8	—	1 1/2	—	31/32
1.875	1 7/8	2 13/16	—	—	—	2 13/16	—	1 39/64	—	1 1/32
2.000	2	3	—	1 11/32	1 7/32	3	—	1 23/32	—	1 3/32
2.250	2 1/4	3 3/8	—	1 1/2	1 3/8	3 3/8	—	1 59/64	—	1 13/64
2.500	2 1/2	3 3/4	—	1 21/32	1 17/32	3 3/4	—	2 9/64	—	1 23/64
2.750	2 3/4	4 1/8	—	1 13/16	1 11/16	4 1/8	—	2 23/64	—	1 37/64
3.000	3	4 1/2	—	2	1 7/8	4 1/2	—	2 37/64	—	1 45/64

*ANSI B18.2.1–1965; ANSI B18.2.2–1965.
Thread-bolts: coarse thread series, class 2A.
Thread-nuts: unfinished; coarse series, class 2B: semifinished; coarse, fine, or 8-pitch series.
†Square bolts in 1/4–1 1/2 in. sizes (nominal) only.

Table 14 □ American national standard finished hexagon castle nuts*

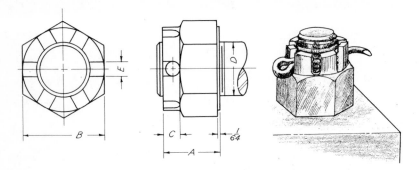

Nominal size diameter D		Threads per inch		Thickness A	Width across flats B	Slot		Diameter of cylindrical part (min)
		UNC	UNF			Depth C	Width E	
.250	¼	20	28	⁹⁄₃₂	⁷⁄₁₆	.094	.078	.371
.3125	⁵⁄₁₆	18	24	²¹⁄₆₄	½	.094	.094	.425
.375	⅜	16	24	¹³⁄₃₂	⁹⁄₁₆	.125	.125	.478
.4375	⁷⁄₁₆	14	20	²⁹⁄₆₄	¹¹⁄₁₆	.156	.125	.582
.500	½	13	20	⁹⁄₁₆	¾	.156	.156	.637
.5625	⁹⁄₁₆	12	18	³⁹⁄₆₄	⅞	.188	.156	.744
.625	⅝	11	18	²³⁄₃₂	¹⁵⁄₁₆	.219	.188	.797
.750	¾	10	16	¹³⁄₁₆	1⅛	.250	.188	.941
.875	⅞	9	14	²⁹⁄₃₂	1⁵⁄₁₆	.250	.188	1.097
1.000	1	8	12	1	1½	.281	.250	1.254
1.125	1⅛	7	12	1⁵⁄₃₂	1¹¹⁄₁₆	.344	.250	1.411

*ANSI B18.2.2–1965.
Thread may be coarse- or fine-thread series, class 2B tolerance; unless otherwise specified fine-thread series will be furnished.

Table 15 □ American national standard set screws*

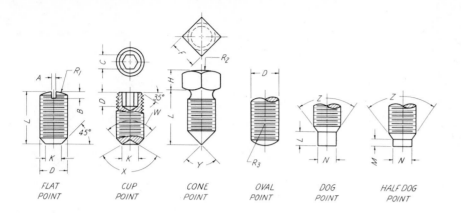

FLAT POINT CUP POINT CONE POINT OVAL POINT DOG POINT HALF DOG POINT

Nominal size diameter D		Slotted headless			Hexagonal socket (min)		Square-head (F-max)			Points								
										Cup and flat (K-max)	Cone			Oval	Full- and half-dog (N-max)			
D		A	B	R_1	C	D†	F	H	R_2	K	W	X	Y	R_3	L	M	N	Z
.125	5	.023	.031	.125	1/16	.050	—	—	—	.067				.094	.060	.030	.083	
.138	6	.025	.035	.138	1/16	.050	—	—	—	.074				.109	.070	.035	.092	
.164	8	.029	.041	.164	5/64	.062	—	—	—	.087				.125	.080	.040	.109	
.190	10	.032	.048	.190	3/32	.075	.1875	9/64	15/32	.102				.141	.090	.045	.127	
.216	12	.036	.054	.216	3/32	.075	.216	5/32	35/64	.115				.156	.110	.055	.144	
.250	1/4	.045	.063	.250	1/8	.100	.250	3/16	5/8	.132				.188	.125	.063	.156	
.3125	5/16	.051	.078	.313	5/32	.125	.3125	15/64	25/32	.172				.234	.156	.078	.203	
.375	3/8	.064	.094	.375	3/16	.150	.375	9/32	15/16	.212				.281	.188	.094	.250	
.4375	7/16	.072	.109	.438	7/32	.175	.4375	21/64	1 3/32	.252				.328	.219	.109	.297	
.500	1/2	.081	.125	.500	1/4	.200	.500	3/8	1 1/4	.291				.375	.250	.125	.344	
.5625	9/16	.091	.141	.563	1/4	.200	.5625	27/64	1 13/32	.332				.422	.281	.140	.391	
.625	5/8	.102	.156	.625	5/16	.250	.625	15/32	1 9/16	.371				.469	.313	.156	.469	
.750	3/4	.129	.188	.750	3/8	.300	.750	9/16	1 7/8	.450				.563	.375	.188	.563	
.875	7/8	—	—	—	1/2	.400	.875	21/32	2 3/16	—				—	—	—	—	
1.000	1	—	—	—	9/16	.450	1.000	3/4	2 1/2	—				—	—	—	—	
1.125	1 1/8	—	—	—	9/16	.450	1.125	27/32	2 13/16	—				—	—	—	—	
1.250	1 1/4	—	—	—	5/8	.500	1.250	15/16	3 1/8	—				—	—	—	—	
1.375	1 3/8	—	—	—	5/8	.500	1.375	1 1/32	3 7/16	—				—	—	—	—	
1.500	1 1/2	—	—	—	3/4	.600	1.500	1 1/8	3 3/4	—				—	—	—	—	

Cone column notes:
$W = 80°$–$90°$ (Draw as $90°$.)
$X = 118° \pm 5°$
Y: When L equals nominal diameter or less, $Y = 118° \pm 2°$. When L exceeds nominal diameter, $Y = 90° \pm 2°$.
$Z = 100°$–$110°$

*ANSI B18.6.2–1956; ANSI B18.3–1969.
†Dimensions apply to cup and flat-point screws 1 diam in length or longer. For screws shorter than 1 diam in length, and for other types of points, socket to be as deep as practicable.

Table 16 □ American national standard machine screws*

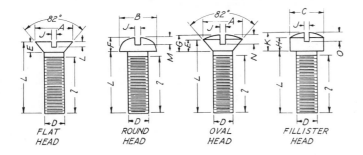

FLAT HEAD ROUND HEAD OVAL HEAD FILLISTER HEAD

Size (number) and threads per inch				Standard dimensions (max)													
Nominal size diameter D		Thread		Head diameter			Height dimensions					Slot width	Slot depth			
		Coarse	Fine	A	B	C	E	F	G	H	K	J	L	M	N	O
.060	0	—	80	.119	.113	.096	.035	.053	.056	.045	.059	.023	.015	.039	.030	.025
.073	1	64	72	.146	.138	.118	.043	.061	.068	.053	.071	.026	.019	.044	.038	.031
.086	2	56	64	.172	.162	.140	.051	.069	.080	.062	.083	.031	.023	.048	.045	.037
.099	3	48	56	.199	.187	.161	.059	.078	.092	.070	.095	.035	.027	.053	.052	.043
.112	4	40	48	.225	.211	.183	.067	.086	.104	.079	.107	.039	.030	.058	.059	.048
.125	5	40	44	.252	.236	.205	.075	.095	.116	.088	.120	.043	.034	.063	.067	.054
.138	6	32	40	.279	.260	.226	.083	.103	.128	.096	.132	.048	.038	.068	.074	.060
.164	8	32	36	.332	.309	.270	.100	.120	.152	.113	.156	.054	.045	.077	.088	.071
.190	10	24	32	.385	.359	.313	.116	.137	.176	.130	.180	.060	.053	.087	.103	.083
.216	12	24	28	.438	.408	.357	.132	.153	.200	.148	.205	.067	.060	.096	.117	.094
.250	¼	20	28	.507	.472	.414	.153	.175	.232	.170	.237	.075	.070	.109	.136	.109
.3125	⁵⁄₁₆	18	24	.635	.590	.518	.191	.216	.290	.211	.295	.084	.088	.132	.171	.137
.375	⅜	16	24	.762	.708	.622	.230	.256	.347	.253	.355	.094	.106	.155	.206	.164
.4375	⁷⁄₁₆	14	20	.812	.750	.625	.223	.328	.345	.265	.368	.094	.103	.196	.210	.170
.500	½	13	20	.875	.813	.750	.223	.355	.354	.297	.412	.106	.103	.211	.216	.190
.5625	⁹⁄₁₆	12	18	1.000	.938	.812	.260	.410	.410	.336	.466	.118	.120	.242	.250	.214
.625	⅝	11	18	1.125	1.000	.875	.298	.438	.467	.375	.521	.133	.137	.258	.285	.240
.750	¾	10	16	1.375	1.250	1.000	.372	.547	.578	.441	.612	.149	.171	.320	.353	.281

*ANSI B18.6.3–1962.
Thread length—screws 2 in. in length or less are threaded as close to the head as practicable. Screws longer than 2 in. should have a minimum thread length of 1¾ in.

Table 17 □ American national standard cap screws*

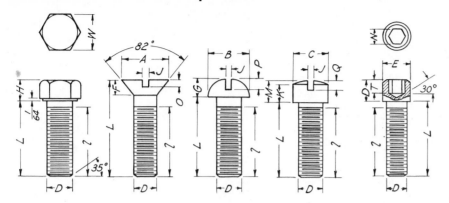

Nominal size diameter		Head diameter					Height dimensions					Slot width	Slot depth			Socket dimensions	
		A	B	C	E	W	F av	G	H nom	K	M	J	O	P	Q	N min	T min
.250	¼	.500	.437	.375	⅜	⁷⁄₁₆	.140	.191	⁵⁄₃₂	.172	.216	.075	.068	.117	.097	³⁄₁₆	.120
.3125	⁵⁄₁₆	.625	.562	.437	⁷⁄₁₆	½	.177	.245	¹³⁄₆₄	.203	.253	.084	.086	.151	.115	⁷⁄₃₂	.151
.375	⅜	.750	.625	.562	⁹⁄₁₆	⁹⁄₁₆	.210	.273	¹⁵⁄₆₄	.250	.314	.094	.103	.168	.142	⁵⁄₁₆	.182
.4375	⁷⁄₁₆	.8125	.750	.625	⅝	⅝	.210	.328	⁹⁄₃₂	.297	.368	.094	.103	.202	.168	⁵⁄₁₆	.213
.500	½	.875	.812	.750	¾	¾	.210	.354	⁵⁄₁₆	.328	.413	.106	.103	.218	.193	⅜	.245
.5625	⁹⁄₁₆	1.000	.937	.812	¹³⁄₁₆	¹³⁄₁₆	.244	.409	²³⁄₆₄	.375	.467	.118	.120	.252	.213	⅜	.276
.625	⅝	1.125	1.000	.875	⅞	¹⁵⁄₁₆	.281	.437	²⁵⁄₆₄	.422	.521	.133	.137	.270	.239	½	.307
.750	¾	1.375	1.250	1.000	1	1⅛	.352	.546	¹⁵⁄₃₂	.500	.612	.149	.171	.338	.283	⁹⁄₁₆	.370
.875	⅞	1.625	—	1.125	1⅛	1⁵⁄₁₆	.423	—	³⁵⁄₆₄	.594	.720	.167	.206	—	.334	⁹⁄₁₆	.432
1.000	1	1.875	—	1.312	1⁵⁄₁₆	1½	.494	—	³⁹⁄₆₄	.656	.803	.188	.240	—	.371	⅝	.495
1.125	1⅛	2.062	—	—	1½	1¹¹⁄₁₆	.529	—	¹¹⁄₁₆	—	—	.196	.257	—	—	¾	.557
1.250	1¼	2.312	—	—	1¾	1⅞	.600	—	²⁵⁄₃₂	—	—	.211	.291	—	—	¾	.620
1.375	1⅜	2.562	—	—	1⅞	2¹⁄₁₆	.665	—	²⁷⁄₃₂	—	—	.226	.326	—	—	¾	.682
1.500	1½	2.812	—	—	2	2¼	.742	—	¹⁵⁄₁₆	—	—	.258	.360	—	—	1	.745

*ANSI B18.3–1969. ANSI B18.6.2–1956.

Bold type indicates products unified dimensionally with British and Canadian standards.

Basically, threads may be coarse, fine, or 8-thread series; class 2A for plain (unplated) cap screws.

Minimum thread length will b 2D + ¼ in. for lengths up to and including 6 in.

Socket-head cap screws—thread coarse or fine, class 3A. Thread length: coarse, 2D + ½ in; fine, 1½D + ½ in.

Standard dimensions are maximum except as noted.

Table 18 □ American national standard plain washers (type A)*

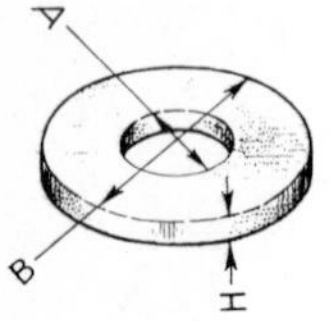

Washer size nominal		Light-SAE (N = narrow)			Standard-plate (W = wide)		
		A ID	B OD	H Thickness	A ID	B OD	H Thickness
.250	¼	.281	.625	.065	.312	.734	.065
.312	⁵⁄₁₆	.344	.688	.065	.375	.875	.083
.375	⅜	.406	.812	.065	.438	1.000	.083
.438	⁷⁄₁₆	.469	.922	.065	.500	1.250	.083
.500	½	.531	1.062	.095	.562	1.375	.109
.562	⁹⁄₁₆	.594	1.156	.095	.625	1.469	.109
.625	⅝	.656	1.312	.095	.688	1.750	.134
.750	¾	.812	1.469	.134	.812	2.000	.148
.875	⅞	.938	1.750	.134	.938	2.250	.165
1.000	1	1.062	2.000	.134	1.062	2.500	.165
1.125	1⅛	1.250	2.250	.134	1.250	2.750	.165
1.250	1¼	1.375	2.500	.165	1.375	3.000	.165
1.375	1⅜	1.500	2.750	.165	1.500	3.250	.180
1.500	1½	1.625	3.000	.165	1.625	3.500	.180

*ANSI B27.2–1965.
Plain washers are specified as follows: ID × OD × thickness − .375 × .875 × .083 PLAIN WASHER.

Table 19 □ American national standard lock washers (selected sizes)

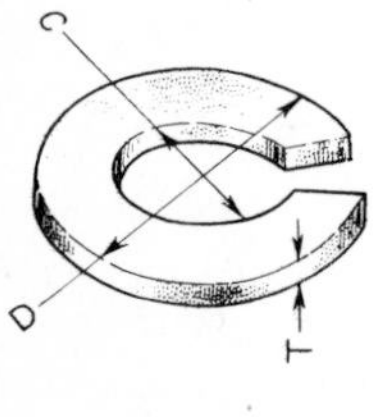

Washer size nominal		Regular (light)			Extra-duty (heavy)		
		C ID (min)	D OD (max)	T Thickness (min)	C ID (min)	D OD (max)	T Thickness (min)
.250	¼	.255	.489	.062	Same	.535	.084
.312	⁵⁄₁₆	.318	.586	.078	as for	.622	.108
.375	⅜	.382	.683	.094	regular	.741	.123
.438	⁷⁄₁₆	.446	.779	.109	lock	.839	.143
.500	½	.509	.873	.125	washers	.939	.162
.562	⁹⁄₁₆	.572	.971	.141		1.041	.182
.625	⅝	.636	1.079	.156		1.157	.202
.750	¾	.763	1.271	.188		1.361	.241
.875	⅞	.890	1.464	.219		1.576	.285
1.000	1	1.017	1.661	.250		1.799	.330
1.125	1⅛	1.144	1.853	.281		2.019	.375
1.250	1¼	1.271	2.045	.312		2.231	.417
1.375	1⅜	1.398	2.239	.344		2.439	.458
1.500	1½	1.525	2.430	.375		2.638	.496

*ANSI B27.1–1965.
Lock washers are specified by giving nominal size and series (⅜ regular lock washer).

Table 20 □ American national standard cotter pins

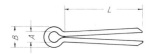

Pin diameter				Eye diameter		Drill size recommended hole/diameter	Clevis pin or shaft diameter
Nominal		Max	Min	A—inside	B—outside		
.031	1/32	.032	.028	1/32	1/16	3/64—.0469	1/8
.047	3/64	.048	.044	3/64	3/32	1/16—.0625	3/16
.062	1/16	.060	.056	1/16	1/8	5/64—.0781	1/4
.078	5/64	.076	.072	5/64	5/32	3/32—.0938	5/16
.094	3/32	.090	.086	3/32	3/16	7/64—.1094	3/8
.125	1/8	.120	.116	1/8	1/4	9/64—.1406	1/2
.156	5/32	.150	.146	5/32	5/16	11/64—.1719	5/8
.188	3/16	.176	.172	3/16	3/8	13/64—.2031	—
.219	7/32	.207	.202	7/32	7/16	15/64—.2344	—
.250	1/4	.225	.220	1/4	1/2	17/64—.2656	—

*ANSI B5.20–1958.

For shafts up to 5/8 in. diam select a cotter pin that is approximately equal to one-fourth of the shaft diameter. For larger shaft sizes use a cotter pin that is from one-fourth to one-sixth of the shaft diameter.

Table 21 □ American national standard machine screw and stove bolt nuts

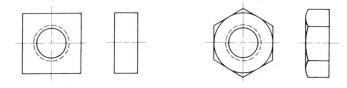

Diameter		Nominal size												
		0	1	2	3	4	5	6	8	10	12	1/4	5/16	3/8
Across flats (nominal)		5/32	5/32	3/16	3/16	1/4	5/16	5/16	11/32	3/8	7/16	7/16	9/16	5/8
Across corners (min)	Hexagonal	.171	.171	.205	.205	.275	.344	.344	.378	.413	.482	.482	.621	.692
	Square	.206	.206	.247	.247	.331	.415	.415	.456	.497	.581	.581	.748	.833
Thickness (nom)		3/64	3/64	1/16	1/16	3/32	7/64	7/64	1/8	1/8	5/32	3/16	7/32	1/4

*ANSI B18.6.3–1962.
Dimensions in inches.
Square nuts—Threads are UNC, Class 2B.
Hexagonal nuts—Threads are UNC or UNF, Class 2B.

Table 22 □ American national standard small rivets*

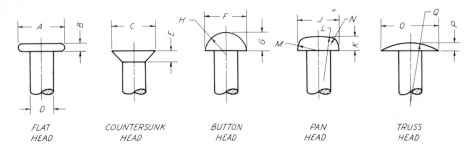

FLAT HEAD COUNTERSUNK HEAD BUTTON HEAD PAN HEAD TRUSS HEAD

Rivet diam D		Flat Diam A max	Height B max	Csk. Diam C max	Height E max	Button Diam F max	Height G max	Rad H	Pan Diam J max	Height K max	Rad L	Rad M	Rad N	Truss Diam O max	Height P max	Rad Q
.062	⅟₁₆	.140	.027	.118	.027	.122	.052	.055	.118	.040	.217	.052	.019	—	—	—
.094	³⁄₃₂	.200	.038	.176	.040	.182	.077	.084	.173	.060	.326	.080	.030	.226	.038	.239
.125	⅛	.260	.048	.235	.053	.235	.100	.111	.225	.078	.429	.106	.039	.297	.048	.314
.156	⁵⁄₃₂	.323	.059	.293	.066	.290	.124	.138	.279	.096	.535	.133	.049	.368	.059	.392
.188	³⁄₁₆	.387	.069	.351	.079	.348	.147	.166	.334	.114	.641	.159	.059	.442	.069	.470
.219	⁷⁄₃₂	.453	.080	.413	.094	.405	.172	.195	.391	.133	.754	.186	.069	.515	.080	.555
.250	¼	.515	.091	.469	.106	.460	.196	.221	.444	.151	.858	.213	.079	.590	.091	.628
.281	⁹⁄₃₂	.579	.103	.528	.119	.518	.220	.249	.499	.170	.963	.239	.088	.661	.103	.706
.313	⁵⁄₁₆	.641	.113	.588	.133	.572	.243	.276	.552	.187	1.070	.266	.098	.732	.113	.784
.344	¹¹⁄₃₂	.705	.124	.646	.146	.630	.267	.304	.608	.206	1.176	.292	.108	.806	.124	.862
.375	⅜	.769	.135	.704	.159	.684	.291	.332	.663	.225	1.286	.319	.118	.878	.135	.942
.406	¹³⁄₃₂	.834	.146	.763	.172	.743	.316	.358	.719	.243	1.392	.345	.127	.949	.145	1.028
.438	⁷⁄₁₆	.896	.157	.823	.186	.798	.339	.387	.772	.261	1.500	.372	.137	1.020	.157	1.098

*ANSI B18.1–1965.
The length of a rivet is measured from the underside (bearing surface) of the head to the end of the shank except in the case of a rivet with a countersunk head. The length of a counter-sunk-head rivet is measured from the top of the head to the end of the shank.

Table 23 □ American national standard square and flat keys*

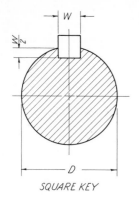

SQUARE KEY

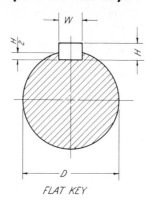

FLAT KEY

Shaft Diam Nominal		Square stock	Rectangular stock	Shaft diam nominal		Square stock	Rectangular stock
Over	To (inclusive)	W	W × H	Over	To (inclusive)	W	W × H
$\frac{5}{16}$	$\frac{7}{16}$	$\frac{3}{32}$	—	$1\frac{3}{4}$	$2\frac{1}{4}$	$\frac{1}{2}$	$\frac{1}{2} \times \frac{3}{8}$
$\frac{7}{16}$	$\frac{9}{16}$	$\frac{1}{8}$	$\frac{1}{8} \times \frac{3}{32}$	$2\frac{1}{4}$	$2\frac{3}{4}$	$\frac{5}{8}$	$\frac{5}{8} \times \frac{7}{16}$
$\frac{9}{16}$	$\frac{7}{8}$	$\frac{3}{16}$	$\frac{3}{16} \times \frac{1}{8}$	$2\frac{3}{4}$	$3\frac{1}{4}$	$\frac{3}{4}$	$\frac{3}{4} \times \frac{1}{2}$
$\frac{7}{8}$	$1\frac{1}{4}$	$\frac{1}{4}$	$\frac{1}{4} \times \frac{3}{16}$	$3\frac{1}{4}$	$3\frac{3}{4}$	$\frac{7}{8}$	$\frac{7}{8} \times \frac{5}{8}$
$1\frac{1}{4}$	$1\frac{3}{8}$	$\frac{5}{16}$	$\frac{5}{16} \times \frac{1}{4}$	$3\frac{3}{4}$	$4\frac{1}{2}$	1	$1 \times \frac{3}{4}$
$1\frac{3}{8}$	$1\frac{3}{4}$	$\frac{3}{8}$	$\frac{3}{8} \times \frac{1}{4}$	$4\frac{1}{2}$	$5\frac{1}{2}$	$1\frac{1}{4}$	$1\frac{1}{4} \times \frac{7}{8}$

*ANSI B17.1–1967.
All dimensions given in inches.

Table 24 □ American national standard plain taper and gib-head keys**

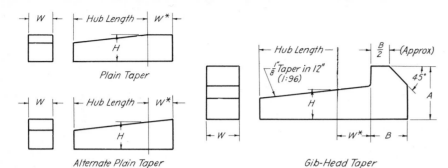

Plain Taper

Alternate Plain Taper

Gib-Head Taper

Plain and Gib-Head Taper Keys Have a ⅛" Taper in 12"

Plain taper and gib-head keys square and rectangular				Gib-head			
Shaft diam nominal		Square type W = H	Rectangular type W × H	Square		Rectangular	
				Head height A	Length B	Head height A	Length B
Over	To (inclusive)						
⁵⁄₁₆	⁷⁄₁₆	—	—	—	—	—	—
⁷⁄₁₆	⁹⁄₁₆	⅛	⅛ × ³⁄₃₂	¼	¼	³⁄₁₆	⅛
⁹⁄₁₆	⅞	³⁄₁₆	³⁄₁₆ × ⅛	⁵⁄₁₆	⁵⁄₁₆	¼	¼
⅞	1¼	¼	¼ × ³⁄₁₆	⁷⁄₁₆	⅜	⁵⁄₁₆	⁵⁄₁₆
1¼	1⅜	⁵⁄₁₆	⁵⁄₁₆ × ¼	½	⁷⁄₁₆	⁷⁄₁₆	⅜
1⅜	1¾	⅜	⅜ × ¼	⅝	½	⁷⁄₁₆	⅜
1¾	2¼	½	½ × ⅜	⅞	⅝	⅝	½
2¼	2¾	⅝	⅝ × ⁷⁄₁₆	1	¾	¾	⁹⁄₁₆
2¾	3¼	¾	¾ × ½	1¼	⅞	⅞	⅝
3¼	3¾	⅞	⅞ × ⅝	1⅜	1	1	¾

**ANSI B17.1–1967.
*For locating position of dimension H.
For longer sizes see standard.
All dimensions given in inches.

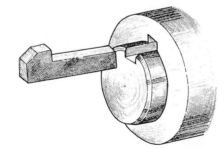

Table 25 □ American national standard Woodruff keys*

Key no.	Nominal size $A \times B$	Height of key C_{max}	Height of key D_{max}	Distance below center E	Depth of keyseat in shaft $+.005$ $-.000$
202	1/16 × 1/4	.109	.109	1/64	.0728
202.5	1/16 × 5/16	.140	.140	1/64	.1038
203	1/16 × 3/8	.172	.172	1/64	.1358
204	1/16 × 1/2	.203	.194	3/64	.1668
302.5	3/32 × 5/16	.140	.140	1/64	.0882
303	3/32 × 3/8	.172	.172	1/64	.1202
304	3/32 × 1/2	.203	.194	3/64	.1511
305	3/32 × 5/8	.250	.240	1/16	.1981
403	1/8 × 3/8	.172	.172	1/64	.1045
404	1/8 × 1/2	.203	.194	3/64	.1355
405	1/8 × 5/8	.250	.240	1/16	.1825
406	1/8 × 3/4	.313	.303	1/16	.2455
505	5/32 × 5/8	.250	.240	1/16	.1669
506	5/32 × 3/4	.313	.303	1/16	.2299
507	5/32 × 7/8	.375	.365	1/16	.2919
605	3/16 × 5/8	.250	.240	1/16	.1513
606	3/16 × 3/4	.313	.303	1/16	.2143
607	3/16 × 7/8	.375	.365	1/16	.2763
608	3/16 × 1	.438	.428	1/16	.3393
609	3/16 × 1 1/8	.484	.475	5/64	.3853
610	3/16 × 1 1/4	.547	.537	5/64	.4483
707	7/32 × 7/8	.375	.365	1/16	.2607
708	7/32 × 1	.438	.428	1/16	.3237
709	7/32 × 1 1/8	.484	.475	5/64	.3697
710	7/32 × 1 1/4	.547	.537	5/64	.4327
806	1/4 × 3/4	.313	.303	1/16	.1830
807	1/4 × 7/8	.375	.365	1/16	.2450
808	1/4 × 1	.438	.428	1/16	.3080
809	1/4 × 1 1/8	.484	.475	5/64	.3540
810	1/4 × 1 1/4	.547	.537	5/64	.4170
811	1/4 × 1 3/8	.594	.584	3/32	.4640
812	1/4 × 1 1/2	.641	.631	7/64	.5110
1008	5/16 × 1	.438	.428	1/16	.2768
1009	5/16 × 1 1/8	.484	.475	5/64	.3228
1010	5/16 × 1 1/4	.547	.537	5/64	.3858
1011	5/16 × 1 3/8	.594	.584	3/32	.4328
1012	5/16 × 1 1/2	.641	.631	7/64	.4798
1208	3/8 × 1	.438	.428	1/16	.2455
1210	3/8 × 1 1/4	.547	.537	5/64	.3545
1211	3/8 × 1 3/8	.594	.584	3/32	.4015
1212	3/8 × 1 1/2	.641	.631	7/64	.4485

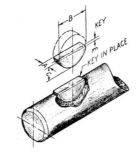

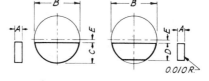

*ANSI B17.2–1967.

All dimensions in inches. Key numbers indicate the nominal key dimensions. The last two digits give the nominal diameter in eighths of an inch and the digits preceding the last two give the nominal width in thirty-seconds of an inch.

Examples: No. 204 indicates a key 2/32 × 4/8 or 1/16 × 1/2.

No. 808 indicates a key 8/32 × 8/8 or 1/4 × 1.

Table 26 □ Pratt and Whitney keys

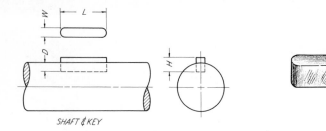

SHAFT & KEY

Key no.	L*	W	H	D	Key no.	L*	W	H	D
1	½	1/16	3/32	1/16	22	1⅜	¼	⅜	¼
2	½	3/32	9/64	3/32	23	1⅜	5/16	15/32	5/16
3	½	⅛	3/16	⅛	F	1⅜	⅜	9/16	⅜
4	⅝	3/32	9/64	3/32	24	1½	¼	⅜	¼
5	⅝	⅛	3/16	⅛	25	1½	5/16	15/32	5/16
6	⅝	5/32	15/64	5/32	G	1½	⅜	9/16	⅜
7	¾	⅛	3/16	⅛	51	1¾	¼	⅜	¼
8	¾	5/32	15/64	5/32	52	1¾	5/16	15/32	5/16
9	¾	3/16	9/32	3/16	53	1¾	⅜	9/16	⅜
10	⅞	5/32	15/64	5/32	26	2	3/16	9/32	3/16
11	⅞	3/16	9/32	3/16	27	2	¼	⅜	¼
12	⅞	7/32	21/64	7/32	28	2	5/16	15/32	5/16
A	⅞	¼	⅜	¼	29	2	⅜	9/16	⅜
13	1	3/16	9/32	3/16	54	2¼	¼	⅜	¼
14	1	7/32	21/64	7/32	55	2¼	5/16	15/32	5/16
15	1	¼	⅜	¼	56	2¼	⅜	9/16	⅜
B	1	5/16	15/32	5/16	57	2¼	7/16	21/32	7/16
16	1⅛	3/16	9/32	3/16	58	2½	5/16	15/32	5/16
17	1⅛	7/32	21/64	7/32	59	2½	⅜	9/16	⅜
18	1⅛	¼	⅜	¼	60	2½	7/16	21/32	7/16
C	1⅛	5/16	15/32	5/16	61	2½	½	¾	½
19	1¼	3/16	9/32	3/16	30	3	⅜	9/16	⅜
20	1¼	7/32	21/64	7/32	31	3	7/16	21/32	7/16
21	1¼	¼	⅜	¼	32	3	½	¾	½
D	1¼	5/16	15/32	5/16	33	3	9/16	27/32	9/16
E	1¼	⅜	9/16	⅜	34	3	⅝	15/16	⅝

*The length L may vary but should always be at least 2W.

Table 27 □ American national standard taper pins*

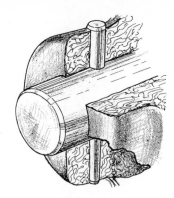

Number of pin†	Diam at large end		Maximum length L	Approx. shaft diam‡	Drill size‡
00000	.094	3/32	1	.250	No. 47 (.0785)
0000	.109	7/64	1	.312	No. 42 (.0935)
000	.125	1/8	1	.375	No. 37 (.1040)
00	.141	9/64	1¼	.438	No. 31 (.1200)
0	.156	5/32	1½	.500	No. 28 (.1405)
1	.172	11/64	2	.562	No. 25 (.1495)
2	.193	—	2½	.625	No. 19 (.1660)
3	.219	7/32	3	.750	No. 12 (.1890)
4	.250	¼	3	.812	No. 3 (.2130)
5	.289	—	3	.875	¼ (.2500)
6	.341	—	4	1.000	9/32 (.2812)
7	.409	—	4	1.250	11/32 (.3438)
8	.492	—	4	1.500	13/32 (.4062)

*ANSI B5.20–1958.
† For Nos. 7/0, 6/0, 9, and 10 see the standard. Pins Nos. 11–14 are special sizes.
‡ Suggested sizes; not American National Standard.
Drill size is for reamer. The small diameter of a pin is equal to the large diameter minus $0.02083 \times L$, where L is the length.

Table 28 ◻ American national standard wrought-iron and steel pipe*

Standard weight							Heavy			
					Nominal wall thickness		Nominal wall thickness			
							Extra heavy		Double extra heavy	
Nominal size	Outside diameter (all weights)	Threads per inch	Tap drill sizes†	Distance pipe enters fitting	Wrought iron	Steel	Wrought iron	Steel	Wrought iron	Steel
1/8	.405	27	11/32	5/16	.069	.068	.099	.095	—	—
1/4	.540	18	7/16	7/16	.090	.088	.122	.119	—	—
3/8	.675	18	19/32	7/16	.093	.091	.129	.126	—	—
1/2	.840	14	23/32	9/16	.111	.109	.151	.147	.307	.294
3/4	1.050	14	15/16	9/16	.115	.113	.157	.154	.318	.308
1	1.315	11½	1 5/32	11/16	.136	.133	.183	.179	.369	.358
1¼	1.660	11½	1½	11/16	.143	.140	.195	.191	.393	.382
1½	1.900	11½	1 23/32	11/16	.148	.145	.204	.200	.411	.400
2	2.375	11½	2 3/16	3/4	.158	.154	.223	.218	.447	.436
2½	2.875	8	2 5/8	1 1/16	.208	.203	.282	.276	.567	.552
3	3.500	8	3¼	1 1/8	.221	.216	.306	.300	.615	.600
3½	4.000	8	3¾	1 3/16	.231	.226	.325	.318	—	—
4	4.500	8	4¼	1 3/16	.242	.237	.344	.337	.690	.674
5	5.563	8	5 5/16	1 5/16	.263	.258	.383	.375	.768	.750
6	6.625	8	6 3/8	1 3/8	.286	.280	.441	.432	.884	.864
8	8.625	8	—	—	.329	.322	.510	.500	.895	.875

*ANSI B36.10–1970, ANSI B2.1–1968.
All dimensions in inches.
† Not American Standard. See ANSI B36.10–1970 for sizes larger than 8 in.

Table 29A-1* □ American national standard running and sliding fits‡

Nominal size range (inches)† Over to	Class RC 1			Class RC 2			Class RC 3			Class RC 4		
	Limits of clearance	Standard limits		Limits of clearance	Standard limits		Limits of clearance	Standard limits		Limits of clearance	Standard Limits	
		Hole H5	Shaft g4		Hole H6	Shaft g5		Hole H7	Shaft f6		Hole H8	Shaft f7
0–0.12	.1 .45	+.2 0	−.1 .25	.1 .55	+.25 0	−.1 −.3	.3 .95	+.4 0	−.3 −.55	.3 1.3	+.6 0	−.3 −.7
0.12–0.24	.15 .5	+.2 0	−.15 −.3	.15 .65	+.3 0	−.15 −.35	.4 1.12	+.5 0	−.4 −.7	.4 1.6	+.7 0	−.4 −.9
0.24–0.40	.2 .6	.25 0	−.2 −.35	.2 .85	+.4 0	−.2 −.45	.5 1.5	+.6 0	−.5 −.9	.5 2.0	+.9 0	−.5 −1.1
0.40–0.71	.25 .75	+.3 0	−.25 −.45	.25 .95	+.4 0	−.25 −.55	.6 1.7	+.7 0	−.6 −1.0	.6 2.3	+1.0 0	−.6 −1.3
0.71–1.19	.3 .95	+.4 0	−.3 −.55	.3 1.2	+.5 0	−.3 −.7	.8 2.1	+.8 0	−.8 −1.3	.8 2.8	+1.2 0	−.8 −1.6
1.19–1.97	.4 1.1	+.4 0	−.4 −.7	.4 1.4	+.6 0	−.4 −.8	1.0 2.6	+1.0 0	−1.0 −1.6	1.0 3.6	+1.6 0	−1.0 −2.0
1.97–3.15	.4 1.2	+.5 0	−.4 −.7	.4 1.6	+.7 0	−.4 −.9	1.2 3.1	+1.2 0	−1.2 −1.9	1.2 4.2	+1.8 0	−1.2 −2.4
3.15–4.73	.5 1.5	+.6 0	−.5 −.9	.5 2.0	+.9 0	−.5 −1.1	1.4 3.7	+1.4 0	−1.4 −2.3	1.4 5.0	+2.2 0	−1.4 −2.8
4.73–7.09	.6 1.8	+.7 0	−.6 −1.1	.6 2.3	+1.0 0	−.6 −1.3	1.6 4.2	+1.6 0	−1.6 −2.6	1.6 5.7	+2.5 0	−1.6 −3.2

*Tables 29A-1 through 29E are extracted from ANSI B4.1–1967.

†For diameters greater than those listed in Tables 29A-1 through 29E see Standard.

‡Limits are in thousandths of an inch. Limits for hole and shaft are applied algebraically to the basic size to obtain the limits of size for the parts. Symbols H5, g5, etc., are hole and shaft designations used in ABC system.

Table 29A-2* American national standard running and sliding fits

Nominal size range (inches) Over to	Class RC 5			Class RC 6			Class RC 7			Class RC 8			Class RC 9		
	Limits of clearance	Standard limits		Limits of clearance	Standard limits		Limits of clearance	Standard limits		Limits of clearance	Standard limits		Limits of clearance	Standard limits	
		Hole H8	Shaft e7		Hole H9	Shaft e8		Hole H9	Shaft d8		Hole H10	Shaft c9		Hole H11	Shaft
0–0.12	.6 1.6	+.6 −0	−.6 −1.0	.6 2.2	+1.0 −0	−.6 −1.2	1.0 2.6	+1.0 0	−1.0 −1.6	2.5 5.1	+1.6 0	−2.5 −3.5	4.0 8.1	+2.5 0	−4.0 −5.6
0.12–0.24	.8 2.0	+.7 −0	−.8 −1.3	.8 2.7	+1.2 −0	−.8 −1.5	1.2 3.1	+1.2 0	−1.2 −1.9	2.8 5.8	+1.8 0	−2.8 −4.0	4.5 9.0	+3.0 0	−4.5 −6.0
0.24–0.40	1.0 2.5	+.9 −0	−1.0 −1.6	1.0 3.3	+1.4 −0	−1.0 −1.9	1.6 3.9	+1.4 0	−1.6 −2.5	3.0 6.6	+2.2 0	−3.0 −4.4	5.0 10.7	+3.5 0	−5.0 −7.2
0.40–0.71	1.2 2.9	+1.0 −0	−1.2 −1.9	1.2 3.8	+1.6 −0	−1.2 −2.2	2.0 4.6	+1.6 0	−2.0 −3.0	3.5 7.9	+2.8 0	−3.5 −5.1	6.0 12.8	+4.0 0	−6.0 −8.8
0.71–1.19	1.6 3.6	+1.2 −0	−1.6 −2.4	1.6 4.8	+2.0 −0	−1.6 −2.8	2.5 5.7	+2.0 0	−2.5 −3.7	4.5 10.0	+3.5 0	−4.5 −6.5	7.0 15.5	+5.0 0	−7.0 −10.5
1.19–1.97	2.0 4.6	+1.6 −0	−2.0 −3.0	2.0 6.1	+2.5 −0	−2.0 −3.6	3.0 7.1	+2.5 0	−3.0 −4.6	5.0 11.5	+4.0 0	−5.0 −7.5	8.0 18.0	+6.0 0	−8.0 −12.0
1.97–3.15	2.5 5.5	+1.8 −0	−2.5 −3.7	2.5 7.3	+3.0 −0	−2.5 −4.3	4.0 8.8	+3.0 0	−4.0 −5.8	6.0 13.5	+4.5 0	−6.0 −9.0	9.0 20.5	+7.0 0	−9.0 −13.5
3.15–4.73	3.0 6.6	+2.2 −0	−3.0 −4.4	3.0 8.7	+3.5 −0	−3.0 −5.2	5.0 10.7	+3.5 0	−5.0 −7.2	7.0 15.5	+5.0 0	−7.0 −10.5	10.0 24.0	+9.0 0	−10.0 −15.0
4.73–7.09	3.5 7.6	+2.5 −0	−3.5 −5.1	3.5 10.0	+4.0 −0	−3.5 −6.0	6.0 12.5	+4.0 0	−6.0 −8.5	8.0 18.0	+6.0 0	−8.0 −12.0	12.0 28.0	+10.0 0	−12.0 −18.0

*Limits are in thousandths of an inch. Limits for hole and shaft are applied algebraically to the basic size to obtain the limits of size for the parts. Symbols H8, e7, etc., are hole and shaft designations used in ABC system.

Table 29B-1* □ American national standard locational clearance fits

Nominal size range (inches) Over to	Class LC 1 Limits of clearance	Standard limits Hole H6	Shaft h5	Class LC 2 Limits of clearance	Standard limits Hole H7	Shaft h6	Class LC 3 Limits of clearance	Standard limits Hole H8	Shaft h7	Class LC 4 Limits of clearance	Standard limits Hole H10	Shaft h9	Class LC 5 Limits of clearance	Standard limits Hole H7	Shaft g6
0–0.12	0 .45	+.25 −0	+0 −.2	0 .65	+.4 −0	+0 −.25	0 1	+.6 −0	+0 −.4	0 2.6	+1.6 −0	+0 −1.0	.1 .75	+.4 −0	−.1 −.35
0.12–0.24	0 .5	+.3 −0	+0 −.2	0 .8	+.5 −0	+0 −.3	0 1.2	+.7 −0	+0 −.5	0 3.0	+1.8 −0	+0 −1.2	.15 .95	+.5 −0	−.15 −.45
0.24–0.40	0 .65	+.4 −0	+0 −.25	0 1.0	+.6 −0	+0 −.4	0 1.5	+.9 −0	+0 −.6	0 3.6	+2.2 −0	+0 −1.4	.2 1.2	+.6 −0	−.2 −.6
0.40–0.71	0 .7	+.4 −0	+0 −.3	0 1.1	+.7 −0	+0 −.4	0 1.7	+1.0 −0	+0 −.7	0 4.4	+2.8 −0	+0 −1.6	.25 1.35	+.7 −0	−.25 −.65
0.71–1.19	0 .9	+.5 −0	+0 −.4	0 1.3	+.8 −0	+0 −.5	0 2	+1.2 −0	+0 −.8	0 5.5	+3.5 −0	+0 −2.0	.3 1.6	+.8 −0	−.3 −.8
1.19–1.97	0 1.0	+.6 −0	+0 −.4	0 1.6	+1.0 −0	+0 −.6	0 2.6	+1.6 −0	+0 −1	0 6.5	+4.0 −0	+0 −2.5	.4 2.0	+1.0 −0	−.4 −1.0
1.97–3.15	0 1.2	+.7 −0	+0 −.5	0 1.9	+1.2 −0	+0 −.7	0 3	+1.8 −0	+0 −1.2	0 7.5	+4.5 −0	+0 −3	.4 2.3	+1.2 −0	−.4 −1.1
3.15–4.73	0 1.5	+.9 −0	+0 −.6	0 2.3	+1.4 −0	+0 −.9	0 3.6	+2.2 −0	+0 −1.4	0 8.5	+5.0 −0	+0 −3.5	.5 2.8	+1.4 −0	−.5 −1.4
4.73–7.09	0 1.7	+1.0 −0	+0 −.7	0 2.6	+1.6 −0	+0 −1.0	0 4.1	+2.5 −0	+0 −1.6	0 10	+6.0 −0	+0 −4	.6 3.2	+1.6 −0	−.6 −1.6

*Limits are in thousandths of an inch. Limits for hole and shaft are applied algebraically to the basic size to obtain the limits of size for parts. Symbols H6, h5, etc., are hole and shaft designations used in ABC system.

Table 29B-2* □ American national standard locational clearance fits

Nominal size range (inches) Over to	Class LC 6			Class LC 7			Class LC 8			Class LC 9			Class LC 10			Class LC 11		
	Limits of clearance	Standard limits		Limits of clearance	Standard limits		Limits of clearance	Standard limits		Limits of clearance	Standard limits		Limits of clearance	Standard limits		Limits of clearance	Standard limits	
		Hole H9	Shaft f8		Hole H10	Shaft e9		Hole H10	Shaft d9		Hole H11	Shaft c10		Hole H12	Shaft		Hole H13	Shaft
0–0.12	.3 1.9	+1.0 0	−.3 −.9	.6 3.2	+1.6 0	−.6 −1.6	1.0 3.6	+1.6 −0	−1.0 −2.0	2.5 6.6	+2.5 −0	−2.5 −4.1	4 12	+4 −0	−4 −8	5 17	+6 −0	−5 −11
0.12–0.24	.4 2.3	+1.2 0	−.4 −1.1	.8 3.8	+1.8 0	−.8 −2.0	1.2 4.2	+1.8 −0	−1.2 −2.4	2.8 7.6	+3.0 −0	−2.8 −4.6	4.5 14.5	+5 −0	−4.5 −9.5	6 20	+7 −0	−6 −13
0.24–0.40	.5 2.8	+1.4 0	−.5 −1.4	1.0 4.6	+2.2 0	−1.0 −2.4	1.6 5.2	+2.2 −0	−1.6 −3.0	3.0 8.7	+3.5 −0	−3.0 −5.2	5 17	+6 −0	−5 −11	7 25	+9 −0	−7 −16
0.40–0.71	.6 3.2	+1.6 0	−.6 −1.6	1.2 5.6	+2.8 0	−1.2 −2.8	2.0 6.4	+2.8 −0	−2.0 −3.6	3.5 10.3	+4.0 −0	−3.5 −6.3	6 20	+7 −0	−6 −13	8 28	+10 −0	−8 −18
0.71–1.19	.8 4.0	+2.0 0	−.8 −2.0	1.6 7.1	+3.5 0	−1.6 −3.6	2.5 8.0	+3.5 −0	−2.5 −4.5	4.5 13.0	+5.0 −0	−4.5 −8.0	7 23	+8 −0	−7 −15	10 34	+12 −0	−10 −22
1.19–1.97	1.0 5.1	+2.5 0	−1.0 −2.6	2.0 8.5	+4.0 0	−2.0 −4.5	3.0 9.5	+4.0 −0	−3.0 −5.5	5 15	+6 −0	−5 −9	8 28	+10 −0	−8 −18	12 44	+16 −0	−12 −28
1.97–3.15	1.2 6.0	+3.0 0	−1.2 −3.0	2.5 10.0	+4.5 0	−2.5 −5.5	4.0 11.5	+4.5 −0	−4.0 −7.0	6 17.5	+7 −0	−6 −10.5	10 34	+12 −0	−10 −22	14 50	+18 −0	−14 −32
3.15–4.73	1.4 7.1	+3.5 0	−1.4 −3.6	3.0 11.5	+5.0 0	−3.0 −6.5	5.0 13.5	+5.0 −0	−5.0 −8.5	7 21	+9 −0	−7 −12	11 39	+14 −0	−11 −25	16 60	+22 −0	−16 −38
4.73–7.09	1.6 8.1	+4.0 0	−1.6 −4.1	3.5 13.5	+6.0 0	−3.5 −7.5	6 16	+6 −0	−6 −10	8 24	+10 −0	−8 −14	12 44	+16 −0	−12 −28	18 68	+25 −0	−18 −43

*Limits are in thousandths of an inch. Limits for hole and shaft are applied algebraically to the basic size to obtain the limits of size for the parts. Symbols H9, f8, etc., are hole and shaft designations used in ABC system.

Table 29C* □ American national standard locational transition fits

Nominal size range (inches) Over to	Class LT 1 Fit	Standard limits Hole H7	Shaft js6	Class LT 2 Fit	Standard limits Hole H8	Shaft js7	Class LT 3 Fit	Standard limits Hole H7	Shaft k6	Class LT 4 Fit	Standard limits Hole H8	Shaft k7	Class LT 5 Fit	Standard limits Hole H7	Shaft n6	Class LT 6 Fit	Standard limits Hole H7	Shaft n7
0–0.12	−.10 +.50	+.4 −0	+.10 −.10	−.2 +.8	+.6 −0	+.2 −.2							−.5 +.15	+.4 −0	+.5 +.25	−.65 +.15	+.4 −0	+.65 +.25
0.12–0.24	−.15 +.65	+.5 −0	+.15 −.15	−.25 +.95	+.7 −0	+.25 −.25							−.6 +.2	+.5 −0	+.6 −.3	−.8 +.2	+.5 −0	+.8 +.3
0.24–0.40	−.2 +.8	+.6 −0	+.2 −.2	−.3 +1.2	+.9 −0	+.3 −.3	−.5 +.5	+.6 −0	+.5 +.1	−.7 +.8	+.9 −0	+.7 +.1	−.8 +.2	+.6 −0	+.8 +.4	−1.0 +.2	+.6 −0	+1.0 +.4
0.40–0.71	−.2 +.9	+.7 −0	+.2 −.2	−.35 +1.35	+1.0 −0	+.35 −.35	−.5 +.6	+.7 −0	+.5 +.1	−.8 +.9	+1.0 −0	+.8 +.1	−.9 +.2	+.7 −0	+.9 +.5	−1.2 +.2	+.7 −0	+1.2 +.5
0.71–1.19	−.25 +1.05	+.8 −0	+.25 −.25	−.4 +1.6	+1.2 −0	+.4 −.4	−.6 +.7	+.8 −0	+.6 +.1	−.9 +1.1	+1.2 −0	+.9 +.1	−1.1 +.2	+.8 −0	+1.1 +.6	−1.4 +.2	+.8 −0	+1.4 +.6
1.19–1.97	−.3 +1.3	+1.0 −0	+.3 −.3	−.5 +2.1	+1.6 −0	+.5 −.5	−.7 +.9	+1.0 −0	+.7 +.1	−1.1 +1.5	+1.6 −0	+1.1 +.1	−1.3 +.3	+1.0 −0	+1.3 +.7	−1.7 +.3	+1.0 −0	+1.7 +.7
1.97–3.15	−.3 +1.5	+1.2 −0	+.3 −.3	−.6 +2.4	+1.8 −0	+.6 −.6	−.8 +1.1	+1.2 −0	+.8 +.1	−1.3 +1.7	+1.8 −0	+1.3 +.1	−1.5 +.4	+1.2 −0	+1.5 +.8	−2.0 +.4	+1.2 −0	+2.0 +.8
3.15–4.73	−.4 +1.8	+1.4 −0	+.4 −.4	−.7 +2.9	+2.2 −0	+.7 −.7	−1.0 +1.3	+1.4 −0	+1.0 +.1	−1.5 +2.1	+2.2 −0	+1.5 +.1	−1.9 +.4	+1.4 −0	+1.9 +1.0	−2.4 +.4	+1.4 −0	+2.4 +1.0
4.73–7.09	−.5 +2.1	+1.6 −0	+.5 −.5	−.8 +3.3	+2.5 −0	+.8 −.8	−1.1 +1.5	+1.6 −0	+1.1 +.1	−1.7 +2.4	+2.5 −0	+1.7 +.1	−2.2 +.4	+1.6 −0	+2.2 +1.2	−2.8 +.4	+1.6 −0	+2.8 +1.2

*Limits are in thousandths of an inch. Limits for hole and shaft are applied algebraically to the basic size to obtain the limits of size for the mating parts. "Fit" represents the maximum interference (minus values) and the maximum clearance (plus values). Symbols H7, js6, etc., are hole and shaft designations used in ABC system.

Table 29D* □ American national standard locational interference fits

Nominal size range (inches) Over to	Class LN 1 Limits of interference	Standard limits Hole H6	Shaft n5	Class LN 2 Limits of interference	Standard limits Hole H7	Shaft p6	Class LN 3 Limits of interference	Standard limits Hole H7	Shaft r6
0–0.12	0 .45	+.25 −0	+.45 +.25	0 .65	+.4 −0	+.65 +.4	.1 .75	+.4 −0	+.75 +.5
0.12–0.24	0 .5	+.3 −0	+.5 +.3	0 .8	+.5 −0	+.8 +.5	.1 .9	+.5 −0	+.9 +.6
0.24–0.40	0 .65	+.4 −0	+.65 +.4	0 1.0	+.6 −0	+1.0 +.6	.2 1.2	+.6 −0	+1.2 +.8
0.40–0.71	0 .8	+.4 −0	+.8 +.4	0 1.1	+.7 −0	+1.1 +.7	.3 1.4	+.7 −0	+1.4 +1.0
0.71–1.19	0 1.0	+.5 −0	+1.0 +.5	0 1.3	+.8 −0	+1.3 +.8	.4 1.7	+.8 −0	+1.7 +1.2
1.19–1.97	0 1.1	+.6 −0	+1.1 +.6	0 1.6	+1.0 −0	+1.6 +1.0	.4 2.0	+1.0 −0	+2.0 +1.4
1.97–3.15	.1 1.3	+.7 −0	+1.3 +.7	.2 2.1	+1.2 −0	+2.1 +1.4	.4 2.3	+1.2 −0	+2.3 +1.6
3.15–4.73	.1 1.6	+.9 −0	+1.6 +1.0	.2 2.5	+1.4 −0	+2.5 +1.6	.6 2.9	+1.4 −0	+2.9 +2.0
4.73–7.09	.2 1.9	+1.0 −0	+1.9 +1.2	.2 2.8	+1.6 −0	+2.8 +1.8	.9 3.5	+1.6 −0	+3.5 +2.5

*Limits are in thousandths of an inch. Limits for hole and shaft are applied algebraically to the basic size to obtain the limits of size for the parts. Symbols H7, p6, etc., are hole and shaft designations used in ABC system.

Table 29E* □ American national standard force and shrink fits

Nominal size range (inches) Over to	Class FN 1 Limits of interference	Standard limits Hole H6	Shaft	Class FN 2 Limits of interference	Standard limits Hole H7	Shaft s6	Class FN 3 Limits of interference	Standard limits Hole H7	Shaft t6	Class FN 4 Limits of interference	Standard limits Hole H7	Shaft u6	Class FN 5 Limits of interference	Standard limits Hole H8	Shaft x7
0–0.12	.05	+.25	+.5	.2	+.4	+.85				.3	+.4	+.95	.3	+.6	+1.3
	.5	−0	+.3	.85	−0	+.6				.95	−0	+.7	1.3	−0	+.9
0.12–0.24	.1	+.3	+.6	.2	+.5	+1.0				.4	+.5	+1.2	.5	+.7	+1.7
	.6	−0	+.4	1.0	−0	+.7				1.2	−0	+.9	1.7	−0	+1.2
0.24–0.40	.1	+.4	+.75	.4	+.6	+1.4				.6	+.6	+1.6	.5	+.9	+2.0
	.75	−0	+.5	1.4	−0	+1.0				1.6	−0	+1.2	2.0	−0	+1.4
0.40–0.56	.1	+.4	+.8	.5	+.7	+1.6				.7	+.7	+1.8	.6	+1.0	+2.3
	.8	−0	+.5	1.6	−0	+1.2				1.8	−0	+1.4	2.3	−0	+1.6
0.56–0.71	.2	+.4	+.9	.5	+.7	+1.6				.7	+.7	+1.8	.8	+1.0	+2.5
	.9	−0	+.6	1.6	−0	+1.2				1.8	−0	+1.4	2.5	−0	+1.8
0.71–0.95	.2	+.5	+1.1	.6	+.8	+1.9				.8	+.8	+2.1	1.0	+1.2	+3.0
	1.1	−0	+.7	1.9	−0	+1.4				2.1	−0	+1.6	3.0	−0	+2.2
0.95–1.19	.3	+.5	+1.2	.6	+.8	+1.9	.8	+.8	+2.1	1.0	+.8	+2.3	1.3	+1.2	+3.3
	1.2	−0	+.8	1.9	−0	+1.4	2.1	−0	+1.6	2.3	−0	+1.8	3.3	−0	+2.5
1.19–1.58	.3	+.6	+1.3	.8	+1.0	+2.4	1.0	+1.0	+2.6	1.5	+1.0	+3.1	1.4	+1.6	+4.0
	1.3	−0	+.9	2.4	−0	+1.8	2.6	−0	+2.0	3.1	−0	+2.5	4.0	−0	+3.0
1.58–1.97	.4	+.6	+1.4	.8	+1.0	+2.4	1.2	+1.0	+2.8	1.8	+1.0	+3.4	2.4	+1.6	+5.0
	1.4	−0	+1.0	2.4	−0	+1.8	2.8	−0	+2.2	3.4	−0	+2.8	5.0	−0	+4.0
1.97–2.56	.6	+.7	+1.8	.8	+1.2	+2.7	1.3	+1.2	+3.2	2.3	+1.2	+4.2	3.2	+1.8	+6.2
	1.8	−0	+1.3	2.7	−0	+2.0	3.2	−0	+2.5	4.2	−0	+3.5	6.2	−0	+5.0
2.56–3.15	.7	+.7	+1.9	1.0	+1.2	+2.9	1.8	+1.2	+3.7	2.8	+1.2	+4.7	4.2	+1.8	+7.2
	1.9	−0	+1.4	2.9	−0	+2.2	3.7	−0	+3.0	4.7	−0	+4.0	7.2	−0	+6.0
3.15–3.94	.9	+.9	+2.4	1.4	+1.4	+3.7	2.1	+1.4	+4.4	3.6	+1.4	+5.9	4.8	+2.2	+8.4
	2.4	−0	+1.8	3.7	−0	+2.8	4.4	−0	+3.5	5.9	−0	+5.0	8.4	−0	+7.0
3.94–4.73	1.1	+.9	+2.6	1.6	+1.4	+3.9	2.6	+1.4	+4.9	4.6	+1.4	+6.9	5.8	+2.2	+9.4
	2.6	−0	+2.0	3.9	−0	+3.0	4.9	−0	+4.0	6.9	−0	+6.0	9.4	−0	+8.0

*Limits are in thousandths of an inch. Limits for hole and shaft are applied algebraically to the basic size to obtain the limits of size for the parts. Symbols H7, s6, etc., are hole and shaft designations used in ABC system.

Table 30 □ Twist drill sizes*

Number sizes

No. size	Decimal equivalent	No. size	Decimal equivalent
1	.2280	41	.0960
2	.2210	42	.0935
3	.2130	43	.0890
4	.2090	44	.0860
5	.2055	45	.0820
6	.2040	46	.0810
7	.2010	47	.0785
8	.1990	48	.0760
9	.1960	49	.0730
10	.1935	50	.0700
11	.1910	51	.0670
12	.1890	52	.0635
13	.1850	53	.0595
14	.1820	54	.0550
15	.1800	55	.0520
16	.1770	56	.0465
17	.1730	57	.0430
18	.1695	58	.0420
19	.1660	59	.0410
20	.1610	60	.0400
21	.1590	61	.0390
22	.1570	62	.0380
23	.1540	63	.0370
24	.1520	64	.0360
25	.1495	65	.0350
26	.1470	66	.0330
27	.1440	67	.0320
28	.1405	68	.0310
29	.1360	69	.0292
30	.1285	70	.0280
31	.1200	71	.0260
32	.1160	72	.0250
33	.1130	73	.0240
34	.1110	74	.0225
35	.1100	75	.0210
36	.1065	76	.0200
37	.1040	77	.0180
38	.1015	78	.0160
39	.0995	79	.0145
40	.0980	80	.0135

Letter sizes

Size letter	Decimal equivalent
A	.234
B	.238
C	.242
D	.246
E	.250
F	.257
G	.261
H	.266
I	.272
J	.277
K	.281
L	.290
M	.295
N	.302
O	.316
P	.323
Q	.332
R	.339
S	.348
T	.358
U	.368
V	.377
W	.386
X	.397
Y	.404
Z	.413

*Fraction-size drills range in size from one-sixteenth to 4 in. and over in diameter—by sixty-fourths.

Table 31 □ Standard wire and sheet-metal gages*

Gage number	(A) Brown & Sharpe or American	(B) American Steel & Wire Co.	(C) Piano wire	(E) U.S. ST'D.	Gage number
0000000	.6513	.4900	—	.5000	0000000
000000	.5800	.4615	.004	.4688	000000
00000	.5165	.4305	.005	.4375	00000
0000	.4600	.3938	.006	.4063	0000
000	.4096	.3625	.007	.3750	000
00	.3648	.3310	.008	.3438	00
0	.3249	.3065	.009	.3125	0
1	.2893	.2830	.010	.2813	1
2	.2576	.2625	.011	.2656	2
3	.2294	.2437	.012	.2500	3
4	.2043	.2253	.013	.2344	4
5	.1819	.2070	.014	.2188	5
6	.1620	.1920	.016	.2031	6
7	.1443	.1770	.018	.1875	7
8	.1285	.1620	.020	.1719	8
9	.1144	.1483	.022	.1563	9
10	.1019	.1350	.024	.1406	10
11	.0907	.1205	.026	.1250	11
12	.0808	.1055	.029	.1094	12
13	.0720	.0915	.031	.0938	13
14	.0641	.0800	.033	.0781	14
15	.0571	.0720	.035	.0703	15
16	.0508	.0625	.037	.0625	16
17	.0453	.0540	.039	.0563	17
18	.0403	.0475	.041	.0500	18
19	.0359	.0410	.043	.0438	19
20	.0320	.0348	.045	.0375	20
21	.0285	.0317	.047	.0344	21
22	.0253	.0286	.049	.0313	22
23	.0226	.0258	.051	.0281	23
24	.0201	.0230	.055	.0250	24
25	.0179	.0204	.059	.0219	25
26	.0159	.0181	.063	.0188	26
27	.0142	.0173	.067	.0172	27
28	.0126	.0162	.071	.0156	28
29	.0113	.0150	.075	.0141	29
30	.0100	.0140	.080	.0125	30
31	.0089	.0132	.085	.0109	31
32	.0080	.0128	.090	.0102	32
33	.0071	.0118	.095	.0094	33
34	.0063	.0104	.100	.0086	34
35	.0056	.0095	.106	.0078	35
36	.0050	.0090	.112	.0070	36
37	.0045	.0085	.118	.0066	37
38	.0040	.0080	.124	.0063	38
39	.0035	.0075	.130	—	39
40	.0031	.0070	.138	—	40

*Dimensions in decimal parts of an inch.
(A) Standard in United States for sheet metal and wire (except steel and iron).
(B) Standard for iron and steel wire (U.S. Steel Wire Gage).
(C) American Steel and Wire Company's music (or piano) wire gage sizes. Recognized by U.S. Bureau of Standards.
(E) U.S. Standard for iron and steel plate. However, plate is now generally specified by its thickness in decimals of an inch.

Symbols, abbreviations, ANSI standards

Graphical symbols for electrical diagrams

AMPLIFIER General	**DEVICE, VISUAL SIGNALING** Annunciator	**MICROPHONE**
ANTENNA General	**ELEMENT, THERMAL** Thermal Cutout, Flasher	**PATH, TRANSMISSION** General
Loop	Thermal Relay or	Crossing, Not Connected
		Junction
ARRESTER General Multigap		Junction, Connected Paths
	FUSE General	Pair
BATTERY Multicell	Fusible Element	Assembled Conductors; Cable Coaxial
BREAKER, CIRCUIT General	**GROUND**	2-Conductor Cable
CAPACITOR General		Grouping Leads or
COIL Blowout (Broken line not part of symbol) Operating	**HANDSET** General	**RECEIVER** General Headset
	INDUCTOR WINDING General or	**RECTIFIER** General
CONNECTOR Female Contact Male Contact	**LAMP** Ballast Lamp Incandescent	**REPEATER** 1-Way
CONTACT, ELECTRIC Fixed ∘ or Locking Nonlocking Rotating Closed or Open or	**MACHINE, ROTATING** Basic Generator Motor Wound Rotor Armature	**RESISTOR** General
		SWITCH General Single Throw Double Throw Knife
CORE Magnetic (General) Magnet or Relay	Winding Symbols 1-Phase 2-Phase 3-Phase Wye 3-Phase Delta	**THERMOCOUPLE** Temperature-Measuring
COUPLER, DIRECTIONAL General		**TRANSFORMER** Magnetic Core Shielded

*ANSI Y32.2.

Abbreviations and symbols*

alternating current	AC	engineer	ENGR	pitch	P
aluminum	AL	external	EXT	pitch diameter	PD
American Standard	AMER STD	fabricate	FAB	plate	PL
approved	APPD	fillister	FIL	pound	# or LB
average	AVG	finish	FIN.	pounds per square foot	PSF
Babbitt	BAB	foot	′ or FT	pounds per square inch	PSI
ball bearing	BB	gallon	GAL	Pratt & Whitney	P & W
brass	BRS	galvanized iron	GI	quantity	QTY
Brinell hardness number	BHN	grind	GRD	radius	R or RAD
bronze	BRZ	harden	HDN	required	REQD
Brown & Sharpe	B & S	hexagon	HEX	revolution per minute	RPM
cast iron	CI	horsepower	HP	right hand	RH
center line	CL or ℄	hour	HR	round	RD
center to center	C to C	impregnate	IMPREG	round bar	φ
centimeter	CM	inch	″ or IN.	screw	SCR
chemical	CHEM	inside diameter	ID	second (time)	SEC
circular	CIR	internal	INT	second (angular measure)	″
circular pitch	CP	left hand	LH	section	SECT
copper	COP	lateral	LAT	Society of Automotive Engineers	SAE
cold-rolled steel	CRS	long	LG	square	SQ
counterbore	CBORE	longitude	LONG.	square inch	SQ IN
countersink	CSK	linear	LIN	square foot	SQ FT
cubic	CU	machine	MACH	standard	STD
cubic inch	CU IN	malleable iron	MI	steel	STL
cubic foot	CU FT	material	MATL	steel casting	STL CSTG
cubic yard	CU YD	maximum	MAX	thousand	M
cylinder	CYL	meter	M	ton	TON
degree	DEG or °	miles	MI	thread	THD
diameter	DIA	millimeter	MM	traced	TR
diagonal	DIAG	miles per hour	MPH	volt	V
diametral pitch	DP	minimum	MIN	watt	W
direct current	DC	minute (angular measure)	′ or MIN	weight	WT
drawing	DWG	minute (time)	MIN	Woodruff	WDF
drawn	DR	outside diameter	OD	wrought iron	WI
detail drawing	DET DWG	pattern	PATT	yard	YD
effective	EFF	phosphor bronze	PH BRZ	year	YR
electric	ELEC	piece	PC		

*Only those abbreviations that are commonly used on engineering drawings have been given in the list above. These have been selected from the long and comprehensive list given in ANSI Z32.13–1950. The professional draftsman should have this standard close at hand for use when needed.

American national standards*

A few of the more than 500 standards approved by the American National Standards Institute are listed below. Copies may be obtained from the American Society of Mechanical Engineers at 345 East 47th Street, New York, N.Y. 10017

B1.1–1960	Unified and American Screw Threads for Screws, Bolts, Nuts, and Other Threaded Parts
B1.5–1952	Acme Screw Threads
B1.8–1952	Stub Acme Screw Threads
B1.9–1953	Buttress Screw Threads
B2.1–1968	Pipe Threads
B2.4–1966	Hose Coupling Screw Threads
B4.1–1967	Preferred Limits and Fits for Cylindrical Parts
B5.10–1963	Machine Tapers, Self-Holding and Steep Taper Series
B5.15–1960	Involute Splines, Side Bearing
B5.20–1958	Machine Pins
B16.1–1967	Cast-Iron Pipe Flanges and Flanged Fittings, Class 25, 125, 250, and 800 lb
B16.3–1963	Malleable-Iron Screwed Fittings, 150 lb
B16.4–1963	Cast-Iron Screwed Fittings for Maximum WSP of 125 and 250 lb
B16.5–1968	Steel Pipe Flanges and Flanged Fittings
B16.9–1964	Steel Butt-Welding Fittings
B17.1–1967	Shafting and Stock Keys
B17.2–1967	Woodruff Keys and Keyseats
B18.1–1965	Small Solid Rivets
B18.2.1–1965	Square and Hex Bolts and Screws
B18.2.2–1965	Square and Hex Nuts
B18.3–1969	Socket Set Screws and Socket-Head Cap Screws
B18.4–1960	Large Rivets
B18.5–1959	Round-Head Bolts
B18.6.2–1956	Hexagon-Head Cap Screws, Slotted-Head Cap Screws, Square-Head Set Screws, and Slotted Headless Set Screws
B36.1–1956	Welded and Seamless Pipe (ASTM A53–44)
B36.2–1956	Welded Wrought-Iron Pipe
B36.10–1970	Wrought-Iron and Wrought-Steel Pipe
B48.1–1933	Inch-Millimeter Conversion for Industrial Use
B94.6–1966	Knurling

*Formerly the ASA and USASI Standards.

Y14.1–1957	Size and Format
Y14.2–1973	Line Conventions, Sectioning, and Lettering
Y14.3–1957	Projections
Y14.4–1957	Pictorial Drawing
Y14.5–1973	Dimensioning and Notes
Y14.6–1957	Screw Threads
Y14.7–1958	Gears, Splines, and Serrations
Y14.7.1–1971	For Spur, Helical, Double Helical and Rack
Y14.8– —	Castings (In preparation)
Y14.9–1958	Forgings
Y14.10–1959	Metal Stampings
Y14.11–1958	Plastics
Y14.12– —	Die Castings } (In preparation)
Y14.13– —	Springs, Helical and Flat }
Y14.14–1961	Mechanical Assemblies
Y14.15–1966	Electrical and Electronics Diagrams
Y14.15a–1971	Supplement-Interconnecting Diagrams
Y14.17–1966	Fluid Power Diagrams
Y14.18– —	Drawings for Optical Parts
Y14.22– —	Drawing Titles and Notes
Y14.24– —	Classifications of Drawings } (In preparation)
Y14.25– —	Guide for Numbering Systems
Y14.26– —	Computer Graphics

Graphical and letter symbols*

Y1.1–1972	Abbreviations for Use on Drawings and in Text (Graphical)*
Y10.2–1958	Hydraulics (Letter)
Y10.3–1968	Mechanics for Solid Bodies (Letter)
Y10.4–1957	Heat and Thermodynamics (Letter)
Y10.7–1954	Aeronautical Sciences (Letter)
Y10.8–1962	Structural Analysis (Letter)*
Y10.9–1953	Radio (Letter)
Y10.11–1959	Acoustics (Letter)
Y10.12–1961	Chemical Engineering (Letter)
Y10.14–1959	Rocket Propulsion (Letter)
Y10.15–1958	Petroleum Reservoir Engineering and Electric Logging (Letter)
Y10.16–1964	Shell Theory (Letter)

*Y is the new letter assigned to standards for abbreviations, charts and graphs, drawings, graphical symbols, and letter symbols. The Z will be changed to Y as the standards are revised and reaffirmed.

Y10.17–1961	Guide for Selecting Greek Letters Used as Letter Symbols for Engineering Mathematics (Letter)
Y15.2–1960	Time-Series Charts, Manual of Design
Y32.2–1970	Electrical and Electronics Diagrams (Graphical)
Y32.3–1969	Welding (Graphical)
Y32.4–1955	Plumbing (Graphical)
Y32.7–1972	Use on Railroad Maps and Profiles (Graphical)
Y32.10–1967	Fluid Power Diagrams (Graphical)
Y32.11–1961	Process Flow Diagrams in Petroleum and Chemical Industries (Graphical)
Y32.12–1960	Metallizing Symbols (Graphical)
Y32.17–1962	Nondestructive Testing Symbols (Graphical)
Z10.1–1941	Abbreviations for Scientific and Engineering Terms (Letter)*
Z10.6–1948	Physics (Letter)*
Z15.3–1943	Engineering and Scientific Graphs for Publications
Z32.2.3–1953	Pipe Fittings, Valves, and Piping (Graphical)*
Z32.2.4–1953	Heating, Ventilating, and Air Conditioning (Graphical)*
Z32.2.6–1956	Heat-Power Apparatus (Graphical)*

Appendix G

Bibliography of engineering drawing and allied subjects

Aeronautical drafting and engineering

Anderson, N. H., *Aircraft Layout and Detail Design*, 2nd ed., New York: McGraw-Hill.

Katz, H. H., *Aircraft Drafting*, New York: Macmillan.

LeMaster, C. A., *Aircraft Sheet Metal Work*, Chicago: American Technical Society.

Castings

Campbell, H. I., *Metal Castings*, New York: Wiley.

Morris, J. L., *Metal Castings*, Englewood Cliffs, N.J.: Prentice-Hall.

Catalogs (instruments)

Boston Gear Works, Inc., Chicago, Ill. (Gears, etc).

Crane Co., Chicago, Ill. (Pipe Fittings)

New Departure, Bristol, Conn. (Ball-Bearing Handbook)

Timken Roller Bearing Co., Canton, Ohio. (Roller Bearings)

Computer graphics films

DAC-1 (color), General Motors Corporation, Public Relations Staff, Film Library, Detroit, Michigan 48202.

Graphic Data Processing, International Business Machines Corp., available through local IBM representative.

Sketchpad, Lincoln Laboratories, Massachusetts Institute of Technology, Lexington, Massachusetts.

Computers-data processing systems

Bernstein, Jeremy, *The Analytical Engine*, New York: Random House.

Englebardt, S. L., *Computers*, New York: Pyramid.

Fetter, W. A., *Computer Graphics in Communication*, New York: McGraw-Hill.

Hagg, J. N., *Comprehensive Fortran Programming*, New York: Hayden.

International Business Machines Corp., *1130 Presentation System*.

———, *Introduction to Data Processing Systems*.

Lohberg, Rolf and Theo Lutz, *Electronic Brains*, New York: Sterling.

Thornhill, R. B., *Engineering Graphics and Numerical Control*, New York: McGraw-Hill.

Descriptive geometry

Johnson, L. O., and I. Wladaver, *Elements of Descriptive Geometry*, Englewood Cliffs, N.J.: Prentice-Hall.

Paré, E. G., R. O. Loving, and I. L. Hill, *Descriptive Geometry*, 2nd ed., New York: Macmillan.

Slaby, S. M., *Three-Dimensional Descriptive Geometry*, New York: Harcourt.

Street, W. E., *Technical Descriptive Geometry*, New York: Van Nostrand.

Warner, F. M., and M. McNeary, *Applied Descriptive Geometry*, 5th ed., New York: McGraw-Hill.

Watts, E. F., and J. T. Rule, *Descriptive Geometry*, Englewood Cliffs, N.J.: Prentice-Hall.

Wellman, B. L., *Technical Descriptive Geometry*, 2nd ed., New York: McGraw-Hill.

Descriptive geometry problems

Paré, E. G., and others, *Descriptive Geometry Worksheets*, New York: Macmillan.

Wiedhaas, E. R., *Applied Descriptive Geometry Problems*, New York: McGraw-Hill.

Die casting

Chase, H., *Die Castings*, New York: Wiley.

Electrical drawing

Baer, C. J., *Electrical and Electronic Drawing*, New York: McGraw-Hill.

Bishop, C. C., C. T. Gilliam, and others, *Electrical Drafting and Design*, 3rd ed., New York: McGraw-Hill.

Carini, L. F. D., *Drafting for Electronics*, New York: McGraw-Hill.

Kuller, K. K., *Electronics Drafting*, New York: McGraw-Hill.

Mark, D., *How to Read Schematic Diagrams*, New York: Rider.

Raskhodoff, N. M., *Electronic Drafting Handbook*, New York: Macmillan.

Shiers, G., *Electronic Drafting*, Englewood Cliffs, N.J.: Prentice-Hall.

———, *Electronic Drafting Techniques and Exercises*, Englewood Cliffs, N.J.: Prentice-Hall.

Van Gieson, D. W., *Electrical Drafting*, New York: McGraw-Hill.

Engineering design

Alger, J. R., and C. V. Hayes, *Creative Synthesis in Design*, Englewood Cliffs, N.J.: Prentice-Hall.

Asimow, Morris, *Introduction to Design*, Englewood Cliffs, N.J.: Prentice-Hall.

Bartee, E. M., *Engineering Experimental Design Fundamentals*, Englewood Cliffs, N.J.: Prentice-Hall.

Crede, C. E., *Shock and Vibration Concepts in Engineering Design*, Englewood Cliffs, N.J.: Prentice-Hall.

Damon, Albert, H. W. Stoudt, and R. A. McFarland, *The Human Body in Equipment Design*, Cambridge, Mass.: Harvard University Press.

Dixon, J. R., *Design Engineering: Inventiveness, Analysis, and Decision-Making*, New York: McGraw-Hill.

Dreyfuss, Henry, *The Measure of Man*, New York: Whitney Library of Design.

Edel, D. H., *Introduction to Creative Design*, Englewood Cliffs, N.J.: Prentice-Hall.

Gibson, J. E., *Introduction to Engineering Design*, New York: Holt, Rinehart and Winston.

Graves, M. E., *The Art of Color and Design*, New York: McGraw-Hill.

Greenwood, D. C., *Mechanical Details for Product Design*, New York: McGraw-Hill.

Hill, P. H., *The Science of Engineering Design*, New York: Holt, Rinehart and Winston.

Klemm, Friedrich, *A History of Western Technology*, Cambridge, Mass.: M.I.T. Press.

Krick, E. V., *An Introduction to Engineering and Engineering Design*, New York: Wiley.

Mason, O. T., *The Origins of Invention*, Cambridge, Mass.: M.I.T. Press.

McCormick, E. J., *Human Factors Engineering*, New York: McGraw-Hill.

Middendorf, W. H., *Engineering Design*, Boston: Allyn and Bacon.

Mischke, C. R., *An Introduction to Computer-aided Design*, Englewood Cliffs, N.J.: Prentice-Hall.

Ray, W. S., *An Introduction to Experimental Design*, New York: Macmillan.

Ruskin, A. M., *Materials Considerations in Design*, Englewood Cliffs, N.J.: Prentice-Hall.

Spotts, M. F., *Design Engineering Projects*, Englewood Cliffs, N.J.: Prentice-Hall.

Starr, M. K., *Product Design and Decision Theory*, Englewood Cliffs, N.J.: Prentice-Hall.

Tuska, C. D., *An Introduction to Patents for Inventors and Engineers*, New York: Dover Publications.

U.S. Patent Office, *Patent Laws*, Washington, D.C.: Superintendent of Documents, Dept. of Commerce.

U.S. Patent Office, *Patents and Inventions: An Information Aid for Inventors*, Washington, D.C.: Superintendent of Documents, Dept. of Commerce.

Von Fange, Eugene, *Professional Creativity*, Englewood Cliffs, N.J.: Prentice-Hall.

White, W. J. and Solomom Schneyer, *Pocket Data for Human Factor Engineering*, Buffalo, N.Y.: Cornell University.

Engineering graphics

Black, E. D., *Graphical Communication*, New York: McGraw-Hill.

Earle, J. H., *Engineering Design Graphics*, Reading, Mass.: Addison-Wesley.

French, T. E., and C. J. Vierck, *Engineering Drawing*, 10th ed., New York: McGraw-Hill.

Giesecke, F. E., and others, *Technical Drawing*, 6th ed., New York: Macmillan.

Hammond, R. H., and others, *Engineering Graphics*, New York: Ronald.

Hoelscher, R. P., and C. H. Springer, *Engineering Drawing and Geometry*, 2nd ed., New York: Wiley.

Hornung, W. J., *Mechanical Drafting*, Englewood Cliffs, N.J.: Prentice-Hall.

Levens, A. S., *Graphics in Engineering and Science*, New York: Wiley.

Luzadder, W. J., *Basic Graphics for Design, Analysis, Communications, and the Computer*, 2nd ed., Englewood Cliffs, N.J.: Prentice-Hall.

———, *Fundamentals of Engineering Drawing for Design, Communication, and Numerical Control*, 6th ed., Englewood Cliffs, N.J.: Prentice-Hall.

———, *Fundamentos de Dibujo Para Ingenieros*, 2nd ed., Mexico, D.F.: Compañia Editorial Continental.

——— *Graphics for Engineers*, India: Prentice-Hall.

——— *Technical Drafting Essentials*, 2nd ed., Englewood Cliffs, N.J.: Prentice-Hall.

Rule, J. T., and S. A. Coons, *Graphics*, New York: McGraw-Hill.

Schneerer, W. F., *Programmed Graphics*, New York: McGraw-Hill.

Engineering graphics films (see text films)

Engineering graphics problems

Earle, J. H., *Engineering Graphics and Design Problems* (Series 1, 2 and 3), Reading, Mass.: Addison-Wesley.

Giesecke, F. E., and others, *Technical Drawing Problems*, 3rd ed., New York: Macmillan.

Johnson, L. O., and I. Wladaver, *Engineering Drawing Problems*, Englewood Cliffs, N.J.: Prentice-Hall.

Levens, A. S., and A. E. Edstrom, *Problems in Mechanical Drawing*, New York: McGraw-Hill.

Luzadder, W. J., K. E. Botkin, and F. H. Thompson, *Problems in Engineering Drawing*, 6th ed., Englewood Cliffs, N.J.: Prentice-Hall.

———, K. E. Botkin and C. J. Rogers, *Engineering Graphics Problems for Design, Analysis and Communications*, Englewood Cliffs, N.J.: Prentice-Hall.

———, and R. E. Bolles, *Problems in Drafting Fundamentals*, Englewood Cliffs, N.J.: Prentice-Hall.

Spencer, H. C., and I. L. Hill, *Technical Drawing Problems*, New York: Macmillan.

Vierck, C. J., and R. I. Hang, *Engineering Drawing Problems*, New York: McGraw-Hill.

Graphical representation and computation

Davis, D. S., *Empirical Equations and Nomography*, New York: McGraw-Hill.

Douglass, R. D., and D. P. Adams, *Elements of Nomography*, New York: McGraw-Hill.

Levens, A. S., *Nomography*, 2nd ed., New York: Wiley.

Lipka, J., *Graphical and Mechanical Computation*, New York: Wiley.

Robinson, A. H., *Elements of Cartography*, New York: Wiley.

Handbooks

American Institute of Steel Construction, *Steel Construction*, New York.

Crocker, S., *Piping Handbook*, New York: McGraw-Hill.

Kent, W., *Mechanical Engineer's Handbook*, 12th ed., New York: Wiley.

Knowlton, A. E., *Standard Handbook for Electrical Engineers*, 9th ed., New York: McGraw-Hill.

Le Grand, R., Ed., *New American Machinist's Handbook*, New York: McGraw-Hill.

McNeese, D. C., and A. L. Hoag, *Engineering and Technical Handbook*, Englewood Cliffs, N.J.: Prentice-Hall.

Marks, L. S., *Mechanical Engineers' Handbook*, 6th ed., New York: McGraw-Hill.

Oberg, E., and F. D. Jones, *Machinery's Handbook*, 17th ed., New York: Industrial Press.

O'Rourke, C. E., *General Engineering Handbook*, 2nd ed., New York: McGraw-Hill.

Perry, J. H., *Chemical Engineers' Handbook*, 4th ed., New York: McGraw-Hill.

Wilson, F. W., Ed., *Tool Engineers' Handbook*, 2nd ed., New York: McGraw-Hill.

Kinematics—machine design

Faires, V. M., *Design of Machine Elements*, 3rd ed., New York: Macmillan.

Hinkle, R. T., *Kinematics of Machines*, Englewood Cliffs, N.J.: Prentice-Hall.

Lent, D., *Analysis and Design of Mechanisms*, Englewood Cliffs, N.J.: Prentice-Hall.

Maleev, V. L., *Machine Design*, 3rd ed., Scranton, Pa.: International.

Norman, C. A., E. S. Ault, and I. F. Zarobsky, *Fundamentals of Machine Design*, New York: Macmillan.

Spotts, M. F., *Design of Machine Elements*, 2nd ed., Englewood Cliffs, N.J.: Prentice-Hall.

Machine drawing

Svensen, C. L., *Machine Drawing*, 3rd ed., New York: Van Nostrand-Reinhold.

Map and topographic drawing

Sloane, R. C., and I. M. Montz, *Elements of Topographic Drawing*, 2nd ed., New York: McGraw-Hill.

Metric System (Système International d'Unités-SI)

American Society For Testing Materials, *Metric Practice Guide*, New York.

Beloit Tool Co., *Discover Why Metrics*, Roscoe, Ill.: Swani Publishing Co.

———, *U.S.A. Goes Metric*, Roscoe, Ill.: Swani Publishing Co.

Society Of Automotive Engineers, *Dual-Dimensioning* (SAE J390), New York.

Numerical control

International Business Machines Corp., *Automatic Programming of Machine Tools*, Poughkeepsie, N.Y.

Patent drawing (See Engineering Design)

Publications of U.S. Government Printing Office, Washington, D.C.:
Guide for Patent Draftsmen
Rules of Practice of the U.S. Patent Office

Perspective

Lawson, P. J., *Practical Perspective Drawing*, New York: McGraw-Hill.

Turner, W. W., *Simplified Perspective*, New York: Ronald.

Slide rule

Arnold, J. N., *The Slide Rule*, Englewood Cliffs, N.J.: Prentice-Hall.

Structural drafting

American Institute of Steel Construction, *Structural Steel Detailing*, New York.

Bishop, C. T., *Structural Drafting*, New York: Wiley.

Lothers, S. E., *Design in Structural Steel*, Englewood Cliffs, N.J.: Prentice-Hall.

Technical dictionary

Tweney, C. F., and L. E. C. Hughes, *Chamber's Technical Dictionary*, 3rd rev. ed., New York: Macmillan.

Technical illustration

Gibby, J. C., *Technical Illustration*, Chicago: American Technical Society.

Thomas, T. A., *Technical Illustration*, New York: McGraw-Hill.

Text films

Engineering Drawing, 7 films, 6 strips, New York: McGraw-Hill.

Engineering Drawing, 16 films. Purdue University.*

Welding

Lincoln Electric Co., *Procedure Handbook of Arc Welding Design and Practice*, Cleveland.

Morris, J. L., *Welding Processes and Procedures*, Englewood Cliffs, N.J.: Prentice-Hall.

——, *Welding Principles for Engineers*, Englewood Cliffs, N.J.: Prentice-Hall.

*These films are available on a rental basis or they may be purchased from Audio-Visual Center, Stewart Center, Purdue University, West Lafayette, Indiana 47907.

Index